VOLUME ONE HUNDRED AND TWENTY FIVE

VITAMINS AND HORMONES
Glycation

EDITORIAL BOARD

David Alpers, MD
Washington University

Tadgh Begley, PhD
Texas A&M University

Nathan A. Berger, MD
Case Western Reserve University

Daniel J. Bernard, PhD
McGill University

Jonathan Bogan, MD
Yale University

Monica P. Colaiacovo, PhD
Harvard University

Kevin P.M. Currie, PhD
Rowan University

Pierre DeMeyts, MD, PhD
Catholic University of Louvain

Briony Forbes, PhD
Flinders University of South Australia

Peter Fuller, MD
Hudson Institute of Medical Research

Liisa Galea, PhD
Center for Addiction & Mental Health (Toronto)

Ralph Green, MD, PhD
University of California, Davis

Jean-Louis Gueant, MD, DSc
University of Lorraine

Gabor M. Halmos, PhD
Debrecen University

Luciana Hanibal, PhD
University of Freiburg

Mark R. Haussler, PhD
University of Arizona

Carolyn M. Klinge, PhD
University of Louisville

Raj Kumar, PhD
Touro University

Michaela Luconi, PhD
University of Florence

David Lyons, PhD
University of Manchester

Kostas Pantopoulos, PhD
Lady Davis Institute for Medical Research

Trevor M. Penning, PhD
University of Pennsylvania

JoAnne S. Richards, PhD
Baylor College of Medicine

Jacqueline M. Stephens, PhD
Louisiana State University

Robert Unwin, PhD
University College London

Jean-Pierre Vilardaga, PhD
University of Pittsburgh

VOLUME ONE HUNDRED AND TWENTY FIVE

VITAMINS AND HORMONES
Glycation

Series Editor

GERALD LITWACK, PhD
Toluca Lake, North Hollywood, California

Academic Press is an imprint of Elsevier
125 London Wall, London, EC2Y 5AS, United Kingdom
50 Hampshire Street, 5th Floor, Cambridge, MA 02139, United States
525 B Street, Suite 1650, San Diego, CA 92101, United States

First edition 2024

Copyright © 2024 Elsevier Inc. All rights are reserved, including those for text and data mining, AI training, and similar technologies.

Publisher's note: Elsevier takes a neutral position with respect to territorial disputes or jurisdictional claims in its published content, including in maps and institutional affiliations.

No part of this publication may be reproduced or transmitted in any form or by any means, electronic or mechanical, including photocopying, recording, or any information storage and retrieval system, without permission in writing from the publisher. Details on how to seek permission, further information about the Publisher's permissions policies and our arrangements with organizations such as the Copyright Clearance Center and the Copyright Licensing Agency, can be found at our website: www.elsevier.com/permissions.

This book and the individual contributions contained in it are protected under copyright by the Publisher (other than as may be noted herein).

Notices
Knowledge and best practice in this field are constantly changing. As new research and experience broaden our understanding, changes in research methods, professional practices, or medical treatment may become necessary.

Practitioners and researchers must always rely on their own experience and knowledge in evaluating and using any information, methods, compounds, or experiments described herein. In using such information or methods they should be mindful of their own safety and the safety of others, including parties for whom they have a professional responsibility.

To the fullest extent of the law, neither the Publisher nor the authors, contributors, or editors, assume any liability for any injury and/or damage to persons or property as a matter of products liability, negligence or otherwise, or from any use or operation of any methods, products, instructions, or ideas contained in the material herein.

ISBN: 978-0-443-19402-3
ISSN: 0083-6729

For information on all Academic Press publications
visit our website at https://www.elsevier.com/books-and-journals

Publisher: Zoe Kruze
Acquisitions Editor: Leticia M. Lima
Editorial Project Manager: Sneha Apar
Production Project Manager: Maria Shalini
Cover Designer: Greg Harris

Typeset by MPS Limited, India

Former Editors

ROBERT S. HARRIS
Newton, Massachusetts

JOHN A. LORRAINE
University of Edinburgh
Edinburgh, Scotland

PAUL L. MUNSON
University of North Carolina
Chapel Hill, North Carolina

JOHN GLOVER
University of Liverpool
Liverpool, England

GERALD D. AURBACH
Metabolic Diseases Branch
National Institute of
Diabetes and Digestive and
Kidney Diseases
National Institutes of Health
Bethesda, Maryland

KENNETH V. THIMANN
University of California
Santa Cruz, California

IRA G. WOOL
University of Chicago
Chicago, Illinois

EGON DICZFALUSY
Karolinska Sjukhuset
Stockholm, Sweden

ROBERT OLSEN
School of Medicine
State University of New York
at Stony Brook
Stony Brook, New York

DONALD B. MCCORMICK
Department of Biochemistry
Emory University School of
Medicine, Atlanta, Georgia

Dedication

To those who will read what is one of Dr. Litwack's final volumes as editor of this series:

Gerry worked with an abiding commitment to his profession and the value he and others might contribute to its advancement. At the same time, he was objective and humble in his view of scientific achievement regardless of the discipline. His approach to the endeavor of science was creative and instinctive, nearly artistic. Above all else, he was dedicated to his research, his writing, the institutions he so loyally served, his many students and collaborators.

As a non-scientist but an interested bystander and often captive audience, one of the many things he said to me over the years was what I believe to be a deeply spiritual insight.

"Scientists only discover what is already there"

Ellie Litwack

Contents

Contributors   xv
About the editor   xix

1. The dynamic roles of advanced glycation end products   1
Mariyam Khalid and Abdu Adem

1. Introduction   2
2. Sources of AGEs   4
3. AGEs receptors   10
4. Downstream signaling of AGEs   13
5. Perspectives   20
Funding   22
Conflicts of interest   22
References   22

2. Methylglyoxal-induced modification of myoglobin: An insight into glycation mediated protein aggregation   31
Sauradipta Banerjee

1. Protein glycation   32
2. Glycation and protein aggregation   33
3. Structural alterations of myoglobin by methylglyoxal   34
4. Aggregation of myoglobin by methylglyoxal   36
5. Mass spectrometric analysis of MG modified Mb   39
6. Concluding remarks   42
Acknowledgments   43
References   43

3. Glycation in the cardiomyocyte   47
Christine E. Delligatti and Jonathan A. Kirk

1. Glycation and the cardiomyocyte   48
2. Extracellular glycation   52
3. Glycation disrupts cardiomyocyte calcium handling   56
4. Glycation of the cytoskeleton   60
5. Other effects of glycation in the cardiomyocyte   64

6. Therapeutics and the future of cardiomyocyte glycation	66
7. Other future directions and concluding remarks	75
References	77

4. Glycation and drug binding by serum albumin — 89
Anu Jain and Nand Kishore

1. Glycation	90
2. Glycation of transport protein serum albumin	91
3. Role of advanced glycation end products	93
4. Characterisation of glycated protein and glycation end-products	94
5. Glycation and protein conformation	98
6. Calorimetry in glycation studies	99
7. Glycation and drug binding	106
8. Future perspectives	109
References	109

5. Advanced glycation end products and insulin resistance in diabetic nephropathy — 117
Kirti Parwani and Palash Mandal

1. Introduction	118
2. Advanced glycation end products	119
3. Receptors for AGEs	123
4. Role of AGEs in insulin resistance	125
5. AGEs and their involvement in diabetic nephropathy	129
6. Mechanisms of insulin resistance and its impact on the kidney	131
7. Inhibition of the AGE-RAGE axis: Treatments for diabetic nephropathy	134
8. Conclusion and future perspectives	138
References	139

6. The Maillard reactions: Pathways, consequences, and control — 149
Delia B. Rodriguez-Amaya and Jaime Amaya-Farfan

1. Introduction	150
2. Stages of the Maillard reactions	150
3. Influencing factors	167
4. Consequences of the Maillard reactions	170
5. Strategies for controlling the Maillard reactions	173

6. Final considerations	174
References	175

7. Structural changes in hemoglobin and glycation — 183
Amanda Luise Alves Nascimento, Ari Souza Guimarães, Tauane dos Santos Rocha, Marilia Oliveira Fonseca Goulart, Jadriane de Almeida Xavier, and Josué Carinhanha Caldas Santos

1. Introduction	184
2. Glycation of hemoglobin	187
3. Hb glycation: *in vitro* protocols evaluation	195
4. Structural changes in Hb after glycation	199
5. Techniques and methods for assessing structural and conformational changes in glycated Hb	200
6. Trends	215
7. Conclusions	217
Acknowledgements	217
References	217

8. Attenuation of albumin glycation and oxidative stress by minerals and vitamins: An *in vitro* perspective of dual-purpose therapy — 231
Ashwini Dinkar Jagdale, Rahul Shivaji Patil, and Rashmi Santosh Tupe

1. Introduction	232
2. Materials and methods	234
3. Methods	235
4. Results and discussion	239
5. Conclusion and future directions	246
Acknowledgments	246
References	247

9. Non-enzymatic glycation and diabetic kidney disease — 251
Anil K. Pasupulati, Veerababu Nagati, Atreya S.V. Paturi, and G. Bhanuprakash Reddy

1. Kidney architecture	252
2. Kidney dysfunction in diabetes	254
3. Altered glucose homeostasis	256

4.	Non-enzymatic glycation (NEG)- a post-translational modification	258
5.	Fate of AGEs	262
6.	Pathology of AGEs in DKD	264
7.	Strategies to prevent AGEs formation and DKD	272
8.	Summary and future perspective	276
Acknowledgments		278
References		278

10. Nanoparticles in prevention of protein glycation   287
Aruna Sivaram and Nayana Patil

1.	Introduction	288
2.	Chemistry of the glycation reaction	288
3.	Glycation of biological macromolecules	290
4.	DNA glycation	292
5.	RNA glycation	293
6.	Lipid glycation	293
7.	Clinical relevance of glycation and advanced glycation end products	294
8.	Osteoporosis	295
9.	Cancer	296
10.	Neurodegenerative diseases	296
11.	Cardiovascular diseases	296
12.	Prevention of glycation	297
13.	Nanotechnology approach for combating glycation	298
14.	Nanoparticles	299
15.	Gold nanoparticles	299
16.	Silver nanoparticles	301
17.	Zinc nanoparticles	302
18.	Challenges and future perspectives	304
19.	Challenges	304
20.	Future perspectives	305
21.	Conclusion	305
References		306

11. Breath of fresh air: Investigating the link between AGEs, sRAGE, and lung diseases   311
Charlotte Delrue, Reinhart Speeckaert, Joris R. Delanghe, and Marijn M. Speeckaert

1.	Introduction	314
2.	Chronic obstructive pulmonary disease	318

3.	Asthma	327
4.	Lung fibrosis	334
5.	Cystic fibrosis	338
6.	Acute lung injury/acute respiratory distress syndrome	339
7.	Lung cancer	342
8.	Unmasking RAGE's complex role in lung cancer progression	343
9.	Unlocking the impact of AGER polymorphisms on lung cancer	345
10.	sRAGE: Illuminating lung cancer diagnosis and beyond	346
11.	Conclusions and future directions	347
	References	348

12. Antioxidant and antibrowning properties of Maillard reaction products in food and biological systems   367

Majid Nooshkam and Mehdi Varidi

1.	Introduction	368
2.	Chemistry of the Maillard reaction	369
3.	MRPs as antioxidants	371
4.	MRPs as antibrowning agents	387
5.	Deleterious effect of MRPs as glycation end products in biological systems	391
6.	Conclusions	392
	Acknowledgments	392
	References	392

13. Vitamin B6 and diabetes and its role in counteracting advanced glycation end products   401

F. Vernì

1.	Vitamin B6	402
2.	Vitamin B6 and diabetes	409
3.	Advanced glycation end products	415
4.	Vitamin B6, diabetic complications and AGEs	421
5.	Conclusions	427
	References	428

Contributors

Abdu Adem
Department of Pharmacology and Therapeutics, College of Medicine and Health Sciences, Khalifa University, Abu Dhabi, United Arab Emirates

Jaime Amaya-Farfan
School of Food Engineering, University of Campinas, Campinas, SP, Brazil

Sauradipta Banerjee
George College, MAKAUT

Joris R. Delanghe
Department of Diagnostic Sciences, Ghent University, Ghent, Belgium

Christine E. Delligatti
Department of Cell and Molecular Physiology, Loyola University Chicago Stritch School of Medicine, Maywood, IL, United States

Charlotte Delrue
Department of Nephrology, Ghent University Hospital, Ghent, Belgium

Marilia Oliveira Fonseca Goulart
Federal University of Alagoas, Institute of Chemistry and Biotechnology, Campus A. C. Simões, Maceió, Alagoas, Brazil

Ari Souza Guimarães
Federal University of Alagoas, Institute of Chemistry and Biotechnology, Campus A. C. Simões, Maceió, Alagoas, Brazil

Ashwini Dinkar Jagdale
Symbiosis School of Biological Sciences (SSBS), Symbiosis International (Deemed University) (SIU), Pune, Maharashtra, India

Anu Jain
Department of Chemistry, Indian Institute of Technology Bombay, Mumbai, India

Mariyam Khalid
Department of Pharmacology and Therapeutics, College of Medicine and Health Sciences, Khalifa University, Abu Dhabi, United Arab Emirates

Jonathan A. Kirk
Department of Cell and Molecular Physiology, Loyola University Chicago Stritch School of Medicine, Maywood, IL, United States

Nand Kishore
Department of Chemistry, Indian Institute of Technology Bombay, Mumbai, India

Palash Mandal
Department of Biological Sciences, P. D. Patel Institute of Applied Sciences, Charotar University of Science & Technology, Gujarat, India

Veerababu Nagati
Department of Biochemistry, University of Hyderabad, Hyderabad, India

Amanda Luise Alves Nascimento
Federal University of Alagoas, Institute of Chemistry and Biotechnology, Campus A. C. Simões, Maceió, Alagoas, Brazil

Majid Nooshkam
Department of Food Science and Technology, Faculty of Agriculture, Ferdowsi University of Mashhad, Mashhad, Iran

Kirti Parwani
Department of Biological Sciences, P. D. Patel Institute of Applied Sciences, Charotar University of Science & Technology, Gujarat, India

Anil K. Pasupulati
Department of Biochemistry, University of Hyderabad, Hyderabad, India

Nayana Patil
School of Bioengineering Sciences and Research, MIT ADT University, Pune, India

Rahul Shivaji Patil
Vascular Biology Center, Medical College of Georgia, Augusta University, Augusta, GA, United States

Atreya S.V. Paturi
Department of Biochemistry, University of Hyderabad, Hyderabad, India

G. Bhanuprakash Reddy
Department of Biochemistry, ICMR-National Institute of Nutrition, Hyderabad, India

Tauane dos Santos Rocha
Federal University of Alagoas, Institute of Chemistry and Biotechnology, Campus A. C. Simões, Maceió, Alagoas, Brazil

Delia B. Rodriguez-Amaya
School of Food Engineering, University of Campinas, Campinas, SP, Brazil

Josué Carinhanha Caldas Santos
Federal University of Alagoas, Institute of Chemistry and Biotechnology, Campus A. C. Simões, Maceió, Alagoas, Brazil

Aruna Sivaram
School of Bioengineering Sciences and Research, MIT ADT University, Pune, India

Marijn M. Speeckaert
Department of Nephrology, Ghent University Hospital, Ghent; Research Foundation-Flanders (FWO), Brussels, Belgium

Reinhart Speeckaert
Department of Dermatology, Ghent University Hospital, Ghent, Belgium

Rashmi Santosh Tupe
Symbiosis School of Biological Sciences (SSBS), Symbiosis International (Deemed University) (SIU), Pune, Maharashtra, India

Mehdi Varidi
Department of Food Science and Technology, Faculty of Agriculture, Ferdowsi University of Mashhad, Mashhad, Iran

F. Vernì
Department of Biology and Biotechnology "Charles Darwin" Sapienza University of Rome, Rome, Italy

Jadriane de Almeida Xavier
Federal University of Alagoas, Institute of Chemistry and Biotechnology, Campus A. C. Simões, Maceió, Alagoas, Brazil

About the editor

Gerald (Gerry) Litwack was educated at Hobart College (BA) with graduate work at the University of Wisconsin, Madison (MS and PhD in biochemistry). After spending a summer at the University of Wisconsin as an instructor in a course on enzymes, he sailed to Paris (Liberté) where he spent a year as a postdoctoral fellow of the National Foundation for Infantile Paralysis at the Laboratoire de Chimie Biologique of the Sorbonne. He then became Assistant Professor of Biochemistry at Rutgers University, later becoming Associate Professor at the Graduate School of Medicine of the University of Pennsylvania. Future posts took him to the Fels Institute for Cancer Research and Molecular Biology at the Temple University School of Medicine where he was the Laura H. Carnell Professor of Biochemistry and the Deputy Director of the institute. Later, he became the Chairman of the Department of Biochemistry and Molecular Pharmacology, Professor of Biochemistry, Vice Dean for Research, and the Director of the Institute for Apoptosis at Thomas Jefferson University Kimmel Medical College with future appointments as Visiting Scholar at the Geffen School of Medicine at UCLA, the Founding Chair of the Department of Basic Science of the Geisinger Commonwealth School of Medicine, and, as his final position, Professor of Molecular and Cellular Medicine, and Associate Director of the Institute for Regenerative Medicine at Texas A&M University School of Medicine. Dr. Litwack was Sabbatical Visitor at the University of California, Berkeley; the University of California, San Francisco; the Courtauld Institute of Biochemistry (London); and the Wistar Institute. He was appointed Emeritus Professor and/or Chairman at Rutgers University, Thomsas Jefferson University, and Geisinger Commonwealth School of Medicine. He has published more than 350 papers in scientific journals, is named on more than 20 patents, and has held numerous editorial positions on biochemical and cancer journals. He has authored 3 textbooks in the areas of biochemistry or endocrinology and has edited more than 65 books. Following his retirement, he lives with his family in Los Angeles where he continues his work as an author and editor and paints in watercolor during his leisure time.

CHAPTER ONE

The dynamic roles of advanced glycation end products

Mariyam Khalid and Abdu Adem[*]

Department of Pharmacology and Therapeutics, College of Medicine and Health Sciences, Khalifa University, Abu Dhabi, United Arab Emirates
[*]Corresponding author. e-mail address: abdu.adem@ku.ac.ae

Contents

1. Introduction	2
2. Sources of AGEs	4
2.1 Endogenous AGEs	4
2.2 Exogenous AGEs	6
3. AGEs receptors	10
3.1 AGEs receptor complex	10
3.2 Scavenger receptors	11
3.3 RAGE	11
4. Downstream signaling of AGEs	13
4.1 AGEs/RAGE signaling	13
4.2 Pathophysiology of AGEs/RAGE axis	16
4.3 AGEs/RAGE axis and retinopathy	18
4.4 AGEs/RAGE axis and nephropathy	18
4.5 AGEs/RAGE axis and Cardiovascular disorders	19
4.6 AGEs/RAGE axis and neurodegeneration	20
5. Perspectives	20
Funding	22
Conflicts of interest	22
References	22

Abstract

Advanced glycation end products (AGEs) are a heterogeneous group of potentially harmful molecules that can form as a result of a non-enzymatic reaction between reducing sugars and proteins, lipids, or nucleic acids. The total body pool of AGEs reflects endogenously produced AGEs as well as exogenous AGEs that come from sources such as diet and the environment. Engagement of AGEs with their cellular receptor, the receptor for advanced glycation end products (RAGE), which is expressed on the surface of various cell types, converts a brief pulse of cellular activation to sustained cellular dysfunction and tissue destruction. The AGEs/RAGE interaction triggers a cascade of intracellular signaling pathways such as mitogen-activated protein kinase/extracellular signal-regulated kinase, phosphoinositide 3-

kinases, transforming growth factor beta, c-Jun N-terminal kinases (JNK), and nuclear factor kappa B, which leads to the production of pro-inflammatory cytokines, chemokines, adhesion molecules, and oxidative stress. All these events contribute to the progression of several chronic diseases. This chapter will provide a comprehensive understanding of the dynamic roles of AGEs in health and disease which is crucial to develop interventions that prevent and mitigate the deleterious effects of AGEs accumulation.

1. Introduction

The AGEs, also known as glycotoxins, are the product of the nonenzymatic reaction between reducing sugars and free amino groups of proteins, lipids, or nucleic acids known as the Maillard or browning reaction (Kuzan, 2021; Uribarri et al., 2010). Maillard reaction was first described in the early 1900s when it was discovered that heating amino acids in the presence of reducing sugars develop a characteristic yellowish-brown color (Ellis, 1912; Singh, Barden, Mori, & Beilin, 2001). It is a complex series of reactions that constitute stable irreversible compounds termed AGEs that accumulate over the lifetime of the protein (Thomas, 2011; Yamagishi, 2019).

Under normal physiological conditions, glycation is relatively a slow naturally occurring reaction and a normal part of the aging process resulting in modifications primarily in long-lived macromolecules. The long-lived proteins constituting the extracellular matrix (ECM) and vascular basement membranes including collagen and elastin are highly susceptible to AGEs-mediated crosslinks (Khalid, Petroianu, & Adem, 2022; Vlassara & Palace, 2002). As glucose is the primary fuel source for the brain, sustained glucose levels are required for optimal brain function, so the majority of the proteins include AGEs related modifications to a certain extent. However, certain factors as in hyperglycemic state in diabetes mellitus or impaired renal clearance exacerbate glycation process and AGEs can arise on short-lived molecules as well (Abdel-Wahab, O'Harte, Boyd, Barnett, & Flatt, 1997; Brings et al., 2017; Singh et al., 2001; Vlassara & Palace, 2002).

AGEs modification of proteins and lipids renders irreversible damage to the biological macro and micro molecules, disrupts their structural and functional integrity, alters molecular conformation, effect enzymatic activity, reduces degradative capacity, promotes crosslinking of circulatory protein, and results in abnormal recognition and clearance by receptors (Singh, Bali, Singh, & Jaggi, 2014; Vlassara & Palace, 2002). To date, numerous AGEs compounds have been identified so far in dietary items, human blood, and tissues. Some of the important AGEs have been mentioned in Fig. 1.

Over the last twenty years, there has been an increasing amount of research investigating the role of AGEs in disease development. There is evidence from several studies in animal models and humans to suggest that the accumulation of AGEs in the body could play a role in the aging process and also in the development of chronic diseases (Chaudhuri et al., 2018; Kuzan, 2021). It can contribute to the development of diabetes and its related complications by promoting insulin resistance and impairing insulin secretion (Vlassara & Palace, 2002). AGEs are also involved in neurodegenerative diseases as they can accumulate in the brain and promote the formation of amyloid plaques, a hallmark of Alzheimer's disease (Li, Liu, Sun, Lu, & Zhang, 2012). Besides these AGEs also participate in the pathophysiology of cardiovascular disease (CVD), sarcopenia, rheumatoid arthritis, renal disease, and other degenerative disorders (Luevano-Contreras & Chapman-Novakofski, 2010; Singh et al., 2014).

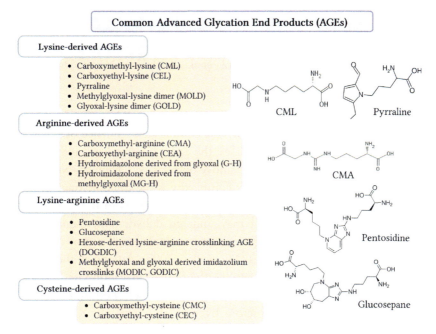

Fig. 1 Common Advanced Glycation End Products (AGEs). Figure created with BioRender.com

There are two distinct methods through which AGEs exert their deleterious effects. The non-receptor-mediated mechanism involves the direct chemical modification of proteins, lipids, and nucleic acids by AGEs leading to impaired cellular function and tissue damage (Saito & Marumo, 2015; Thomas, 2011). The receptor-mediated mechanism involves the binding of AGEs to specific receptors on the cell surface such as RAGE which triggers a cascade of cellular events that can lead to inflammation, oxidative stress, and cellular dysfunction. (Kuzan, 2021; Luevano-Contreras & Chapman-Novakofski, 2010).

Due to the ability of AGEs to increase the risk of chronic degenerative diseases, AGEs are becoming an increasingly popular subject of research in recent years. In this chapter, our focus is on the chemistry and kinetics of the Maillard reaction, different types of AGEs, glycation pathways that lead to the formation of the main identified AGEs, specific receptors of AGEs, and the role of AGEs in the pathogenesis of chronic diseases. We suggest several approaches with a potential therapeutic value that can be targeted to prevent or reduce AGEs accumulation in the body.

2. Sources of AGEs

Recently many studies have suggested that the body pool of AGEs reflects endogenously produced AGEs as well as exposure to glycation products formed exogenously (Mastrocola et al., 2020; Van Dongen et al., 2022).

2.1 Endogenous AGEs

The formation of endogenous AGEs is a complicated sequence of events and occurs in distinguishable three phases: the early, intermediate, and advanced stages (John & Lamb, 1993; Luevano-Contreras & Chapman-Novakofski, 2010). One of the reasons for this complexity is the multiple fragmentation reactions of the sugar moiety. The sugar moiety refers to the carbohydrate component of a molecule that may undergo dehydration, oxidation, and fragmentation reactions and give rise to multiple smaller molecules with diverse properties and structures (Thornalley, 2005). Among all the naturally occurring reducing sugars, glucose has the slowest glycation rate because of the rate-limiting step of glycolysis and its natural selection as an ideal fuel source for cells. Although fructose bypasses this step, but the absorptive capacity for fructose is lower than glucose or sucrose. (Cho, Roman, Yeboah, & Konishi, 2007; Singh et al., 2001).

The initiation of Maillard reaction starts with the condensation of a carbonyl group from a reducing sugar with a free amino acid of proteins, lipids, and nucleic acids in a non-enzymatic way to form an imine, a Schiff base (John & Lamb, 1993). The Schiff base formation is a crucial step in the Maillard reaction. Schiff bases are early, unstable, and reversible compounds that can be further transformed into more complex compounds as the Maillard reaction proceeds (Aragno & Mastrocola, 2017; John & Lamb, 1993; Khalid et al., 2022). The rate of AGEs formation is influenced by several factors, such as the concentration of reducing sugars and amino acids, pH, temperature, and the presence of metal ions (Van Dongen et al., 2022). These factors can affect the chemical reactivity of the reactants and the stability of the intermediate compounds. Higher concentrations of reducing sugars and amino acids lead to a higher rate of reaction as an increased number of reactants are available to participate in the reaction (Martins, Jongen, & Van Boekel, 2000). During the second phase, the Schiff base undergoes a well-known rearrangement, called the Amadori rearrangement over a period which produces a more stable isomer of the Schiff base known as early glycation products or Amadori products (John & Lamb, 1993; Van Dongen et al., 2022). Further modification of the Amadori product is dependent on the pH of the system. At pH 7 or below, the transformation of the sugar moiety of Amadori products into α-dicarbonyl compounds such as 3-deoxyglucosone which will degrade further to generate methylglyoxal and glyceraldehyde (Cho et al., 2007). The 3-deoxyglucosone reacts rapidly with protein amino groups to form AGEs such as N-ε-carboxymethyllysine (CML), pyrraline, imidazolone, glucosepane, and pentosidine (Chaudhuri et al., 2018). At pH above 7, the degradation of the Amadori compound takes place by 2,3 enolisation, and involves the formation of reductones, and a variety of fission products, including acetol, pyruvaldehyde, and diacetyl (Martins et al., 2000).

All these intermediate compounds formed during Amadori rearrangement are highly reactive and take part in further reactions. These compounds known as α- dicarbonyls or oxoaldehydes are key precursors of AGEs formation. They have been considered as an important focal point and biomarker in production of AGEs as they are formed from all stages of glycation (Luevano-Contreras & Chapman-Novakofski, 2010; Singh et al., 2001). These reactive carbonyls are potent glycating agents since these intermediates are up to 20,000 times more reactive than glucose in glycation reactions (Thornalley, 2005). The imbalance between generation & accumulation of these intermediates and efficiency of scavenger pathway results in carbonyl stress (Singh et al., 2001; Turk, 2010). The phenomenon of carbonyl stress has been extensively documented in complicated diabetes, vascular damage, and uremia (Thornalley, 2005; Turk, 2010).

In the final step, Amadori products undergo further chemical rearrangements such as cyclization, oxidation reductions, and dehydration, over a period of time leading to crosslinking with proteins, and formation of irreversible AGEs as shown in Fig. 2. The final glycation products are very stable and accumulate inside and outside the cells and interfere with protein function (Cho et al., 2007; John & Lamb, 1993; Luevano-Contreras & Chapman-Novakofski, 2010).

2.2 Exogenous AGEs

Exogenous AGEs and their precursors are present in a wide variety of food items and tobacco smoking. High-temperature treatment of food to attain desirable taste, aroma, safety, bioavailability, and shelf life generate increased levels of AGEs (Luevano-Contreras & Chapman-Novakofski, 2010; Mastrocola et al., 2020). Modern diets exposed to thermal processing such as dry heat, grilling, roasting, broiling, or frying propagate a plentiful source of exogenous AGEs. However, the AGEs content of a diet depends on the processing method used and nutrient composition as animal-origin foods that are high in fat and protein have the highest AGEs content and are prone to new AGEs formation during cooking. (Uribarri et al., 2010;

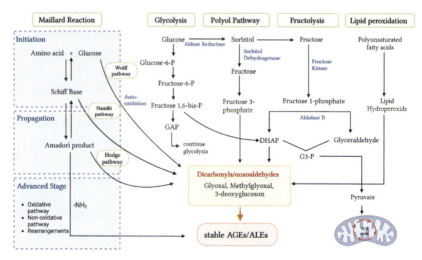

Fig. 2 Endogenous pathways of AGEs formation. Figure created with BioRender.com

Van Dongen et al., 2022). In the modern food industry, Maillard reaction is very important owing to its profound effect on sensory properties that influence the consumer's preferences and increase food consumption. The Maillard reaction has been used commonly in the food industry, especially in baking, coffee roasting, processed food products, sodas & juices, and for caramel production (Luevano-Contreras & Chapman-Novakofski, 2010; Prasad, Imrhan, Marotta, Juma, & Vijayagopal, 2014; Uribarri et al., 2015). However, the harmful potential of dietary AGEs has been controversial. Doubts have been raised regarding the contribution of exogenous AGEs to the systemic burden of AGEs and their association with adverse effects on human health (Van Dongen et al., 2022; Xu et al., 2003). Recently numerous animal and human studies revealed the association of excessive dietary AGEs consumption with inflammation, oxidative stress, and other diabetes-related complications. The animal study by Xu et al. used labeled AGEs and present a direct link between exogenous AGEs and vascular change resembling that of diabetic retinopathy in non-diabetic rats (Xu et al., 2003). While administration of advanced glycation inhibitor prevents the progression of diabetic nephropathy in chronic diabetic rats through inhibition of AGEs-derived cross-linking (Nakamura et al., 1997). The study conducted on mice indicated that glycation of insulin significantly reduced insulin's biological activity and compromised its mechanism of action, thus contributing to insulin resistance and glucose intolerance (Abdel-Wahab et al., 1997). High levels of diet-derived AGEs significantly delay wound healing in diabetic *db/db* (+/+) mice (Peppa, Brem, et al., 2003), while blockade of AGEs receptor, RAGE can accelerate wound repair (Goova et al., 2001). AGEs-enriched diet plays an important role in the pathogenesis of type 1 diabetes by enhancing T-cell-mediated injury in NOD mice (Peppa, He, et al., 2003) while restricted AGEs intake leads to improved insulin sensitivity in db/db mice (Hofmann et al., 2002).

Moreover, the pathogenic role of oral glycotoxins has been correlated with impaired cognition, dementia, and Alzheimer's disease due to specific neurotoxic AGEs derivatives (Cai et al., 2014; Pucci et al., 2021). Another noteworthy finding demonstrates that diet-originating AGEs elevates inflammatory markers and lipid abnormalities linked to diabetes and vascular dysfunction in diabetic subjects, while restriction of AGEs intake result in suppression of these effects (Vlassara & Palace, 2002). In a short-term study, restriction of dietary glycotoxins proves to be effective to reduce the risk of cardiovascular-associated mortality in renal failure patients (Uribarri et al., 2003). Meals high in AGEs have been shown to affect the postprandial ghrelin, glucose responses, and oxidative stress, in healthy overweight individuals (Poulsen et al., 2014). The consumption of a diet low in AGEs through modulation of cooking methods for 4 weeks have reported to significantly improve insulin resistance in overweight women (Mark et al., 2014). Another study conducted on obese human subjects reported reduced serum AGEs levels and indices of body fat as beneficial effects of exercise and an AGEs-restricted diet without caloric restriction for 12 weeks (Cai et al., 2008). One study investigated the AGEs content by measuring CML concentrations in two-hundred fifty foods from different hospital cafeterias and local eating establishments in Mount Sinai and reported high content of AGEs in commonly consumed food (Goldberg et al., 2004).

Besides dietary AGEs, previous studies have reported that glycotoxins inhaled through cigarette smoke are highly reactive, they can interact with ECM proteins, low-density lipoprotein, and lens protein to form stable covalent adducts and produce AGEs (Nicholl & Bucala, 1998; Prasad et al., 2014). Tobacco-derived AGEs can cause an increase in serum levels of AGEs (Nicholl & Bucala, 1998) which is associated with heart stiffness in patients with type 1 diabetes (Berg et al., 1999). Numerous animal and human studies have shown that high dietary intake of exogenous AGEs contributes to the total pool of circulating AGEs in the body, independent of age and hyperglycemia as explained in Fig. 3. This is because exogenous AGEs are absorbed intact and can directly contribute to the circulating pool of AGEs. Additionally, some of the absorbed dicarbonyl compounds can act as seeds for the formation of additional AGEs in the body (Uribarri et al., 2007). The dietary AGEs associated with increased protein-linked tissue deposition of methylgloxal and CML, are considered diabetogenic AGEs and are associated with collagen-associated AGEs accumulation (Peppa, Brem, et al., 2003), increased metalloproteinase production (Daoud et al., 2001), delayed wound healing and impaired growth factor activity (Duraisamy et al., 2001).

Limited clinical studies are available regarding the absorption, biodistribution, and elimination of dietary AGEs due to the complex nature and heterogeneity of AGEs that can influence their absorption rate (Sergi, Boulestin, Campbell, & Williams, 2021). Earlier investigations indicated that only 10% of ingested AGEs are transported into the systemic circulation (Koschinsky et al., 1997), and only ~30% of that is eliminated through the renal route within 3 days from ingestion (He, Sabol, Mitsuhashi, & Vlassara, 1999; Snelson & Coughlan, 2019). However, a marked reduction in urinary

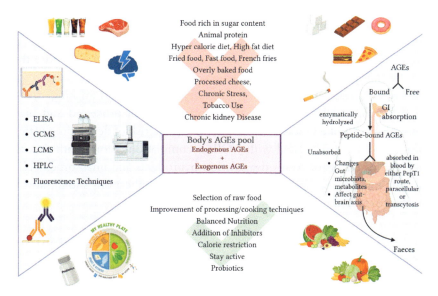

Fig. 3 Exogenous AGEs sources, their absorption, and detection techniques. Figure created with BioRender.com

clearance of orally absorbed AGEs is observed in diabetic nephropathy patients, which potentially aggregates the adverse effects inflicted by these agents (He et al., 1999; Koschinsky et al., 1997). Dietary AGEs can exit either in free form or peptide-bound AGEs which are hydrolyzed into absorbable or unabsorbable fractions during their passage through the gastrointestinal tract as demonstrated in Fig. 3. The individual variations in gut microbiota determine the extent to which ingested AGEs are absorbed into the bloodstream (Snelson & Coughlan, 2019; Zhao et al., 2017). The subsequent fate of peptide-bound AGEs in the intestinal microflora is not yet fully understood and is important for future research to understand this process and counteract its deleterious effects on health.

In future studies, it is important to differentiate between free AGEs and bound AGEs in food. Protein-bound AGEs are mainly hydrolyzed into peptide-bound AGEs, a major form of bound AGEs after gastrointestinal digestion (Zhao et al., 2017, 2019). Free AGEs enter the circulatory system by diffusion, which is less efficient than the transportation of peptide-bound AGEs by PepT1. After absorption, peptide-bound AGEs are thought to bind more stably to RAGEs than free AGEs. The renal clearance rate of peptide-bound AGEs is lower than that of free AGEs (Yuan et al., 2021; Zhao et al., 2019). These differences suggest that free and bound AGEs in food may have different impacts on health (Cai et al., 2012, 2014). Furthermore, during gastrointestinal digestion, protein-bound AGEs and longer peptide-bound AGEs (>2 amino acids) have to undergo enzymatic hydrolysis before they can be transported by PepT 1 (Zhao et al., 2019).

Overall, reducing the intake of dietary AGEs, and making dietary modifications such as reducing the consumption of processed and high-temperature cooked foods may help to lower the risk of developing various health problems associated with AGEs accumulation. Along with this, the deleterious effects of AGEs degradation byproducts and unabsorbed AGEs on gut microbiota, and their potential health implications should also be further explored.

3. AGEs receptors

The AGEs are a heterogeneous group of compounds, that interact with several cellular receptors to stimulate signal transduction through diverse pathways (Ott et al., 2014). The most studied and well-characterized receptor for AGEs is known as RAGE, a pattern-recognizing, multiligand receptor with different isoforms (Zeng, Li, Ma, Niu, & Tay, 2019). In recent years, several other AGEs receptors have been discovered which include AGEs receptor complex (AGE-R1, AGE-R2, AGE-R3/ galectin-3), scavenger receptors including class A macrophage scavenger receptor (SR-A), class B scavenger receptors such as cluster of differentiation 36 (SR-B/CD36), scavenger receptor class B type I and II (SR-BI, SR-BII), class E scavenger receptor (SR-E/LOX-1), FEEL-1 and FEEL-2 receptors (Ott et al., 2014; Rungratanawanich, Qu, Wang, Essa, & Song, 2021; Tamura et al., 2003). The expression of these AGEs receptors varies among different cell and tissue types and is regulated by multiple factors including inflammation, oxidative stress, and metabolic changes such as hyperglycemia, aging, hyperlipidemia, etc (Ott et al., 2014; Vlassara, 2001).

3.1 AGEs receptor complex

Evidence indicates that AGEs receptor complex comprises of AGE-R1 (OST-48, 48 kDa) and AGE-R3/galectin-3 (26 kDa) have high affinity for AGEs binding and are involved in endocytic uptake and degradation of AGEs-modified proteins (Sourris, Watson, & Jandeleit-Dahm, 2021). AGER1, encoded by the gene DDOST was the first AGEs clearance receptor identified (Lu et al., 2004; Yamagata et al., 1997). It is a type 1 integral plasma membrane protein, present in most cells. AGER1 plays a central part in the clearance of AGEs by sequestering and degrading AGEs, blocking ROS generation, and suppressing new AGEs promoted either by receptor or nonreceptor-based mechanisms (Uribarri et al., 2011; Vlassara & Striker, 2011). The precise biological role of AGER2 (80 K-H, 90 kDa) remains unclear. However, research has shown that this protein undergoes phosphorylation at tyrosine residues and belong to the inflammatory receptors. It is involved in AGEs signal transduction, the formation of a complex with PKC, and has a role in the intracellular signaling of several receptors including the FGF receptor (Ott et al., 2014; Vlassara, 2001).

3.2 Scavenger receptors

The term "scavenger receptor" is used to describe a large and diverse group of cell surface receptors that have the ability to scavenge and degrade modified proteins, thus preventing their accumulation in tissues (Miyazaki, Nakayama, & Horiuchi, 2002; Ott et al., 2014). The scavenger receptors SR-A I/II (77 kDa), SR-B/CD36 (88 kDa), and Fasciclin, EGF-like, laminin-type EGF-like and link domain-containing scavenger receptors 1 and 2 (FEEL-1 and FEEL-2) drive the degradation of AGEs-modified molecules into smaller fragments and play part in host defense mechanism against AGEs accumulation (Sourris et al., 2021; Tamura et al., 2003). The SR-BI (76 kDa) and the lectin-like oxidized LDL receptor-1 (SR-E/LOX-1) (50 kDa) have multiple ligand binding sites and can bind with various ligands. SR-BI is primarily involved in the uptake of high-density lipoprotein cholesterol while LOX-1 is identified as a novel scavenger receptor for oxidized low-density lipoprotein (ox-LDL) that mediates endocytic uptake and subsequent lysosomal degradation of ox-LDL (Jono et al., 2002; Miyazaki et al., 2002; Ott et al., 2014).

3.3 RAGE

The RAGE receptor belongs to the immunoglobulin (Ig) superfamily (Chavakis, Bierhaus, & Nawroth, 2004; Khalid et al., 2022). This patternrecognizing, multiligand receptor is widely expressed in various cell types as endothelial cells, astrocytes, neurons, macrophages/monocytes, lymphocytes, neutrophils, and several organs such as the lung, pancreas, brain, skeletal muscles, heart, kidney, and liver (Singh et al., 2001; Zeng et al., 2019).

The RAGE receptor is composed of three Ig-like domains, including one variable (V) domain and two constant (C) domains, a short hydrophobic transmembrane domain, and a 43-amino acid cytoplasmic tail. The V domain is responsible for ligand binding, while the cytosolic tail is involved in signal transduction (Bierhaus et al., 2005; Chavakis et al., 2004). Due to alternative splicing multiple variants of RAGE have been identified in humans and mice. These isoforms differ in their extracellular domains, which can affect their ligand binding properties and signaling capabilities (Chavakis et al., 2004; Ding & Keller, 2005). Among these, there are at least 3 major forms of RAGE which can be defined as the membrane-bound full-length RAGE (fl-RAGE), dominant negative RAGE (dnRAGE), and soluble/secretory RAGE (sRAGE) as demonstrated in Fig. 4 (Ding & Keller, 2005). The most abundant RAGE isoform is fl-RAGE encoded by gene hRAGE, which is involved in the activation of several pathways by binding to its ligands such as AGEs, S100/calgranulins, amyloid-beta peptide, and ECM proteins (Bierhaus et al., 2005; Scavello et al., 2019).

The variant hRAGE_v1 encodes for the secreted esRAGE, while proteolytic cleavage of fl-RAGE generates cleaved RAGE (cRAGE). Both esRAGE and cRAGE are collectively known as sRAGE which is a truncated form of the receptor (Oliveira et al., 2013; Scavello et al., 2019). The sRAGE comprises the extracellular domain of RAGE but lacks the transmembrane region and the cytosolic tail with similar ligand affinity as that of membrane-bound RAGE. It functions as an endogenous protective decoy receptor and suppresses RAGE-mediated signaling by sequestering RAGE ligands (Bierhaus et al., 2005; Ding & Keller, 2005). Evidence

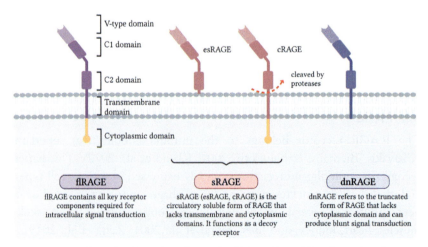

Fig. 4 Receptor for AGEs (RAGE) isoforms. RAGE is a 50–55 kDa glycosylated protein that contains an extracellular region consisting of a variable (V) immunoglobulin (Ig) domain (residues 23–116), and two constant Ig domains, C1 (residues 124–221) & C2 (residues 227–317); a hydrophobic transmembrane (residues 343–363); and a short cytoplasmic (residues 363–404) tail. Figure created with BioRender.com.

suggests circulating sRAGE as a potential biomarker for several pathological conditions. Clinical studies have shown that higher plasma levels of sRAGE are associated with a reduced risk of metabolic syndrome, Alzheimer's disease, cardiovascular disorders, and obesity (Geroldi, Falcone, & Emanuele, 2006). The dn-RAGE is the least understood form of RAGE receptor. It lacks only the short intracellular domain and competes for the ligand binding to prevent membrane-bound RAGE activation. Although more studies are required to identify the role of dnRAGE as a suppressor of fl-RAGE signal transduction (Ding & Keller, 2005).

4. Downstream signaling of AGEs

As discussed above, the formation of AGEs depends on multiple factors, including the amount of AGEs precursors, the concentration and reactivity of glucose, and the availability of free amino groups (Perrone, Giovino, Benny, & Martinelli, 2020; Yamagishi, 2019). The interaction of AGEs with several cellular receptors induces sustained post-receptor signaling, including activation of inflammatory pathways and promotion of oxidative stress (Vlassara & Palace, 2002).

AGEs can activate toll-like receptors (TLRs) which are involved in the recognition of pathogen-associated molecular patterns and damage-associated molecular patterns in order to enable the innate immune system in response to endogenous danger molecules (Gąsiorowski, Brokos, Echeverria, Barreto, & Leszek, 2018; Van Beijnum, Buurman, & Griffioen, 2008). AGEs and TLRs interaction can lead to the production of pro-inflammatory cytokines, chemokines, neutrophil recruitment, and adhesion molecules (Van Zoelen et al., 2009). AGEs can interact with integrins, which are cell surface receptors that mediate cell adhesion, migration, and signaling. The binding of AGEs to integrins can lead to the activation of intracellular signaling pathways, such as the focal adhesion kinase pathway, and contribute to various pathologies (Pullerits, Brisslert, Jonsson, & Tarkowski, 2006). Among all the receptors of AGEs, RAGE is considered the major cellular receptor for AGEs and plays a crucial role in mediating the harmful effects of AGEs on cellular processes (Chavakis et al., 2004).

4.1 AGEs/RAGE signaling

Upon binding to cell surface receptor RAGE, AGEs trigger the activation of various downstream effectors including activation of NADPH oxidase, extracellular signal-regulated kinases 1 and 2, phosphoinositide 3-kinase, protein kinase B (AKT), and nuclear factor-κB (NF-κB) as shown in Fig. 5 (Bierhaus et al., 2005; Bopp et al., 2008). This array of signaling events which usually increases oxidative stress, and inflammation promotes RAGE expression in a positive feed-forward loop. The oxidative damage can further contribute to the formation of AGEs and perpetuate a cycle of cellular damage and inflammation (Burgos-Morón et al., 2019; Tobon-Velasco, Cuevas, & Torres-Ramos, 2014).

In resting cells, the transcription factor NF-κB resides in the cytoplasm in an inactive form, complexed with inhibitory molecules called inhibitors of κB (IκBs). The most common form of the NF-κB complex in resting cells is the p50/p65 heterodimer bound to IκBα (Ramasamy, Yan, & Schmidt, 2012; Vlassara & Striker, 2011). RAGE activates downstream signaling pathways, including the activation of the IκB kinase (IKK) complex. Upon activation, IKK complex phosphorylates IκBα and IκBβ, leading to their ubiquitination and proteasomal degradation, resulting in the release and translocation of NF-κB into the nucleus (Bierhaus et al., 2005; Vlassara & Striker, 2011). The NF-κB activation promotes transcription of NF-κB regulated target genes, such as proinflammatory cytokines, RAGE expression, prothrombotic and vasoconstrictive gene products, adhesion molecules, and de novo synthesis of NF-κB p65 mRNA (Ramasamy et al., 2012; Vlassara, 2001; Yamagishi, 2019). The upregulation of NF-κB p65 expression by RAGE can create a positive feedback loop causing over-expression of RAGE which in turn perpetuates vicious cycles of oxidative stress and contribute to systemic inflammation (Khalid, Alkaabi, Khan, & Adem, 2021; Tobon-Velasco et al., 2014).

AGEs can induce mitochondrial dysfunction and oxidative stress by stimulating mitochondrial ROS production and activating the c-Jun N-terminal protein kinase (JNK) and NADPH oxidase pathways. JNK is a stress-activated protein kinase (Hirosumi et al., 2002). The JNK pathway can regulate gene expression and cellular functions by phosphorylating various target proteins, including c-Jun, ATF2, and p53 (Bierhaus et al., 2005; Hirosumi et al., 2002; Solinas & Becattini, 2017). The dysregulation of the JNK pathway, NADPH oxidase activity, and ROS production induce mitochondrial dysfunction and endoplasmic reticulum (ER) stress. The induction of ER stress can downregulate sirtuin1 which is a NAD+-dependent deacetylase and plays a critical role in regulating cellular metabolism, oxidative stress, and inflammation (Cai et al., 2012; Vlassara & Striker, 2011). Sirtuin1 and the cell-surface AGER1 (also known as DDOST) provide a protective host defense mechanism by

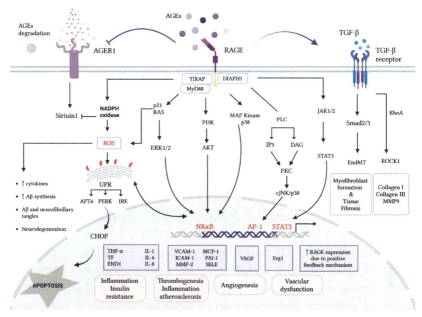

Fig. 5 Intracellular RAGE signaling. RAGE interacts with a diverse spectrum of extracellular ligands and multiple signal transduction pathways, including PI3K, p38MAPK, ERK1/2, PKC, JNK, JAK and TGFβ pathways, leading to activation of transcription factors NF-κB, AP-1, and STAT3. This leads to changes in gene expression and alterations in cellular functions, including inflammation, oxidative stress, angiogenesis, thrombogenesis, insulin resistance and upregulation of RAGE expression. At the same time excess AGEs can directly generate reactive oxygen species (ROS) through multiple pathways, including the activation of NADPH oxidase and the inhibition of antioxidant enzymes that further amplifies AGEs formation and endoplasmic reticulum (ER) stress upon stimulation of unfolded protein response (UPR). AKT, protein kinase B; AP-1, activator protein-1; ATF6, activating transcription factor 6; Aβ, amyloid beta; CHOP, CCAT/enhancer-binding protein homologous protein; DAG, diacylglycerol; DIAPH1, diaphanous 1; EndMT, endothelial to mesenchymal transition; EGR-1, early growth response protein 1; IRE1, serine/threonine-protein kinase/endoribonuclease; IP3, inositol trisphosphate; JAK2, janus kinase 2; JNK, c-Jun N-terminal kinase; MyD88, myeloid differentiating factor 88; MAPK, mitogen-activated protein kinase; NF-κB, nuclear factor kappa B; PERK, protein kinase RNA-like endoplasmic reticulum kinase; PI3-K, phosphoinositide 3-kinase; ROCK1, Rho-associated kinase 1; STAT1, signal transducer and activator of transcription-1; TIRAP, Toll/interleukin-1 receptor domain-containing adapter protein; TGFβ, transforming growth factor beta. Figure created with BioRender.com.

suppressing AGEs and AGEs-related ROS and inflammation (Uribarri et al., 2011). Sustained prolonged supply of exogenous and endogenous AGEs, depletes AGER1 expression (Aragno & Mastrocola, 2017). Recent studies have suggested a potential protective synergism between AGER1 and sirtuin1.

AGER1 has the capability to sequester AGEs and prevent them from binding to RAGE. It can also regulate sirtuin1 activity through modulation of its protein levels (Oliveira et al., 2013; Uribarri et al., 2011). Sirtuin1 via its deacetylase activity inhibits NFκB p65 hyperacetylation and suppresses enhanced transcription of inflammatory genes (Vlassara & Striker, 2011). Sirtuin1 is known to regulate various downstream targets, including peroxisome proliferator-activated receptor γ co-activator 1α (PGC-1α), forkhead box transcription factors (FOXO), and p53, which play important roles in cellular metabolism, oxidative stress, and inflammation. AGEs can suppress sirtuin1 activity via changes in the NAD+/NADH ratio. Besides AGEs, the intracellular NAD+/NADH ratio can be affected by various factors, including oxidative stress, mitochondrial dysfunction, and overnutrition (Cai et al., 2012; Uribarri et al., 2011; Vlassara & Striker, 2011).

Overall, AGEs/RAGE axis appears to play a central role in the pathogenesis of various diseases, including cancer, diabetes, diabetic complication, atherosclerosis, vascular dysfunction, and neurodegenerative disorders through prolonged activation and promotion of signaling mechanisms inducing the continuous generation of inflammation and oxidative stress as shown in Fig. 5.

4.2 Pathophysiology of AGEs/RAGE axis

The interaction between AGEs and RAGE is a key contributor to the development and progression of many chronic diseases. The sustained postreceptor signaling allows AGEs/RAGE/NFκB to propagate cellular dysfunction in several pathophysiological relevant situations.

4.2.1 AGEs/RAGE axis and progression of diabetes mellitus

AGEs and diabetes have a complex relationship, the AGEs can be the cause and effect of diabetes mellitus. The high glucose levels in diabetes mellitus due to defective insulin signaling or secretion can lead to the formation of AGEs (Vlassara & Uribarri, 2014). Chronic hyperglycemia has been shown to increase the formation of AGEs, serum AGEs level, and their accumulation in various tissues (Rungratanawanich et al., 2021; Yao & Brownlee, 2010). AGEs can accumulate in pancreatic beta cells and bind to RAGE receptors, leading to the activation of various signaling pathways that contribute to beta cell dysfunction. The downstream effects of AGEs/RAGE signaling include oxidative stress, inflammation, and apoptosis in pancreatic beta cells (Coughlan et al., 2011; Khalid et al., 2022). The overload of ROS formation exceeds the pancreatic antioxidant defense mechanisms (Lin, Zhang, & Su, 2012). Furthermore, RAGE is expressed on beta cells and promotes aggregation of islet amyloid polypeptide (IAPP) into amyloid fibrils that deposit in pancreatic islets, contributing to beta cell dysfunction and death (Iki & Pour, 2007; Khalid et al., 2022). Overall, the interaction between AGEs/RAGE axis and IAPP contributes to the progression of diabetes by causing insulin secretory defects, impairment of mitochondrial function, and decrease in beta cell regeneration & survival (Cerf, 2013; Gerber & Rutter, 2017). Studies have shown that AGEs contribute to pancreatic beta-cell apoptosis via the interaction of AGE/RAGE signaling, while inhibition of RAGE can effectively protect beta-cells against AGE-induced apoptosis (Lee et al., 2010; Zhu et al., 2011).

The AGEs alter insulin action by direct modification of insulin through covalent crosslinking (Fedintsev & Moskalev, 2020; Unoki & Yamagishi, 2008). The AGEs/RAGE axis has been implicated in the development of insulin resistance, a hallmark of type 2 diabetes mellitus (Nin et al., 2011). The underlying molecular mechanisms leading to AGEs-induced insulin resistance comprise activation of protein kinase C, p38, IKK/NF-κB, JAK/STAT, JNK pathways, ER stress, and redox imbalance which results in dysregulation of intracellular signaling (Burgos-Morón et al., 2019; Solinas & Becattini, 2017; Tobon-Velasco et al., 2014; Unoki & Yamagishi, 2008; Vlassara & Uribarri, 2014). These pathways work either independently or synergistically and can influence insulin sensitivity in major insulin-sensitive organs. The skeletal muscle is the largest organ and principal tissue responsible for glucose uptake and utilization through glucose transporters (GLUTs) to maintain whole body glucose homeostasis (Abdul-Ghani & DeFronzo, 2010; Brons & Grunnet, 2017). The glucose transporter protein 4 (GLUT4) encoded by *Slc2a4* gene is the major isoform expressed in skeletal muscles (Gannon, Conn, & Vaughan, 2015). The activated NFκB specifically its p65 subunit binds into the promoter region of *Slc2a4* gene and represses its transcription. The decreased GLUT4 expression impairs glucose homeostasis in skeletal muscles and is considered the major mechanism contributing to insulin resistance (Deshmukh, 2016; Pinto-Junior et al., 2018). Moreover, AGEs have also a direct impact on mitochondrial energy metabolism via the downregulation of AGER1 and sirtuin1. Sirtuin1 is responsible for the activation of PGC 1α, the master regulator of mitochondrial metabolism and biogenesis (Sergi et al., 2021). Thus AGEs/RAGE axis contributes to insulin resistance and diabetes progression through direct and indirect actions via numerous pathways including induction of oxidative stress generation, chronic inflammation, and causing defects in the insulin signaling cascade.

4.3 AGEs/RAGE axis and retinopathy

Diabetics are more likely to develop micro and macrovascular complications which are the chief cause of morbidity and mortality in diabetes (Vlassara, 2001). The excess of AGEs contributes to the metabolic memory that provides the clinical link between diabetic complications and their pathogenesis (Ramasamy, Yan, & Schmidt, 2011). The ECM proteins modified by AGEs may severely affect cell function and plays a crucial role in the development of diabetic complications. Modification of collagen with glyoxal and methylglyoxal impairs collagen degradation, stimulates myofibroblast differentiation, and facilitates fibrosis (Pedchenko et al., 2005; Ramasamy et al., 2011; Saito & Marumo, 2015). Fibrosis leads to increased stiffness and reduced elasticity of tissues which impair their normal function. Glycation of other ECM proteins including laminin or fibronectin has shown impaired interactions with other matrix components as well as weakened cell-matrix interactions (Ramasamy et al., 2011; Singh et al., 2014). Regarding diabetic retinopathy, AGEs can affect retinal cells through both receptor-dependent and receptor-independent pathways. Receptor-dependent pathways involve the AGEs/RAGE interaction while receptor-independent pathways involve the non-enzymatic glycation of ECMs (Fukami, Yamagishi, & Okuda, 2014; Zong, Ward, & Stitt, 2011). One study conducted on primary cultures of bovine retinal pericytes demonstrated that modification of fibronectin by α-dicarbonyl compounds like methylglyoxal and glyoxal triggers apoptosis and contribute to loss of pericytes in diabetic retinopathy (Liu, Bhat, Padival, Smith, & Nagaraj, 2004).

4.4 AGEs/RAGE axis and nephropathy

Consistent with the above-mentioned examples, another study demonstrated that AGEs-induced cross-links and structural alterations of ECM proteins correlate with the severity of diabetic nephropathy (Forbes, Cooper, Oldfield, & Thomas, 2003). RAGE expression has been detected in several types of cells in the kidney, including podocytes, mesangial cells, glomerular endothelial cells, tubular epithelial cells, and endothelial cells. RAGE expression is normally low in the kidney, but an upregulated RAGE expression is observed in both animal and human nephropathies (D'agati & Schmidt, 2010). RAGE activation by AGEs amplifies inflammatory processes, activates oxidative stress dependent mechanisms, and induces glomerular stress (Sanajou, Haghjo, Argani, & Aslani, 2018). All these processes have been implicated in the pathogenesis of diabetic nephropathy and other kidney diseases as these culminate in the gradual loss of kidney architecture and function (D'agati & Schmidt, 2010). Studies have shown that treatment with anti-RAGE antibodies can improve renal function and reduce markers of kidney damage in mouse models of diabetes (Jensen et al., 2006; Sanajou et al., 2018).

4.5 AGEs/RAGE axis and Cardiovascular disorders

AGEs/RAGE interactions play a significant role in the pathogenesis of CVD through multiple mechanisms. AGEs-crosslinks in the ECM of the vasculature can cause structural modification and functional alteration of the ECM proteins as well as intracellular signaling molecules (Zhao, Randive, & Stewart, 2014). Increased ECM deposition and accumulation may result in vessel stiffening, impaired arterial compliance, and an increase in blood pressure which contribute to the development of hypertension and other CVDs (Burr & Stewart, 2020; Zhao et al., 2014). AGEs can also promote oxidative stress, inflammation, and endothelial dysfunction, leading to the development of atherosclerosis and other CVDs (Hyun & Shim, 2011). Atherosclerosis is the major cause of most ischemic heart diseases in which the blood supply to the heart is often reduced due to plaque formation in the arterial wall (Nowotny, Jung, Höhn, Weber, & Grune, 2015). Atherosclerosis is characterized by the deposition of plaques on the insides of arterial walls, restricted blood flow to vital organs and tissues, and eventually myocardial infarction (Ahmed, 2005; Bucciarelli et al., 2002). Studies have shown that intracellular AGEs deposition in aortic atherosclerosis lesions retrieved from human subjects (Kume et al., 1995). Studies have demonstrated that the absence of apolipoprotein E in diabetic db/db mice leads to the upregulation of RAGE-dependent gene expression of several pro-inflammatory molecules such as vascular cell adhesion molecule-1, tissue factor, and matrix metalloproteinase-9 in comparison to non-diabetic animals. This enhances the formation of atherosclerotic plaques at the aortic sinus, which eventually occludes blood flow and leads to myocardial infarction (Fowlkes et al., 2013; Fukami et al., 2014). It has been discussed already that AGEs/RAGE signaling can induce proinflammatory cytokine expression and oxidative stress via the NF-κB pathway (Nowotny et al., 2015). Another chief pathway found to be impacted by RAGE is the ROCK1 branch of the TGF-β signaling pathway (Bu et al., 2010). Activation of this pathway can lead to increased production of ECM proteins, vascular smooth muscle cell proliferation, and endothelial dysfunction, all of which contribute to the development and progression of CVD (Ramasamy et al., 2012). Nitric Oxide (NO) is a key mediator of vascular homeostasis, regulating vasodilation, platelet aggregation, and leukocyte adhesion to the endothelium. However, AGEs can react with and scavenge NO, leading to decreased bioavailability of NO and impaired vascular function. This can contribute to the development and progression of atherosclerosis and other CVD

(Nowotny et al., 2015). Taken together, AGEs/RAGE interaction is linked to mechanisms of atherosclerosis, and other CVDs while findings from animal and human studies support the potential benefits of RAGE blockade for cardioprotection (Fukami et al., 2014; Hyun & Shim, 2011; Ramasamy et al., 2011).

4.6 AGEs/RAGE axis and neurodegeneration

The major pathological feature of neurodegeneration including Alzheimer's disease, Parkinson's disease, Huntington's disease, multiple sclerosis, and amyotrophic lateral sclerosis, is the presence of a large number of neuroinflammatory plaques (Li et al., 2012). The amyloid beta protein is the main component of these neuroinflammatory plaques in the cerebral cortex, hippocampus, certain subcortical nuclei, and thalamus (Dong, Zhang, Huang, & Deng, 2022). Both AGEs and amyloid beta protein are able to bind and activate RAGE signaling and induce the expression of pro-inflammatory cytokines through NF-κB dependent pathways (Tobon-Velasco et al., 2014). High expression of RAGE accelerates the accumulation of amyloid beta, neuroinflammatory plaques formation, and functions as a signal transducer for AGEs-induced synaptic dysfunction (Dong et al., 2022; Li et al., 2012; Pintana et al., 2017). This chronic neuroinflammation contributes to neuropathological and biochemical changes, causes neuronal damage and loss, leading to the cognitive and motor impairments, spatial learning deficits, and memory impairment observed in neurodegenerative diseases (Pintana et al., 2017; Tobon-Velasco et al., 2014). Studies have shown that the blockade of RAGE signaling in neurons or microglia of Alzheimer's disease mice attenuates amyloid beta-induced deterioration and prevents spatial learning deficits, memory impairment, and neuropathological and biochemical changes (Ding et al., 2020). Genetic deletion of RAGE using novel transgenic neuronal RAGE knockout mice offers protection against synaptic injury induced by AGEs and leads to an improvement in cognitive decline. (Zhang et al., 2014) .

5. Perspectives

In summary, the AGEs/RAGE axis provides a fundamental cue in understanding the development of chronic diseases that takes years to develop. The accumulation of AGEs over time plays a critical role in the pathogenesis of several diseases. Thus, future strategies aimed at preventing or reducing AGEs production should be targeted. We have mentioned some mechanistic strategies against AGEs production, consumption, toxicity, and accumulation in Table 1.

Table 1 Defense Mechanisms to counteract AGEs toxicity.

Defense Mechanisms

Strategies	Explanation	References
Lifestyle Changes	Clinical studies have shown that lifestyle modification by reducing dietary AGEs can help to modulate the AGEs/RAGE axis and can be effective in preventing the development of metabolic syndrome. Physical activity has been shown to increase the activity of enzymes that break down AGEs precursors and can help to prevent AGEs accumulation.	Santos-Bezerra et al. (2018), Magalhães et al. (2008), Hull et al. (2005)
AGEs Inhibitors	Clinical evidence supports sevelamer carbonate, a nonabsorbable oral drug can be used to prevent the absorption of diet AGEs. Other studies suggest that AST-120, an oral adsorbent drug may also be capable of binding to AGEs, such as carboxymethyl lysine (CML). This phenomenon could potentially impair their absorption into the bloodstream. Moreover, a recent human trial showed that L-Carnosine supplementation attenuates fasting glucose and serum AGEs.	Houjeghani et al. (2018), Vlassara et al. (2012), Yamagishi et al. (2007)
Antioxidants	Antioxidants can help to scavenge free radicals and prevent oxidative stress, which contributes to AGEs formation and accumulation. An efficient way for AGEs detoxification is the early degradation of AGEs by chemo-preventive enzymes and antioxidants. Several studies have shown the efficacy of Glyoxalase 1 (GLO-1) to prevent carbonyl stress. GLO-1 detoxifies reactive dicarbonyls and reduces AGEs levels and oxidative stress in mouse mesangial cells under diabetic conditions. While impairment of GLO-1 is associated in AGEs accumulation in the cartilage of osteoarthritis patients. The use of phytochemicals having potent antioxidant and anti-inflammatory activity is another strategy to overcome AGEs load. Studies have shown the beneficial effects of several phytochemicals like quercetin, sulforaphane, iridoids, and curcumin on AGEs formation as well as RAGE-mediated signaling pathways in various cell types and organs.	Costa et al. (2020), Trellu et al. (2019), Yamagishi et al. (2017), Alfarano et al. (2018), Miller et al. (2010)
RAGE Antagonist	RAGE inhibition strategies by using a neutralizing monoclonal anti-RAGE antibody, sRAGE, and RAGE blockers have a great therapeutic impact to counteract AGE/RAGE-associated pathophysiology. Initial data from recent studies evaluating the RAGE inhibition effects of FPS-ZM1 and RAGE-aptamers showed promising results.	Sanajou et al. (2018), Hong et al. (2016), Christaki et al. (2011), Bro et al. (2008)

Although recent studies have provided important insights into the AGEs/RAGE axis and structural and functional properties of RAGE, there are critical issues that remain to be addressed. In this context, determining the mechanisms of AGEs toxicity, intervention to break preexisting AGEs crosslinks, AGEs/RAGE axis associated signaling pathways, the impact of RAGE blockage, metabolic memory-related pathophysiology, and techniques to determine a variety of AGEs need to be further elucidated.

Funding

M.K. is supported by the internal award of Khalifa University of Science and Technology under Grant [8474000360, FSU-2021–020], and [8474000478, ESIG−2023–007].

Conflicts of interest

The authors have no conflict of interest.

References

Abdel-Wahab, Y., O'Harte, F., Boyd, A., Barnett, C., & Flatt, P. (1997). Glycation of insulin results in reduced biological activity in mice. *Acta Diabetologica, 34*(4), 265–270.
Abdul-Ghani, M. A., & DeFronzo, R. A. (2010). Pathogenesis of insulin resistance in skeletal muscle. *Journal of Biomedicine and Biotechnology, 2010*.
Ahmed, N. (2005). Advanced glycation endproducts—Role in pathology of diabetic complications. *Diabetes Research and Clinical Practice, 67*(1), 3–21.
Alfarano, M., Pastore, D., Fogliano, V., Schalkwijk, C. G., & Oliviero, T. (2018). The effect of sulforaphane on glyoxalase I expression and activity in peripheral blood mononuclear cells. *Nutrients, 10*(11), 1773.
Aragno, M., & Mastrocola, R. (2017). Dietary sugars and endogenous formation of advanced glycation endproducts: Emerging mechanisms of disease. *Nutrients, 9*(4), 385.
Berg, T. J., Snorgaard, O., Faber, J., Torjesen, P. A., Hildebrandt, P., Mehlsen, J., & Hanssen, K. F. (1999). Serum levels of advanced glycation end products are associated with left ventricular diastolic function in patients with type 1 diabetes. *Diabetes care, 22*(7), 1186–1190.
Bierhaus, A., Humpert, P. M., Morcos, M., Wendt, T., Chavakis, T., Arnold, B., ... Nawroth, P. P. (2005). Understanding RAGE, the receptor for advanced glycation end products. *Journal of Molecular Medicine, 83*(11), 876–886.
Bopp, C., Bierhaus, A., Hofer, S., Bouchon, A., Nawroth, P. P., Martin, E., & Weigand, M. A. (2008). Bench-to-bedside review: The inflammation-perpetuating pattern-recognition receptor RAGE as a therapeutic target in sepsis. *Critical Care, 12*(1), 1–8.
Brings, S., Fleming, T., Freichel, M., Muckenthaler, M. U., Herzig, S., & Nawroth, P. P. (2017). Dicarbonyls and advanced glycation end-products in the development of diabetic complications and targets for intervention. *International Journal of Molecular Sciences, 18*(5), 984.
Bro, S., Flyvbjerg, A., Binder, C. J., Bang, C. A., Denner, L., Olgaard, K., & Nielsen, L. B. (2008). A neutralizing antibody against receptor for advanced glycation end products (RAGE) reduces atherosclerosis in uremic mice. *Atherosclerosis, 201*(2), 274–280.
Brons, C., & Grunnet, L. G. (2017). Skeletal muscle lipotoxicity in insulin resistance and type 2 diabetes: A causal mechanism or an innocent bystander. *European Journal of Endocrinology/European Federation of Endocrine Societies, 176*(2), R67–R78.

Bu, D., Rai, V., Shen, X., Rosario, R., Lu, Y., D'Agati, V., ... Schmidt, A. M. (2010). Activation of the ROCK1 branch of the transforming growth factor-β pathway contributes to RAGE-dependent acceleration of atherosclerosis in diabetic ApoE-null mice. *Circulation Research, 106*(6), 1040–1051.

Bucciarelli, L. G., Wendt, T., Qu, W., Lu, Y., Lalla, E., Rong, L. L., ... Lee, D. C. (2002). RAGE blockade stabilizes established atherosclerosis in diabetic apolipoprotein E–null mice. *Circulation, 106*(22), 2827–2835.

Burgos-Morón, E., Abad-Jiménez, Z., Martinez De Maranon, A., Iannantuoni, F., Escribano-López, I., López-Domènech, S., ... Roldan, I. (2019). Relationship between oxidative stress, ER stress, and inflammation in type 2 diabetes: The battle continues. *Journal of Clinical Medicine, 8*(9), 1385.

Burr, S. D., & Stewart, Jr. J. A. (2020). Extracellular matrix components isolated from diabetic mice alter cardiac fibroblast function through the AGE/RAGE signaling cascade. *Life Sciences, 250*, 117569.

Cai, W., He, J. C., Zhu, L., Chen, X., Zheng, F., Striker, G. E., & Vlassara, H. (2008). Oral glycotoxins determine the effects of calorie restriction on oxidant stress, age-related diseases, and lifespan. *The American Journal of Pathology, 173*(2), 327–336.

Cai, W., Ramdas, M., Zhu, L., Chen, X., Striker, G. E., & Vlassara, H. (2012). Oral advanced glycation endproducts (AGEs) promote insulin resistance and diabetes by depleting the antioxidant defenses AGE receptor-1 and sirtuin 1. *Proceedings of the National Academy of Sciences, 109*(39), 15888–15893.

Cai, W., Uribarri, J., Zhu, L., Chen, X., Swamy, S., Zhao, Z., ... Schnaider-Beeri, M. (2014). Oral glycotoxins are a modifiable cause of dementia and the metabolic syndrome in mice and humans. *Proceedings of the National Academy of Sciences, 111*(13), 4940–4945.

Cerf, M. E. (2013). Beta cell dysfunction and insulin resistance. *Frontiers in Endocrinology, 4*, 37.

Chaudhuri, J., Bains, Y., Guha, S., Kahn, A., Hall, D., Bose, N., ... Kapahi, P. (2018). The role of advanced glycation end products in aging and metabolic diseases: Bridging association and causality. *Cell Metabolism, 28*(3), 337–352.

Chavakis, T., Bierhaus, A., & Nawroth, P. P. (2004). RAGE (receptor for advanced glycation end products): A central player in the inflammatory response. *Microbes and Infection, 6*(13), 1219–1225.

Cho, S.-J., Roman, G., Yeboah, F., & Konishi, Y. (2007). The road to advanced glycation end products: A mechanistic perspective. *Current Medicinal Chemistry, 14*(15), 1653–1671.

Christaki, E., Opal, S. M., Keith, J. C., Jr, Kessimian, N., Palardy, J. E., Parejo, N. A., ... Pittman, D. D. (2011). A monoclonal antibody against RAGE alters gene expression and is protective in experimental models of sepsis and pneumococcal pneumonia. *Shock, 35*(5), 492–498.

Costa, A. F., Campos, D., Reis, C. A., & Gomes, C. (2020). Targeting glycosylation: a new road for cancer drug discovery. *Trends in Cancer, 6*(9), 757–766.

Coughlan, M. T., Yap, F. Y., Tong, D. C., Andrikopoulos, S., Gasser, A., Thallas-Bonke, V., ... Slattery, R. M. (2011). Advanced glycation end products are direct modulators of β-cell function. *Diabetes, 60*(10), 2523–2532.

D'agati, V., & Schmidt, A. M. (2010). RAGE and the pathogenesis of chronic kidney disease. *Nature Reviews Nephrology, 6*(6), 352–360.

Daoud, S., Schinzel, R., Neumann, A., Loske, C., Fraccarollo, D., Diez, C., & Simm, A. (2001). Advanced glycation end products: activators of cardiac remodelling in primary fibroblasts from adult rat hearts. *Molecular Medicine, 7*, 543–551.

Deshmukh, A. S. (2016). Insulin-stimulated glucose uptake in healthy and insulin-resistant skeletal muscle. *Hormone Molecular Biology and Clinical Investigation, 26*(1), 13–24.

Ding, B., Lin, C., Liu, Q., He, Y., Ruganzu, J. B., Jin, H., ... Yang, W. (2020). Tanshinone IIA attenuates neuroinflammation via inhibiting RAGE/NF-κB signaling pathway in vivo and in vitro. *Journal of Neuroinflammation, 17*(1), 1–17.

Ding, Q., & Keller, J. N. (2005). Evaluation of rage isoforms, ligands, and signaling in the brain. *Biochimica et Biophysica Acta (BBA)-Molecular Cell Research, 1746*(1), 18–27.

Dong, H., Zhang, Y., Huang, Y., & Deng, H. (2022). Pathophysiology of RAGE in inflammatory diseases. *Frontiers in Immunology, 13*.

Duraisamy, Y., Slevin, M., Smith, N., Bailey, J., Zweit, J., Smith, C., ... Gaffney, J. (2001). Effect of glycation on basic fibroblast growth factor induced angiogenesis and activation of associated signal transduction pathways in vascular endothelial cells: possible relevance to wound healing in diabetes. *Angiogenesis, 4*, 277–288.

Ellis, H. (1912). Psychopathic Pains [Des Différents Espèces de Douleurs Psychopathiques]. (L'Encéphale, Sept. 10th, 1911.) Mallard, etc. *Journal of Mental Science, 58*(240), 122–123.

Fedintsev, A., & Moskalev, A. (2020). Stochastic non-enzymatic modification of long-lived macromolecules-A missing hallmark of aging. *Ageing Research Reviews, 62*, 101097.

Forbes, J. M., Cooper, M. E., Oldfield, M. D., & Thomas, M. C. (2003). Role of advanced glycation end products in diabetic nephropathy. *Journal of the American Society of Nephrology, 14*(suppl 3), S254–S258.

Fowlkes, V., Clark, J., Fix, C., Law, B. A., Morales, M. O., Qiao, X., ... Murray, D. B. (2013). Type II diabetes promotes a myofibroblast phenotype in cardiac fibroblasts. *Life Sciences, 92*(11), 669–676.

Fukami, K., Yamagishi, S., & Okuda, S. (2014). Role of AGEs-RAGE system in cardiovascular disease. *Current Pharmaceutical Design, 20*(14), 2395–2402.

Gannon, N. P., Conn, C. A., & Vaughan, R. A. (2015). Dietary stimulators of GLUT4 expression and translocation in skeletal muscle: A mini-review. *Molecular Nutrition & Food Research, 59*(1), 48–64.

Gąsiorowski, K., Brokos, B., Echeverria, V., Barreto, G. E., & Leszek, J. (2018). RAGE-TLR crosstalk sustains chronic inflammation in neurodegeneration. *Molecular Neurobiology, 55*(2), 1463–1476.

Gerber, P. A., & Rutter, G. A. (2017). The role of oxidative stress and hypoxia in pancreatic beta-cell dysfunction in diabetes mellitus. *Antioxidants & Redox Signaling, 26*(10), 501–518.

Geroldi, D., Falcone, C., & Emanuele, E. (2006). Soluble receptor for advanced glycation end products: From disease marker to potential therapeutic target. *Current Medicinal Chemistry, 13*(17), 1971–1978.

Goldberg, T., Cai, W., Peppa, M., Dardaine, V., Baliga, B. S., Uribarri, J., & Vlassara, H. (2004). Advanced glycoxidation end products in commonly consumed foods. *Journal of the American Dietetic Association, 104*(8), 1287–1291.

Goova, M. T., Li, J., Kislinger, T., Qu, W., Lu, Y., Bucciarelli, L. G., ... Yan, S. F. (2001). Blockade of receptor for advanced glycation end-products restores effective wound healing in diabetic mice. *The American Journal of Pathology, 159*(2), 513–525.

He, C., Sabol, J., Mitsuhashi, T., & Vlassara, H. (1999). Dietary glycotoxins: Inhibition of reactive products by aminoguanidine facilitates renal clearance and reduces tissue sequestration. *Diabetes, 48*(6), 1308–1315.

Hirosumi, J., Tuncman, G., Chang, L., Görgün, C. Z., Uysal, K. T., Maeda, K., ... Hotamisligil, G. S. (2002). A central role for JNK in obesity and insulin resistance. *Nature, 420*(6913), 333–336.

Hofmann, S. M., Dong, H.-J., Li, Z., Cai, W., Altomonte, J., Thung, S. N., ... Vlassara, H. (2002). Improved insulin sensitivity is associated with restricted intake of dietary glycoxidation products in the db/db mouse. *Diabetes, 51*(7), 2082–2089.

Hong, C. O., Nam, M. H., Oh, J. S., Lee, J. W., Kim, C. T., Park, K. W., ... Lee, K. W. (2016). Pheophorbide a from Capsosiphon fulvescens inhibits advanced glycation end products mediated endothelial dysfunction. *Planta medica, 82*(01/02), 46–57.

Houjeghani, S., Kheirouri, S., Faraji, E., & Jafarabadi, M. A. (2018). L-Carnosine supplementation attenuated fasting glucose, triglycerides, advanced glycation end products, and tumor necrosis factor–α levels in patients with type 2 diabetes: a double-blind placebo-controlled randomized clinical trial. *Nutrition Research, 49*, 96–106.

Hull, G. L., Woodside, J. V., Ames, J. M., & Cuskelly, G. J. (2012). Nε-(carboxymethyl) lysine content of foods commonly consumed in a Western style diet. *Food Chemistry, 131*(1), 170–174.

Hyun, S. P. S.-J. Y., & Shim, J. T. C. Y. (2011). RAGE and cardiovascular disease. *Frontiers in Bioscience, 16*, 486–497.

Iki, K., & Pour, P. M. (2007). Distribution of pancreatic endocrine cells including IAPP-expressing cells in non-diabetic and type 2 diabetic cases. *Journal of Histochemistry & Cytochemistry, 55*(2), 111–118.

Jensen, L. J., Denner, L., Schrijvers, B. F., Tilton, R. G., Rasch, R., & Flyvbjerg, A. (2006). Renal effects of a neutralising RAGE-antibody in long-term streptozotocin-diabetic mice. *Journal of Endocrinology, 188*(3), 493–501.

John, W. G., & Lamb, E. J. (1993). The Maillard or browning reaction in diabetes. *Eye (London, England), 7*(2), 230–237.

Jono, T., Miyazaki, A., Nagai, R., Sawamura, T., Kitamura, T., & Horiuchi, S. (2002). Lectin-like oxidized low density lipoprotein receptor-1 (LOX-1) serves as an endothelial receptor for advanced glycation end products (AGE). *FEBS Letters, 511*(1–3), 170–174.

Khalid, M., Alkaabi, J., Khan, M. A., & Adem, A. (2021). Insulin signal transduction perturbations in insulin resistance. *International Journal of Molecular Sciences, 22*(16), 8590.

Khalid, M., Petroianu, G., & Adem, A. (2022). Advanced Glycation End Products and Diabetes Mellitus: Mechanisms and Perspectives. *Biomolecules, 12*(4), 542.

Koschinsky, T., He, C.-J., Mitsuhashi, T., Bucala, R., Liu, C., Buenting, C., ... Vlassara, H. (1997). Orally absorbed reactive glycation products (glycotoxins): An environmental risk factor in diabetic nephropathy. *Proceedings of the National Academy of Sciences, 94*(12), 6474–6479.

Kume, S., Takeya, M., Mori, T., Araki, N., Suzuki, H., Horiuchi, S., ... Takahashi, K. (1995). Immunohistochemical and ultrastructural detection of advanced glycation end products in atherosclerotic lesions of human aorta with a novel specific monoclonal antibody. *The American journal of pathology, 147*(3), 654.

Kuzan, A. (2021). Toxicity of advanced glycation end products. *Biomedical Reports, 14*(5), 1–8.

Lee, B.-W., Chae, H. Y., Kwon, S. J., Park, S. Y., Ihm, J., & Ihm, S.-H. (2010). RAGE ligands induce apoptotic cell death of pancreatic β-cells via oxidative stress. *International Journal of Molecular Medicine, 26*(6), 813–818.

Li, J., Liu, D., Sun, L., Lu, Y., & Zhang, Z. (2012). Advanced glycation end products and neurodegenerative diseases: Mechanisms and perspective. *Journal of the Neurological Sciences, 317*(1–2), 1–5.

Lin, N., Zhang, H., & Su, Q. (2012). Advanced glycation end-products induce injury to pancreatic beta cells through oxidative stress. *Diabetes & Metabolism, 38*(3), 250–257.

Liu, B., Bhat, M., Padival, A. K., Smith, D. G., & Nagaraj, R. H. (2004). Effect of dicarbonyl modification of fibronectin on retinal capillary pericytes. *Investigative Ophthalmology & Visual Science, 45*(6), 1983–1995.

Lu, C., He, J. C., Cai, W., Liu, H., Zhu, L., & Vlassara, H. (2004). Advanced glycation endproduct (AGE) receptor 1 is a negative regulator of the inflammatory response to AGE in mesangial cells. *Proceedings of the National Academy of Sciences, 101*(32), 11767–11772.

Luevano-Contreras, C., & Chapman-Novakofski, K. (2010). Dietary advanced glycation end products and aging. *Nutrients, 2*(12), 1247–1265.

Magalhães, P. M., Appell, H. J., & Duarte, J. A. (2008). Involvement of advanced glycation end products in the pathogenesis of diabetic complications: the protective role of regular physical activity. *European Review of Aging and Physical Activity, 5*, 17–29.

Mark, A. B., Poulsen, M. W., Andersen, S., Andersen, J. M., Bak, M. J., Ritz, C., ... Dragsted, L. O. (2014). Consumption of a diet low in advanced glycation end products

for 4 weeks improves insulin sensitivity in overweight women. *Diabetes Care, 37*(1), 88–95.
Martins, S. I., Jongen, W. M., & Van Boekel, M. A. (2000). A review of Maillard reaction in food and implications to kinetic modelling. *Trends in Food Science & Technology, 11*(9–10), 364–373.
Mastrocola, R., Collotta, D., Gaudioso, G., Le Berre, M., Cento, A. S., Ferreira Alves, G., ... Manig, F. (2020). Effects of exogenous dietary advanced glycation end products on the cross-talk mechanisms linking microbiota to metabolic inflammation. *Nutrients, 12*(9), 2497.
Miller, A. G., Tan, G., Binger, K. J., Pickering, R. J., Thomas, M. C., Nagaraj, R. H., ... Wilkinson-Berka, J. L. (2010). Candesartan attenuates diabetic retinal vascular pathology by restoring glyoxalase-I function. *Diabetes, 59*(12), 3208–3215.
Miyazaki, A., Nakayama, H., & Horiuchi, S. (2002). Scavenger receptors that recognize advanced glycation end products. *Trends in Cardiovascular Medicine, 12*(6), 258–262.
Nakamura, S., Makita, Z., Ishikawa, S., Yasumura, K., Fujii, W., Yanagisawa, K., ... Koike, T. (1997). Progression of nephropathy in spontaneous diabetic rats is prevented by OPB-9195, a novel inhibitor of advanced glycation. *Diabetes, 46*(5), 895–899.
Nicholl, I., & Bucala, R. (1998). Advanced glycation endproducts and cigarette smoking. *Cellular and Molecular Biology (Noisy-Le-Grand, France), 44*(7), 1025–1033.
Nin, J. W., Jorsal, A., Ferreira, I., Schalkwijk, C. G., Prins, M. H., Parving, H.-H., ... Stehouwer, C. D. (2011). Higher plasma levels of advanced glycation end products are associated with incident cardiovascular disease and all-cause mortality in type 1 diabetes: A 12-year follow-up study. *Diabetes Care, 34*(2), 442–447.
Nowotny, K., Jung, T., Höhn, A., Weber, D., & Grune, T. (2015). Advanced glycation end products and oxidative stress in type 2 diabetes mellitus. *Biomolecules, 5*(1), 194–222.
Oliveira, M. I. A., Souza, E. M. De, Pedrosa, F. de O., Réa, R. R., Alves, A. Da. S. C., Picheth, G., & Rego, F. G. De. M. (2013). RAGE receptor and its soluble isoforms in diabetes mellitus complications. *Jornal Brasileiro de Patologia e Medicina Laboratorial, 49*, 97–108.
Ott, C., Jacobs, K., Haucke, E., Santos, A. N., Grune, T., & Simm, A. (2014). Role of advanced glycation end products in cellular signaling. *Redox Biology, 2*, 411–429.
Pedchenko, V. K., Chetyrkin, S. V., Chuang, P., Ham, A.-J. L., Saleem, M. A., Mathieson, P. W., ... Voziyan, P. A. (2005). Mechanism of perturbation of integrin-mediated cell-matrix interactions by reactive carbonyl compounds and its implication for pathogenesis of diabetic nephropathy. *Diabetes, 54*(10), 2952–2960.
Peppa, M., Brem, H., Ehrlich, P., Zhang, J.-G., Cai, W., Li, Z., ... Vlassara, H. (2003). Adverse effects of dietary glycotoxins on wound healing in genetically diabetic mice. *Diabetes, 52*(11), 2805–2813.
Peppa, M., He, C., Hattori, M., McEvoy, R., Zheng, F., & Vlassara, H. (2003). Fetal or neonatal low-glycotoxin environment prevents autoimmune diabetes in NOD mice. *Diabetes, 52*(6), 1441–1448.
Perrone, A., Giovino, A., Benny, J., & Martinelli, F. (2020). Advanced glycation end products (AGEs): Biochemistry, signaling, analytical methods, and epigenetic effects. *Oxidative Medicine and Cellular Longevity, 2020.*
Pintana, H., Apaijai, N., Kerdphoo, S., Pratchayasakul, W., Sripetchwandee, J., Suntornsaratoon, P., ... Chattipakorn, S. C. (2017). Hyperglycemia induced the Alzheimer's proteins and promoted loss of synaptic proteins in advanced-age female Goto-Kakizaki (GK) rats. *Neuroscience Letters, 655*, 41–45.
Pinto-Junior, D. C., Silva, K. S., Michalani, M. L., Yonamine, C. Y., Esteves, J. V., Fabre, N. T., ... Seraphim, P. M. (2018). Advanced glycation end products-induced insulin resistance involves repression of skeletal muscle GLUT4 expression. *Scientific Reports, 8*(1), 8109.

Poulsen, M. W., Bak, M. J., Andersen, J. M., Monošík, R., Giraudi-Futin, A. C., Holst, J. J., ... Bügel, S. (2014). Effect of dietary advanced glycation end products on postprandial appetite, inflammation, and endothelial activation in healthy overweight individuals. *European Journal of Nutrition, 53*, 661–672.

Prasad, C., Imrhan, V., Marotta, F., Juma, S., & Vijayagopal, P. (2014). Lifestyle and advanced glycation end products (AGEs) burden: Its relevance to healthy aging. *Aging and Disease, 5*(3), 212.

Pucci, M., Aria, F., Premoli, M., Maccarinelli, G., Mastinu, A., Bonini, S., ... Abate, G. (2021). Methylglyoxal affects cognitive behaviour and modulates RAGE and Presenilin-1 expression in hippocampus of aged mice. *Food and Chemical Toxicology, 158*, 112608.

Pullerits, R., Brisslert, M., Jonsson, I., & Tarkowski, A. (2006). Soluble receptor for advanced glycation end products triggers a proinflammatory cytokine cascade via β2 integrin Mac-1. *Arthritis & Rheumatism: Official Journal of the American College of Rheumatology, 54*(12), 3898–3907.

Ramasamy, R., Yan, S. F., & Schmidt, A. M. (2011). Receptor for AGE (RAGE): Signaling mechanisms in the pathogenesis of diabetes and its complications. *Annals of the New York Academy of Sciences, 1243*, 88.

Ramasamy, R., Yan, S. F., & Schmidt, A. M. (2012). Advanced glycation endproducts: From precursors to RAGE: Round and round we go. *Amino Acids, 42*, 1151–1161.

Rungratanawanich, W., Qu, Y., Wang, X., Essa, M. M., & Song, B.-J. (2021). Advanced glycation end products (AGEs) and other adducts in aging-related diseases and alcohol-mediated tissue injury. *Experimental & Molecular Medicine, 53*(2), 168–188.

Saito, M., & Marumo, K. (2015). Effects of collagen crosslinking on bone material properties in health and disease. *Calcified Tissue International, 97*(3), 242–261.

Sanajou, D., Haghjo, A. G., Argani, H., & Aslani, S. (2018). AGE-RAGE axis blockade in diabetic nephropathy: Current status and future directions. *European Journal of Pharmacology, 833*, 158–164.

Sanajou, D., Ghorbani Haghjo, A., Argani, H., Roshangar, L., Ahmad, S. N. S., Jigheh, Z. A., ... Mesgari Abbasi, M. (2018). FPS-ZM1 and valsartan combination protects better against glomerular filtration barrier damage in streptozotocin-induced diabetic rats. *Journal of physiology and biochemistry, 74*, 467–478.

Santos-Bezerra, D. P., Machado-Lima, A., Monteiro, M. B., Admoni, S. N., Perez, R. V., Machado, C. G., ... Corrêa-Giannella, M. L. (2018). Dietary advanced glycated endproducts and medicines influence the expression of SIRT1 and DDOST in peripheral mononuclear cells from long-term type 1 diabetes patients. *Diabetes and Vascular Disease Research, 15*(1), 81–89.

Scavello, F., Zeni, F., Tedesco, C. C., Mensà, E., Veglia, F., Procopio, A. D., ... Raucci, A. (2019). Modulation of soluble receptor for advanced glycation end-products (RAGE) isoforms and their ligands in healthy aging. *Aging (Albany NY), 11*(6), 1648.

Sergi, D., Boulestin, H., Campbell, F. M., & Williams, L. M. (2021). The role of dietary advanced glycation end products in metabolic dysfunction. *Molecular Nutrition & Food Research, 65*(1), 1900934.

Singh, R., Barden, A., Mori, T., & Beilin, L. (2001). Advanced glycation end-products: A review. *Diabetologia, 44*(2), 129–146.

Singh, V. P., Bali, A., Singh, N., & Jaggi, A. S. (2014). Advanced glycation end products and diabetic complications. *The Korean Journal of Physiology & Pharmacology: Official Journal of the Korean Physiological Society and the Korean Society of Pharmacology, 18*(1), 1.

Snelson, M., & Coughlan, M. T. (2019). Dietary advanced glycation end products: Digestion, metabolism and modulation of gut microbial ecology. *Nutrients, 11*(2), 215.

Solinas, G., & Becattini, B. (2017). JNK at the crossroad of obesity, insulin resistance, and cell stress response. *Molecular Metabolism, 6*(2), 174–184.

Sourris, K. C., Watson, A., & Jandeleit-Dahm, K. (2021). Inhibitors of advanced glycation end product (AGE) formation and accumulation. *Reactive Oxygen Species: Network Pharmacology and Therapeutic Applications, 395*–423.
Tamura, Y., Adachi, H., Osuga, J., Ohashi, K., Yahagi, N., Sekiya, M., ... Shimano, H. (2003). FEEL-1 and FEEL-2 are endocytic receptors for advanced glycation end products. *Journal of Biological Chemistry, 278*(15), 12613–12617.
Tobon-Velasco, C. J., Cuevas, E., & Torres-Ramos, A. M. (2014). Receptor for AGEs (RAGE) as mediator of NF-kB pathway activation in neuroinflammation and oxidative stress. *CNS & Neurological Disorders-Drug Targets (Formerly Current Drug Targets-CNS & Neurological Disorders), 13*(9), 1615–1626.
Trellu, S., Courties, A., Jaisson, S., Gorisse, L., Gillery, P., Kerdine-Römer, S., ... Sellam, J. (2019). Impairment of glyoxalase-1, an advanced glycation end-product detoxifying enzyme, induced by inflammation in age-related osteoarthritis. *Arthritis Research & Therapy, 21*, 1–12.
Thomas, M. C. (2011). Advanced glycation end products. *Diabetes and the Kidney, 170*, 66–74.
Thornalley, P. J. (2005). Dicarbonyl intermediates in the Maillard reaction. *Annals of the New York Academy of Sciences, 1043*(1), 111–117.
Turk, Z. (2010). Glycotoxines, carbonyl stress and relevance to diabetes and its complications. *Physiological Research, 59*(2).
Unoki, H., & Yamagishi, S. (2008). Advanced glycation end products and insulin resistance. *Current Pharmaceutical Design, 14*(10), 987–989.
Uribarri, J., Cai, W., Peppa, M., Goodman, S., Ferrucci, L., Striker, G., & Vlassara, H. (2007). Circulating glycotoxins and dietary advanced glycation endproducts: Two links to inflammatory response, oxidative stress, and aging. *The Journals of Gerontology Series A: Biological Sciences and Medical Sciences, 62*(4), 427–433.
Uribarri, J., Cai, W., Ramdas, M., Goodman, S., Pyzik, R., Chen, X., ... Vlassara, H. (2011). Restriction of advanced glycation end products improves insulin resistance in human type 2 diabetes: Potential role of AGER1 and SIRT1. *Diabetes Care, 34*(7), 1610–1616.
Uribarri, J., del Castillo, M. D., de la Maza, M. P., Filip, R., Gugliucci, A., Luevano-Contreras, C., ... Menini, T. (2015). Dietary advanced glycation end products and their role in health and disease. *Advances in Nutrition, 6*(4), 461–473.
Uribarri, J., Peppa, M., Cai, W., Goldberg, T., Lu, M., He, C., & Vlassara, H. (2003). Restriction of dietary glycotoxins reduces excessive advanced glycation end products in renal failure patients. *Journal of the American Society of Nephrology, 14*(3), 728–731.
Uribarri, J., Woodruff, S., Goodman, S., Cai, W., Chen, X., Pyzik, R., ... Vlassara, H. (2010). Advanced glycation end products in foods and a practical guide to their reduction in the diet. *Journal of the American Dietetic Association, 110*(6), 911–916.
Van Beijnum, J. R., Buurman, W. A., & Griffioen, A. W. (2008). Convergence and amplification of toll-like receptor (TLR) and receptor for advanced glycation end products (RAGE) signaling pathways via high mobility group B1 (HMGB1). *Angiogenesis, 11*, 91–99.
Van Dongen, K. C., Kappetein, L., Estruch, I. M., Belzer, C., Beekmann, K., & Rietjens, I. M. (2022). Differences in kinetics and dynamics of endogenous versus exogenous advanced glycation end products (AGEs) and their precursors. *Food and Chemical Toxicology,* 112987.
Van Zoelen, M. A., Yang, H., Florquin, S., Meijers, J. C., Akira, S., Arnold, B., ... Van Der Poll, T. (2009). Role of Toll-like receptors 2 and 4, and the receptor for advanced glycation end products in high-mobility group box 1–induced inflammation in vivo. *Shock (Augusta, Ga.), 31*(3), 280.
Vlassara, H. (2001). The AGE-receptor in the pathogenesis of diabetic complications. *Diabetes/Metabolism Research and Reviews, 17*(5), 436–443.
Vlassara, H., & Palace, M. (2002). Diabetes and advanced glycation endproducts. *Journal of Internal Medicine, 251*(2), 87–101.

Vlassara, H., & Striker, G. E. (2011). AGE restriction in diabetes mellitus: A paradigm shift. *Nature Reviews Endocrinology, 7*(9), 526.

Vlassara, H., & Uribarri, J. (2014). Advanced glycation end products (AGE) and diabetes: Cause, effect, or both? *Current Diabetes Reports, 14*, 1–10.

Vlassara, H., Uribarri, J., Cai, W., Goodman, S., Pyzik, R., Post, J., ... Striker, G. E. (2012). Effects of sevelamer on HbA1c, inflammation, and advanced glycation end products in diabetic kidney disease. *Clinical Journal of the American Society of Nephrology, 7*(6), 934–942.

Xu, X., Li, Z., Luo, D., Huang, Y., Zhu, J., Wang, X., ... Patrick, C. (2003). Exogenous advanced glycosylation end products induce diabetes-like vascular dysfunction in normal rats: A factor in diabetic retinopathy. *Graefe's Archive for Clinical and Experimental Ophthalmology, 241*, 56–62.

Yamagata, T., Tsuru, T., Momoi, M. Y., Suwa, K., Nozaki, Y., Mukasa, T., ... Momoi, T. (1997). Genome organization of human 48-kDa oligosaccharyltransferase (DDOST). *Genomics, 45*(3), 535–540.

Yamagishi, S. (2019). Role of advanced glycation endproduct (AGE)-receptor for advanced glycation endproduct (RAGE) axis in cardiovascular disease and its therapeutic intervention. *Circulation Journal, 83*(9), 1822–1828.

Yamagishi, S. I., Matsui, T., Ishibashi, Y., Isami, F., Abe, Y., Sakaguchi, T., & Higashimoto, Y. (2017). Phytochemicals against advanced glycation end products (AGEs) and the receptor system. *Current Pharmaceutical Design, 23*(8), 1135–1141.

Yamagishi, S. I., Ueda, S., & Okuda, S. (2007). Food-derived advanced glycation end products (AGEs): a novel therapeutic target for various disorders. *Current pharmaceutical design, 13*(27), 2832–2836.

Yao, D., & Brownlee, M. (2010). Hyperglycemia-induced reactive oxygen species increase expression of the receptor for advanced glycation end products (RAGE) and RAGE ligands. *Diabetes, 59*(1), 249–255.

Yuan, X., Nie, C., Liu, H., Ma, Q., Peng, B., Zhang, M., ... Li, J. (2021). Comparison of metabolic fate, target organs, and microbiota interactions of free and bound dietary advanced glycation end products. *Critical Reviews in Food Science and Nutrition*, 1–22.

Zeng, C., Li, Y., Ma, J., Niu, L., & Tay, F. R. (2019). Clinical/translational aspects of advanced glycation end-products. *Trends in Endocrinology & Metabolism, 30*(12), 959–973.

Zhang, H., Wang, Y., Yan, S., Du, F., Wu, L., Yan, S., & Yan, S. S. (2014). Genetic deficiency of neuronal RAGE protects against AGE-induced synaptic injury. *Cell Death & Disease, 5*(6), e1288.

Zhao, D., Li, L., Le, T. T., Larsen, L. B., Su, G., Liang, Y., & Li, B. (2017). Digestibility of glyoxal-glycated β-casein and β-lactoglobulin and distribution of peptide-bound advanced glycation end products in gastrointestinal digests. *Journal of Agricultural and Food Chemistry, 65*(28), 5778–5788.

Zhao, D., Sheng, B., Wu, Y., Li, H., Xu, D., Nian, Y., ... Zhou, G. (2019). Comparison of free and bound advanced glycation end products in food: A review on the possible influence on human health. *Journal of Agricultural and Food Chemistry, 67*(51), 14007–14018.

Zhao, J., Randive, R., & Stewart, J. A. (2014). Molecular mechanisms of AGE/RAGE-mediated fibrosis in the diabetic heart. *World Journal of Diabetes, 5*(6), 860.

Zhu, Y., Shu, T., Lin, Y., Wang, H., Yang, J., Shi, Y., & Han, X. (2011). Inhibition of the receptor for advanced glycation endproducts (RAGE) protects pancreatic β-cells. *Biochemical and Biophysical Research Communications, 404*(1), 159–165.

Zong, H., Ward, M., & Stitt, A. W. (2011). AGEs, RAGE, and diabetic retinopathy. *Current Diabetes Reports, 11*, 244–252.

CHAPTER TWO

Methylglyoxal-induced modification of myoglobin: An insight into glycation mediated protein aggregation

Sauradipta Banerjee*
George College, MAKAUT
*Corresponding author. e-mail address: dipta.saura@gmail.com

Contents

1. Protein glycation	32
2. Glycation and protein aggregation	33
3. Structural alterations of myoglobin by methylglyoxal	34
3.1 Gel electrophoresis	34
3.2 Spectroscopic studies	34
4. Aggregation of myoglobin by methylglyoxal	36
5. Mass spectrometric analysis of MG modified Mb	39
6. Concluding remarks	42
Acknowledgments	43
References	43

Abstract

Post-translational modification of proteins by Maillard reaction, known as glycation, is thought to be the root cause of different complications, particularly in diabetes mellitus and age-related disorders. Methylglyoxal (MG), a reactive α-oxoaldehyde, increases in diabetic condition and reacts with the proteins to form advanced glycation end products (AGEs) following a Maillard-like reaction. In a time-dependent reaction study of MG with the heme protein myoglobin (Mb), MG was found to induce significant structural alterations of the heme protein, such as heme loss, changes in tryptophan fluorescence, and decrease of α-helicity with increased β-sheet content. These changes were found to occur gradually with increasing period of incubation. Incubation of Mb with MG induced the formation of several AGE adducts, including, carboxyethyllysine at Lys-16, carboxymethyllysine at Lys-87, carboxyethyllysine or pyrraline-carboxymethyllysine at Lys-133, carboxyethyllysine at Lys-42 and hydroimidazolone or argpyrimidine at Arg-31 and Arg-139. MG induced amyloid-like aggregation of Mb was detected at a longer period of incubation. MG-derived AGEs, therefore, appear to have an important role as the precursors of protein aggregation, which, in turn, may be associated with pathophysiological complications.

1. Protein glycation

Protein glycation is an irreversible, non-enzymatic post-translational modification of protein amino groups (N-terminal and arginine/lysine side chains) by carbonyl compounds leading to the formation of advanced glycation end-products (AGEs). The reaction initiates via Schiff base formation, followed by Amadori rearrangement and AGE formation, collectively known as the Maillard reaction (Cohen & Wu, 1994; Giardino, Edelstein, & Brownlee, 1994). AGEs are thought to be the root cause of different complications in diabetes mellitus (Rosen et al., 2001), which may be physiologically significant due to the increased levels of several active carbonyl compounds. Besides glucose, other glycating agents include fructose (Ghosh Moulick, Bhattacharya, Roy, Basak, & Dasgupta, 2007), glyoxal (Iram et al., 2013a), methylglyoxal (Oliveira, Lages, Gomes, Neves, & Familia, 2011), 3-deoxyglucosone (Mera et al., 2010). Glycation of hemoglobin by glucose (Kar & Chakraborti, 2001; Sen, Kar, Roy, & Chakraborti, 2005), fructose (Bose & Chakraborti, 2008) and methylglyoxal (Bose, Bhattacherjee, Banerjee, & Chakraborti, 2013) was found to promote iron release and free radical-mediated oxidative reactions.

Methylglyoxal (MG), a highly reactive α-oxoaldehyde, is mainly derived from triose phosphates D-glyceraldehyde-3-phosphate and dihydroxyacetone phosphate during glycolysis in eukaryotic cells, and its blood level increases in both type 1 and type 2 diabetes mellitus (Lapolla et al., 2003; Lu et al., 2011; Ramasamy, Yan, & Schmidt, 2006). It is estimated that an adult human produces approximately 3 mmol MG per day but only about 0.3% of this, forms glycation adducts—the remaining 99.7% is metabolized, mostly by glyoxalase 1 (Rabbani & Thornalley, 2012). The median concentration of MG is increased by 5–6-fold and 2–3-fold in blood samples of diabetic patients with Type 1 and Type 2 diabetes mellitus, respectively (McLellan, Thornalley, Benn, & Sonksen, 1994), and the formation of MG-derived AGEs is increased accordingly. MG-derived AGEs (MAGEs) have been reported with different proteins namely insulin (Oliveira et al., 2011), human serum albumin (Ahmed, Dobler, Dean, & Thornalley, 2005), cytochrome c (Oliviera et al., 2013), α-synuclein (Lee, Park, Paik, & Choi, 2009) and superoxide dismutase (Khan et al., 2014). MG also interacts with hemoglobin to modify the arginine residues forming hydroimidazolones (MG-H1) (Chen, Ahmed, & Thornalley, 2005; Gao & Wang, 2006), leading to the changes in structural and functional properties of the heme protein (Bose et al., 2013).

Monomeric heme protein myoglobin (Mb) is released into the circulation due to muscle damage by various factors including vigorous repeated exercises (David, 2000). Although free Mb has a short half-life, it may come in contact with MG to form modified Mb, which may be damaging, particularly in the diabetic condition with increased level of the α-oxoaldehyde. Considering these, it is of clinical relevance to investigate the interaction study of Mb with MG, which is about 20,000 times more reactive than glucose (Ahmed et al., 2005). Mb has also been used for studying in vitro glycation reactions with different sugars namely glucose (Roy, Sil, & Chakraborti, 2010), fructose (Bhattacherjee & Chakraborti, 2011) and ribose (Bokiej, Livermore, Harris, Onishi, & Sandwick, 2011). Glyoxal-derived AGE adducts has been reported to induce the structural alterations of Mb (Banerjee & Chakraborti, 2014; Banerjee, 2017).

2. Glycation and protein aggregation

Partial or total unfolding of the native protein is required to establish a new intermolecular interaction and promote amyloid aggregation. Covalent modification of the lysine and arginine residues may induce partial unfolding and aggregation of proteins. AGE adducts may influence the overall structure of proteins leading to aggregation (Obrenovich & Monnier, 2004; Shaikh & Nicholson, 2008; Vitek, Bhattacharya, Glendening, Stopa, & Vlassara, 1994) and pathological complications. AGE-modified amyloid plaques and fibrils have been reported in the brain of Alzheimer's disease patients (Vitek et al., 1994) and islets of Langerhans of diabetic patients (Kapurniotu, Bernhagen, Greenfield, Al-Abed, & Teichberg, 1998), respectively. In several in vitro studies, AGE-mediated protein aggregation has been reported. MG-induced modifications of a single arginine residue of insulin (Arg-46) or cytochrome c (Arg-92) have been reported to cause conformational changes and aggregation of the proteins (Oliveira et al., 2011, 2013). Glucose, fructose and glyoxal induced the aggregation of hemoglobin and hen egg white lysozyme on long-term (21–180 days) incubation (Fazili, Bhat, & Naeem, 2014; Ghosh et al., 2013; Ghosh Moulick et al., 2007; Iram et al., 2013a). On the other hand, short-term (24 h to 7 days) incubation of several proteins namely insulin, cytochrome c and histone H2A with high concentrations of MG (1–10 mM) induces their aggregation (Mir, Uddin, Alam, & Ali, 2014; Oliveira et al., 2011, 2013). Thus, both concentrations of the glycating agent and incubation period of the reaction mixture appear to control the extent and nature of modification. In our study, we found that MG-derived adducts induced aggregation of the heme protein at a longer period of incubation. The concentration of MG used was much higher than the plasma concentration of the dicarbonyl in diabetic patients (McLellan et al., 1994). However, as the vast majority (approximately 90%) of MG is bound to the cysteinyl, lysyl and arginyl residues of proteins (Chaplin, Fahl, & Cameron, 1996), much higher concentration of total MG in vivo is suggested. Moreover, in several studies, quite high concentration of MG (1–10 mM) has been used for in the vitro reactions with proteins, namely, insulin, cytochrome c and histone H2A (Mir et al., 2014; Oliveira et al., 2011, 2013).

3. Structural alterations of myoglobin by methylglyoxal

3.1 Gel electrophoresis

Monomeric Mb appears mostly as a single major protein band in native PAGE followed by Coomassie blue staining. Long-term incubation of Mb with MG, result in the appearance of high molecular weight band, indicating MG-induced aggregation of the heme protein. Furthermore, SDS resistant property of the band (following analysis in SDS PAGE), indicated that major covalent interactions are involved in the formation of high molecular weight aggregate species of Mb. MG-mediated covalent cross-linking has been reported to induce the formation of high molecular aggregates of α-crystallin (Kumar, Mrudula, Mitra, & Reddy, 2004), as well as histone H2A (Mir et al., 2014). Increased mobility of Mb band observed at the shorter periods of incubation with MG was due to the loss of positive charge(s) on MG induced modification and AGE formation, gradually leading to aggregation of the protein.

3.2 Spectroscopic studies

In comparison with the control Mb, MG-incubated Mb samples exhibited a gradual decrease in Soret absorbance, indicating heme loss and increased absorbance in the 250–300 nm region. No shift in the Soret absorbance or appearance of Q bands (absorbance peaks at 540 and 580 nm), characteristic of heme reduction (Fe^{3+} to Fe^{2+}) and formation of oxy form of Mb, was observed in the absorbance spectra of MG-treated Mb. According to Sen et al. (2005) and Kar, Roy, Bose & Chakraborti (2006), glycated hemoglobin

(HbA1$_c$) exhibits a weaker heme-globin linkage and is more susceptible to the heme loss compared to normal hemoglobin (HbA$_0$). Increase in absorbance in the 250–300 nm range may be due to the formation of covalent adducts between MG and the protein amino groups (Mukhopadhyay, Kar, & Das, 2010). Bokiej et al. (2011) have shown that ribose modification also causes increased absorbance in the 270–320 nm region of Mb due to the AGE formation.

The wavelength of maximum fluorescence emission intensity (λ_{max}) of control Mb was found to be around 332 nm, when excited at 295 nm. With increasing time of incubation with MG, an increase in fluorescence intensity, as well as a significant red shift of λ_{max} was observed for the heme protein. The results indicated a considerable structural change of Mb with increasing time of incubation with MG. Red shift of λ_{max} suggested the exposure of tryptophan residues to a more polar solvent (Chaudhuri, Haldar, & Chattopadhyay, 2010). Increase in the intensity of fluorescence emission may be associated with reduced fluorescence energy transfer to the heme group due to an increase in distance between the heme moiety and tryptophan residues (Jian-Zhong, 2007). MG thus appeared to unfold Mb in course of incubation, which may lead to its subsequent aggregation. Loss of heme group due to the weakening of heme-globin linkage and unfolding has been found to be associated with the aggregation of hemoglobin in an earlier study (Iram & Naeem, 2013). Fluorescence emission exhibited by the fluorescent AGEs in the region 400–500 nm may also contribute to the increased emission of MG-treated Mb samples at the lower wavelengths.

MG-mediated secondary structural alteration of Mb was studied by far UV CD spectroscopy. The CD spectrum of control Mb indicated that it is predominantly α-helical protein, with characteristic negative absorbance minima around 208 and 222 nm. The α-helicity was found to be about 84%. As the period of incubation with MG increased, a decrease in the α-helicity of Mb was observed as evident from the decrease in minima at 208 and 222 nm. The reduction in α-helicity of the protein was compensated by an increase of β-sheet. CD spectra recorded in the visible region (350–450 nm) indicated the Soret peak position of control Mb and MG-treated Mb samples at around 410 nm, characteristic of met Mb (Fe^{3+}). The treated samples showed a gradual decrease in the peak height without any peak shift, as also found in the absorption spectral pattern, indicating heme loss without reduction of the heme moiety of Mb by MG.

4. Aggregation of myoglobin by methylglyoxal

Increase of β-sheet has been found to be associated with protein aggregation in several studies (Bakhti, Habibi-Rezaei, Moosavi-Movahedi, & Khazaei, 2007; Ghosh et al., 2013; Ghosh Moulick et al., 2007; Iram et al., 2013a). This is because β-sheet generally provides a better environment for the intermolecular interactions required for aggregation (Creighton, 1993). Intermolecular β-sheet structure is normally associated with the amyloid aggregation of proteins. The cationic benzothiazole dye, thioflavin T (ThT) is a specific fluorescent marker for intermolecular cross β-sheets and is often used for the detection and characterization of amyloid aggregation (Ghosh et al., 2013; Iram et al., 2013a; Oliveira et al., 2011). On binding to the amyloid aggregates, ThT normally exhibits intense green fluorescence signal (Pandey, Ghosh, & Dasgupta, 2013). Long-term incubation of Mb with MG exhibited strong a ThT fluorescence emission of the heme protein, indicating the presence of amyloid cross β-sheet structure (Fig. 1A). ThT fluorescence microscopic imaging studies further indicate MG-incubated Mb sample to exhibit intense fluorescence signal (Fig. 1B). The strong binding affinity of ThT suggests the possible presence of intermolecular cross β-sheet structure in MG incubated Mb for long term.

Amyloid aggregates exhibit a characteristic intrinsic fluorescence, which is independent of the fluorescence of aromatic amino acid residues (Chan et al., 2013; Pinotsi, Buell, Dobson, Schierle, & Kaminski, 2013). This property of the cross β-sheet structures is probably due to the presence of extensive arrays of hydrogen bonds that cause the delocalization of peptide electrons, giving rise to optically excitable energy states (Chan et al., 2013). Thus, the electronic transition may account for the characteristic UV-A fluorescence emission. MG incubated Mb exhibited intense intrinsic fluorescence characteristic of amyloidal nature of aggregation as observed in the fluorescence and confocal microscopic imaging studies (Fig. 1C and D).

Fourier transform infrared (FTIR) spectroscopy was used to find the presence of intermolecular cross β-sheet structure in the aggregated Mb. The spectrum of the control Mb exhibited a sharp peak at 1654.59 cm^{-1} in the amide I region, corresponding to mainly α-helical structure (Fig. 2A). Additionally, a low intensity peak around at 1548.25 cm^{-1} in the amide II region was also observed. In the spectrum of MG-incubated Mb, the amide I peak was found to shift from 1654.59 to 1638.22 cm^{-1} (Fig. 2B). Moreover, the peak at 1548.25 cm^{-1} was found to disappear. In one study, unfolding

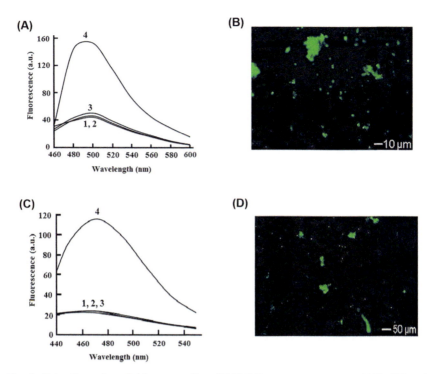

Fig. 1 Detection of amyloid aggregation. (A) ThT fluorescence spectra (460–600 nm) of control Mb (1), Mb incubated with MG for 7 days (2), Mb incubated with MG for 14 days (3) and Mb incubated with MG for 18 days (4), recorded with excitation at 435 nm. Fluorescence was measured after incubating the samples (6 μM each) with a fixed concentration of ThT (15 μM) overnight at 37 °C. (B) Fluorescence microscopic image of ThT-stained protein sample (Mb incubated with MG for 18 days). (C) Amyloid intrinsic fluorescence spectra (440–540 nm) of control Mb (1), Mb incubated with MG for 7 days (2), Mb incubated with MG for 14 days (3) and Mb incubated with MG for 18 days (4), recorded with excitation at 405 nm. Protein concentration was adjusted to 10 μM. (D) Confocal microscopic image for the detection of intrinsic fluorescence in protein sample (Mb incubated with MG for 18 days).

and aggregation of Mb have been associated with a drastic decrease in peak area of the amide II peak (around 1545 cm^{-1}) along with the formation of intermolecular β-sheet structure (Calabro & Magazu, 2013).

The amide I band was subjected to Fourier self-deconvolution analysis. Deconvolution of the amide I band of Mb again exhibited a sharp signal at 1655.07 cm^{-1}, indicating a dominant α-helical structure (Fig. 2C). A weak signal was obtained at 1637.88 cm^{-1} in the deconvoluted spectrum, indicating a minor contribution from the β-pleated sheet structure. As shown in

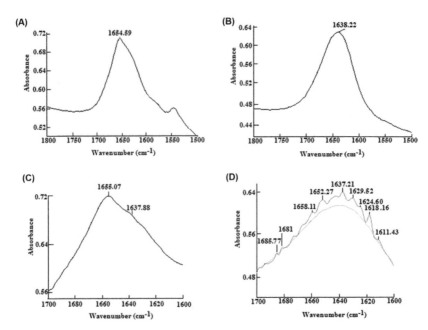

Fig. 2 FTIR studies. Representative infrared spectra of (A) control Mb and (B) Mb incubated with MG for 18 days in the wave number range 1800–1500 cm^{-1}. Fourier self-deconvolution analysis applied to the infrared spectra of (C) control Mb and (D) Mb incubated with MG for 18 days in the amide I region. The peak positions are denoted in the spectra.

the deconvoluted amide I band spectrum of MG incubated Mb (Fig. 2D), the peaks corresponding to 1652.27 and 1658.11 cm^{-1} indicated α-helical structure. The peak at 1637.21 cm^{-1} indicated the presence of β-pleated sheet structure. The peaks in the range 1611–1630 cm^{-1} (1611.43, 1618.16, 1624.60, and 1629.52 cm^{-1}) denoted the presence of cross β-sheet structure (Lara, Gourdin-Bertin, Adamcik, Bolisetty, & Mezzenga, 2012). The peaks around 1680 cm^{-1} region (1681 and 1685.77 cm^{-1}) were also indicative of intermolecular β-sheet conformation (Iram et al., 2013a; Iram, Amani, Furkan, & Naeem, 2013b). The spectral pattern thus suggested the presence of intermolecular cross β-sheet structure, indicating amyloid nature of aggregation in MG-incubated Mb.

X-ray diffraction (XRD) is a powerful technique for confirming the presence of amyloid cross-β structure and is used for determining the nature of protein aggregation (Iram & Naeem, 2013; Naeem & Amani, 2013). XRD pattern of MG incubated Mb exhibited two peaks, corresponding to the d-spacings of 4.7 Å (denoted by peak of 2θ around 20°)

and 10.4 Å (denoted by peak of 2θ around 10°), indicating the respective inter-strand and inter-sheet distances in a cross-β structure (Fig. 3) (Lara et al., 2012). These peaks were absent in the XRD pattern of control Mb. Calculations for d-spacings were carried out using the Bragg's law of diffraction. Thermally aggregated BSA has been found to exhibit the characteristic XRD pattern of amyloid cross-β structure, as reported in an earlier study (Holma et al., 2007). The result further confirms the amyloid nature of aggregation in MG incubated Mb and strengthens the findings obtained by other experimental methods.

Scanning electron microscopy (SEM) is used for studying the morphology of protein aggregates (Borana, Mishra, Pissurlenkar, Hosur, & Ahmad, 2014; Chan et al., 2013; Iram et al., 2013b). While SEM imaging of control Mb sample did not show the presence of any aggregate, MG incubated Mb revealed the presence of non-fibrillar amorphous aggregates (Fig. 4A). A magnified protein aggregate is shown in Fig. 4B. Hemoglobin forms amorphous aggregates when incubated with 70% glyoxal for 4 h, and fibrils when incubated with 30% glyoxal for 20 days (Iram et al., 2013a). It appears that the formation of fibrillar or non-fibrillar protein aggregates largely depend on the conditions of in vitro reaction, type and concentration of glycating/modifying agent etc.

5. Mass spectrometric analysis of MG modified Mb

In a time-dependent reaction study of Mb with MG, the following AGE adducts were detected and identified: Pyrraline-carboxymethyllysine (Pyr-CML) and carboxyethyllysine (CEL) at Lys-133, CEL at Lys-16 and

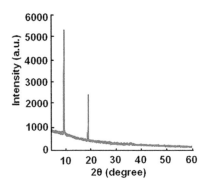

Fig. 3 XRD analysis of proteins. XRD pattern of Mb incubated with MG for 18 days.

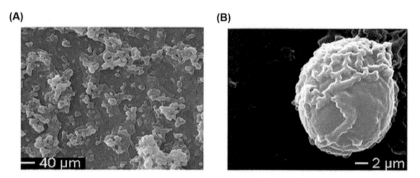

Fig. 4 SEM imaging study for the detection of aggregate morphology. SEM image of (A) Mb incubated with MG for 18 days and (B) magnified (view) of MG-modified Mb aggregate.

Lys-42, CML at Lys-87, hydroimidazolone (MG-H1) and argpyrimidine (ArgP) at each of the arginine residue, Arg-31 and Arg-139. The sites (amino acid residues) and nature of AGE modifications in MG modified Mb are summarized in Table 1. ArgP is a fluorescent AGE and may contribute to the enhanced fluorescence emission of MG-treated samples. As evident from mass spectrometric analysis, two different types of AGEs were found at Lys-133, Arg-31 and Arg-139. Formation of different AGEs on the same residue indicates that modification of Mb by MG is not highly specific under in vitro conditions. Such heterogeneity has also been reported for MG-induced modifications of insulin (Oliveira et al., 2011) and yeast enolase (Gomes et al., 2008). Reaction of Mb with glyoxal resulted in increased band mobility in the native PAGE due to AGE modifications at the lysine and arginine residues (Banerjee & Chakraborti, 2014). Enhanced electrophoretic mobility of other proteins namely hemoglobin (Bose et al., 2013) and ceruloplasmin (Kang, 2006) due to modification of the positively charged residues by MG have been reported.

All modifications were confirmed by MS/MS fragmentation of the peptides using collision-induced dissociation (CID) to determine their sequences and identify the modified amino acid residues. In CID fragmentation technique, the peptide bond is fragmented to generate b-ions (charge retained by the amino-terminal fragment) and y ions (charge retained by the carboxy-terminal fragment). Thus, if an amino acid residue is modified, the particular y and b fragment ions containing the modified amino acid residue, should have the particular amino acid mass value plus mass increment due to the modification.

Table 1 Assignment of modified amino acid residues.

Observed mass (Da)	Theoretical mass (Da)	Peptide sequence	Mass increase (Da)	AGE identified	Modified residue
2304	2232	HPGDFGADAQGAMTK*ALELFR (119–139)	72	CEL	Lys-133
3475	3403	GLSDGEWQQVLNVWGK*VEAD IAGHGQEVLIR (1–31)	72	CEL	Lys-16
2272	2232	HPGDFGADAQGAMTK*ALELFR (119–139)	40	Pyr-CML	Lys-133
2177	2119	GHHEAELK*PLAQSHATKHK (80–98)	58	CML	Lys-87
1733	1661	LFTGHPETLEK*FDK(32–45)	72	CEL	Lys-42
1660	1606	VEADIAGHGQEVLIR*(17–31)	54	MG-H1	Arg-31
2939	2859	VEADIAGHGQEVLIR*LFTG HPETLEK (17–42)	80	ArgP	Arg-31
1414	1360	ALELFR*NDIAAK (134–145)	54	MG-H1	Arg-139
1440	1360	ALELFR*NDIAAK (134–145)	80	ArgP	Arg-139
2913	2859	VEADIAGHGQEVLIR*LFTG HPETLEK (17–42)	54	MG-H1	Arg-31

The specific AGEs are indicated in the table and the modified amino acid residues are marked (*).

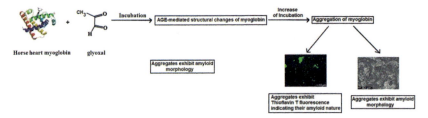

Fig. 5 A schematic presentation of MG-mediated aggregation of Mb.

A schematic representation of MG-derived AGE mediated aggregation of Mb in a time-dependent incubation study is shown in Fig. 5.

6. Concluding remarks

Several reports on MG-mediated amyloid aggregation of proteins are known. MG induces amyloid aggregation of cystatin, characterized by fibrillar structure and extensive β-sheet content (Bhat, Bhat, Khaki, & Bano, 2015). It also promotes amyloid-like aggregation of fibrinogen (Xu, Qiang, Zhang, Liu, & He, 2012) and enhances the oligomerization and formation of amyloid β-sheets of amyloid β-peptides (Chen, Maley, & Yu, 2006). Protein glycation including MG-induced CML formation has been reported to induce protein cross-linking in several studies (Chellan & Nagaraj, 1999; Ghosh Moulick et al., 2007; Kleter, Damen, Buijs, & Cate, 1998). The existing reports thus strengthen our findings on MG-induced AGE modification and amyloid-like aggregation of Mb. MG-induced aggregates of Mb, however, are found to exhibit nonfibrillar, amorphous morphology, in spite of their amyloid nature, an observation consistent with the other reports (Iram et al., 2013a, 2013b).

In conclusion, MG modifies several lysine and arginine residues of Mb in a time-dependent manner. MG-derived AGE adducts induce significant structural alterations of the heme protein, including heme loss, protein unfolding, reduction of native α-helical content and increase of β-sheet structure. The intermolecular cross β-sheet leads to the amyloid-like aggregation of Mb. Our findings as well as existing reports suggest that the AGE adducts derived from the glycating agents appear to play an important role as the precursors of protein aggregation, which may have important clinical implications in the design of therapeutics for the treatment of AGE-induced protein conformational disorders.

Acknowledgments
None.

References

Ahmed, N., Dobler, D., Dean, M., & Thornalley, P. J. (2005). Peptide mapping identifies hotspot site of modification in human serum albumin by methylglyoxal involved in ligand binding and esterase activity. *The Journal of Biological Chemistry, 280*, 5724–5732.

Bakhti, M., Habibi-Rezaei, M., Moosavi-Movahedi, A. A., & Khazaei, M. R. (2007). Consequential alterations in hemoglobin structure upon glycation with fructose: Prevention by acetylsalicylic acid. *Journal of Biochemistry, 141*, 827–833.

Banerjee, S., & Chakraborti, A. S. (2014). Structural alterations of hemoglobin and myoglobin by glyoxal: A comparative study. *International Journal of Biological Macromolecules, 66*, 311–318.

Banerjee, S. (2017). Formation of pentosidine cross-linking in myoglobin by glyoxal: Detection of fluorescent advanced glycation end product. *Journal of Fluorescence, 27*, 1213–1219.

Bhattacherjee, A., & Chakraborti, A. S. (2011). Fructose-induced modifications of myoglobin: Change of structure from met (Fe^{3+}) to oxy (Fe^{2+}) form. *International Journal of Biological Macromolecules, 48*, 202–209.

Bhat, W. F., Bhat, S. A., Khaki, P. S. S., & Bano, B. (2015). Employing in vitro analysis to test the potency of methylglyoxal in inducing the formation of amyloid-like aggregates of caprine brain cystatin. *Amino Acids, 47*, 135–146.

Bokiej, M., Livermore, A. T., Harris, A. W., Onishi, A. C., & Sandwick, R. K. (2011). Ribose sugars generate internal glycation cross-links in horse heart myoglobin. *Biochemical and Biophysical Research Communications, 407*, 191–196.

Borana, M. S., Mishra, P., Pissurlenkar, R. R. S., Hosur, R. V., & Ahmad, B. (2014). Curcumin and kaempferol prevent lysozyme fibril formation by modulating aggregation kinetic parameters. *Biochimica et Biophysica Acta, 1844*, 670–680.

Bose, T., Bhattacherjee, A., Banerjee, S., & Chakraborti, A. S. (2013). Methylglyoxal-induced modifications of hemoglobin: Structural and functional characteristics. *Archives of Biochemistry and Biophysics, 529*, 99–104.

Bose, T., & Chakraborti, A. S. (2008). Fructose-induced structural and functional modifications of hemoglobin: Implication for oxidative stress in diabetes mellitus. *Biochimica et Biophysica Acta, 1780*, 800–808.

Calabro, E., & Magazu, S. (2013). Unfolding and aggregation of myoglobin can be induced by three hours' exposure to mobile phone microwaves: A FTIR spectroscopy study. *Spectroscopy Letters, 46*, 586–589.

Chan, F. T., Kaminski Schierle, G. S., Kumita, J. R., Bertoncini, C. W., Dobson, C. M., & Kaminski, C. F. (2013). Protein amyloids develop an intrinsic fluorescence signature during aggregation. *Analyst, 138*, 2156–2162.

Chaplin, F. W. R., Fahl, W. E., & Cameron, D. C. (1996). Effect of endogenous methylglyoxal on Chinese hamster ovary cells growth in culture. *Cytotechnology, 22*, 33–42.

Chaudhuri, A., Haldar, S., & Chattopadhyay, A. (2010). Organization and dynamics of tryptophans in the molten globule state of bovine α-lactalbumin utilizing wavelength selective fluorescence approach: Comparisons with native and denatured states. *Biochemical and Biophysical Research Communications, 394*, 1082–1086.

Chellan, P., & Nagaraj, R. H. (1999). Protein crosslinking by the Maillard reaction: Dicarbonylderived imidazolium crosslinks in aging and diabetes. *Archives of Biochemistry and Biophysics, 368*(1), 98–104.

Chen, Y., Ahmed, N., & Thornalley, P. J. (2005). Peptide mapping of human hemoglobin modified minimally by methylglyoxal in vitro. *Annals of the New York Academy of Sciences, 1043*, 905.

Chen, K., Maley, J., & Yu, P. H. (2006). Potential implications of endogenous aldehydes in β-amyloid misfolding, oligomerization and fibrillogenesis. *Journal of Neurochemistry, 99*, 1413–1424.

Cohen, M. P., & Wu, V. (1994). Purification of glycated hemoglobin. *Methods in Enzymology, 231*, 65–75.

Creighton, T. E. (1993). *Proteins: Structures and molecular properties*. New York: W. H. Freeman and Company.

David, W. S. (2000). Myoglobinuria. *Neurologic Clinics, 18*, 215–243.

Fazili, N. A., Bhat, W. F., & Naeem, A. (2014). Induction of amyloidogenicity in wild type HEWL by a dialdehyde: Analysis involving multi dimensional approach. *International Journal of Biological Macromolecules, 64*, 36–44.

Gao, Y., & Wang, Y. (2006). Site-selective modifications of arginine residues in human haemoglobin induced by methylglyoxal. *Biochemistry, 45*, 15654–15660.

Ghosh Moulick, R., Bhattacharya, J., Roy, S., Basak, S., & Dasgupta, A. K. (2007). Compensatory secondary structure alterations in protein glycation. *Biochimica et Biophysica Acta, 1774*, 233–242.

Ghosh, S., Pandey, N. K., Roy, A. S., Tripathy, D. R., Dinda, A. K., & Dasgupta, S. (2013). Prolonged glycation of hen egg white lysozyme generates non amyloidal structures. *PLoS One, 8*, e74336.

Giardino, I., Edelstein, D., & Brownlee, M. (1994). Nonenzymatic glycosylation in vitro in bovine endothelial cells alters basic fibroblast growth factor. *The Journal of Clinical Investigation, 94*, 110–117.

Gomes, R. A., Oliveira, L. M. A., Silva, M., Ascenso, C., Quintas, A., Costa, G., ... Cordeiro, C. (2008). Protein glycation in vivo: Functional and structural effects on yeast enolase. *The Biochemical Journal, 416*, 317–326.

Holma, N. K., Jespersen, S. K., Thomassen, L. V., Wolff, T. Y., Sehgal, P., Thomsen, L. A., ... Otzen, D. E. (2007). Aggregation and fibrillation of bovine serum albumin. *Biochimica et Biophysica Acta, 1774*, 1128–1138.

Iram, A., Alam, T., Khan, J. M., Khan, T. A., Khan, R. H., & Naeem, A. (2013a). Molten globule of hemoglobin proceeds into aggregates and advanced glycated end products. *PLoS One, 8*, e72075.

Iram, A., Amani, S., Furkan, M., & Naeem, A. (2013b). Equilibrium studies of cellulase aggregates in presence of ascorbic and boric acid. *International Journal of Biological Macromolecules, 52*, 286–295.

Iram, A., & Naeem, A. (2013). Detection and analysis of protofibrils and fibrils of hemoglobin: Implications for the pathogenesis and cure of heme loss related maladies. *Archives of Biochemistry and Biophysics, 533*, 69–78.

Jian-Zhong, L., & Wang, M. (2007). Improvement of activity and stability of chloroperoxidase by chemical modification. *BMC Biotechnology, 7*, 1–8.

Kang, J. H. (2006). Oxidative modification of human ceruloplasmin by methylglyoxal: an in vitro study. *Journal of Biochemistry and Molecular Biology, 39*, 335–338.

Kapurniotu, A., Bernhagen, J., Greenfield, N., Al-Abed, Y., & Teichberg, S. (1998). Contribution of advanced glycosylation to the amyloidogenicity of islet amyloid polypeptide. *European Journal of Biochemistry/FEBS, 251*, 208–216.

Kar, M., & Chakraborti, A. S. (2001). Effect of glycosylation on iron-mediated free radical reactions of hemoglobin. *Current Science, 80*, 770–773.

Kar, M., Roy, A., Bose, T., & Chakraborti, A. S. (2006). Effect of glycation of hemoglobin on its interaction with trifluoperazine. *The Protein Journal, 25*, 202–211.

Khan, M. A., Anwar, S., Aljarbou, A. N., Al-Orainy, M., Aldebasi, Y. H., Islam, S., & Younus, H. (2014). Protective effect of thymoquinone on glucose or methylglyoxal-induced glycation of superoxide dismutase. *International Journal of Biological Macromolecules, 65*, 16–20.

Kleter, G. A., Damen, J. J. M., Buijs, M. J., & Cate, J. M. T. (1998). Modification of amino acid residues in carious dentin matrix. *Journal of Dental Research, 77*, 488–495.

Kumar, M. S., Mrudula, T., Mitra, N., & Reddy, G. B. (2004). Enhanced degradation and decreased stability of eye lens α-crystallin upon methylglyoxal modification. *Experimental Eye Research, 79*, 577–583.

Lapolla, A., Flamini, R., Vedova, A. D., Senesi, A., Reitano, R., Fedele, D., ... Traldi, P. (2003). Glyoxal and methylglyoxal levels in diabetic patients: Quantitative determination by a new GC/MS method. *Clinical Chemistry and Laboratory Medicine: CCLM/FESCC, 41*, 1166–1173.

Lara, C., Gourdin-Bertin, S., Adamcik, J., Bolisetty, S., & Mezzenga, R. (2012). Self-assembly of ovalbumin into amyloid and non-amyloid fibrils. *Biomacromolecules, 13*, 4213–4221.

Lee, D., Park, C. W., Paik, S. R., & Choi, K. Y. (2009). The modification of α-synuclein by dicarbonyl compounds inhibits its fibril-forming process. *Biochimica et Biophysica Acta, 1794*, 421–430.

Lu, J., Randell, E., Han, Y., Adeli, K., Krhan, J., & Meng, Q. H. (2011). Increased plasma methylglyoxal level, inflammation, and vascular endothelial dysfunction in diabetic nephropathy. *Clinical Biochemistry, 44*, 307–311.

McLellan, A. C., Thornalley, P. J., Benn, J., & Sonksen, P. H. (1994). Glyoxalase system in clinical diabetes mellitus and correlation with diabetic complications. *Clinical Science, 87*, 21–29.

Mera, K., Takeo, K., Izumi, M., Maruyama, T., Nagai, R., & Otagiri, M. (2010). Effect of reactive aldehydes on the modification and dysfunction of human serum albumin. *Journal of Pharmaceutical Sciences, 99*, 1614–1625.

Mir, A. R., Uddin, M., Alam, K., & Ali, A. A. (2014). Methylglyoxal mediated conformational changes in histone H2A—Generation of carboxyethylated advanced glycation end products. *International Journal of Biological Macromolecules, 69*, 260–266.

Mukhopadhyay, S., Kar, M., & Das, K. P. (2010). Effect of methylglyoxal modification of human α-crystallin on the structure, stability and chaperone function. *The Protein Journal, 29*, 551–566.

Naeem, A., & Amani, S. (2013). Deciphering structural intermediates and genotoxic fibrillar aggregates of albumins: A molecular mechanism underlying for degenerative diseases. *PLoS One, 8*, e54061.

Obrenovich, M. E., & Monnier, V. M. (2004). Glycation stimulates amyloid formation. *Science of Aging Knowledge Environment, 2004*, pe3.

Oliveira, L. M. A., Lages, A., Gomes, R. A., Neves, H., & Familia, C. (2011). Insulin glycation by methylglyoxal results in native-like aggregation and inhibition of fibril formation. *BMC Biochemistry, 12*, 41.

Oliviera, L. M. A., Gomes, R. A., Yang, D., Dennison, S. R., Familia, C., Lages, A., ... Quintas, A. (2013). Insights into the molecular mechanism of protein native-like aggregation upon glycation. *Biochimica et Biophysica Acta, 1834*, 1010–1022.

Pandey, N. K., Ghosh, S., & Dasgupta, S. (2013). Fructose restrains fibrillogenesis in human serum albumin. *International Journal of Biological Macromolecules, 61*, 424–432.

Pinotsi, D., Buell, A. K., Dobson, C. M., Schierle, G. S. K., & Kaminski, C. F. (2013). A label-free quantitative assay of amyloid fibril growth based on intrinsic fluorescence. *Chembiochem: A European Journal of Chemical Biology, 14*, 846–850.

Rabbani, N., & Thornalley, P. J. (2012). Methylglyoxal, glyoxalase 1 and the dicarbonyl proteome. *Amino Acids, 42*, 1133–1142.

Ramasamy, R., Yan, S. F., & Schmidt, A. M. (2006). Methylglyoxal comes of AGE. *Cell, 124*, 258–260.
Rosen, P., Nawroth, P. P., King, G., Moller, W., Tritschler, H. J., & Packer, L. (2001). The role of oxidative stress in the onset and progression of diabetes and its complications: A summary of Congress Series sponsored by UNESCO-MCBN, the American Diabetes Association and the German Diabetes Society. *Diabetes/Metabolism Research and Reviews, 17*, 189–212.
Roy, A., Sil, R., & Chakraborti, A. S. (2010). Non-enzymatic glycation induces structural modifications of myoglobin. *Molecular and Cellular Biochemistry, 338*, 105–114.
Sen, S., Kar, M., Roy, A., & Chakraborti, A. S. (2005). Effect of nonenzymatic glycation on functional and structural properties of hemoglobin. *Biophysical Chemistry, 113*, 289–298.
Shaikh, S., & Nicholson, L. F. (2008). Advanced glycation end products induce in vitro crosslinking of alpha-synuclein and accelerate the process of intracellular inclusion body formation. *Journal of Neuroscience Research, 86*, 2071–2082.
Vitek, M. P., Bhattacharya, K., Glendening, J. M., Stopa, E., & Vlassara, H. (1994). Advanced glycation end products contribute to amyloidosis in Alzheimer disease. *Proceedings of the National Academy of Sciences, 91*, 4766–4770.
Xu, Y., Qiang, M., Zhang, J., Liu, Y., & He, R. (2012). Reactive carbonyl compounds (RCCs) cause aggregation and dysfunction of fibrinogen. *Protein Cell, 3*, 627–640.

CHAPTER THREE

Glycation in the cardiomyocyte

Christine E. Delligatti and Jonathan A. Kirk*

Department of Cell and Molecular Physiology, Loyola University Chicago Stritch School of Medicine, Maywood, IL, United States
*Corresponding author. e-mail address: jkirk2@luc.edu

Contents

1. Glycation and the cardiomyocyte	48
1.1 Glycation and the glyoxalase cycle	48
1.2 The cardiomyocyte	52
2. Extracellular glycation	52
2.1 Extracellular matrix glycation	53
2.2 Extracellular glycation in cardiovascular diseases	54
2.3 Future directions	55
3. Glycation disrupts cardiomyocyte calcium handling	56
3.1 RAGE signaling effects on calcium handling	56
3.2 Intracellular protein glycation effects on calcium handling	57
3.3 Glycation induced Ca^{2+} dysregulation in cardiovascular diseases	59
3.4 Future directions	59
4. Glycation of the cytoskeleton	60
4.1 Glycation impairs cardiomyocyte sarcomere function	61
4.2 Myofilament glycation in diabetes	63
4.3 Next steps in understanding cytoskeletal glycation	63
5. Other effects of glycation in the cardiomyocyte	64
5.1 Glycation and the mitochondria	64
5.2 Protein quality control	64
5.3 Future directions in glycation of PQC-related proteins	66
6. Therapeutics and the future of cardiomyocyte glycation	66
6.1 Alagebrium (ALT-711), an AGE-breaker	67
6.2 Aminoguanidine, an AGE-inhibitor	74
6.3 N-terminal MYPBC peptides	74
6.4 Glo1 overexpression and activation	75
7. Other future directions and concluding remarks	75
7.1 PTM crosstalk	75
7.2 Concluding remarks	76
References	77

Abstract

Glycation is a protein post-translational modification that can occur on lysine and arginine residues as a result of a non-enzymatic process known as the Maillard reaction. This modification is irreversible, so the only way it can be removed is by protein degradation and replacement. Small reactive carbonyl species, glyoxal and methylglyoxal, are the primary glycating agents and are elevated in several conditions associated with an increased risk of cardiovascular disease, including diabetes, rheumatoid arthritis, smoking, and aging. Thus, how protein glycation impacts the cardiomyocyte is of particular interest, to both understand how these conditions increase the risk of cardiovascular disease and how glycation might be targeted therapeutically. Glycation can affect the cardiomyocyte through extracellular mechanisms, including RAGE-based signaling, glycation of the extracellular matrix that modifies the mechanical environment, and signaling from the vasculature. Intracellular glycation of the cardiomyocyte can impact calcium handling, protein quality control and cell death pathways, as well as the cytoskeleton, resulting in a blunted contractility. While reducing protein glycation and its impact on the heart has been an active area of drug development, multiple clinical trials have had mixed results and these compounds have not been translated to the clinic—highlighting the challenges of modulating myocyte glycation. Here we will review protein glycation and its effects on the cardiomyocyte, therapeutic attempts to reverse these, and offer insight as to the future of glycation studies and patient treatment.

1. Glycation and the cardiomyocyte

In 1912, Louis Camille Maillard wrote to the journal of *Comptes Rendus* describing a reaction by which amino acids and sugars would react to form a dark product (Maillard, 1912). The rate of this reaction was moderately increased under the presence of heat. Maillard, while attempting to mimic how proteins fold in our cells, had inadvertently discovered a post translational modification (**PTM**) that occurs when proteins are irreversibly and non-enzymatically modified by reducing sugars. Although Maillard's insight into which specific amino acids participated in this reaction and the resulting products was limited due to the technical limitations of the time, the reaction is now known as the Maillard reaction. It is also one of the 'browning' reactions because it is responsible for giving foods such as bread and meat their brown color during the cooking process. However, the reaction later became known as glycation, which is how we will refer to it here.

1.1 Glycation and the glyoxalase cycle

In the century following Maillard's discovery, glycation has been characterized as the reaction between a reducing sugar and an amino acid (typically with a free amino group) or a nucleic acid's amino group. Reducing sugars that participate in this reaction include: glucose, ribose, and fructose (Dyntar et al., 2001; Lester, 1989; Levi & Werman, 1998;

Rabbani & Thornalley, 2014), and aldehyde byproducts of sugar metabolism with highly reactive carbonyl groups, including glyoxal and methylglyoxal (Bidasee et al., 2004; Cai et al., 2014; Moore et al., 2013; Papadaki et al., 2018; Shao et al., 2011). While glycation can also occur on lipids and nucleic acids (Miyazawa, Nakagawa, Shimasaki, & Nagai, 2012; Rabbani & Thornalley, 2014), almost nothing is known about these processes in the cardiomyocyte, therefore this review will only cover protein glycation.

The first step in the nonenzymatic reaction forms an unstable intermediate that is subsequently stabilized, also nonenzymatically, into the irreversible protein modification (Hodge, 1953). After a protein has been glycated, the entire molecule is referred to as an Advanced Glycation End-Product (**AGE**). Once an AGE is formed, it is generally considered irreversible; however, there is evidence that certain adducts may be reversed enzymatically by a protein encoded in the Parkinsonian Associated Deglycase (PARK7) gene, known as DJ-1 (Hodge, 1953; Rabbani & Thornalley, 2014; Richarme et al., 2015), although this is controversial (Mazza et al., 2022).

Glycating compounds are generated from a variety of sources. For example, methylglyoxal is generated as a consequence of glycolysis, from excess dihydroxyacetone phosphate (DHAP) via methylglyoxal synthase in a pathway designed to generate D-Lactate (Yuan & Gracy, 1977). This occurs in all tissues, although at an increased rate in cells with greater metabolic demand such as cardiomyocytes, and in conditions like hyperglycemia and aging (Bidasee et al., 2003; Bou-Teen et al., 2022; Moore et al., 2013; Papadaki et al., 2018, 2022; Ruiz-Meana et al., 2019; Tian et al., 2014). Glycating compounds can also be directly consumed in the diet, including glucose and fructose (Bidasee et al., 2004; Moore et al., 2013). Dietary AGEs constitute a very small percentage of overall AGEs in the body, although they can still influence function of various organs, such as the vasculature (Stirban et al., 2013).

Dicarbonyl aldehydes such as methylglyoxal and glyoxal are reactive carbonyl species (RCS) and the major glycating agents in the cardiomyocyte. Methylglyoxal is responsible for the majority of glycation associated with pathologies such as diabetes (Thornalley et al., 2003). RCS like glyoxal and methylglyoxal react with lysine (K) and arginine (R), two basic amino acids (Bonsignore, Leoncini, Siri, & Ricci, 1973a, 1973b; McLaughlin, Pethig, & Szent-Gyorgyi, 1980). Methylglyoxal glycates arginine residues to form Nδ-(5-hydro-5-methyl4-imidazolon-2-yl)ornithine (**MG-H1**) (Rabbani & Thornalley, 2014); and with lysine residues to form Nε-carboxyethyl-lysine (**CEL**) (Papadaki et al., 2018; Rabbani & Thornalley, 2014; Semba, Bandinelli, Sun, Guralnik, & Ferrucci, 2009). Glyoxal modifies lysine to generate N(ε)-(carboxymethyl)lysine (**CML**) (Rabbani & Thornalley, 2014). These PTMs are the most commonly occurring in the body, but there are other possible glycation modifications these compounds can generate. Reducing sugars such as ribose can cross-link lysine and arginine residues, generating **pentosidine** modifications (Grandhee & Monnier, 1991). Hemoglobin glycated by glucose is called hemoglobin A1c (**HbA1c**) and is used to track "blood glucose" levels over the lifespan of a red blood cell (Cai et al., 2014; Hong et al., 2014; Jiao, Zhang, Peng, & Shen, 2023). Cross-linking AGEs are relevant in cardiovascular pathologies, as they are implicated in the pathogenesis of hypertension risk with age (Aronson, 2003; Kass et al., 2001). The CEL, CML, and pentosidine modifications are shown in Fig. 1A.

Cells detect glycated proteins (AGE) in the serum and extracellular environment via Receptors of AGE (RAGEs) (Liu et al., 2021; Lu et al., 2021; Scavello et al., 2022). RAGE isoforms can localize to plasma membranes, nuclear membranes, and the cytoplasm (Hou et al., 2014; Kumar et al., 2017; Liu, Chen, Luo, & Zheng, 2016), resulting in a variety of signaling cascades, discussed in Section 2.

While there is a baseline level of protein glycation and therefore it is likely involved in physiological signaling, glycation is more often associated with detrimental outcomes and pathology. Thus, cells typically metabolize glycating sugars before they generate AGEs. The primary pathway for this detoxification is the glyoxalase pathway, in which Glyoxalase-1 (**Glo1**) utilizes glutathione (GSH; replenished by Glyoxalase-2 (**Glo2**)) to metabolize methylglyoxal into D-Lactate (Averill-Bates, 2023; Rabbani, Xue, & Thornalley, 2014; Thornalley, 2003a). Dysregulation of Glo1 results in disease (Arai et al., 2010; Rabbani et al., 2014). Manipulations of Glo1 expression and activity are currently under investigation as a therapeutic for diseases where glycation is elevated, such as diabetes (Section 6.3) (Prisco et al., 2022; Rabbani, Xue, Weickert, & Thornalley, 2021).

Methylglyoxal levels increase during the aging process, arising at least partially from a reduction in glyoxalase-dependent clearance. Indeed, aging has shown to reduce glyoxalase cycle efficiency (Ruiz-Meana et al., 2019) and Glo1 expression levels (Li, Zheng, Chen, Liu, & Zhang, 2019). In the cardiomyocyte, CaMKIIδ, the most abundant CaMKII isoform in the heart (Edman & Schulman, 1994), phosphorylates Glo1 at Thr107 and

Glycation in the cardiomyocyte 51

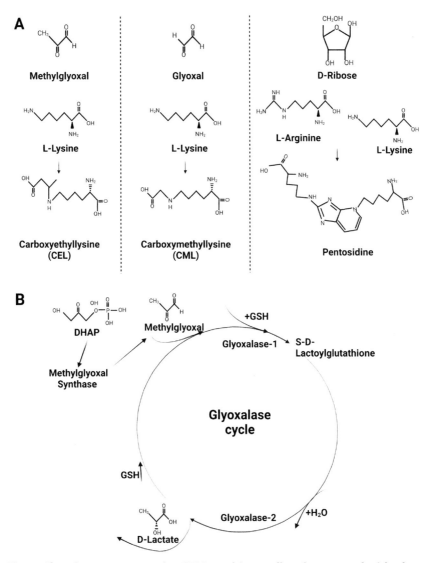

Fig. 1 Glycation targets reactive PTMs and is usually taken care of with glyoxalase cycle. (A) Three examples of glycation: Methylglyoxal and lysine reacting to form CEL (left); methylglyoxal and arginine reacting to form MG-H1 (middle); ribose reacting with lysine and arginine to form pentosidine (right). (B) Diagram of the glyoxalase cycle. Glyoxalase 1 and 2 turn methylglyoxal into D-Lactate, utilizing and regenerating GSH in the process.

increases its maximum catalysis rate and efficiency (Morgenstern et al., 2020). Importantly, CaMKIIδ phosphorylation of Glo1 decreases with age, contributing to the increase in methylglyoxal levels in the heart. Glo1 overexpression supports longer lives and reduced oxidative stress in *C. elegans* (Morcos et al., 2008).

Glo1 has an antioxidant response element (ARE) domain, which responds to high oxidative stress to increase its activity. Oxidative stress can be deleterious to both proteins and nucleic acids (Xue et al., 2012), thus Glo1's role in responding to it is critical to a cell such as a cardiomyocyte, which has high metabolic activity.

1.2 The cardiomyocyte

The pumping action of the heart occurs through the coordinated contraction of individual cardiomyocytes. Thus, the cardiomyocyte contracts continuously throughout the lifespan of the organism, placing it under substantial mechanical and metabolic stress. Although some data suggests cardiomyocytes have a low level of proliferation, this role is minimal (near or below 1% in youth), which is further reduced in aging (Bergmann et al., 2009; Senyo et al., 2013), so these cells are considered quiescent with protective signaling cascades that prevent proliferation (Cho et al., 2023). Typically, a human heart will beat 2.5 billion times or more across the lifespan; therefore, it is important to understand how the terminally differentiated cardiomyocyte responds to stress, such as elevated dicarbonyl stress (Gupta et al., 2016; Martin et al., 2021; Martin, Delligatti, Muntu, Stachowski-Doll, & Kirk, 2022; Townsend et al., 2004).

This continuous cyclic contraction requires the combined efforts of ion channels, cytoskeletal proteins, metabolic activity, protein quality control (**PQC**) mechanisms, and others. Further, the signal to contract is propagated through the heart by structures known as gap junctions which connect the cardiomyocytes, assisting in the tight coordination of movement that keeps the heart pumping synchronously (Grant, 2009; Hesketh, Van Eyk, & Tomaselli, 2009; Jafri, 2012). Each of these components and how glycation affects them will be discussed below.

2. Extracellular glycation

The extracellular environment provides nutrients, structural scaffolding, and signaling to the cardiomyocyte. For example, RAGE on the cardiomyocyte plasma membrane detect glycated albumin (Son et al., 2017), glycated ECM components such as collagen (Alikhani et al., 2005), and other proteins such as s100 proteins in the extracellular environment and nearby vasculature (Hofmann et al., 1999). Often, these can represent dangerous positive feedback signaling, as AGE itself can increase RAGE protein expression in a dose-dependent manner (Hou et al., 2014). Indeed, RAGE signaling in the heart plays a substantial role in response to various stresses and disease states, activating inflammation, gene expression changes, apoptosis, and more. Many of these pathways are common across multiple cell types and will not be covered in detail here, and readers are directed towards a comprehensive review on cardiac RAGE signaling (Ramasamy & Schmidt, 2012).

Importantly, the heart is a unique mechanical environment, so cardiomyocytes also respond to critical outside-in and inside-out mechano-signaling and support. Indeed, cardiomyocytes dynamically interact with and respond to changes in the ECM and vasculature (Rienks, Papageorgiou, Frangogiannis, & Heymans, 2014).

2.1 Extracellular matrix glycation

Changes in the ECM can influence the cardiomyocyte, so the cell can appropriately respond to environmental disruptions such as increased afterload or injury (Rienks et al., 2014). Collagen represents an essential component of the cardiac ECM which is known to influence vital myocyte functions such as calcium handling (Lu et al., 2011) and maturation (Tani et al., 2023) and is often found dysregulated alongside Titin in disease (Roe et al., 2017; Wu, Cazorla, Labeit, Labeit, & Granzier, 2000). Broadly, collagen cross-linking occurs when two non-sequential amino acids are modified in such a way that they covalently link together and is associated with cardiovascular disease (Neff & Bradshaw, 2021). In diabetes and diabetic rodent models, cross-linking of lysine residues of collagen is increased (Hudson, Archer, King, & Eyre, 2018; Monnier, Glomb, Elgawish, & Sell, 1996), and cross-linked collagen can result in elevated stiffness of heart tissue (Lopez, Querejeta, Gonzalez, Larman, & Diez, 2012).

Glycation of lysine residues can result in this cross-linking. Glucose-initiated (fructosyl-lysine) glycation of collagen can cross-link lysine residues in diabetic mouse models (Hudson et al., 2018) and compete for enzymatic, healthy cross-linking in collagen. In atrial fibrillation, where tissue stiffness is increased (Khurram et al., 2016), myocardial AGEs are increased as well (Bi et al., 2021). However, there was no additional impact of AGEs on collagen stiffness in hypertrophic obstructive cardiomyopathy when comparing sinus rhythm to atrial fibrillation patients in this group, although this has not yet been investigated outside of the context of hypertrophic cardiomyopathies (Bi et al., 2021). AGE-breakers and AGE inhibitors, drugs intended to break these cross-links or prevent them from forming, have shown some effectiveness attenuating collagen cross-linking and deposition (Brownlee, Vlassara, Kooney, Ulrich, & Cerami, 1986; Candido et al., 2003) and have been since investigated as treatments for heart disease. These will be discussed in more detail in Section 6.

RAGE activation also increases matrix-metalloprotease-2 (MMP-2) activity (Harja et al., 2008). VCAM-1 and MMP-2 are extracellular matrix (ECM)-associated proteins (Goncalves, Nascimento, Gerlach, Rodrigues, & Prado, 2022; Harja et al., 2008), meaning ECM remodeling may result from this.

2.2 Extracellular glycation in cardiovascular diseases

Diabetes and hyperglycemia are among the most studied contributors to increased AGE levels. Diabetic hemoglobin A1c (HbA1c)—the preferred method of long-term blood sugar analysis due to glycation being irreversible and the longevity of red blood cells—is a measurement of glucose-catalyzed AGE in hemoglobin (Lester, 1989). Elevated HbA1c is a reliable predictor of adverse cardiovascular events (Cai et al., 2014; Hong et al., 2014; Jiao et al., 2023; Ravipati et al., 2006). Recent studies have found glycated albumin results in whole-body insulin resistance and skeletal muscle downregulation of GLUT4 (Pinto-Junior et al., 2018). Also, AGE increases in diabetes reduces an individual's capacity to respond to cardiovascular insults, such as a heart attack (myocardial infarction; MI) (Liu et al., 2021). Smoking also increases AGEs in human serum (Cerami et al., 1997; de Groot et al., 2011; Nicholl & Bucala, 1998). Elevated serum AGE from smoking (1) contributes to diseases that independently cause AGEs to increase (de Groot et al., 2011) and (2) occurs independently of diabetes (Nicholl & Bucala, 1998), suggesting an important, understudied role of AGEs in diseases which impact smokers.

As organisms age, AGEs begin to accumulate (Schlotterer et al., 2009; Semba et al., 2009), which correlates with higher mortality risk, independent of other cardiovascular comorbidities (Semba et al., 2009). Rheumatoid arthritis, which increases cardiovascular disease risk independent of diabetes (van Halm et al., 2009), increases plasma AGE (de Groot et al., 2011; Rabbani & Thornalley, 2014). Although this modest increase was highest in populations such as the elderly and smokers where AGE is already elevated (de Groot et al., 2011), it correlated with atherosclerosis precursor signals in endothelial cells. Therefore, it may serve as a useful biomarker for cardiovascular disease risk in this population (de Groot et al., 2011).

AGEs and their impact may increase with aging and diabetes, but contrary to this, some RAGE receptor isoforms may reduce in expression with aging, RAGE receptor mRNA loss in diabetes may be as much as 50% (Bidasee et al., 2003; Liu et al., 2016; Scavello et al., 2021).

2.3 Future directions

The contribution of RAGE signaling to metabolic outcomes which may impact cardiomyocytes, such as reduced GLUT4 expression in skeletal muscle (Passarelli & Machado, 2021; Pinto-Junior et al., 2018), remains understudied. This could represent an early, dangerous feed-forward cycle in early disease stages of diabetes, where AGE levels are rising, and hyperglycemia and insulin resistance are both worsening.

Smoking of E-cigarettes is also associated with methylglyoxal and glyoxal introduction to airways and induction of RAGE associated pathways (Section 2.1), such as NF-κB activation, in airway epithelial cells (Kwak et al., 2021). Whether this also increases serum AGE as traditional cigarettes remains unknown—however, with the growing popularity of E-cigarettes, particularly among the younger population (Willett et al., 2019), investigating this question is of great importance.

Increased vascular AGE deposition in diabetic aortic stenosis patients has also been shown to correlate with worsened cardiomyocyte passive tension (Falcao-Pires et al., 2011; van Heerebeek et al., 2008). The reason for this correlation remains an open area of research.

The overall impact and role of cardiomyocyte RAGE signaling remains unclear. RAGE activation increases oxidative stress in cardiomyocyte cytosol and mitochondria (Chen et al., 2022), and soluble serum RAGE inversely correlates with hypertension (Liu et al., 2016). However, there is evidence that AGE-RAGE pathways may be required for normal cell function or even involved in protective responses. Historically, there was interest in anti-RAGE therapeutics, but targeting RAGE has not always resulted in positive outcomes. While RAGE antibody administration decreases RAGE-induced arrhythmias in mice (Liu et al., 2021); genetic RAGE knockout increases fibrosis (Scavello et al., 2021) and upregulates genes involved in cardiac remodeling in aging mice (Scavello et al., 2022). Furthermore, nuclear isoforms of RAGE may be involved for DNA repair in some tissues (Kumar et al., 2017). Converse to Scavello et al., other research has observed RAGE signaling prevents cardiac fibroblast migration

(Burr, Harmon, & Stewart, 2020), which could be a positive outcome for patients at risk for increased fibrosis. These studies suggest that *some* RAGE expression is required for normal physiology. Whether targeting RAGE is a viable therapeutic option, and to what extent and in what capacity RAGE are protective or support normal physiology in myocardial ECM and nearby vasculature, remains unknown.

3. Glycation disrupts cardiomyocyte calcium handling

Cardiomyocytes rely on Ca^{2+} signaling to trigger important pathways in the cell, such as facilitating the function of contraction (via binding to RyR and TnC) and CaMKII activity (Beckendorf, van den Hoogenhof, & Backs, 2018; Eisner, Caldwell, Kistamas, & Trafford, 2017; Grant, 2009; Jafri, 2012). For a more extensive review describing the excitation-contraction cycle in cardiac cells, please see (Eisner et al., 2017). Briefly, a depolarization signal opens voltage gated sodium channels, until the membrane has become positive enough to trigger the opening of the L-Type Calcium channels. This small influx of Ca^{2+} then triggers calcium-induced calcium release (**CICR**), where the ryanodine receptor (**RyR**) releases large stores of Ca^{2+} from the endoplasmic reticulum (in muscle cells, sarcoplasmic reticulum, SR). This Ca^{2+} is then able to bind to troponin on the thin filament of the sarcomere (the fundamental functional unit of a muscle cell) to initiate contraction. When the heart is relaxing, sarco/endoplasmic reticulum ATPase (**SERCA**) pumps Ca^{2+} back into the SR, decreasing the cytoplasmic levels causing thin filament de-activation and relaxation of the sarcomere. These carefully regulated processes are central to contraction of the cardiomyocyte, and many of the central proteins are glycated under stress.

3.1 RAGE signaling effects on calcium handling

A major pathway that RAGE signaling impacts is calcium (**Ca^{2+}**) handling in the cardiomyocyte (Fig. 2A and B). It was found that serum AGE positively correlated with premature ventricular contraction (PVC), a type of arrhythmia, incidence in humans as well as other ventricular arrythmias in diabetic, post-MI rats (Liu et al., 2021). The mechanism was through RAGE induced increases in GRP78-PERK mediated ER-stress initiated by aberrant Ca^{2+} signaling, specifically ryanodine receptor (**RyR**) leakiness (Liu et al., 2021), and inhibiting RAGE was able to attenuate ventricular arrhythmias (Liu et al., 2021).

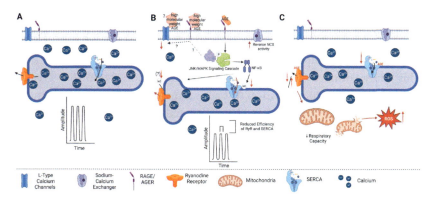

Fig. 2 Calcium handling of cardiomyocytes is disrupted by both intra- and extracellular glycation. (A) Normal calcium handling in the cardiomyocyte. To be compared with B and C. (B) Example of the effect RAGE signaling has on calcium handling in the cardiomyocyte. (C) Example of the effect intracellular AGEs have on the cardiomyocyte.

Indeed, RAGE signaling has been shown to impact many calcium proteins in the cardiomyocyte. In rodent models of diabetes, plasma AGE induced S-nitrosylation of RyR and SERCA in cardiomyocytes that resulted in impaired peak Ca^{2+} release and the rate of Ca^{2+} return (Hegab et al., 2017). However, this dysfunction occurred without altering expression levels of any calcium handling protein (Hegab et al., 2017; Petrova et al., 2002). Overall, whether nitrogenous PTMs on RyR are beneficial or harmful is controversial, as acknowledged by the authors (Gonzalez, Beigi, Treuer, & Hare, 2007; Gonzalez, Treuer, Castellanos, Dulce, & Hare, 2010).

Recently, RAGE activation has been shown to increase reverse activity of the sodium-calcium exchanger (NCX) in AGE-treated myocytes (Chen et al., 2022), which could have a modest effect on overall calcium handling in the cardiomyocyte (more-so in humans than in rodents). Even higher molecular weight AGEs that may not activate the traditional RAGE-JNK/MAPK-NF-κB signaling pathway (Deluyker et al., 2016; Deluyker, Evens, Belien, & Bito, 2019) can impair L-Type Ca^{2+} current in isolated cardiomyocytes, resulting in reduced shortening capacity (Deluyker et al., 2019).

3.2 Intracellular protein glycation effects on calcium handling

Intracellular Ca^{2+}-signaling proteins can also be directly modified by intracellular glycating compounds that modify their function (Fig. 2C). Multiple studies have shown that glycation of SERCA has a significant negative impact on its function. However, the types of SERCA2A-AGEs that are present in a diabetic individual are diverse, with several kinds of glycation on different amino acids increasing in a model of rodent type 1 diabetes when compared to nondiabetic control rats (Shao et al., 2011). This study, done by Shao et al., effectively demonstrated that glycation indeed caused reduced SERCA Ca^{2+} transport by generating mutated SERCA plasmids that had a charge-neutralized lysine at K481 due to pyralline-causing glycation (Shao et al., 2011). Earlier studies done by Bidasee et al. similarly found glycated SERCA2A in rat models of diabetes. This proteomic analysis found a large number of pentosidine and imidazolone modifications and a decrease in total SERCA2A protein (Bidasee et al., 2004). Further, there was an increase in PLB phosphorylation (which would reduce SERCA inhibition—a possible compensatory attempt by the cell) (Bidasee et al., 2004). However, the basal level of PLB expression was also increased by about 40% (Bidasee et al., 2004), and how the pPLB:PLB compares with the control or what glycation may exist on PLB itself was not addressed in this study. In pancreatic cell lines, SERCA is modified by methylglyoxal in a dose-dependent manner with a concurrent decrease in activity level, and overall, SERCA decreases as methylglyoxal increases (Zizkova et al., 2018). However, the concentration of methylglyoxal used (5–40 mM in initial experiments; 1–5 mM in later) is supraphysiological. In other studies, 100 μM methylglyoxal is the highest concentration used to imitate pathological levels in acute experiments (Papadaki et al., 2018).

Similar to SERCA, glycation of RyR is also functionally detrimental (Bidasee et al., 2003). An increase in RyR2-AGEs such as CML occurs concurrently with a greater than 58% decrease in binding of RyR2 to its ligand, ryanodine, and a reduced overall affinity for Ca^{2+} (Bidasee et al., 2003)—critical for the CICR component of excitation-contraction coupling that the RyR participates in. This reduced Ca^{2+} affinity is thought to be responsible for the decreased activity of the RyR in diabetic models that has been noted in some studies (Yaras et al., 2005). In patient groups where AGEs are known to be high, there is an increase in PVCs, which were found to be caused by leaky RyR activity, rather than a decrease in its activity (Liu et al., 2021). This raises the interesting possibility that both extracellular and intracellular glycation contribute to cardiac arrhythmias (Koponen et al., 2020), which are a well-understood consequence of RyR dysfunction. Similar findings elsewhere (Ruiz-Meana et al., 2019) suggest RyR AGEs present in aging result in leaky RyRs and subsequent mitochondrial damage. This again suggests the effect of glycation on this Ca^{2+} handling protein may be disease and glycation-adduct dependent.

3.3 Glycation induced Ca^{2+} dysregulation in cardiovascular diseases

In rodent models of diabetes, RyR mRNA and protein is consistently decreased when compared to 'nondiabetic' mice, and has previously estimated to be at about 66% to as low as 56% the level of expression (Bidasee et al., 2003; Yaras et al., 2005), with aberrant (i.e., slower, leakier) activity (Bidasee et al., 2003; Yaras et al., 2005). L-Type calcium channels do not contribute much to the development of metabolic syndrome in these models (Yaras et al., 2005). However, SERCA glycation does occur in these models and is now thought to contribute to diastolic dysfunction experienced by diabetic individuals (Bidasee et al., 2004; Shao et al., 2011). Rodent models of Type-1 diabetes also show calmodulin, an important endogenous Ca^{2+} chelator (Walsh, 1983), is glycated, and that overall calmodulin is reduced in submandibular salivary glands (Nicolau, de Souza, & Carrilho, 2009).

Aging also impacts Ca^{2+} handling glycation of cardiomyocytes. In old mice compared to young, there is a 30–40% increase in RyR glycation (Ruiz-Meana et al., 2019). Age also impacts the glycation of mitochondrial proteins, and the mitochondria and sarco/endoplasmic reticulum (where Ca^{2+} is stored) are in a delicate balance and must be able to properly communicate (Bou-Teen et al., 2022).

3.4 Future directions

The effect of SERCA glycation and subsequent PLB phosphorylation states (Bidasee et al., 2004; Moore et al., 2013; Shao et al., 2011; Zizkova et al., 2018) have lingering questions. Increased pPLB (Bidasee et al., 2004) is incongruent with other studies which found no change in or decreased pPLB, as noted by the authors in Discussion (Bidasee et al., 2004; Choi et al., 2002; Kim, Ch, Lee, Park, & Kim, 2001). As PLB regulates SERCA, resolving glycation's impact on pPLB is critical to better understand the cellular response to impaired Ca^{2+} cycling under glycation stress. Moreover, the glycation status of many Ca^{2+} handling proteins, such as NCX, CAMKII, and L-Type Ca^{2+} Channels remains completely unknown. AGE-RAGE signaling reverses NCX activity in ventricular cells (Chen et al., 2022); yet, despite conditions of high AGE impacting both the extracellular and intracellular environment, whether NCX is glycated and, if so, the impact on the RAGE-activated outcome is unknown.

A recent study using *ex-vivo* hearts and isolated cardiomyocytes from rats shows methylglyoxal incubation increased intracellular Ca^{2+}, though not in the presence of aminoguanidine (an AGE scavenger, see Section 6) or verapamil (an L-Type Ca^{2+} channel blocker) (Peyret et al., 2024). Although concentrations of methylglyoxal used were supraphysiological (0.5–1 mM), whether it is the Ca^{2+} channel itself which is glycated or interacting proteins merits further investigation as verapamil is already approved for use in humans.

Studies such as (Yaras et al., 2005) have suggested reduced RyR2 activity in diabetes is *not* due to glycation, but rather decreased expression and phosphorylation. Whether glycation or expression is the culprit still remains unknown, and further studies should be done to solidify our understanding of (1) whether glycation is directly responsible for RyR2 dysfunction and, if so, what residues and specific modifications are the cause; (2) whether RyR2 is degraded *as a consequence of* glycation, or because of some other signal; and (3) what the role of RyR2 glycation is on the RyR's ability to communicate with other organelles, like the mitochondria. This last point is of particular interest, as in rodent models of aging, (high glycation, reduced RyR2 activity, no RyR2 expression changes), a loss of connectivity between RyR2 and mitochondria has been proposed to be a mechanism by which mitochondrial dysfunction occurs (Fernandez-Sanz et al., 2014).

The cardiovascular risks from extracellular AGE from smoking traditional and E-Cigarettes due to increased methylglyoxal and glyoxal is known (Cerami et al., 1997; Nicholl & Bucala, 1998; Ruedisueli, Lakhani, Nguyen, Gornbein, & Middlekauff, 2023) (Section 2), however whether smoking impacts cardiomyocyte Ca^{2+} handling remains unstudied. E-Cigarette smoking may result in impaired repolarization compared to nonusers (Ruedisueli et al., 2023), but whether RAGE signaling, glycation of voltage-gated potassium channels, SERCA glycation, or other proteins responsible for maintaining a healthy rate of ventricular cardiomyocyte repolarization are responsible for this is unknown.

4. Glycation of the cytoskeleton

Many of the cytoskeletal proteins in the cardiomyocyte are components of the sarcomere, the repeating lattice structure and fundamental contractile unit in muscle cells (Martin, Thompson, et al., 2022) (Fig. 3A). The sarcomere consists of thick filaments (primarily composed of myosin and myosin binding proteins), and thin filaments (primarily composed of actin, troponin, and tropomyosin) anchored at structural z-disks. The giant

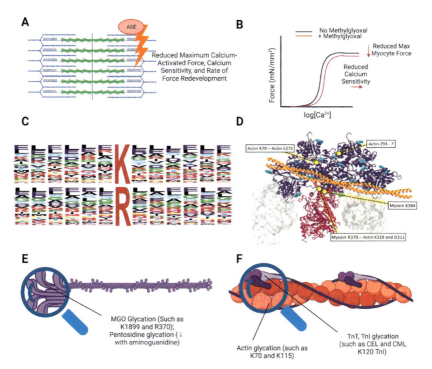

Fig. 3 Glycation of the myofilament is detrimental to cardiomyocyte function. (A,B) Diagram of the sarcomere and the effect AGE generation has on sarcomere function (C) Sequence analysis of glycated peptides indicating no consensus sequence for glycation (D) Glycation of the thick and thin filament proteins, highlighting actin and myosin glycation (E) Glycation of the thick filament (F) Glycation of the thin filament.

protein Titin aids in sarcomere organization and maintaining restorative passive force. These proteins are under tight stoichiometric control in the cardiomyocyte, so function is regulated or dysregulated primarily via PTMs on these proteins, such as glycation.

4.1 Glycation impairs cardiomyocyte sarcomere function

Sarcomeric proteins generally have long half-lives in the cardiomyocyte (days–weeks) compared to many other intracellular proteins. Thus, these proteins may accumulate glycation over a longer period before they are degraded, making the sarcomere a "high-risk" structure for glycation-induced dysfunction (Fig. 3A and B).

In mass spectrometry studies of sarcomeric proteins, we found that glycation occurred in certain reproducible "hot spots", indicating that there is preference for some residues over others (Papadaki et al., 2018, 2022). For example, glycation of K1899 is observed on α-Myosin Heavy Chain (α-MHC), but *not* on the corresponding residue (K1898) on the highly homologous β-MHC (Papadaki et al., 2018). However, this specificity is likely not due to a protein consensus sequence, since we found no obvious patterns in the regions surrounding a glycated arginine or lysine residue (Fig. 3C). This makes sense, given that PTM consensus sequences are usually recognized by the involved enzyme, and glycation is non-enzymatic. The selectivity could arise from steric considerations, since many of the modifications we detected were on the surface of the proteins (Fig. 3D), although given the small size of glyoxal and methylglyoxal, this explanation also seems somewhat unlikely (or at least incomplete). Regardless, the question of specificity remains an open one across all glycation biology, not just in the cardiomyocyte.

Glycation is detrimental to sarcomere function, and thus contractility of the cardiomyocyte. In permeabilized murine myocytes, methylglyoxal incubation impaired maximum Ca^{2+} activated force, Ca^{2+} sensitivity, and the rate of force redevelopment (Papadaki et al., 2018, 2022) (Fig. 3A and B). Specifically, methylglyoxal glycation of actin (Fig. 3D) impeded normal tropomyosin-actin interactions, inhibiting thin filament activation (Papadaki et al., 2022). Moreover, methylglyoxal also directly impairs myosin function, although the molecular mechanism remains unclear. Interestingly, some of the residues on which glycation was observed are also residues on which point mutations are associated with cardiomyopathy. These residues include K1899 and K383 on myosin (Kuang et al., 1996).

Other groups have shown glycation of myosin. For example, diabetic STZ injected rats developed pentosidine modifications of ventricular myosin heavy chains, which was attenuated with aminoguanidine treatment (Shao et al., 2010)—indicating this stress may be successfully reversed by drugs intended to inhibit AGE formation (discussed in Section 6). Further, insulin control was able to almost restore myosin glycation levels completely to that of controls (Shao et al., 2010).

Glycation of the troponin complex has also been observed, particularly on several lysine and arginine residues of TnI and TnT (Janssens et al., 2018; Ruiz-Meana et al., 2019), indicating another pathway through which glycation could interfere with the Ca^{2+} sensitivity of the myofilament in environments that favor glycation. However, other studies do not detect troponin glycation (Papadaki et al., 2018). Therefore, the role troponin glycation plays in healthy and diseased sarcomere function requires further investigation.

Underlining the critical nature of investigating glycation's effect on the cardiac sarcomere is the abundance of lysine-based methylglyoxal and glyoxal glycation in the sarcomere (Papadaki et al., 2018, 2022). Lysine contributes a positive charge to participate in structural interactions and enzymatic catalysis (Holliday, Mitchell, & Thornton, 2009), critical for normal sarcomere function. For example, actin arginine and lysine residues are thought to participate in electrostatic interactions that stabilize tropomyosin's inhibitory position on the thin filament, which can be fine-tuned by PTMs (Schmidt & Cammarato, 2020; Schmidt, Madan, Foster, & Cammarato, 2020). Some lysine residues such as K210 on Troponin T (**TnT**) result in disease (in this case, DCM) when mutated (Clippinger et al., 2019).

There is also evidence that extracellular AGE signaling increases MMP-2 activity. MMP-2 is active within cardiomyocytes and is associated with the degradation of sarcomere proteins including Troponin I (**TnI**) (Wang et al., 2002), which supports the interaction of actin and tropomyosin, preventing inappropriate sarcomere activation (Lehman, Pavadai, & Rynkiewicz, 2021). Thus, aberrant sarcomere degradation could result via this pathway.

4.2 Myofilament glycation in diabetes

Diabetes is associated with an increase in glycation at the cardiac myofilament. In particular, there is a significant increase in methylglyoxal glycation on lysine (CEL modifications) (Papadaki et al., 2018, 2022). Increased levels of pentosidine on α- and β-myosin heavy chain (MHC) have also been observed (Shao et al., 2010). In rat models of diabetes and in isolated pig cells, microtubule glycation in neuronal cells has been assessed and compared between the control and experimental group, but whether tubulin experienced an increase in glycation and/or decrease in stability was not fully clear from the data (McLean, Pekiner, Cullum, & Casson, 1992).

4.3 Next steps in understanding cytoskeletal glycation

Recent work demonstrates myofilament glycation results in functional deficits in the cardiomyocyte (Papadaki et al., 2018, 2022), and that myofilament glycation in disease may be severe (Janssens et al., 2018; Papadaki et al., 2018, 2022; Shao et al., 2010). This has primarily been investigated under the context of diabetes. However, a recent aging study has also shown elevated glycation on myofilament proteins in myocardium of older mice when compared to younger ones, including on several myosins, Titin, and TnI (Ruiz-Meana et al., 2019).

Lysine residues, which can be targets of glyoxal or methylglyoxal glycation, are also critical to the microtubule polymerization (which are critical for structural support and vesicle trafficking in cardiomyocytes) (Caporizzo, Chen, & Prosser, 2019). There has been some investigation into microtubule glycation in neuronal cells (McLean et al., 1992), however microtubule glycation in cardiomyocytes is not well studied. Further, whether kinesin or dynein are glycated in cardiomyocytes is completely unknown. Future work should investigate if cardiomyocyte tubulin glycation (if it occurs) impairs polymerization of microtubules or the transport highways they comprise.

5. Other effects of glycation in the cardiomyocyte
5.1 Glycation and the mitochondria

Cardiomyocyte mitochondria accumulate diet and age induced AGEs (Pamplona, Portero-Otin, Bellmun, Gredilla, & Barja, 2002). The impact of these modifications is varied, but consistently harmful. Within the cardiomyocyte, aging rats displayed increased mitochondrial glycation, which was somewhat mitigated by a change in diet (Pamplona et al., 2002), and other studies have found ATP Synthase to be more highly glycated in older rats when compared to younger ones (Gu et al., 2022), resulting in impaired ATP synthesis (Bou-Teen et al., 2022). Damaged mitochondria also undergo their own death pathway, called mitophagy. Mitophagy is linked to aging and cell death (although it is controversial whether the process promotes aging and cell death or counters them) (Ding & Yin, 2012; Werbner, Tavakoli-Rouzbehani, Fatahian, & Boudina, 2023), and there is evidence that administration of the glycating molecule glycolaldehyde initiates mitochondria-associated cell death pathways in renal mesangial cells (Gu et al., 2022). However, this impact has yet to be demonstrated in cardiomyocytes.

5.2 Protein quality control

Glycation is irreversible, with just a few controversial studies indicating deglycase activity of certain enzymes such as DJ-1 (Mazza et al., 2022; Richarme et al., 2015). Therefore, AGEs must be degraded and replaced, meaning proteins that reduce glycated molecules are the primary intracellular mechanism responsible for reducing glycation once it has occurred. Yet, whether proteins involved in PQC pathways are themselves glycated remains largely unknown. Carbonylation, a similar PTM to glycation that is largely aldehyde related, impacts PQC-related proteins such as the 26S proteasome in HeLa cells (Bollineni, Hoffmann, & Fedorova, 2014) and HSP90 in human colorectal cells (Connor, Marnett, & Liebler, 2011). Studies examining glycation of proteins involved in PQC mechanisms in the cardiomyocyte are limited, but we will cover what is known in this section (Fig. 4).

Incubation of cardiomyocytes with methylglyoxal increases glycation of thioredoxin, impairing its ability to respond to oxidative stress (Wang et al., 2010). This is of note, as diabetes significantly increases the likelihood a patient will experience a myocardial infarction, which results in substantial oxidative stress on the remaining cardiomyocytes (Cui, Liu, Li, Xu, & Liu, 2021; Leon & Maddox, 2015). Alcohol and similar substances which create an oxidative environment result in an increase in carbonylation of PQC-related proteins like HSP90 and the 26S proteasome (Bollineni et al., 2014; Connor et al., 2011). However, it is not known whether these modifications alter the function of these proteins or impact PQC in the cell. High glucose and Glo1 knockdown increases HSP70 and GRP78 protein expression in endothelial cells and generates some PQC protein glycation (Irshad et al., 2019), but whether this plays a role in the cardiomyocyte is unknown. Furthermore, it is not known whether the increased expression of these proteins is to responding to decreased efficiency of PQC proteins from internal glycation, or increased

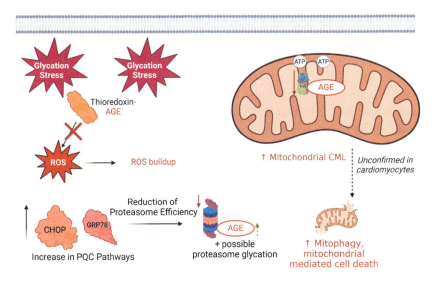

Fig. 4 Diagram of known AGE-induced PQC disruption in cardiomyocytes.

PQC stressed from glycation on other proteins. Neonatal rat ventricular cardiomyocytes incubated with 1 mM methylglyoxal results in increased cell death and cell stress pathways such as cleaved PARP-1 and caspase-3, ATF4, CHOP, and GRP98 (Nam et al., 2015), however whether this is a direct result of glycation of these proteins or as a result of upregulated stress pathways remains unknown.

5.3 Future directions in glycation of PQC-related proteins

The unfolded protein response (UPR) largely starts with activation of ER stress pathways. The 78-kDa glucose-regulated protein (GRP78) is one of these proteins. GRP78 has found to be glycated with extremely active ER stress pathways such as upregulation of C/EBP Homologous Protein (CHOP), a protein highly associated with cell death, but not in cardiomyocytes (Oyadomari & Mori, 2004; Yamabe et al., 2013). Activated Protein C (APC) treatments have shown some efficacy in ameliorating 1 mM methylglyoxal-induced ER stress responses in cardiomyocytes (Nam et al., 2016). Moreover, lysine residues on PQC proteins, including the 26S proteasome, are carbonylated (Bollineni et al., 2014), a PTM similar to methylglyoxal and glyoxal glycation. However, the effect of glycation on PQC mechanisms overall remains unstudied. PQC is critical in the cardiomyocyte (Maejima, 2020). Therefore, better understanding how glycation impacts PQC is of the utmost importance, as deviations to the efficiency of these pathways may have severe disease consequences. Glycation of mitochondrial proteins should also be investigated, as cardiomyoblasts incubated with 625 μM methylglyoxal show decreased mitochondrial respiration (including impaired ATP production) (Prisco et al., 2022), which could exacerbate other cell stressors. This makes this an interesting avenue for future research. Together, these studies demonstrate ER activation and mitochondrial stress occur after RAGE activation, that ER-stress proteins may themselves become AGEs, and respiratory stress is a consequence of mitochondrial AGEs. Further work must elucidate which of these pathways occur in cardiomyocytes and what the functional effect of specific protein glycation is on overall PQC in the heart.

6. Therapeutics and the future of cardiomyocyte glycation

While there are some exceptions (Hodge, 1953; Mazza et al., 2022; Rabbani & Thornalley, 2014; Richarme et al., 2015), most studies agree glycation is largely irreversible. Therefore, many approaches to treating glycation do not attempt to reverse glycation or crosslinking, with the exception of AGE-breakers like Alagebrium (Kass et al., 2001). Investigation into drugs which have anti-glycation properties has been extensive. We will highlight four approaches that have been explored to treat glycation or its effects: ALT-711, an AGE-breaker; aminoguanidine, an AGE-inhibitor; N-Terminal MYBPC treatment; and Glo1 overexpression. Fig. 5 contains a summary of three anti-glycation drugs discussed in Section 6. Table 1 contains a list of clinical trials aimed at mitigating AGE-related damage in the cardiovascular system, including ones that do not fit in any category discussed, such as amino acid-derived supplements and other AGE-Inhibitors (Kennedy, Solano, Meneghini, Lo, & Cohen, 2010; Ranasinghe, Jayawardena, Chandrasena, Noetzel, & Burd, 2019).

6.1 Alagebrium (ALT-711), an AGE-breaker

Alagebrium (ALT-711) (Fig. 5A) is intended to break the cross-linking from AGEs as a treatment for hypertension and heart failure (Kass et al., 2001; Little et al., 2005; Toprak & Yigitaslan, 2019). Whether Alagebrium is a true "AGE-breaker" is yet to be seen, as some evidence suggests it is a chelator of AGE-inducing molecules, like aminoguanidine (Fig. 5B).

Cross-linking AGEs can accumulate in the extracellular matrix and increase the stiffness of the surrounding area, which is thought to contribute to disease in diabetes and age-related cardiovascular complications. Therefore, the purpose of AGE-breakers is to relieve this tissue stress and improve overall cardiovascular function (Candido et al., 2003; Kass et al., 2001). Indeed, in diabetic rats, Alagebrium was able to not only reduce AGEs found in the left ventricle when compared to untreated diabetic rats, but was able to reduce RAGE mRNA in both diabetic and control groups and appeared to attenuate type-III collagen deposition (Candido et al., 2003). However, in models of rodent diabetes no changes in overall HbA1c were seen (Candido et al., 2003). Alagebrium has also been effective in animal models when administered tandem with other medications such as nifedipine, a calcium channel blocker, in treating diabetic hypertension (Zhang et al., 2014). AGE breakers have not just been explored as a treatment for diabetes, as they have also been examined for efficiency in hypertension. Indeed, Alagebrium was found to have a modest, borderline significant, impact on decreasing diastolic blood pressure, and improved arterial compliance (Kass et al., 2001).

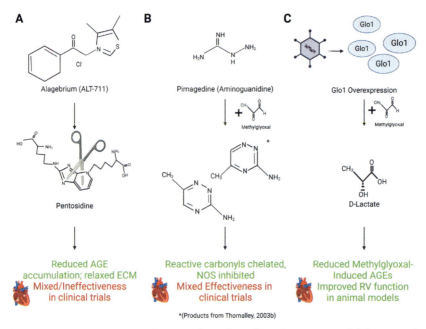

Fig. 5 Attempts at treating glycation have been largely unsuccessful, but remain under investigation. (A) The mechanism of Alagebrium, an AGE-breaker. (B) The mechanism of aminoguanidine, an AGE-inhibitor. (C) The theory behind Glo1 over-expression in reducing glycation.

Despite these early successes, the clinical trials of Alagebrium have had mixed, but generally negative, results. BENEFICIAL was a randomized, placebo-controlled trial which enrolled heart failure with reduced ejection fraction (HFrEF) below 45%. They did not exclude patients with controlled diabetes or hypertension, however patients with severely impaired exercise tolerance were excluded (Willemsen et al., 2010). No differences were noted between placebo and experimental groups for any parameter of cardiovascular function, including LV ejection fraction and NT-proBNP levels, and there was a trending increase in negative cardiovascular outcomes such as rhythm disturbances and cardiac failure (Hartog et al., 2011). As a cardiovascular health preventative, it also failed in elderly patients to improve cardiovascular health or lifetime risk beyond those on exercise plans (1-year or lifetime), and Alagebrium + 1-year exercise did not act synergistically for a greater impact (Carrick-Ranson et al., 2016; Oudegeest-Sander et al., 2013). However, it was seen to potentially decrease LV stiffness (Fujimoto et al., 2013). Other clinical trials in elderly patients with heart failure with preserved ejection

Table 1 Clinical trials of anti-glycation drugs.

Trial	Drug	Period	Status	Disease	Primary and secondary measurements	Positive outcomes	Negative outcomes	Tx duration
ACTION II	Aminoguanidine (AGE inhibitor)	1995–98	Completed	Type 2 DM	Serum creatinine; *motality, ESRD, cardiovascular morbidity and mortality, renal function parameters, urinary AGEs*	Unavailable	Unavailable	1 year
ACTION I	Aminoguanidine (AGE inhibitor)		Completed	Type 1 DM	Serum creatinine; *urinary protein, lipids, HgA1C, P, ETDRS score, mortality, ESRD, adverse event incidence*	AGEs reduced	Not effective against diabetic nephropathy	2–4 years
DIAMOND (NC-T00043836)	Alagebrium (AGE-Breaker)	2002–09	Completed	Diastolic HF, dyspnea, pulmonary edema in elderly	O2 consumption in exercise; quality of llife, cardiac echocardiography parameters; pulmonary function in exercise, aortic distensibility	↓ LV mass (↓ hypertrophy), ↑ diastolic function, ↑ quality of life	No change in 16 weeks aortic distention	
SAPPHIRE (NC-T00045981)	Alagebrium (AGE-Breaker)	2002–10	Completed	Hypertension (>50 years old)	Unavailable	Unavailable	Unavailable	6 months

(continued)

Table 1 Clinical trials of anti-glycation drugs. (cont'd)

Trial	Drug	Period	Status	Disease	Primary and secondary measurements	Positive outcomes	Negative outcomes	Tx duration
SILVER (NCT00045994)	Alagebrium (AGE-Breaker)	2002–09	Completed	Hypertension, LV hypertrophy (>50 years old)	Unavailable	Unavailable	Unavailable	6 months
SPECTRA (NCT00089713)	Alagebrium (AGE-Breaker)	2004–10	Terminated	Hypertension (>45 years old)	Unavailable	Unavailable	Unavailable	17–20 weeks
STRETCH (NCT00302250)	Alagebrium (AGE-Breaker)	2006	Completed	Vasoreactivity and hypertension (>50 years old)	Unavailable	Unavailable	Unavailable	8 weeks
STRETCH (NCT00277875)	Alagebrium (AGE-Breaker)	2006–09	Completed	Vasoreactivity and hypertension (>50 years old)	Endothelial function; distensibility; arterial stiffening; vascular inflammation; collagen; augmentation index	↔ HR, pulse pressure; ↓ collagen, VCAM1, TGFβ, PINP, PIIINP	No change in BP, MAP, brachial artery distensibility	8 weeks
BENEFICIAL (NCT00516646)	Alagebrium (AGE-Breaker)	2007–10	Completed	HFrEF	Aerobic capacity (VO2); *systolic and diastolic fxn; AGEs; heart failure scores; global assessments; NT-pro-BNP after 9 months*	None	No improvement - trending ↑ in negative cardiovascular events	9 months

Trial ID	Drug	Years	Status	Condition	Outcomes	Results	Duration
NCT00557518	Alagebrium (AGE-Breaker)	2007–09	Terminated	Type 1 DM with nephropathy	Change in albumin excretion; Albumin:creatinine, plasma renin, collagen and AGE markers, blood pressure	Unavailable	24 weeks
NCT00544921	GLY-230	2007–16	Completed	Type 1 or 2 diabetic adult males	Glycated albumin; urine albumin; EKGs; HbA1c	↓ glycated albumin in serum and urine	↔ HbA1c	2 weeks
BENEFICIAL (NCT00739687)	Alagebrium (AGE-Breaker)	2008–09	Terminated	HFrEF	Aerobic capacity (VO2)	Unavailable	9 months
BREAK-DHF-I (NCT00662116)	Alagebrium (AGE-Breaker)	2008–09	Terminated	Diastolic HF	Exercise tolerance (6 min walk); quality of life, risk assesment; negative cardiovascular outcomes and hospitalizations	Unavailable	24 weeks
AGE (NCT01014572)	Alagebrium (AGE-Breaker)	2009–20	Completed	Ventricular dysfunction in elderly population	LV compliance; distensibility	Slight ↓ LV tissue stiffness	↔ LV function, exercise capacity	1 year

(continued)

Table 1 Clinical trials of anti-glycation drugs. (cont'd)

Trial	Drug	Period	Status	Disease	Primary and secondary measurements	Positive outcomes	Negative outcomes	Tx duration
NCT01417663	Alagebrium (AGE-Breaker)	2011–13	Completed	CV disease development and endothelial dysfunction in aging	Endothelial function; cardiac *echocardiography parameters*; *cerebral perfusion and cognitive function*	No difference in improvement compared to exercise alone, Alagebrium tolerated relatively well	No improvement in CV health beyond exercise alone; no alleviation of CV risk	1 year
NCT01456026	Glycated/Non-glycated beta-lactoglobulins (Dietary AGE)	2011	Completed	Type 2 DM	Pre-to-Postprandial FMD change, *Pre-to-Postprandial microcirculation change*	Vascular effect of dietary AGEs are vascular at acute time points	Dietary AGEs affect vasculature	Single administration
BREAK-DHF-1 (NCT01913301)	Alagebrium (AGE-Breaker)	2013	Terminated	Diastolic HF	Exercise tolerance (6 min walk)	Unavailable	Unavailable	24 weeks
HATFF (NCT02095873)	Trans-resveratrol & hesperetin (Glo1 Activator)	2014–17	Completed	Glucose intolerance, aortic stiffness, vasodilation	AUC for gluose intolerance; *capillary density, flow-mediated dilation*, aortal pulse wave velocity	↓ insulin resistance	No information on cardiovascular health	22 weeks

NCT02768220	Empagliflozin, Linagliptin (SGLT2 Inhibitor)	2016–18	Withdrawn	Type 2 DM without kidney disease	Serum and urinary AGE levels, including CML, free methylglyoxal; *BP, weight*	Unavailable	30 days
NCT04234581	Lysulin (AGE inhibitor)	2020	Terminated	Type 2 DM	Fasting glucose, HbA1c, *Fasting C-Peptide, CML, CEL, GH1, MGH1, 3DGH1, MetSO, 2-AAA (baseline vs 12 weeks)*	Unavailable	12 weeks
AGLIANICO (NCT05112731)	Biomarker discovery	2021–	Recruiting	STEMI (all adults)	Identification of biomarkers (inflammatory, glycation) to ischemic events and outcomes immediately after; *Identification of biomarkers including glycation biomarkers for complications and outcomes after 3 months*	Unavailable	3 months post STEMI

fraction (HFpEF) demonstrated an efficacy of the drug in this patient population (Little et al., 2005), indicating the utility of Alagebrium as a therapeutic may simply require further investigation.

6.2 Aminoguanidine, an AGE-inhibitor

Aminoguanidine (Pimagedine) (Fig. 5B) treatment was initially pursued as a chelator or sponge for reactive carbonyl species like glyoxal, methylglyoxal, and 3-deoxyglucosone to directly remove them from the body. In fact, it is so good at this job that one of its main downsides is how overreactive it is with carbonyl-containing compounds and can inhibit nitric oxide synthase (Thornalley, 2003b). Aminoguanidine has been used in many AGE-related studies to reduce levels of methylglyoxal or other AGE-producing RCS or to rule out RCS/AGE contribution to pathology or phenotype (e.g., Cerami et al., 1997; Dyntar et al., 2001; Gu et al., 2022; Moore et al., 2013; Papadaki et al., 2022; Wang et al., 2010), and aminoguanidine may even be able to prevent AGE-related cross-linking (Cerami et al., 1997). Indeed, the outlook for aminoguanidine as a therapeutic drug was initially good, as rodent studies had noted marked reductions in AGE levels in the hearts of old rats (particularly female rats) versus controls (Li et al., 1996). However, these results were not always consistent *in vivo*; for instance, in F344 rats the reduction was significant in the heart, but not vascular or renal tissue; furthermore, these effects varied by strain (Li et al., 1996). And in models of diabetes, ocular studies of cataract development suggested the window of opportunity with aminoguanidine may be narrow, with 'less sick' rats receiving more relief from aminoguanidine administration than sicker counterparts (Swamy-Mruthinti, Green, & Abraham, 1996).

6.3 N-terminal MYPBC peptides

Methylglyoxal is known to impair the ability of tropomyosin to interact with actin, thus impairing the ability of the cell to contract (Papadaki et al., 2022) (Section 5). Methods of directly targeting AGE have been thus-far unsuccessful, therefore N-Terminal MYBPC peptides containing domains C0-C2 of MYBPC, a protein known to activate the thin filament (Mun et al., 2014), have been investigated as a method of rescuing thin-filament dysregulation in skinned murine myocytes pretreated with 100 μM methylglyoxal (Papadaki et al., 2022). Glycated myocytes treated with this peptide were unable to produce a comparable amount of force to controls after treatment with the peptide, however the treatment did reverse the negative impact on Ca^{2+} sensitivity (Papadaki et al., 2022), indicating thin filament targeting could be a successful treatment for cardiovascular complications caused by sarcomere glycation. However, this study directly treated de-membranated cardiomyocytes with purified peptides. *In vivo* studies will need to be performed to determine whether this peptide treatment can improve contractile function and outcomes in models of diabetes or aging.

6.4 Glo1 overexpression and activation

In rat models of pulmonary hypertension, administering Glo1 AAV-9 (Fig. 5C) fully rescued some parameters of cardiovascular function, including right ventricular (RV) ejection fraction, and RV end diastolic pressure (Prisco et al., 2022). However, hypertension itself was not affected (Prisco et al., 2022), indicating this treatment may be limited in its ability to relieve cardiomyopathy. Glo1 overexpression as a method of treating AGE-related pathologies has been attempted in other cell types, such as mesangial cells, where Glo1 overexpression via plasmid transfection successfully reduced oxidative damage and restored viability of cells treated with high glucose (Kim, Kim, Jung, Lee, & Kim, 2012).

Moreover, recent clinical trials have attempted to increase Glo1 expression using a combination treatment of *trans*-resveratrol and hesperetin (tRES-HESP) (Rabbani et al., 2021). This clinical trial, done in overweight and obese patients, found that tRES-HESP did alleviate some stress associated with methylglyoxal-mediated glycation (Rabbani et al., 2021). However, this study did not examine cardiovascular health.

7. Other future directions and concluding remarks

The atria may be resistant to glycation stress. Disease-induced AGE increases are readily observed in the ventricles (Papadaki et al., 2018, 2022; Shao et al., 2010). However, atria of insulin-resistant mice do not have increased CML (Maria, Campolo, Scherlag, Ritchey, & Lacombe, 2018). Understanding the mechanisms behind this resistance would be informative for the development of future therapeutics and as to the organ-wide effect of cardiomyocyte glycation.

7.1 PTM crosstalk

Lysine is a site for many PTMs, including acetylation and ubiquitination (Donald, Kulp, & DeGrado, 2011; Wang & Cole, 2020). Pan-HDAC inhibitors (i.e., SAHA), which preserve acetylation modifications, show promise as a heart failure therapeutic in feline models (Eaton et al., 2022). Specific HDAC inhibitors also increase myofilament Ca^{2+} sensitivity. (Gupta, Samant, Smith, & Shroff, 2008)—the opposite effect of glycation (Papadaki et al., 2018). Ubiquitination is a lysine-linked PTM that tags proteins for (among other things) degradation (Mattiroli & Sixma, 2014). Glycation is an irreversible process, and sarcomere proteins have incredibly long half-lives, with some early studies indicating half-lives up to 10 days (Thompson & Metzger, 2014). These two factors raise the possibility of lysine-linked glycation (such as methylglyoxal CEL) blocking these other PTMs. However, whether glycation interferes with this aspect of PQC or acetylation remains unstudied and merits further investigation.

7.2 Concluding remarks

In the last century, glycation has become appreciated as a PTM with diverse protein targets and functional consequences that are associated with many pathological conditions that impact the cardiovascular system, such as diabetes, aging, and heart failure. These irreversible PTMs play a pivotal role in cardiovascular disease, by both increasing the risk of developing the disease and worsening outcomes. Indeed, these pathogenic affects underscore the importance of future research to discover (1) AGEs present in high-risk conditions for cardiovascular disease (such as diabetes), and whether these same AGES are also in high abundance once cardiovascular disease has developed (such as a diabetic patient with heart failure), (2) molecular mechanisms underpinning cardiovascular disease in high-AGE populations, and (3) what, if any, necessary physiological role do AGEs play that must be maintained during approaches to reduce pathological AGEs. Glycation research has met challenges, partly because of the lack of success of clinical trials for drugs such as Alagebrium and aminoguanidine. However, recent technological advances, such as improvement in mass spectrometry and genetic approaches, have revolutionized our available tools to understand the impact of, and how to combat, glycation. Combining mass spectrometry studies with genetic approaches such as MyBP-C adenovirus therapy and functional assays can discover the mechanisms and impact of *specific* AGEs. Furthermore, future work should identify how existing therapeutics can be employed in the clinic (e.g., should a patient with high blood methylglyoxal be prescribed Mavacamten or Verapamil?) and how therapeutics can be developed that are better suited to treat the specific functional deficits glycation induces in the cardiomyocyte.

References

Alikhani, Z., Alikhani, M., Boyd, C. M., Nagao, K., Trackman, P. C., & Graves, D. T. (2005). Advanced glycation end products enhance expression of pro-apoptotic genes and stimulate fibroblast apoptosis through cytoplasmic and mitochondrial pathways. *The Journal of Biological Chemistry, 280*(13), 12087–12095. https://doi.org/10.1074/jbc.M406313200.

Arai, M., Yuzawa, H., Nohara, I., Ohnishi, T., Obata, N., Iwayama, Y., ... Itokawa, M. (2010). Enhanced carbonyl stress in a subpopulation of schizophrenia. *Archives of General Psychiatry, 67*(6), 589–597. https://doi.org/10.1001/archgenpsychiatry.2010.62.

Aronson, D. (2003). Cross-linking of glycated collagen in the pathogenesis of arterial and myocardial stiffening of aging and diabetes. *Journal of Hypertension, 21*(1), 3–12. https://doi.org/10.1097/00004872-200301000-00002.

Averill-Bates, D. A. (2023). The antioxidant glutathione. *Vitamins and Hormones, 121*, 109–141. https://doi.org/10.1016/bs.vh.2022.09.002.

Beckendorf, J., van den Hoogenhof, M. M. G., & Backs, J. (2018). Physiological and unappreciated roles of CaMKII in the heart. *Basic Research in Cardiology, 113*(4), 29. https://doi.org/10.1007/s00395-018-0688-8.

Bergmann, O., Bhardwaj, R. D., Bernard, S., Zdunek, S., Barnabe-Heider, F., Walsh, S., ... Frisen, J. (2009). Evidence for cardiomyocyte renewal in humans. *Science (New York, N. Y.), 324*(5923), 98–102. https://doi.org/10.1126/science.1164680.

Bi, X., Song, Y., Song, Y., Yuan, J., Cui, J., Zhao, S., & Qiao, S. (2021). Collagen cross-linking is associated with cardiac remodeling in hypertrophic obstructive cardiomyopathy. *Journal of the American Heart Association, 10*(1), e017752. https://doi.org/10.1161/JAHA.120.017752.

Bidasee, K. R., Nallani, K., Yu, Y., Cocklin, R. R., Zhang, Y., Wang, M., ... Besch, H. R., Jr. (2003). Chronic diabetes increases advanced glycation end products on cardiac ryanodine receptors/calcium-release channels. *Diabetes, 52*(7), 1825–1836. https://doi.org/10.2337/diabetes.52.7.1825.

Bidasee, K. R., Zhang, Y., Shao, C. H., Wang, M., Patel, K. P., Dincer, U. D., & Besch, H. R., Jr. (2004). Diabetes increases formation of advanced glycation end products on Sarco(endo)plasmic reticulum Ca^{2+}-ATPase. *Diabetes, 53*(2), 463–473. https://doi.org/10.2337/diabetes.53.2.463.

Bollineni, R. C., Hoffmann, R., & Fedorova, M. (2014). Proteome-wide profiling of carbonylated proteins and carbonylation sites in HeLa cells under mild oxidative stress conditions. *Free Radical Biology & Medicine, 68*, 186–195. https://doi.org/10.1016/j.freeradbiomed.2013.11.030.

Bonsignore, A., Leoncini, G., Siri, A., & Ricci, D. (1973a). Kinetic behaviour of glyceraldehyde 3-phosphate conversion into methylglyoxal. *The Italian Journal of Biochemistry, 22*(4), 131–140. https://www.ncbi.nlm.nih.gov/pubmed/4784005.

Bonsignore, A., Leoncini, G., Siri, A., & Ricci, D. (1973b). Polymerization of methylglyoxal in the presence of lysine. *The Italian Journal of Biochemistry, 22*(1), 55–63. https://www.ncbi.nlm.nih.gov/pubmed/4750841.

Bou-Teen, D., Fernandez-Sanz, C., Miro-Casas, E., Nichtova, Z., Bonzon-Kulichenko, E., Casos, K., ... Ruiz-Meana, M. (2022). Defective dimerization of FoF1-ATP synthase secondary to glycation favors mitochondrial energy deficiency in cardiomyocytes during aging. *Aging Cell, 21*(3), e13564. https://doi.org/10.1111/acel.13564.

Brownlee, M., Vlassara, H., Kooney, A., Ulrich, P., & Cerami, A. (1986). Aminoguanidine prevents diabetes-induced arterial wall protein cross-linking. *Science (New York, N. Y.), 232*(4758), 1629–1632. https://doi.org/10.1126/science.3487117.

Burr, S. D., Harmon, M. B., & Stewart, J. A., Jr (2020). The impact of diabetic conditions and AGE/RAGE signaling on cardiac fibroblast migration. *Frontiers in Cell and Developmental Biology, 8*, 112. https://doi.org/10.3389/fcell.2020.00112.

Cai, A., Li, G., Chen, J., Li, X., Wei, X., Li, L., & Zhou, Y. (2014). Glycated hemoglobin level is significantly associated with the severity of coronary artery disease in non-diabetic adults. *Lipids in Health and Disease, 13*, 181. https://doi.org/10.1186/1476-511X-13-181.

Candido, R., Forbes, J. M., Thomas, M. C., Thallas, V., Dean, R. G., Burns, W. C., ... Burrell, L. M. (2003). A breaker of advanced glycation end products attenuates diabetes-induced myocardial structural changes. *Circulation Research, 92*(7), 785–792. https://doi.org/10.1161/01.RES.0000065620.39919.20.

Caporizzo, M. A., Chen, C. Y., & Prosser, B. L. (2019). Cardiac microtubules in health and heart disease. *Experimental Biology and Medicine (Maywood), 244*(15), 1255–1272. https://doi.org/10.1177/1535370219868960.

Carrick-Ranson, G., Fujimoto, N., Shafer, K. M., Hastings, J. L., Shibata, S., Palmer, M. D., ... Levine, B. D. (2016). The effect of 1 year of Alagebrium and moderate-intensity exercise training on left ventricular function during exercise in seniors: A randomized controlled trial. *Journal of Applied Physiology (1985), 121*(2), 528–536. https://doi.org/10.1152/japplphysiol.00021.2016.

Cerami, C., Founds, H., Nicholl, I., Mitsuhashi, T., Giordano, D., Vanpatten, S., ... Cerami, A. (1997). Tobacco smoke is a source of toxic reactive glycation products. *Proceedings of the National Academy of Sciences of the United States of America, 94*(25), 13915–13920. https://doi.org/10.1073/pnas.94.25.13915.

Chen, Y. C., Lu, Y. Y., Wu, W. S., Lin, Y. K., Chen, Y. A., Chen, S. A., & Chen, Y. J. (2022). Advanced glycation end products modulate electrophysiological remodeling of right ventricular outflow tract cardiomyocytes: A novel target for diabetes-related ventricular arrhythmogenesis. *Physiological Reports, 10*(21), e15499. https://doi.org/10.14814/phy2.15499.

Cho, K. W., Andrade, M., Bae, S., Kim, S., Eyun Kim, J., Jang, E. Y., ... Yoon, Y. S. (2023). Polycomb group protein CBX7 represses cardiomyocyte proliferation through modulation of the TARDBP/RBM38 axis. *Circulation*. https://doi.org/10.1161/CIRCULATIONAHA.122.061131.

Choi, K. M., Zhong, Y., Hoit, B. D., Grupp, I. L., Hahn, H., Dilly, K. W., ... Matlib, M. A. (2002). Defective intracellular Ca(2+) signaling contributes to cardiomyopathy in Type 1 diabetic rats. *American Journal of Physiology. Heart and Circulatory Physiology, 283*(4), H1398–H1408. https://doi.org/10.1152/ajpheart.00313.2002.

Clippinger, S. R., Cloonan, P. E., Greenberg, L., Ernst, M., Stump, W. T., & Greenberg, M. J. (2019). Disrupted mechanobiology links the molecular and cellular phenotypes in familial dilated cardiomyopathy. *Proceedings of the National Academy of Sciences of the United States of America, 116*(36), 17831–17840. https://doi.org/10.1073/pnas.1910962116.

Connor, R. E., Marnett, L. J., & Liebler, D. C. (2011). Protein-selective capture to analyze electrophile adduction of hsp90 by 4-hydroxynonenal. *Chemical Research in Toxicology, 24*(8), 1275–1282. https://doi.org/10.1021/tx200157t.

Cui, J., Liu, Y., Li, Y., Xu, F., & Liu, Y. (2021). Type 2 diabetes and myocardial infarction: Recent clinical evidence and perspective. *Frontiers in Cardiovascular Medicine, 8*, 644189. https://doi.org/10.3389/fcvm.2021.644189.

de Groot, L., Hinkema, H., Westra, J., Smit, A. J., Kallenberg, C. G., Bijl, M., & Posthumus, M. D. (2011). Advanced glycation endproducts are increased in rheumatoid arthritis patients with controlled disease. *Arthritis Research & Therapy, 13*(6), R205. https://doi.org/10.1186/ar3538.

Deluyker, D., Evens, L., Belien, H., & Bito, V. (2019). Acute exposure to glycated proteins reduces cardiomyocyte contractile capacity. *Experimental Physiology, 104*(7), 997–1003. https://doi.org/10.1113/EP087127.

Deluyker, D., Ferferieva, V., Noben, J. P., Swennen, Q., Bronckaers, A., Lambrichts, I., ... Bito, V. (2016). Cross-linking versus RAGE: How do high molecular weight advanced glycation products induce cardiac dysfunction? *International Journal of Cardiology, 210,* 100–108. https://doi.org/10.1016/j.ijcard.2016.02.095.

Ding, W. X., & Yin, X. M. (2012). Mitophagy: Mechanisms, pathophysiological roles, and analysis. *Biological Chemistry, 393*(7), 547–564. https://doi.org/10.1515/hsz-2012-0119.

Donald, J. E., Kulp, D. W., & DeGrado, W. F. (2011). Salt bridges: Geometrically specific, designable interactions. *Proteins, 79*(3), 898–915. https://doi.org/10.1002/prot.22927.

Dyntar, D., Eppenberger-Eberhardt, M., Maedler, K., Pruschy, M., Eppenberger, H. M., Spinas, G. A., & Donath, M. Y. (2001). Glucose and palmitic acid induce degeneration of myofibrils and modulate apoptosis in rat adult cardiomyocytes. *Diabetes, 50*(9), 2105–2113. https://doi.org/10.2337/diabetes.50.9.2105.

Eaton, D. M., Martin, T. G., Kasa, M., Djalinac, N., Ljubojevic-Holzer, S., Von Lewinski, D., ... Wallner, M. (2022). HDAC inhibition regulates cardiac function by increasing myofilament calcium sensitivity and decreasing diastolic tension. *Pharmaceutics, 14*(7), https://doi.org/10.3390/pharmaceutics14071509.

Edman, C. F., & Schulman, H. (1994). Identification and characterization of delta B-CaM kinase and delta C-CaM kinase from rat heart, two new multifunctional Ca^{2+}/calmodulin-dependent protein kinase isoforms. *Biochimica et Biophysica Acta, 1221*(1), 89–101. https://doi.org/10.1016/0167-4889(94)90221-6.

Eisner, D. A., Caldwell, J. L., Kistamas, K., & Trafford, A. W. (2017). Calcium and excitation-contraction coupling in the heart. *Circulation Research, 121*(2), 181–195. https://doi.org/10.1161/CIRCRESAHA.117.310230.

Falcao-Pires, I., Hamdani, N., Borbely, A., Gavina, C., Schalkwijk, C. G., van der Velden, J., ... Paulus, W. J. (2011). Diabetes mellitus worsens diastolic left ventricular dysfunction in aortic stenosis through altered myocardial structure and cardiomyocyte stiffness. *Circulation, 124*(10), 1151–1159. https://doi.org/10.1161/CIRCULATIONAHA.111.025270.

Fernandez-Sanz, C., Ruiz-Meana, M., Miro-Casas, E., Nunez, E., Castellano, J., Loureiro, M., ... Garcia-Dorado, D. (2014). Defective sarcoplasmic reticulum-mitochondria calcium exchange in aged mouse myocardium. *Cell Death & Disease, 5*(12), e1573. https://doi.org/10.1038/cddis.2014.526.

Fujimoto, N., Hastings, J. L., Carrick-Ranson, G., Shafer, K. M., Shibata, S., Bhella, P. S., ... Levine, B. D. (2013). Cardiovascular effects of 1 year of alagebrium and endurance exercise training in healthy older individuals. *Circulation: Heart Failure, 6*(6), 1155–1164. https://doi.org/10.1161/CIRCHEARTFAILURE.113.000440.

Goncalves, P. R., Nascimento, L. D., Gerlach, R. F., Rodrigues, K. E., & Prado, A. F. (2022). Matrix metalloproteinase 2 as a pharmacological target in heart failure. *Pharmaceuticals (Basel), 15*(8), https://doi.org/10.3390/ph15080920.

Gonzalez, D. R., Beigi, F., Treuer, A. V., & Hare, J. M. (2007). Deficient ryanodine receptor S-nitrosylation increases sarcoplasmic reticulum calcium leak and arrhythmogenesis in cardiomyocytes. *Proceedings of the National Academy of Sciences of the United States of America, 104*(51), 20612–20617. https://doi.org/10.1073/pnas.0706796104.

Gonzalez, D. R., Treuer, A. V., Castellanos, J., Dulce, R. A., & Hare, J. M. (2010). Impaired S-nitrosylation of the ryanodine receptor caused by xanthine oxidase activity contributes to calcium leak in heart failure. *The Journal of Biological Chemistry, 285*(37), 28938–28945. https://doi.org/10.1074/jbc.M110.154948.

Grandhee, S. K., & Monnier, V. M. (1991). Mechanism of formation of the Maillard protein cross-link pentosidine. Glucose, fructose, and ascorbate as pentosidine precursors. *The Journal of Biological Chemistry, 266*(18), 11649–11653. https://www.ncbi.nlm.nih.gov/pubmed/1904866.

Grant, A. O. (2009). Cardiac ion channels. *Circulation: Arrhythmia and Electrophysiology, 2*(2), 185–194. https://doi.org/10.1161/CIRCEP.108.789081.

Gu, M. J., Hyon, J. Y., Lee, H. W., Han, E. H., Kim, Y., Cha, Y. S., & Ha, S. K. (2022). Glycolaldehyde, an advanced glycation end products precursor, induces apoptosis via ROS-mediated mitochondrial dysfunction in renal mesangial cells. *Antioxidants (Basel), 11*(5), https://doi.org/10.3390/antiox11050934.

Gupta, M. K., Tahrir, F. G., Knezevic, T., White, M. K., Gordon, J., Cheung, J. Y., ... Feldman, A. M. (2016). GRP78 Interacting Partner Bag5 responds to ER stress and protects cardiomyocytes from ER stress-induced apoptosis. *Journal of Cellular Biochemistry, 117*(8), 1813–1821. https://doi.org/10.1002/jcb.25481.

Gupta, M. P., Samant, S. A., Smith, S. H., & Shroff, S. G. (2008). HDAC4 and PCAF bind to cardiac sarcomeres and play a role in regulating myofilament contractile activity. *The Journal of Biological Chemistry, 283*(15), 10135–10146. https://doi.org/10.1074/jbc.M710277200.

Harja, E., Bu, D. X., Hudson, B. I., Chang, J. S., Shen, X., Hallam, K., ... Schmidt, A. M. (2008). Vascular and inflammatory stresses mediate atherosclerosis via RAGE and its ligands in apoE-/- mice. *The Journal of Clinical Investigation, 118*(1), 183–194. https://doi.org/10.1172/JCI32703.

Hartog, J. W., Willemsen, S., van Veldhuisen, D. J., Posma, J. L., van Wijk, L. M., Hummel, Y. M., ... Investigators, B. (2011). Effects of alagebrium, an advanced glycation endproduct breaker, on exercise tolerance and cardiac function in patients with chronic heart failure. *European Journal of Heart Failure: Journal of the Working Group on Heart Failure of the European Society of Cardiology, 13*(8), 899–908. https://doi.org/10.1093/eurjhf/hfr067.

Hegab, Z., Mohamed, T. M. A., Stafford, N., Mamas, M., Cartwright, E. J., & Oceandy, D. (2017). Advanced glycation end products reduce the calcium transient in cardiomyocytes by increasing production of reactive oxygen species and nitric oxide. *FEBS Open Bio, 7*(11), 1672–1685. https://doi.org/10.1002/2211-5463.12284.

Hesketh, G. G., Van Eyk, J. E., & Tomaselli, G. F. (2009). Mechanisms of gap junction traffic in health and disease. *Journal of Cardiovascular Pharmacology, 54*(4), 263–272. https://doi.org/10.1097/FJC.0b013e3181ba0811.

Hodge, J. (1953). Dehydrated foods, chemistry of browning reactions in model systems. *Journal of Agricultural and Food Chemistry, 1*(15), 928–943.

Hofmann, M. A., Drury, S., Fu, C., Qu, W., Taguchi, A., Lu, Y., ... Schmidt, A. M. (1999). RAGE mediates a novel proinflammatory axis: A central cell surface receptor for S100/calgranulin polypeptides. *Cell, 97*(7), 889–901. https://doi.org/10.1016/s0092-8674(00)80801-6.

Holliday, G. L., Mitchell, J. B., & Thornton, J. M. (2009). Understanding the functional roles of amino acid residues in enzyme catalysis. *Journal of Molecular Biology, 390*(3), 560–577. https://doi.org/10.1016/j.jmb.2009.05.015.

Hong, L. F., Li, X. L., Guo, Y. L., Luo, S. H., Zhu, C. G., Qing, P., ... Li, J. J. (2014). Glycosylated hemoglobin A1c as a marker predicting the severity of coronary artery disease and early outcome in patients with stable angina. *Lipids in Health and Disease, 13*, 89. https://doi.org/10.1186/1476-511X-13-89.

Hou, X., Hu, Z., Xu, H., Xu, J., Zhang, S., Zhong, Y., ... Wang, N. (2014). Advanced glycation endproducts trigger autophagy in cadiomyocyte via RAGE/PI3K/AKT/mTOR pathway. *Cardiovascular Diabetology, 13*, 78. https://doi.org/10.1186/1475-2840-13-78.

Hudson, D. M., Archer, M., King, K. B., & Eyre, D. R. (2018). Glycation of type I collagen selectively targets the same helical domain lysine sites as lysyl oxidase-mediated cross-linking. *The Journal of Biological Chemistry, 293*(40), 15620–15627. https://doi.org/10.1074/jbc.RA118.004829.

Irshad, Z., Xue, M., Ashour, A., Larkin, J. R., Thornalley, P. J., & Rabbani, N. (2019). Activation of the unfolded protein response in high glucose treated endothelial cells is mediated by methylglyoxal. *Scientific Reports, 9*(1), 7889. https://doi.org/10.1038/s41598-019-44358-1.

Jafri, M. S. (2012). Models of excitation-contraction coupling in cardiac ventricular myocytes. *Methods in Molecular Biology, 910*, 309–335. https://doi.org/10.1007/978-1-61779-965-5_14.

Janssens, J. V., Ma, B., Brimble, M. A., Van Eyk, J. E., Delbridge, L. M. D., & Mellor, K. M. (2018). Cardiac troponins may be irreversibly modified by glycation: Novel potential mechanisms of cardiac performance modulation. *Scientific Reports, 8*(1), 16084. https://doi.org/10.1038/s41598-018-33886-x.

Jiao, X., Zhang, Q., Peng, P., & Shen, Y. (2023). HbA1c is a predictive factor of severe coronary stenosis and major adverse cardiovascular events in patients with both type 2 diabetes and coronary heart disease. *Diabetology & Metabolic Syndrome, 15*(1), 50. https://doi.org/10.1186/s13098-023-01015-y.

Kass, D. A., Shapiro, E. P., Kawaguchi, M., Capriotti, A. R., Scuteri, A., deGroof, R. C., & Lakatta, E. G. (2001). Improved arterial compliance by a novel advanced glycation end-product crosslink breaker. *Circulation, 104*(13), 1464–1470. https://doi.org/10.1161/hc3801.097806.

Kennedy, L., Solano, M. P., Meneghini, L., Lo, M., & Cohen, M. P. (2010). Anti-glycation and anti-albuminuric effects of GLY-230 in human diabetes. *American Journal of Nephrology, 31*(2), 110–116. https://doi.org/10.1159/000259897.

Khurram, I. M., Maqbool, F., Berger, R. D., Marine, J. E., Spragg, D. D., Ashikaga, H., ... Zimmerman, S. L. (2016). Association between left atrial stiffness index and atrial fibrillation recurrence in patients undergoing left atrial ablation. *Circulation: Arrhythmia and Electrophysiology, 9*(3), https://doi.org/10.1161/CIRCEP.115.003163.

Kim, H. W., Ch, Y. S., Lee, H. R., Park, S. Y., & Kim, Y. H. (2001). Diabetic alterations in cardiac sarcoplasmic reticulum Ca^{2+}-ATPase and phospholamban protein expression. *Life Sciences, 70*(4), 367–379. https://doi.org/10.1016/s0024-3205(01)01483-7.

Kim, K. M., Kim, Y. S., Jung, D. H., Lee, J., & Kim, J. S. (2012). Increased glyoxalase I levels inhibit accumulation of oxidative stress and an advanced glycation end product in mouse mesangial cells cultured in high glucose. *Experimental Cell Research, 318*(2), 152–159. https://doi.org/10.1016/j.yexcr.2011.10.013.

Koponen, M., Marjamaa, A., Tuiskula, A. M., Viitasalo, M., Nallinmaa-Luoto, T., Leinonen, J. T., ... Swan, H. (2020). Genealogy and clinical course of catecholaminergic polymorphic ventricular tachycardia caused by the ryanodine receptor type 2 P2328S mutation. *PLoS One, 15*(12), e0243649. https://doi.org/10.1371/journal.pone.0243649.

Kuang, S. Q., Yu, J. D., Lu, L., He, L. M., Gong, L. S., Chen, S. J., & Chen, Z. (1996). Identification of a novel missense mutation in the cardiac beta-myosin heavy chain gene in a Chinese patient with sporadic hypertrophic cardiomyopathy. *Journal of Molecular and Cellular Cardiology, 28*(9), 1879–1883. https://doi.org/10.1006/jmcc.1996.0180.

Kumar, V., Fleming, T., Terjung, S., Gorzelanny, C., Gebhardt, C., Agrawal, R., ... Nawroth, P. P. (2017). Homeostatic nuclear RAGE-ATM interaction is essential for efficient DNA repair. *Nucleic Acids Research, 45*(18), 10595–10613. https://doi.org/10.1093/nar/gkx705.

Kwak, S., Choi, Y. S., Na, H. G., Bae, C. H., Song, S. Y., & Kim, Y. D. (2021). Glyoxal and methylglyoxal as E-cigarette vapor ingredients-induced pro-inflammatory cytokine and mucins expression in human nasal epithelial cells. *American Journal of Rhinology & Allergy, 35*(2), 213–220. https://doi.org/10.1177/1945892420946968.

Lehman, W., Pavadai, E., & Rynkiewicz, M. J. (2021). C-terminal troponin-I residues trap tropomyosin in the muscle thin filament blocked-state. *Biochemical and Biophysical Research Communications, 551*, 27–32. https://doi.org/10.1016/j.bbrc.2021.03.010.

Leon, B. M., & Maddox, T. M. (2015). Diabetes and cardiovascular disease: Epidemiology, biological mechanisms, treatment recommendations and future research. *World Journal of Diabetes, 6*(13), 1246–1258. https://doi.org/10.4239/wjd.v6.i13.1246.

Lester, E. (1989). The clinical value of glycated haemoglobin and glycated plasma proteins. *Annals of Clinical Biochemistry, 26*(Pt 3), 213–219. https://doi.org/10.1177/000456328902600301.

Levi, B., & Werman, M. J. (1998). Long-term fructose consumption accelerates glycation and several age-related variables in male rats. *The Journal of Nutrition, 128*(9), 1442–1449. https://doi.org/10.1093/jn/128.9.1442.

Li, H., Zheng, L., Chen, C., Liu, X., & Zhang, W. (2019). Brain senescence caused by elevated levels of reactive metabolite methylglyoxal on D-galactose-induced aging mice. *Frontiers in Neuroscience, 13*, 1004. https://doi.org/10.3389/fnins.2019.01004.

Li, Y. M., Steffes, M., Donnelly, T., Liu, C., Fuh, H., Basgen, J., ... Vlassara, H. (1996). Prevention of cardiovascular and renal pathology of aging by the advanced glycation inhibitor aminoguanidine. *Proceedings of the National Academy of Sciences of the United States of America, 93*(9), 3902–3907. https://doi.org/10.1073/pnas.93.9.3902.

Little, W. C., Zile, M. R., Kitzman, D. W., Hundley, W. G., O'Brien, T. X., & Degroof, R. C. (2005). The effect of alagebrium chloride (ALT-711), a novel glucose cross-link breaker, in the treatment of elderly patients with diastolic heart failure. *Journal of Cardiac Failure, 11*(3), 191–195. https://doi.org/10.1016/j.cardfail.2004.09.010.

Liu, Q., Chen, H. B., Luo, M., & Zheng, H. (2016). Serum soluble RAGE level inversely correlates with left ventricular hypertrophy in essential hypertension patients. *Genetics and Molecular Research: GMR, 15*(2), https://doi.org/10.4238/gmr.15028414.

Liu, Z., Zhang, Y., Pan, S., Qiu, C., Jia, H., Wang, Y., & Zhu, H. (2021). Activation of RAGE-dependent endoplasmic reticulum stress associates with exacerbated post-myocardial infarction ventricular arrhythmias in diabetes. *American Journal of Physiology. Endocrinology and Metabolism, 320*(3), E539–E550. https://doi.org/10.1152/ajpendo.00450.2020.

Lopez, B., Querejeta, R., Gonzalez, A., Larman, M., & Diez, J. (2012). Collagen cross-linking but not collagen amount associates with elevated filling pressures in hypertensive patients with stage C heart failure: Potential role of lysyl oxidase. *Hypertension, 60*(3), 677–683. https://doi.org/10.1161/HYPERTENSIONAHA.112.196113.

Lu, P., Xiao, S., Chen, S., Fu, Y., Zhang, P., Yao, Y., & Chen, F. (2021). LncRNA SNHG12 downregulates RAGE to attenuate hypoxia-reoxygenation-induced apoptosis in H9c2 cells. *Bioscience, Biotechnology, and Biochemistry, 85*(4), 866–873. https://doi.org/10.1093/bbb/zbaa090.

Lu, Y. Y., Chen, Y. C., Kao, Y. H., Wu, T. J., Chen, S. A., & Chen, Y. J. (2011). Extracellular matrix of collagen modulates intracellular calcium handling and electrophysiological characteristics of HL-1 cardiomyocytes with activation of angiotensin II type 1 receptor. *Journal of Cardiac Failure, 17*(1), 82–90. https://doi.org/10.1016/j.cardfail.2010.10.002.

Maejima, Y. (2020). The critical roles of protein quality control systems in the pathogenesis of heart failure. *Journal of Cardiology, 75*(3), 219–227. https://doi.org/10.1016/j.jjcc.2019.09.019.

Maillard, L. (1912). Action des acides aminés sur les sucres; formation des mélanoïdes par voie méthodique. *Académie des Sciences: Comptes Redeus, 154*, 66–68.

Maria, Z., Campolo, A. R., Scherlag, B. J., Ritchey, J. W., & Lacombe, V. A. (2018). Dysregulation of insulin-sensitive glucose transporters during insulin resistance-induced atrial fibrillation. *Biochimica et Biophysica Acta (BBA)—Molecular Basis of Disease, 1864* (4 Pt A), 987–996. https://doi.org/10.1016/j.bbadis.2017.12.038.

Martin, A. A., Thompson, B. R., Hahn, D., Angulski, A. B. B., Hosny, N., Cohen, H., & Metzger, J. M. (2022). Cardiac sarcomere signaling in health and disease. *International Journal of Molecular Sciences, 23*(24), https://doi.org/10.3390/ijms232416223.

Martin, T. G., Delligatti, C. E., Muntu, N. A., Stachowski-Doll, M. J., & Kirk, J. A. (2022). Pharmacological inhibition of BAG3-HSP70 with the proposed cancer therapeutic JG-98 is toxic for cardiomyocytes. *Journal of Cellular Biochemistry, 123*(1), 128–141. https://doi.org/10.1002/jcb.30140.

Martin, T. G., Myers, V. D., Dubey, P., Dubey, S., Perez, E., Moravec, C. S., ... Kirk, J. A. (2021). Cardiomyocyte contractile impairment in heart failure results from reduced BAG3-mediated sarcomeric protein turnover. *Nature Communications, 12*(1), 2942. https://doi.org/10.1038/s41467-021-23272-z.

Mattiroli, F., & Sixma, T. K. (2014). Lysine-targeting specificity in ubiquitin and ubiquitin-like modification pathways. *Nature Structural & Molecular Biology, 21*(4), 308–316. https://doi.org/10.1038/nsmb.2792.

Mazza, M. C., Shuck, S. C., Lin, J., Moxley, M. A., Termini, J., Cookson, M. R., & Wilson, M. A. (2022). DJ-1 is not a deglycase and makes a modest contribution to cellular defense against methylglyoxal damage in neurons. *Journal of Neurochemistry, 162*(3), 245–261. https://doi.org/10.1111/jnc.15656.

McLaughlin, J. A., Pethig, R., & Szent-Gyorgyi, A. (1980). Spectroscopic studies of the protein-methylglyoxal adduct. *Proceedings of the National Academy of Sciences of the United States of America, 77*(2), 949–951. https://doi.org/10.1073/pnas.77.2.949.

McLean, W. G., Pekiner, C., Cullum, N. A., & Casson, I. F. (1992). Posttranslational modifications of nerve cytoskeletal proteins in experimental diabetes. *Molecular Neurobiology, 6*(2-3), 225–237. https://doi.org/10.1007/BF02780555.

Miyazawa, T., Nakagawa, K., Shimasaki, S., & Nagai, R. (2012). Lipid glycation and protein glycation in diabetes and atherosclerosis. *Amino Acids, 42*(4), 1163–1170. https://doi.org/10.1007/s00726-010-0772-3.

Monnier, V. M., Glomb, M., Elgawish, A., & Sell, D. R. (1996). The mechanism of collagen cross-linking in diabetes: A puzzle nearing resolution. *Diabetes, 45*(Suppl 3), S67–S72. https://doi.org/10.2337/diab.45.3.s67.

Moore, C. J., Shao, C. H., Nagai, R., Kutty, S., Singh, J., & Bidasee, K. R. (2013). Malondialdehyde and 4-hydroxynonenal adducts are not formed on cardiac ryanodine receptor (RyR2) and sarco(endo)plasmic reticulum Ca^{2+}-ATPase (SERCA2) in diabetes. *Molecular and Cellular Biochemistry, 376*(1-2), 121–135. https://doi.org/10.1007/s11010-013-1558-1.

Morcos, M., Du, X., Pfisterer, F., Hutter, H., Sayed, A. A., Thornalley, P., ... Nawroth, P. P. (2008). Glyoxalase-1 prevents mitochondrial protein modification and enhances lifespan in Caenorhabditis elegans. *Aging Cell, 7*(2), 260–269. https://doi.org/10.1111/j.1474-9726.2008.00371.x.

Morgenstern, J., Katz, S., Krebs-Haupenthal, J., Chen, J., Saadatmand, A., Cortizo, F. G., ... Fleming, T. (2020). Phosphorylation of T107 by camkiidelta regulates the detoxification efficiency and proteomic integrity of glyoxalase 1. *Cell Reports, 32*(12), 108160. https://doi.org/10.1016/j.celrep.2020.108160.

Mun, J. Y., Previs, M. J., Yu, H. Y., Gulick, J., Tobacman, L. S., Beck Previs, S., ... Craig, R. (2014). Myosin-binding protein C displaces tropomyosin to activate cardiac thin filaments and governs their speed by an independent mechanism. *Proceedings of the National Academy of Sciences of the United States of America, 111*(6), 2170–2175. https://doi.org/10.1073/pnas.1316001111.

Nam, D. H., Han, J. H., Kim, S., Shin, Y., Lim, J. H., Choi, H. C., & Woo, C. H. (2016). Activated protein C prevents methylglyoxal-induced endoplasmic reticulum stress and cardiomyocyte apoptosis via regulation of the AMP-activated protein kinase signaling pathway. *Biochemical and Biophysical Research Communications, 480*(4), 622–628. https://doi.org/10.1016/j.bbrc.2016.10.106.

Nam, D. H., Han, J. H., Lee, T. J., Shishido, T., Lim, J. H., Kim, G. Y., & Woo, C. H. (2015). CHOP deficiency prevents methylglyoxal-induced myocyte apoptosis and cardiac dysfunction. *Journal of Molecular and Cellular Cardiology, 85*, 168–177. https://doi.org/10.1016/j.yjmcc.2015.05.016.

Neff, L. S., & Bradshaw, A. D. (2021). Cross your heart? Collagen cross-links in cardiac health and disease. *Cellular Signalling, 79*, 109889. https://doi.org/10.1016/j.cellsig.2020.109889.

Nicholl, I. D., & Bucala, R. (1998). Advanced glycation endproducts and cigarette smoking. *Cellular and Molecular Biology (Noisy-le-grand), 44*(7), 1025–1033. https://www.ncbi.nlm.nih.gov/pubmed/9846884.

Nicolau, J., de Souza, D. N., & Carrilho, M. (2009). Increased glycated calmodulin in the submandibular salivary glands of streptozotocin-induced diabetic rats. *Cell Biochemistry and Function, 27*(4), 193–198. https://doi.org/10.1002/cbf.1555.

Oudegeest-Sander, M. H., Olde Rikkert, M. G., Smits, P., Thijssen, D. H., van Dijk, A. P., Levine, B. D., & Hopman, M. T. (2013). The effect of an advanced glycation endproduct crosslink breaker and exercise training on vascular function in older individuals: A randomized factorial design trial. *Experimental Gerontology, 48*(12), 1509–1517. https://doi.org/10.1016/j.exger.2013.10.009.

Oyadomari, S., & Mori, M. (2004). Roles of CHOP/GADD153 in endoplasmic reticulum stress. *Cell Death and Differentiation, 11*(4), 381–389. https://doi.org/10.1038/sj.cdd.4401373.

Pamplona, R., Portero-Otin, M., Bellmun, M. J., Gredilla, R., & Barja, G. (2002). Aging increases Nepsilon-(carboxymethyl)lysine and caloric restriction decreases Nepsilon-(carboxyethyl)lysine and Nepsilon-(malondialdehyde)lysine in rat heart mitochondrial proteins. *Free Radical Research, 36*(1), 47–54. https://doi.org/10.1080/10715760210165.

Papadaki, M., Holewinski, R. J., Previs, S. B., Martin, T. G., Stachowski, M. J., Li, A., ... Kirk, J. A. (2018). Diabetes with heart failure increases methylglyoxal modifications in the sarcomere, which inhibit function. *JCI Insight, 3*(20), https://doi.org/10.1172/jci.insight.121264.

Papadaki, M., Kampaengsri, T., Barrick, S. K., Campbell, S. G., von Lewinski, D., Rainer, P. P., ... Kirk, J. A. (2022). Myofilament glycation in diabetes reduces contractility by inhibiting tropomyosin movement, is rescued by cMyBPC domains. *Journal of Molecular and Cellular Cardiology, 162*, 1–9. https://doi.org/10.1016/j.yjmcc.2021.08.012.

Passarelli, M., & Machado, U. F. F. (2021). AGEs-induced and endoplasmic reticulum stress/inflammation-mediated regulation of GLUT4 expression and atherogenesis in diabetes mellitus. *Cells, 11*(1), https://doi.org/10.3390/cells11010104.

Petrova, R., Yamamoto, Y., Muraki, K., Yonekura, H., Sakurai, S., Watanabe, T., ... Yamamoto, H. (2002). Advanced glycation endproduct-induced calcium handling impairment in mouse cardiac myocytes. *Journal of Molecular and Cellular Cardiology, 34*(10), 1425–1431. https://doi.org/10.1006/jmcc.2002.2084.

Peyret, H., Konecki, C., Terryn, C., Dubuisson, F., Millart, H., Feliu, C., & Djerada, Z. (2024). Methylglyoxal induces cardiac dysfunction through mechanisms involving altered intracellular calcium handling in the rat heart. *Chemico-Biological Interactions, 394*, 110949. https://doi.org/10.1016/j.cbi.2024.110949.

Pinto-Junior, D. C., Silva, K. S., Michalani, M. L., Yonamine, C. Y., Esteves, J. V., Fabre, N. T., ... Machado, U. F. (2018). Advanced glycation end products-induced insulin resistance involves repression of skeletal muscle GLUT4 expression. *Scientific Reports, 8*(1), 8109. https://doi.org/10.1038/s41598-018-26482-6.

Prisco, S. Z., Hartweck, L., Keen, J. L., Vogel, N., Kazmirczak, F., Eklund, M., ... Prins, K. W. (2022). Glyoxylase-1 combats dicarbonyl stress and right ventricular dysfunction in rodent pulmonary arterial hypertension. *Frontiers in Cardiovascular Medicine, 9*, 940932. https://doi.org/10.3389/fcvm.2022.940932.

Rabbani, N., & Thornalley, P. J. (2014). Dicarbonyl proteome and genome damage in metabolic and vascular disease. *Biochemical Society Transactions, 42*(2), 425–432. https://doi.org/10.1042/BST20140018.

Rabbani, N., Xue, M., & Thornalley, P. J. (2014). Activity, regulation, copy number and function in the glyoxalase system. *Biochemical Society Transactions, 42*(2), 419–424. https://doi.org/10.1042/BST20140008.

Rabbani, N., Xue, M., Weickert, M. O., & Thornalley, P. J. (2021). Reversal of insulin resistance in overweight and obese subjects by trans-resveratrol and hesperetin combination-link to dysglycemia, blood pressure, dyslipidemia, and low-grade inflammation. *Nutrients, 13*(7), https://doi.org/10.3390/nu13072374.

Ramasamy, R., & Schmidt, A. M. (2012). Receptor for advanced glycation end products (RAGE) and implications for the pathophysiology of heart failure. *Current Heart Failure Reports, 9*(2), 107–116. https://doi.org/10.1007/s11897-012-0089-5.

Ranasinghe, P., Jayawardena, R., Chandrasena, L., Noetzel, V., & Burd, J. (2019). Effects of Lysulin supplementation on pre-diabetes: Study protocol for a randomized controlled trial. *Trials, 20*(1), 171. https://doi.org/10.1186/s13063-019-3269-8.

Ravipati, G., Aronow, W. S., Ahn, C., Sujata, K., Saulle, L. N., & Weiss, M. B. (2006). Association of hemoglobin A(1c) level with the severity of coronary artery disease in patients with diabetes mellitus. *The American Journal of Cardiology, 97*(7), 968–969. https://doi.org/10.1016/j.amjcard.2005.10.039.

Richarme, G., Mihoub, M., Dairou, J., Bui, L. C., Leger, T., & Lamouri, A. (2015). Parkinsonism-associated protein DJ-1/Park7 is a major protein deglycase that repairs methylglyoxal- and glyoxal-glycated cysteine, arginine, and lysine residues. *The Journal of Biological Chemistry, 290*(3), 1885–1897. https://doi.org/10.1074/jbc.M114.597815.

Rienks, M., Papageorgiou, A. P., Frangogiannis, N. G., & Heymans, S. (2014). Myocardial extracellular matrix: An ever-changing and diverse entity. *Circulation Research, 114*(5), 872–888. https://doi.org/10.1161/CIRCRESAHA.114.302533.

Roe, A. T., Aronsen, J. M., Skardal, K., Hamdani, N., Linke, W. A., Danielsen, H. E., ... Louch, W. E. (2017). Increased passive stiffness promotes diastolic dysfunction despite improved Ca^{2+} handling during left ventricular concentric hypertrophy. *Cardiovascular Research, 113*(10), 1161–1172. https://doi.org/10.1093/cvr/cvx087.

Ruedisueli, I., Lakhani, K., Nguyen, R., Gornbein, J., & Middlekauff, H. R. (2023). Electronic cigarettes prolong ventricular repolarization in people who smoke tobacco cigarettes: Implications for harm reduction. *American Journal of Physiology. Heart and Circulatory Physiology, 324*(6), H821–H832. https://doi.org/10.1152/ajpheart.00057.2023.

Ruiz-Meana, M., Minguet, M., Bou-Teen, D., Miro-Casas, E., Castans, C., Castellano, J., ... Garcia-Dorado, D. (2019). Ryanodine receptor glycation favors mitochondrial damage in the senescent heart. *Circulation, 139*(7), 949–964. https://doi.org/10.1161/CIRCULATIONAHA.118.035869.

Scavello, F., Piacentini, L., Castiglione, S., Zeni, F., Macri, F., Casaburo, M., ... Raucci, A. (2022). Effects of RAGE deletion on the cardiac transcriptome during aging. *International Journal of Molecular Sciences, 23*(19), https://doi.org/10.3390/ijms231911130.

Scavello, F., Zeni, F., Milano, G., Macri, F., Castiglione, S., Zuccolo, E., ... Raucci, A. (2021). Soluble receptor for advanced glycation end-products regulates age-associated cardiac fibrosis. *International Journal of Biological Sciences, 17*(10), 2399–2416. https://doi.org/10.7150/ijbs.56379.

Schlotterer, A., Kukudov, G., Bozorgmehr, F., Hutter, H., Du, X., Oikonomou, D., ... Morcos, M. (2009). C. elegans as model for the study of high glucose- mediated life span reduction. *Diabetes, 58*(11), 2450–2456. https://doi.org/10.2337/db09-0567.

Schmidt, W., & Cammarato, A. (2020). The actin 'A-triad's' role in contractile regulation in health and disease. *The Journal of Physiology, 598*(14), 2897–2908. https://doi.org/10.1113/JP276741.

Schmidt, W., Madan, A., Foster, D. B., & Cammarato, A. (2020). Lysine acetylation of F-actin decreases tropomyosin-based inhibition of actomyosin activity. *The Journal of Biological Chemistry, 295*(46), 15527–15539. https://doi.org/10.1074/jbc.RA120.015277.

Semba, R. D., Bandinelli, S., Sun, K., Guralnik, J. M., & Ferrucci, L. (2009). Plasma carboxymethyl-lysine, an advanced glycation end product, and all-cause and cardiovascular disease mortality in older community-dwelling adults. *Journal of the American Geriatrics Society, 57*(10), 1874–1880. https://doi.org/10.1111/j.1532-5415.2009.02438.x.

Senyo, S. E., Steinhauser, M. L., Pizzimenti, C. L., Yang, V. K., Cai, L., Wang, M., ... Lee, R. T. (2013). Mammalian heart renewal by pre-existing cardiomyocytes. *Nature, 493*(7432), 433–436. https://doi.org/10.1038/nature11682.

Shao, C. H., Capek, H. L., Patel, K. P., Wang, M., Tang, K., DeSouza, C., ... Bidasee, K. R. (2011). Carbonylation contributes to SERCA2a activity loss and diastolic dysfunction in a rat model of type 1 diabetes. *Diabetes, 60*(3), 947–959. https://doi.org/10.2337/db10-1145.

Shao, C. H., Rozanski, G. J., Nagai, R., Stockdale, F. E., Patel, K. P., Wang, M., ... Bidasee, K. R. (2010). Carbonylation of myosin heavy chains in rat heart during diabetes. *Biochemical Pharmacology, 80*(2), 205–217. https://doi.org/10.1016/j.bcp.2010.03.024.

Son, M., Kang, W. C., Oh, S., Bayarsaikhan, D., Ahn, H., Lee, J., ... Lee, B. (2017). Advanced glycation end-product (AGE)-albumin from activated macrophage is critical in human mesenchymal stem cells survival and post-ischemic reperfusion injury. *Scientific Reports, 7*(1), 11593. https://doi.org/10.1038/s41598-017-11773-1.

Stirban, A., Kotsi, P., Franke, K., Strijowski, U., Cai, W., Gotting, C., & Tschoepe, D. (2013). Acute macrovascular dysfunction in patients with type 2 diabetes induced by ingestion of advanced glycated beta-lactoglobulins. *Diabetes Care, 36*(5), 1278–1282. https://doi.org/10.2337/dc12-1489.

Swamy-Mruthinti, S., Green, K., & Abraham, E. C. (1996). Inhibition of cataracts in moderately diabetic rats by aminoguanidine. *Experimental Eye Research, 62*(5), 505–510. https://doi.org/10.1006/exer.1996.0061.

Tani, H., Kobayashi, E., Yagi, S., Tanaka, K., Kameda-Haga, K., Shibata, S., ... Tohyama, S. (2023). Heart-derived collagen promotes maturation of engineered heart tissue. *Biomaterials, 299*, 122174. https://doi.org/10.1016/j.biomaterials.2023.122174.

Thompson, B. R., & Metzger, J. M. (2014). Cell biology of sarcomeric protein engineering: disease modeling and therapeutic potential. *The Anatomical Record (Hoboken), 297*(9), 1663–1669. https://doi.org/10.1002/ar.22966.

Thornalley, P. J. (2003a). Glyoxalase I—Structure, function and a critical role in the enzymatic defence against glycation. *Biochemical Society Transactions, 31*(Pt 6), 1343–1348. https://doi.org/10.1042/bst0311343.

Thornalley, P. J. (2003b). Use of aminoguanidine (Pimagedine) to prevent the formation of advanced glycation endproducts. *Archives of Biochemistry and Biophysics, 419*(1), 31–40. https://doi.org/10.1016/j.abb.2003.08.013.

Thornalley, P. J., Battah, S., Ahmed, N., Karachalias, N., Agalou, S., Babaei-Jadidi, R., & Dawnay, A. (2003). Quantitative screening of advanced glycation endproducts in cellular and extracellular proteins by tandem mass spectrometry. *The Biochemical Journal, 375*(Pt 3), 581–592. https://doi.org/10.1042/BJ20030763.

Tian, C., Alomar, F., Moore, C. J., Shao, C. H., Kutty, S., Singh, J., & Bidasee, K. R. (2014). Reactive carbonyl species and their roles in sarcoplasmic reticulum Ca^{2+} cycling defect in the diabetic heart. *Heart Failure Reviews, 19*(1), 101–112. https://doi.org/10.1007/s10741-013-9384-9.

Toprak, C., & Yigitaslan, S. (2019). Alagebrium and complications of diabetes mellitus. *The Eurasian Journal of Medicine, 51*(3), 285–292. https://doi.org/10.5152/eurasianjmed.2019.18434.

Townsend, P. A., Cutress, R. I., Carroll, C. J., Lawrence, K. M., Scarabelli, T. M., Packham, G., ... Latchman, D. S. (2004). BAG-1 proteins protect cardiac myocytes from simulated ischemia/reperfusion-induced apoptosis via an alternate mechanism of cell survival independent of the proteasome. *The Journal of Biological Chemistry, 279*(20), 20723–20728. https://doi.org/10.1074/jbc.M400399200.

van Halm, V. P., Peters, M. J., Voskuyl, A. E., Boers, M., Lems, W. F., Visser, M., ... Nurmohamed, M. T. (2009). Rheumatoid arthritis versus diabetes as a risk factor for cardiovascular disease: A cross-sectional study, the CARRE Investigation. *Annals of the Rheumatic Diseases, 68*(9), 1395–1400. https://doi.org/10.1136/ard.2008.094151.

van Heerebeek, L., Hamdani, N., Handoko, M. L., Falcao-Pires, I., Musters, R. J., Kupreishvili, K., ... Paulus, W. J. (2008). Diastolic stiffness of the failing diabetic heart: Importance of fibrosis, advanced glycation end products, and myocyte resting tension. *Circulation, 117*(1), 43–51. https://doi.org/10.1161/CIRCULATIONAHA.107.728550.

Walsh, M. P. (1983). Calmodulin and its roles in skeletal muscle function. *Canadian Anaesthetists' Society Journal, 30*(4), 390–398. https://doi.org/10.1007/BF03007862.

Wang, W., Schulze, C. J., Suarez-Pinzon, W. L., Dyck, J. R., Sawicki, G., & Schulz, R. (2002). Intracellular action of matrix metalloproteinase-2 accounts for acute myocardial ischemia and reperfusion injury. *Circulation, 106*(12), 1543–1549. https://doi.org/10.1161/01.cir.0000028818.33488.7b.

Wang, X. L., Lau, W. B., Yuan, Y. X., Wang, Y. J., Yi, W., Christopher, T. A., ... Ma, X. L. (2010). Methylglyoxal increases cardiomyocyte ischemia-reperfusion injury via glycative inhibition of thioredoxin activity. *American Journal of Physiology. Endocrinology and Metabolism, 299*(2), E207–E214. https://doi.org/10.1152/ajpendo.00215.2010.

Wang, Z. A., & Cole, P. A. (2020). The chemical biology of reversible lysine post-translational modifications. *Cell Chemical Biology, 27*(8), 953–969. https://doi.org/10.1016/j.chembiol.2020.07.002.

Werbner, B., Tavakoli-Rouzbehani, O. M., Fatahian, A. N., & Boudina, S. (2023). The dynamic interplay between cardiac mitochondrial health and myocardial structural remodeling in metabolic heart disease, aging, and heart failure. *The Journal of Cardiovascular Aging, 3*(1), https://doi.org/10.20517/jca.2022.42.

Willemsen, S., Hartog, J. W., Hummel, Y. M., Posma, J. L., van Wijk, L. M., van Veldhuisen, D. J., & Voors, A. A. (2010). Effects of alagebrium, an advanced glycation end-product breaker, in patients with chronic heart failure: Study design and baseline characteristics of the BENEFICIAL trial. *European Journal of Heart Failure: Journal of the Working Group on Heart Failure of the European Society of Cardiology, 12*(3), 294–300. https://doi.org/10.1093/eurjhf/hfp207.

Willett, J. G., Bennett, M., Hair, E. C., Xiao, H., Greenberg, M. S., Harvey, E., ... Vallone, D. (2019). Recognition, use and perceptions of JUUL among youth and young adults. *Tobacco Control, 28*(1), 115–116. https://doi.org/10.1136/tobaccocontrol-2018-054273.

Wu, Y., Cazorla, O., Labeit, D., Labeit, S., & Granzier, H. (2000). Changes in titin and collagen underlie diastolic stiffness diversity of cardiac muscle. *Journal of Molecular and Cellular Cardiology, 32*(12), 2151–2162. https://doi.org/10.1006/jmcc.2000.1281.

Xue, M., Rabbani, N., Momiji, H., Imbasi, P., Anwar, M. M., Kitteringham, N., ... Thornalley, P. J. (2012). Transcriptional control of glyoxalase 1 by Nrf2 provides a stress-responsive defence against dicarbonyl glycation. *The Biochemical Journal, 443*(1), 213–222. https://doi.org/10.1042/BJ20111648.

Yamabe, S., Hirose, J., Uehara, Y., Okada, T., Okamoto, N., Oka, K., ... Mizuta, H. (2013). Intracellular accumulation of advanced glycation end products induces apoptosis via endoplasmic reticulum stress in chondrocytes. *The FEBS Journal, 280*(7), 1617–1629. https://doi.org/10.1111/febs.12170.

Yaras, N., Ugur, M., Ozdemir, S., Gurdal, H., Purali, N., Lacampagne, A., ... Turan, B. (2005). Effects of diabetes on ryanodine receptor Ca release channel (RyR2) and Ca^{2+} homeostasis in rat heart. *Diabetes, 54*(11), 3082–3088. https://doi.org/10.2337/diabetes.54.11.3082.

Yuan, P. M., & Gracy, R. W. (1977). The conversion of dihydroxyacetone phosphate to methylglyoxal and inorganic phosphate by methylglyoxal synthase. *Archives of Biochemistry and Biophysics, 183*(1), 1–6. https://doi.org/10.1016/0003-9861(77)90411-8.

Zhang, B., He, K., Chen, W., Cheng, X., Cui, H., Zhong, W., ... Wang, L. (2014). Alagebrium (ALT-711) improves the anti-hypertensive efficacy of nifedipine in diabetic-hypertensive rats. *Hypertension Research: Official Journal of the Japanese Society of Hypertension, 37*(10), 901–907. https://doi.org/10.1038/hr.2014.98.

Zizkova, P., Viskupicova, J., Heger, V., Rackova, L., Majekova, M., & Horakova, L. (2018). Dysfunction of SERCA pumps as novel mechanism of methylglyoxal cytotoxicity. *Cell Calcium, 74*, 112–122. https://doi.org/10.1016/j.ceca.2018.06.003.

CHAPTER FOUR

Glycation and drug binding by serum albumin

Anu Jain and Nand Kishore*

Department of Chemistry, Indian Institute of Technology Bombay, Mumbai, India
*Corresponding author. e-mail address: nandk@chem.iitb.ac.in

Contents

1. Glycation	90
2. Glycation of transport protein serum albumin	91
3. Role of advanced glycation end products	93
4. Characterisation of glycated protein and glycation end-products	94
5. Glycation and protein conformation	98
6. Calorimetry in glycation studies	99
6.1 Differential scanning calorimetry	99
6.2 Isothermal titration calorimetry	104
7. Glycation and drug binding	106
8. Future perspectives	109
References	109

Abstract

Accumulation of glycation products in patients with hyperglycaemic conditions can lead to their reaction with the proteins in the human system such as serum albumin, haemoglobin, insulin, plasma lipoproteins, lens proteins and collagen among others which have important biological functions. Therefore, it is important to understand if glycation of these proteins affects their normal action not only qualitatively, but also importantly quantitatively. Glycation of human serum albumin can easily be carried out over period of weeks and its drug transportability may be examined, in addition to characterisation of the amadori products. A combination of ultrasensitive isothermal titration calorimetry, differential scanning calorimetry, spectroscopy and chromatography provides structure–property–energetics correlations which are important to obtain mechanistic aspects of drug recognition, conformation of the protein, and role of amadori products under conditions of glycation. The role of advance glycation end products is important in recognition of antidiabetic drugs. Further, the extent of glycation of the protein and its implication on drug transportability investigated by direct calorimetric methods enables unravelling mechanistic insights into role of functionality on drug molecules in the binding process, and hinderance in the recognition process, if any, as a result of glycation. It is possible that the drug binding ability of the protein under glycation conditions may not be adversely affected, or may even lead to strengthened ability. Rigorous studies on such systems with diverse functionality on the drug molecules is required which is essential in deriving guidelines for improvements in the existing drugs or in the synthesis of new molecular entities directed towards addressing diabetic conditions.

1. Glycation

Non-enzymatic glycosylation (glycation) can occur both in-vitro as well as in-vivo. It is a chemical process which occurs naturally in the body, in which a sugar molecule (like glucose or fructose) combines with a protein or lipid molecule. This process can be accelerated by various factors like oxidative stress, inflammation or high blood sugar levels. The carbonyl group of a reducing sugar can undergo reaction with α-amino group of N-terminal amino acid, ε-amino group of lysine, arginine and thiol group of cysteine present in most of the proteins under physiological conditions (Billaud et al., 2005; Mohamadi-Nejad et al., 2002; Nomi & Otsuka, 2020; Sattarahmady et al., 2007; Wang et al., 2020). Schiff base results from Maillard reaction which can equilibrate with amadori products and reactants (Lapolla et al., 1999, 2005; Sattarahmady et al., 2007). Following this the glycation products found as intermediates further rearrange to form advance glycation end products (AGEs), through a process which is irreversible. These AGEs can be overproduced under conditions of increased glucose concentration in the blood of an organism during the process of hyperglycaemia (de La Rochette et al., 2003; Flier et al., 1988; Furusyo & Hayashi, 2013; Glenn & Stitt, 2009; Lapolla et al., 1999, 2005; Westwood & Thornalley, 1995; Yamagishi et al., 2012). The AGEs can subsequently bind to the receptors at an enhanced rate in the patients with diabetic condition and lead to complications like kidney disease, blindness, osteoarthritis and cardiovascular diseases (Barile & Schmidt, 2007; Glenn & Stitt, 2009; Hudson & Schmidt, 2004; Swamy-Mruthinti, 2001; Yamagishi et al., 2012). The amadori compound which works as a clinical marker of glucose control before 1–2 months in diabetic patients is glycohemoglobin A1c (HbA1c) (Sherwani et al., 2016). The proteins in the human system (haemoglobin, serum albumin, insulin, plasma lipoprotein, peripheral nerve myelin, lens proteins, elastin, collagen and selected intrinsically disordered protein) can undergo reaction with the glycation end products which accumulate in patients with such hyperglycaemic conditions (Anguizola et al., 2013; Sattarahmady et al., 2007; Winterhalter, 1985). This interaction can affect biological function of proteins and hence contribute to above mentioned detrimental health complications. Therefore, it is extremely important to obtain mechanistic insights into the effects of such interactions on the role of these proteins both quantitatively and qualitatively.

2. Glycation of transport protein serum albumin

Amongst various proteins in the human system, serum albumin (Fig. 1) has drawn major attention since it is an abundant carrier protein and offers binding sites to a variety of exogenous and endogenous ligands (Ghosh et al., 2019; Ghosh & Kishore, 2020). It is also established that serum albumin can undergo modifications when it is exposed to higher concentrations of glycating agents (Anguizola et al., 2013; Furusyo & Hayashi, 2013; Nakajou et al., 2003; Rondeau et al., 2010; Sattarahmady et al., 2007). Several studies aimed at modulation of binding affinity of new

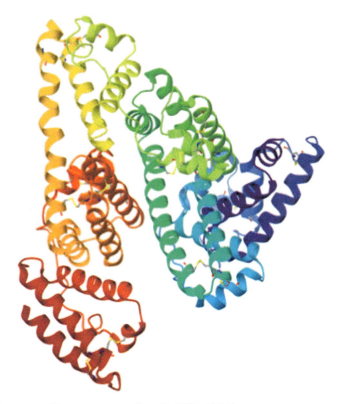

Fig. 1 Structure of human serum albumin (PDB- 1AO6).

molecular entities have been performed with bovine serum albumin and human serum albumin (HSA) because of significant structural similarities between the two (Gelamo & Tabak, 2000; Ossowicz et al., 2020). Specifically, HSA has N-terminal aspartic acid, lysine, and arginine residues in large number in addition to a cysteine residue at position 34, having a free -SH group, which makes it prone to glycation in patients with hyperglycaemia conditions (Bou-Abdallah et al., 2016; Furusyo & Hayashi, 2013; Maiti et al., 2006, 2008; Meloun et al., 1975; Sen et al., 2009; Varshney et al., 2010; Westwood & Thornalley, 1995).

Predominant glycation sites and extent of glycation of HSA both in-vivo and in-vitro have been reported by diverse approaches including isotopic labelling and enzymatic digestion (Barnaby et al., 2010; Stefanowicz et al., 2010). Specifically, lysine residues at the positions 12, 51, 159, 199, 205, 233, 281, 286, 317, 351, 439, 525, 534 and 538; and arginine residues at positions at 160, 222, and 472, have been reported to be susceptible to glycation in HSA molecule. Several groups have been able to separate and characterise glycation products such as amadori products reactive dicarbonyl and AGEs successfully (BIllaud et al., 2005; Dyer et al., 1991; GhoshMoulick et al., 2007; Glenn & Stitt, 2009; Lapolla et al., 1999, 2005; Linetsky et al., 2008; Malik et al., 2015; Nakajou et al., 2003; Nomi & Otsuka, 2020; Sattarahmady et al., 2007; Schmitt et al., 2005; Singh et al., 2002; Wang et al., 2020; Westwood & Thornalley, 1995; Yamagishi et al., 2012). Upon glycation of HSA for a longer period, a molten globule like state, which has poorly defined tertiary structure but strengthened or intact secondary structure, has also been reported (Kumar et al., 2010; Sattarahmady et al., 2007). It must be kept in mind that such intermediate structures can associate to form amyloid fibrils structures which can also lead to neurodegenerative diseases such as Alzheimer's and Parkinson's (Iannuzzi et al., 2014; Sirangelo & Iannuzzi, 2021; Yan et al., 1995). A variety of inhibitors have also been reported which can inhibit Maillard reaction (Anis & Sreerama, 2020; Favre et al., 2020; Hartmann et al., 2019; Nabi et al., 2019; Rebollo-Hernanz et al., 2019; Schmidt et al., 2000; Sharma et al., 2019).

The protein HSA can undergo chemical and conformational transformation depending upon the stage and extent of glycation (Shental-Bechor & Levy, 2008). These transformations can affect the binding abilities of HSA and hence, affect the drug transportability properties of the protein. HSA has two major drug binding sites known as Sudlow sites I and II (Abou-Zied & Al-Lawatia, 2011; Lee & Wu, 2015), which may be impacted as a result of modification of amino acid residues in the glycated protein. This can affect the drug binding ability of this carrier protein and hence may require an improved engineering of binding affinity under such conditions. Obtaining mechanistic insights into the effect of glycation on the extent of binding has direct implications in managing diabetes mellitus and its treatment.

There can be two approaches in addressing the questions raised above: qualitative and quantitative. Even though significant efforts have been made in qualitative understanding (Jud & Sourij, 2019; Nowotny et al., 2015; Van Putte et al., 2016), quantitative aspects still remain largely unanswered. Some of the questions which need a deeper understanding are whether glycation makes HSA unstable, induces aggregation, or leads to loss of binding affinity. A quantitative understanding of the effect of glycation on drug transportability of HSA and on its aggregation, propensity can further lead to identification of functionalities which play important role in these processes. Spectroscopic and high sensitivity calorimetric methods have often provided valuable quantitative and mechanistic insights into ligand binding reactions with biologically important macromolecules (Ghosh et al., 2020; Marjani et al., 2022; Naik & Seetharamappa, 2023; Ovung et al., 2022; Shamsi et al., 2020).

3. Role of advanced glycation end products

The production of AGE product increases in patients suffering from hyperglycaemia which acts on extracellular matrix, lipids and enhances vascular calcification, and can also damage endothelial cells and promote smooth cell proliferation (Han et al., 2023). The production and accumulation of AGEs has been reported and reviewed by several researchers in recent past (Furusyo & Hayashi, 2013; Jud & Sourij, 2019; Nowotny et al., 2015; Oshitari, 2023; Yamagishi et al., 2012).

One of the most widely used technique for quantitative understanding of drug binding with serum albumins is ultrasensitive isothermal titration calorimetry (ITC) (Ghosh et al., 2019; Karonen et al., 2015; Migliore et al., 2022; Ràfols et al., 2018). It measures heats absorbed or released upon interaction of species in aqueous solutions in a series of interacting events. This technique is not only very sensitive but also has provided role of functionality in drug association process along with guidelines for rational drug design and modification of recently discovered new molecular entities

(Linkuvienė et al., 2016; Su & Xu, 2018; Ward & Holdgate, 2001). The uniqueness of this method is that in one single experiment it is possible to extract all necessary thermodynamic signatures accompanying the binding process such as binding constant, change in standard molar enthalpy, Gibbs free energy of interaction, and entropy upon interaction. As an example, these thermodynamic signatures not only reflect on the strength of ligand binding to the protein but also assist in unravelling underlying mechanistic details and the role of surrounding solvents in the recognition process. The detailed experimentation should also focus on the role of non-AGEs and AGEs on the drug transportability of glycated serum albumin. The outcome of all such studies using ITC, spectroscopy and chromatography individually or in combination can enable correlating thermodynamic signatures with conformation and binding ability of the protein and hence enable obtaining relevant mechanistic details of the associated process.

The mechanistic and physicochemical insights into the glycation process and drug binding ability of the transport protein serum albumin under these conditions can be obtained by a rigorous combination of characterisation, conformational analysis, size determination, extent of glycation, dynamic light scattering process, thermal stability assessment, and mass determination of the glycated protein and energetics of its interaction with a wide variety of drugs.

4. Characterisation of glycated protein and glycation end-products

The first step is to carry out the glycation of the protein which can be obtained by incubating it with the glycating agents such as glucose, even at 37 °C in dark environment. A usual method adopted is to collect the protein samples every week and characterise them by using diverse techniques. These methodologies can include employing UV–visible spectroscopy, fluorescence spectroscopy, circular dichroism spectroscopy, dynamic light scattering, time-correlated single photon counting, size exclusion chromatography, nuclear magnetic resonance spectroscopy, SDS–PAGE, matrix assisted laser desorption, ionisation time of flight mass spectroscopy, high-performance liquid chromatography mass spectrometry, ITC, and differential scanning calorimetry (DSC). A detailed characterisation of the modified protein and the side products of glycation is very important. Since, the formation of AGEs can affect the structure and properties of the protein.

Absorption spectroscopy can be conveniently used since it is a simple technique which permits monitoring glycation by comparing the absorption spectra of protein as a function of time. Glycation can distinctively alter the absorption spectra as compared to that of the unglycated protein (Ghosh & Kishore, 2022).

The absorption of aromatic residues of HSA can undergo reduction or increase as a result of glycation, which is a qualitative indicator that the glycation of the protein has occurred (Fig. 2). It must be kept in mind that glycation of the protein can also lead to conformational changes and hence can affect the integrity of binding sites of the protein.

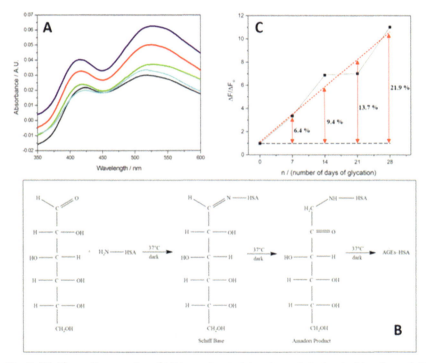

Fig. 2 (A) Absorption spectra of HSA: unglycated (▬), week 1 glycated (▬), week 2 glycated (▬), week 3 glycated (▬) and week 4 glycated (▬) protein at 25 °C in NBT reaction. (B) The non-enzymatic glycosylation (Maillard reaction) scheme generating Schiff base (in GS-Week 2), Amadori product (in GS-Week 3) and AGEs (in GS-Week 3 and GS-Week 4) is given. The formation of AGEs—HSA is explained in details in MALDI-TOF MS section. (C) Extent of glycation is shown by plotting $\Delta F/\Delta F_o$ against number of days of glycation (n). The black solid line represents the extent of glycation and the red dotted line (obtained by linear fitting of the plot) represents the percent glycation (%). Percent glycation has been determined on the basis of molecular weight measurement by using MALDI-TOF MS. (Reproduced with permission from Elsevier with licence number 5584290665482 dated July 08, 2023).

Changes in the tryptophan microenvironment of the protein can also be simultaneously monitored through intrinsic fluorescence of the protein in its glycated and unglycated forms. Fluorescence spectroscopy has been routinely used in monitoring conformational changes in the protein based upon changes in the intrinsic fluorescence emission spectra. Such an approach has provided very useful information about the tryptophan microenvironment and also about the formation of glycation end products as a result of interaction of the amino groups of lysine and arginine residues with the glycating agents (Dyer et al., 1991; Flier et al., 1988; Lapolla et al., 1999, 2005; Sattarahmady et al., 2007; Singh et al., 2002).

Another important step is to identify the amadori products and glycated end products as a result of glycation of protein. In this context it is important to mention malliard reaction which comprises of a set of chemical reactions involving amines of the protein and carbonyl compounds of the sugar (Lapolla et al., 1999, 2005; Sattarahmady et al., 2007). The amadori products exhibit fluorescence emission at 331 nm, when the glycated HSA is excited at 280 nm (Fig. 3). The extent of glycation of protein can also result in its fluorescence quenching which may or may not be accompanied by significant conformational changes which can be monitored in terms of shift in the value of λ_{max}. These shifts in the fluorescence emission profile and quenching can also suggest the presence of amadori products as a result of glycation.

Identification of AGEs can also be conveniently done employing steady state fluorescence measurement by exciting the glycated protein at 370 nm or even above and monitoring the emission in the wavelength range of 440–460 nm (Mohamadi-Nejad et al., 2002; Sattarahmady et al., 2007; Servetnick et al., 1996). The maxima of bands in this emission range corresponds to formation of AGEs, whereas those below this wavelength range correspond to pre-AGEs which are also called early glycation products. Therefore, the emission wavelength range of 440–450 nm can be selected to distinguish between AGEs and non-AGEs.

To examine the formation of amadori products (ketoamines) a usual reaction employed is *para*-nitoglutetrazolium chloride (NBT) (Johnson et al., 1983; Syrový, 1994). Upon reaction with NBT the ketoamines form formazan which can be monitored by measuring absorbance at 530 nm (Johnson et al., 1983; Kirchhoff, 1993; Syrový, 1994). Such an approach has recently been employed to monitor the formation of ketoamines in glycated HSA as a function of number of weeks of glycation. The fluorescence emission can also be employed to examine the percentage increase of glycation as a function of time (Ghosh & Kishore, 2022).

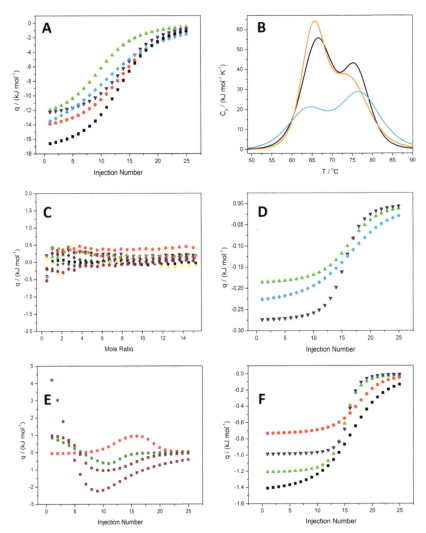

Fig. 3 (A) Far and (B) near UV CD spectra of unglycated HSA at 25 °C (■), 30 °C (■) and 37 °C (■). (C) Far and (D) near UV CD spectra of HSA: unglycated (■), week 1 glycated (■), week 2 glycated (■), week 3 glycated (■) and week 4 glycated (■) protein at 25 °C. (Reproduced with permission from Elsevier with licence number 5584290665482 dated July 08, 2023).

Another alternate method employed to monitor glycation in HSA is by DNS-PBA assay. Here, the change in fluorescence emission intensity is an indicator of the extent of glycation. This can be coupled with an increase in the molecular weight of the protein as a result of glycation which can be easily monitored by employing MALDI-TOF MS. Glycation can also lead to conformational changes in the protein for which circular dichroism spectroscopy can be employed. Recent results demonstrate that the secondary and tertiary structure of the protein lead to reduction in the α-helical content of the protein HSA, compared to the unglycated sample. Further, no significant changes in the tertiary structure were observed in HSA as a result of glycation.

The extent of glycation can also be commented upon based on size determination by dynamic light scattering. As an example, unglycated HSA has a hydrodynamic diameter of around 6.8 nm. The glycation of HSA over a period of four weeks did not lead to noticeable differences in their hydrodynamic radii. Therefore, it was concluded that glycation doesn't lead to major change in hydrodynamic radii of HSA. This observation was well supported by insignificant changes in tertiary structure of the protein as observed in circular dichroism spectroscopic signatures. Absence in changes of size of the protein also suggested that glycation doesn't lead to aggregation of the protein in a reasonable period of time.

5. Glycation and protein conformation

The thermal stability of a protein can be monitored either based upon spectroscopic signatures as a function of temperature or by ultrasensitive DSC which will be discussed ahead. It is reported that thermal unfolding of HSA is not affected appreciably by modifications of the lysine and arginine residues as a result of glycation over a period of two weeks (Ghosh & Kishore, 2022). However, beyond this glycation period the thermal stability of the protein gets compromised as indicated by reduction in cooperative signatures of the thermal unfolding profiles. These results also indicated that non-AGEs such as Schiff base and amadori products do not significantly alter the unfolding nature of the protein. However, AGEs affect cooperative nature of HSA thermal transition. Further, quantitative comparisons can be made by comparing their values of enthalpies of unfolding.

The effect of glycation on microenvironment of tryptophan in the protein can be assessed by fluorescence lifetime measurement. The observations made in literature suggest that the fluorescence decay profiles for week three and week four, glycated HSA are different than those of week one and week two glycated samples. Such results compliment the steady state fluorescence and circular dichroism spectroscopic observations.

Further, characterisation of the effect of glycation on the protein can be monitored based on mass determination by HR-LCMS. This helps in determination of the molecular weight of the glycated protein along with other products if any, in addition to that of the unglycated protein sample. Such studies help in examining the presence of modified and unmodified protein AGEs and other products which may not be attached to the protein.

An alternate approach for characterisation can also be size exclusion chromatography and SDS–PAGE which can provide evidences for monomeric or multimeric state of the protein. As a result of glycation HSA can undergo increase in molecular weight depending upon the extent of glycation. These observations can further be complimented employing matrix assisted laser desorption ionisation time of flight mass spectroscopy (MALDI-TOF MS measurements). The number of glucose molecules can be determined from the difference in molecular weights of the protein in the pre- and post-glycated forms. By using such methodology, it was observed that there are 85 glycation centres in HSA which involve 59 lysine, 24 arginine and one cysteine residues. Another significant technique which can enable an understanding of how glucose can affect the structure and conformation of protein is proton NMR. It is reported that insignificant changes in 1D ^{1}H NMR spectrum does not change even after four weeks of glycation suggesting insignificant structural and conformational changes in the protein which can be further complimented by using CD spectroscopy and more direct calorimetric methods.

6. Calorimetry in glycation studies
6.1 Differential scanning calorimetry

DSC has proved to be an excellent technique which not only has provided information about thermal stability of the protein under given environmental condition but also over a period of time has enabled an understanding of various forces that keep the 3D structure of the protein in its native form (Johnson, 2013; Sanchez-Ruiz, 1995). With the advancement of technology, DSC has become a method of choice for assessing thermodynamic behaviour of the folded and partially folded conformations of the protein with affordable minimum amount of the samples. A typical DSC consist of a pair of matching samples and reference cells which are capillary coiled shaped of 300 μL volume. Such a shape of cell minimises the chances of the protein molecules to aggregate. HSA upon glycation may or may not undergo changes in its conformational and thermal stability. Therefore, a precise quantitative assessment obtained by DSC can prove to be valuable addition to this understanding. Just like mutations of the proteins and subsequent DSC investigation have provide the role of specific amino acid residues in maintaining the thermal and conformational stability of the protein (Durowoju et al., 2017; Sooram et al., 2023), effect of glycation can also be monitored in a similar manner. Typically, a concentration corresponding to $1\,\text{mg}\,\text{mL}^{-1}$ should be enough to obtain a reasonably good unfolding profile in terms of a plot of heat capacity versus temperature at a pressurised heating scan rate. Usually, a scan rate of $1\,\text{K}\,\text{min}^{-1}$ provides acceptable conditions to examine the thermal stability of protein (Kim et al., 2012). Curve deconvolution of reversible DSC profiles can enable evaluation of thermodynamic signatures such a transition temperature ($T_{1/2}$), enthalpy of unfolding (ΔH), heat capacity of unfolding (ΔC_p), entropy change (ΔS) and Gibbs free energy change (ΔG) as a function of temperature. Unlike spectroscopic results, the DSC measurements can enable assessment of cooperativity in thermal unfolding in a more precise manner. Further, spectroscopic methods provide values of enthalpy change employing van't Hoff equation which are model dependent. However, calorimetric method provides the value of ΔH directly from the area under the curve within the selected pre- and post-transition temperature which can be termed as true calorimetric enthalpy and is model independent. In fact, a comparison of calorimetric versus van't Hoff enthalpies can provide valuable information about the nature of folding and unfolding process including the presence of intermediate states, if any.

Thermal stability of HSA with and without glycation has been assessed employing both spectroscopic and DSC methods. As pointed out earlier, unlike spectroscopic observations, calorimetric signatures can provide additional details. Sharp cooperativity in the DSC profiles has been observed for HSA compared to UV–visible based spectroscopic unfolding profile. The transition temperature for unglycated HSA is close to 66 °C (Fig. 4) (Ghosh & Kishore, 2022), usually resulting in a single endothermic profile specifically if the protein HSA is fatty acid free. Depending upon the extent of glycation, the thermal stability of HSA gets affected; for example, incubation of HSA at 37 °C in dark did not show significant change in the value of transition temperature which suggested insignificant modification in HSA even after four weeks of incubation. However, four weeks of glycation lead to modification of HSA by AGEs, which reduced the

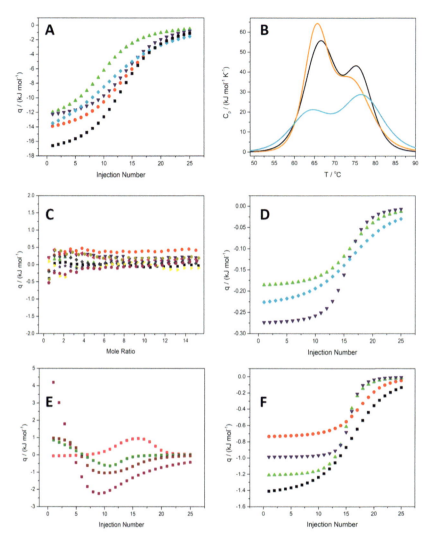

Fig. 4 Binding profiles of the studied drugs with HSA upon glycation. (A) Enthalpy (kJ mol^{-1}) against mole ratio of naproxen to protein plots of HSA: unglycated (▬), week 1 glycated (▬), week 2 glycated (■), week 3 glycated (●) and week 4 glycated (▲) protein at 25 °C. (B) Heat capacity (kJ mol^{-1} K^{-1}) versus temperature (°C) profiles of UGS (–), IS (▼) and GS-Week 4 (◆) at pH = 7.4 obtained with scan rate of 0.5 °C min^{-1}. (C) Enthalpy (kJ mol^{-1}) against mole ratio of aminoguanidine to protein plots of HSA: unglycated at 25 °C (▬) and 30 °C (▬); week 1 glycated at 25 °C (■) and 30 °C (●); week 2 glycated at 25 °C (▬) and 30 °C (▬); week 3 glycated at 25 °C (▬) and 30 °C (●); and week 4 glycated at 25 °C (★) and 30 °C (●). (D) Binding profiles plotting enthalpy (kJ mol^{-1}) against injection number of aminoguanidine to GS-Week 2 (■), GS-Week 3 (●) and GS-Week 4 (■) at 37 °C. (E) Binding profiles plotting enthalpy (kJ mol^{-1}) against injection number of metformin to GS-Week 3 at 20 °C (■)

(Continued)

transition temperature by about 2.5 °C and also reduction in the value of calorimetric enthalpy of unfolding (Ghosh & Kishore, 2022). This result suggested that the formation of AGEs-HSA leads to destabilisation of glycated HSA which is reflected in thermodynamic signature obtained by DSC. It must be noted that calorimetry permits evaluation of thermal unfolding profiles quantitatively even if there are more than one merging thermal transition profiles. Furthermore, the formation of intermediates, aggregation or dissociation in the protein undergoing transition can be assessed from the ratio of calorimetric to van't Hoff enthalpies from the reversible thermal transition.

NMR spectroscopic measurements revealed that HSA retains its sustainability even after glycation at 37 °C for a period of four weeks without degradation (Fig. 5). These observations can also be supported from dynamic light scattering measurements based upon changes in the hydrodynamic property. As discussed earlier, MALDI-TOF MS observations which provide information on molecular weight of protein as a result of glycation, can be a very useful method which can confirm variation in the extent of glycation as a function of time, if any (Hattan et al., 2016). For example, it was observed from steady state fluorescence that there was unusual halt in the change in fluorescence in the third week of HSA glycation. It was further indicated that in the third week of glycation, HSA contains glucosones, which are α-dicarbonyl compounds having 1,2 or 1,3 *cis*-diol structures formed during the process of AGEs formation. Since the modified HSA which accompanied the formation of glycation product did not lead to significant changes in near UV-CD, 1D ^{1}H NMR spectra and thermodynamics signature obtained from DSC, it was inferred that the tertiary structure of HSA is not significantly altered. On the other hand, changes in far UV-CD spectra, absorption of peptide bonds and cooperativity in the thermal unfolding signatures of the protein as a result of glycation suggested secondary structure changes indicating reduction and hydrogen bonding and van der Waal's interactions. The amino acid residues which are affected as a result of glycation can be assessed employing a combination of steady-state fluorescence and TCSPC measurements (Anguizola et al., 2013; Barnaby et al., 2010; Stefanowicz et al., 2010).

Fig. 4—Cont'd and GS-Week 4 at 20 °C (■) and GS-Week 3 at 25 °C (■) and GS-Week 4 (●) at 25 °C. (F) Binding profiles plotting enthalpy (kJ mol^{-1}) against injection number of metformin to GS-Week 1 (▬) and GS-Week 2 (■), GS-Week 3 (●) and GS-Week 4 (▲) at 37 °C. (Reproduced with permission from Elsevier with licence number 5584290665482 dated July 08, 2023 (♦), (◄), (►).

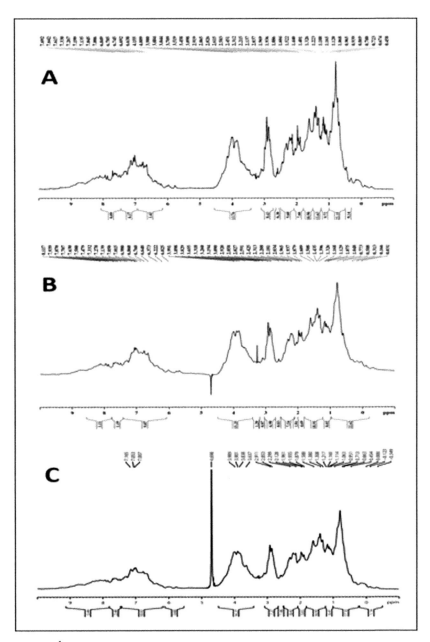

Fig. 5 1D ^{1}H NMR spectra of (A) unglycated (UGS), (B) week 2 glycated (GS-Week 2) and (C) week 4 glycated (GS-Week 4) HSA. (Reproduced with permission from Elsevier with licence number 5584290665482 dated July 08, 2023).

Amongst various techniques employed to monitor changes in the protein as a result of glycation it appears that MALDI-TOF MS measurements provide direct signatures of increase in size of a protein as a result of glycation (Ghosh & Kishore, 2022).

6.2 Isothermal titration calorimetry

As discussed earlier, ITC is a direct method to quantify the binding of ligands with proteins in terms of accompanying thermodynamics signatures. Since, serum albumin is a major transport protein in human system, it is an obvious question to know whether glycation leads to loss of binding affinity on the protein or not. This can be assessed in the terms of changes in the value of dissociation constant and corresponding change in standard molar Gibbs energy at a specified temperature. As the titration of naproxen with glycated and unglycated HSA is exothermic with generation of entropy, it suggested that the binding is both enthalpically and entropically driven. Low exothermic values of ΔH suggest involvement of polar interactions such as hydrogen bonds and van der Waal's interaction at side II of HSA. A binding site of the order of 10^4, thus, has significant component from entropy generation which can be attributed to release of water molecules as a result of association. It was observed that the binding of naproxen with unglycated protein or glycated protein up to four weeks of glycation does not lead to change in the order of binding constant in the temperature range of 25–37 °C. Similar observation is made from the changes in the value of $\Delta H_m^°$ or $\Delta S_m^°$. The exact binding site could be established by using site I blocker warfarin and site II blocker Ibuprofen. Even, nonspecific binding or of pyridoxamine could also be established by using site I or site II blocking agents. Since no significant change in the value of binding constant associated binding of naproxen with glycated or unglycated HSA was observed, it could be inferred that either arginine 410 and lysine 414 do not undergo modification as a result of glycation or the modification of these amino acids residues does not alter the binding of the drug at site II (Ghosh & Kishore, 2022).

In order to generalise observations, it is advisable to study binding of different drugs having varying structures and properties with glycated and unglycated HSA. Such an attempt was made by using aminoguanidine which has been reported to inhibit initial glycation steps in lens protein (Lewis & Harding, 1990). Having structural similarity to L-arginine, aminoguanidine can provide more insights during the progression of glycation. It is reported that at 25 °C aminoguanidine is unable to bind with both the glycated and unglycated HSA in the temperature range of 25 °C to 30 °C. Upon raising the temperature to 37 °C aminoguanidine could established association with glycated HSA obtained after each week of incubation with the glycating agent. The binding of aminoguanidine with HSA strengthened up to week three of glycation followed by a decrease in fourth week of glycation (Ghosh & Kishore, 2022). Literature suggests that during the initial steps of glycation, the formation of amadori products results in their interaction with aminoguanidine involving reactive carbonyl group leading to prevention of early glycation (Lewis & Harding, 1990; Webster et al., 2005). It is further reported that the formation of ketoamines occur from first week of glycation and reaches its highest extent in the third week of glycation. The binding of aminoguanidine which is observed to be entropically driven exhibiting highest value of ΔS after three weeks of glycation suggests a significant role of the solvation process (Ghosh & Kishore, 2022).

The binding of metformin has also been examined with glycated and unglycated HSA (Rahnama et al., 2015). Metformin is a biguanide agent which acts predominantly in the inhibition of hepatic glucose release in type II diabetic patients which is orally administered as antidiabetic hypoglycemic drug (Knowler et al., 2002; Setter et al., 2003). Even though aminoguanidine which has one guanidine group did not associate with unglycated or glycated HSA, metformin consisting of biguanide group established binding with both unglycated and glycated HSA. These results could suggest that AGEs presen offer binding sites for interaction with metformin, a probable reason for the observed multiple binding phenomena. This could be the reason for strengthened association of metformin with glycated HSA as the extent of glycation increased (Ghosh & Kishore, 2022). There are reports that Sudlow site II (subdomain 3A) is a binding domain for metformin. Further, the formation of dihydroxyacetone-AGEs at each stage of glycation is targeted by metformin (Ahmad et al., 2013). Therefore, it was observed that the modification of the proteins with AGEs leads to significant increase in the association of metformin as glycation advances (Ghosh & Kishore, 2022).

ITC was also applied to understand the association of pyridoxamine with glycated and unglycated HSA. Here, the drug binds with unglycated HSA and glycated HSA at all stages of glycation in a non-specific manner. The order of binding constant is observed to increase by a factor of 10, when the protein is modified (Ghosh & Kishore, 2022). It has been observed that oral administration of pyridoxamine led to appreciable reduction of metabolic impairments leading to decrease in AGEs, carboxyethyl lysine and carboxymethyl lysine (Chiazza et al., 2017). Action of pyridoxamine on amadori products and intermediates has also been reported which is correlated to inhibition of glycation products formation by pyridoxamine supplementation at each state. Thermodynamically, the binding of pyridoxamine with HSA both in the glycated and unglycated form is very weakly exothermic associated with predominantly entropically driven process (Ghosh & Kishore, 2022).

As described earlier, DSC is an excellent method to assess effect of ligand binding on thermal stability of the protein. In the case of HSA, the asymmetric transition observed suggested non-two step nature of the thermal unfolding which may lead to protein aggregation above transition temperature. Under such circumstances the thermal transition profiles of the protein are usually calorimetrically irreversible. Further, appearance of the second transition in the DSC profiles of HSA could be associated with the binding of fatty acids to the protein. It was observed that naproxen imparts thermal stability to HSA. The transition temperature ranges from 70.5 °C to 67.7 °C for naproxen-HSA complex with unglycated protein to glycated protein after four weeks of incubation. The transition temperature of unglycated protein in uncomplexed form is 66.4 °C, the transition temperature of which does not change significantly as a result of incubation in dark for a period of four weeks under physiological pH. As a consequence of four weeks of glycation the AGEs lead to a slight reduction in both the transition temperature and calorimetric enthalpy (Ghosh & Kishore, 2022).

7. Glycation and drug binding

Recently effect of HSA glycation on the binding of antidiabetic agent exenatide has been reported based on multi spectroscopic studies. Based on fluorescence measurements the association constant were observed to be the order of 10^4 in the 292–390 K with stoichiometry of binding approximately equal to one. The authors further studied the binding of the drug with the diet supplement such as Ca^{+2}, Cr^{+2}, Zn^{+2} where it was observed that the binding of exenatide with glycated albumin is strengthened in the presence of metal ions. The binding did not lead to appreciable effect on the secondary structure of the protein in the absence of metal ion. The authors also used site specific markers such as warfarin, flufenamic acid, and digitoxin for site I, II, and III, respectively. Based upon displacement studies it was suggested that subdomain II A and III A constitutes the binding site for exenatide (Śliwińska-Hill & Wiglusz, 2023).

Molecular simulation studies have also been recently done on structural and dynamic alteration of glycated serum albumin and Schiff base and amadori adducts (Sittiwanichai et al., 2023). This report highlights damage in the structure and ligand binding ability of HSA upon glycation which is contrary to observations made by us where we experimentally observed that the glycation did not have an appreciable effect on the binding ability of drugs with glycated HSA (Ghosh & Kishore, 2022). The crystal structure of glycated HSA suggests that two glucose isomers are located at Sudlow site I. The author performed molecular dynamics simulation to understand the nature of Schiff base and amadori products. There results suggest that both these forms alter the main protein dynamics leading to broadening of the entrance of Sudlow site I, reduction in the size of nine fatty acid binding pockets and shrinkage of Sudlow site II (Sittiwanichai et al., 2023).

From an extensive study employing calorimetry, spectroscopy, and chromatography, it was observed in general the drug molecule possessing ring structure such as naphthalene moiety in naproxen and pyridine group in pyridoxamine do not experience appreciable change in their association with HSA upon glycation even up to a period of four weeks. However open chain molecules were observed to undergo appreciable strengthening of association with glycated HSA, rather higher in metformin consisting of a greater number of guanidine group. This increase in the extent of association could be attributed to the formation of AGEs-HSA in the modified form, predominantly driven by increase in entropy. These results suggest that the association of metformin is accompanied with significant desolvation around both the drug and AGEs-HSA. The role of guanidine group in the association process was also noted. It was inferred that guanidine group in a drug molecule offers favourable association properties toward binding with HSA in the glycated form (Ghosh & Kishore, 2022).

In general, upon undergoing Maillard reaction, HSA generates amadori products up to second week of glycation followed by AGEs in the third and fourth week of glycation. It may be noted that the AGEs remain bound to the protein. It is also worth noticing that the hydrodynamic radii and tertiary structure of HSA undergo negligible change, though, with an appreciable alteration in the secondary structure upon glycation. 3-(Dansylamino) phenyl boronic acid assay may not always provide extent of glycation, therefore, monitoring by other supporting methods is strongly recommended. The other observations worth noticing are significant changes in the binding affinity of native and glycated HSA but enhancement in the thermal stability of the protein in the presence of naproxen may have implications in association of NSAID to the target site in the patients with diabetes. An important observation that emerges from the result of this study is that association of these drugs with HSA in the glycated form is either unaltered in its order or rather strengthened. These results further suggest that the association properties of these drugs are not affected adversely under diabetic conditions. The noticeable observation that open-chain molecule consisting of guanidine group offers strengthened binding ability towards protein in glycated form, hence, significant consequences in the area of diabetes. Deriving structure–property–energetics correlations in systems consisting variety of drug and HSA or other target protein is extremely important in understanding drug binding abilities in diabetic conditions (Ghosh & Kishore, 2022).

A long-acting glucagon like peptide-I analogue liraglutide has been efficient in managing type-II diabetes and exhibits beneficial effects in patients having cardiovascular diseases. The protracted action of this peptide is reported to be highly dependent on its ability to associate with albumin through its palmitic acid group. Therefore, its therapeutic effect may be adversely affected under diabetic conditions where albumin undergoes glycation. An attempt has been made to determine HSA affinity for liraglutide with two models of glycated albumin using ^{19}F NMR spectroscopy. These studies were done either by glycating HSA in-vitro upon incubation with glucose or methoxy glyoxal and also purified form of in vivo glycated HSA extracted from the plasma of patients having diabetes condition possessing poor glycaemic control. Glycation led to a reduction in binding to liraglutide by seven folds (with G100-HSA) and by five folds (MGO-HSA) in comparison with control. A similar effect was also observed for HSA purified from individuals having diabetic condition in comparison with those in healthy state (Gajahi Soudahome et al., 2018).

It has been reported that second-generation sulfonylurea still has high potential in hypoglycemia treatment. Gliclazide belonging to this group was observed to lose some of its binding ability affinity to the glycated form of BSA compare to that its native form. Under such conditions free drug concentration may be elevated in systemic circulation. An attempt has been made on understanding the conformational behaviour of differently glycated albumin upon association with gliclazide in the presence of another drug combination. This another drug in combination is cilazapril (which is an angiotensin-converting enzyme inhibitor possessing hypotensive activity), atorvastatin (which is a 3-hydroxy-methylglutarl coenzyme A reductase inhibitor), or acetylsalicylic acid (which is an antiplatelet agent responsible for inhibiting cyclooxygenase-1). The glycation of BSA was observed to moderately affect the binding of gliclazide with the protein. However, the binding of gliclazide in the presence of other drugs was dependent on the extent of glycation (Wiglusz et al., 2021).

In another report non-enzymatic glycation of serum albumin by glucose and fructose was studied employing fluorescence spectroscopy. Here, the authors did not observe significant reduction in the value of affinity for ketoprofen with HSA whereas for phenylbutazone it is rather strengthened (Szkudlarek et al., 2016). These observations are in-line with the observations of Ghosh and Kishore (2022).

8. Future perspectives

Drug transportability under diabetic conditions is of prime importance in order for the transport protein to carry out its function. Conflicting reports in literature suggest the need for further investigation in drug binding at different stages of the glycation process on the protein. Obviously, glycation can alter conformation of the protein and modulate its structural integrity, and hence binding sites for the drugs thereby affecting the association. The role of AGEs can be significant in such recognition process. Different groups have used drugs with diverse functionalities and obtained either reduction in the binding affinity or enhancement in the binding affinity of the drug with the transport protein serum albumin. There is a need to obtain both qualitative and quantitative mechanistic insights to unravel the role of functionality in recognition process. Amongst various approaches, microcalorimetry (DSC and ITC) can provide very important guidelines in assessing the suitability of a drug in diabetic conditions depending upon unravelling its thermodynamic signatures in correlation with the extent of association and functionality.

References

Abou-Zied, O. K., & Al-Lawatia, N. (2011). Exploring the drug-binding site sudlow I of human serum albumin: The role of water and Trp214 in molecular recognition and ligand binding. *Chemphyschem: A European Journal of Chemical Physics and Physical Chemistry, 12*(2), 270–274. https://doi.org/10.1002/cphc.201000742.

Ahmad, S., Shahab, U., Baig, M. H., Khan, M. S., Khan, M. S., Srivastava, A. K., ... Moinuddin (2013). Inhibitory effect of metformin and pyridoxamine in the formation of early, intermediate and advanced glycation end-products. *PLoS One, 8*(9), e72128. https://doi.org/10.1371/journal.pone.0072128.

Anguizola, J., Matsuda, R., Barnaby, O. S., Hoy, K. S., Wa, C., DeBolt, E., ... Hage, D. S. (2013). Review: Glycation of human serum albumin. *Clinica Chimica Acta, 425*, 64–76. https://doi.org/10.1016/j.cca.2013.07.013.

Anis, M. A., & Sreerama, Y. N. (2020). Inhibition of protein glycoxidation and advanced glycation end-product formation by barnyard millet (*Echinochloa frumentacea*) phenolics. *Food Chemistry, 315*, 126265. https://doi.org/10.1016/j.foodchem.2020.126265.

Barile, G., & Schmidt, A. (2007). RAGE and its ligands in retinal disease. *Current Molecular Medicine, 7*(8), 758–765. https://doi.org/10.2174/156652407783220778.

Barnaby, O. S., Wa, C., Cerny, R. L., Clarke, W., & Hage, D. S. (2010). Quantitative analysis of glycation sites on human serum albumin using 16O/18O-labeling and matrix-assisted laser desorption/ionization time-of-flight mass spectrometry. *Clinica Chimica Acta, 411*(15–16), 1102–1110. https://doi.org/10.1016/j.cca.2010.04.007.

Billaud, C., Maraschin, C., Peyrat-Maillard, M.-N., & Nicolas, J. (2005). Maillard reaction products derived from thiol compounds as inhibitors of enzymatic browning of fruits and vegetables: The structure-activity relationship. *Annals of the New York Academy of Sciences, 1043*(1), 876–885. https://doi.org/10.1196/annals.1333.099.

Bou-Abdallah, F., Sprague, S. E., Smith, B. M., & Giffune, T. R. (2016). Binding thermodynamics of Diclofenac and Naproxen with human and bovine serum albumins: A calorimetric and spectroscopic study. *The Journal of Chemical Thermodynamics, 103*, 299–309. https://doi.org/10.1016/j.jct.2016.08.020.

Chiazza, F., Cento, A. S., Collotta, D., Nigro, D., Rosa, G., Baratta, F., ... Collino, M. (2017). Protective effects of pyridoxamine supplementation in the early stages of diet-induced kidney dysfunction. *BioMed Research International, 2017*, 1–12. https://doi.org/10.1155/2017/2682861.

de La Rochette, A., Birlouez-Aragon, I., Silva, E., & Morlière, P. (2003). Advanced glycation endproducts as UVA photosensitizers of tryptophan and ascorbic acid: Consequences for the lens. *Biochimica et Biophysica Acta (BBA) – General Subjects, 1621*(3), 235–241. https://doi.org/10.1016/S0304-4165(03)00072-2.

Durowoju, I. B., Bhandal, K. S., Hu, J., Carpick, B., & Kirkitadze, M. (2017). Differential scanning calorimetry — A method for assessing the thermal stability and conformation of protein antigen. *Journal of Visualized Experiments, 121*. https://doi.org/10.3791/55262.

Dyer, D. G., Blackledge, J. A., Thorpe, S. R., & Baynes, J. W. (1991). Formation of pentosidine during nonenzymatic browning of proteins by glucose. Identification of glucose and other carbohydrates as possible precursors of pentosidine in vivo. *Journal of Biological Chemistry, 266*(18), 11654–11660. https://doi.org/10.1016/S0021-9258(18)99007-1.

Favre, L. C., Rolandelli, G., Mshicileli, N., Vhangani, L. N., dos Santos Ferreira, C., van Wyk, J., & Buera, M. del P. (2020). Antioxidant and anti-glycation potential of green pepper (*Piper nigrum*): Optimization of β-cyclodextrin-based extraction by response surface methodology. *Food Chemistry, 316*, 126280. https://doi.org/10.1016/j.foodchem.2020.126280.

Flier, J. S., Underhill, L. H., Brownlee, M., Cerami, A., & Vlassara, H. (1988). Advanced glycosylation end products in tissue and the biochemical basis of diabetic complications. *New England Journal of Medicine, 318*(20), 1315–1321. https://doi.org/10.1056/NEJM198805193182007.

Furusyo, N., & Hayashi, J. (2013). Glycated albumin and diabetes mellitus. *Biochimica et Biophysica Acta (BBA) – General Subjects, 1830*(12), 5509–5514. https://doi.org/10.1016/j.bbagen.2013.05.010.

Gajahi Soudahome, A., Catan, A., Giraud, P., Assouan Kouao, S., Guerin-Dubourg, A., Debussche, X., ... Couprie, J. (2018). Glycation of human serum albumin impairs binding to the glucagon-like peptide-1 analogue liraglutide. *Journal of Biological Chemistry, 293*(13), 4778–4791. https://doi.org/10.1074/jbc.M117.815274.

Gelamo, E. L., & Tabak, M. (2000). Spectroscopic studies on the interaction of bovine (BSA) and human (HSA) serum albumins with ionic surfactants. *Spectrochimica Acta Part A: Molecular and Biomolecular Spectroscopy, 56*(11), 2255–2271. https://doi.org/10.1016/S1386-1425(00)00313-9.

Ghosh, R., Bharathkar, S. K., & Kishore, N. (2019). Anticancer altretamine recognition by bovine serum albumin and its role as inhibitor of fibril formation: Biophysical insights. *International Journal of Biological Macromolecules, 138*, 359–369. https://doi.org/10.1016/j.ijbiomac.2019.07.093.

Ghosh, R., & Kishore, N. (2020). Physicochemical insights into the role of drug functionality in fibrillation inhibition of bovine serum albumin. *The Journal of Physical Chemistry. B, 124*(41), 8989–9008. https://doi.org/10.1021/acs.jpcb.0c06167.

Ghosh, R., & Kishore, N. (2022). Mechanistic physicochemical insights into glycation and drug binding by serum albumin: Implications in diabetic conditions. *Biochimie, 193*, 16–37. https://doi.org/10.1016/j.biochi.2021.10.008.

Ghosh, T., Sarkar, S., Bhattacharjee, P., Jana, G. C., Hossain, M., Pandya, P., & Bhadra, K. (2020). In vitro relationship between serum protein binding to beta-carboline alkaloids: A comparative cytotoxic, spectroscopic and calorimetric assays. *Journal of Biomolecular Structure and Dynamics, 38*(4), 1103–1118. https://doi.org/10.1080/07391102.2019.1595727.

GhoshMoulick, R., Bhattacharya, J., Roy, S., Basak, S., & Dasgupta, A. K. (2007). Compensatory secondary structure alterations in protein glycation. *Biochimica et Biophysica Acta (BBA) – Proteins and Proteomics, 1774*(2), 233–242. https://doi.org/10.1016/j.bbapap.2006.11.018.

Glenn, J. V., & Stitt, A. W. (2009). The role of advanced glycation end products in retinal ageing and disease. *Biochimica et Biophysica Acta (BBA) – General Subjects, 1790*(10), 1109–1116. https://doi.org/10.1016/j.bbagen.2009.04.016.

Han, C., Zhai, L., Shen, H., Wang, J., & Guan, Q. (2023). Advanced glycation end-products (AGEs) promote endothelial cell pyroptosis under cerebral ischemia and hypoxia via HIF-1α-RAGE-NLRP3. *Molecular Neurobiology, 60*(5), 2355–2366. https://doi.org/10.1007/s12035-023-03228-8.

Hartmann, A., Orfanoudaki, M., Blanchard, P., Derbre, S., Schinkovitz, A., Richomme, P., ... Stuppner, H. (2019). Secondary metabolites from marine sources as inhibitors of advanced glycation end products (AGEs) and collagenase. *Planta Medica, 85*(18), 1547. https://doi.org/10.1055/s-0039-3400086.

Hattan, S. J., Parker, K. C., Vestal, M. L., Yang, J. Y., Herold, D. A., & Duncan, M. W. (2016). Analysis and quantitation of glycated hemoglobin by matrix assisted laser desorption/ionization time of flight mass spectrometry. *Journal of the American Society for Mass Spectrometry, 27*(3), 532–541. https://doi.org/10.1007/s13361-015-1316-6.

Hudson, B. I., & Schmidt, A. M. (2004). RAGE: A novel target for drug intervention in diabetic vascular disease. *Pharmaceutical Research, 21*(7), 1079–1086. https://doi.org/10.1023/B:PHAM.0000032992.75423.9b.

Iannuzzi, C., Irace, G., & Sirangelo, I. (2014). Differential effects of glycation on protein aggregation and amyloid formation. *Frontiers in Molecular Biosciences, 1*. https://doi.org/10.3389/fmolb.2014.00009.

Johnson, C. M. (2013). Differential scanning calorimetry as a tool for protein folding and stability. *Archives of Biochemistry and Biophysics, 531*(1–2), 100–109. https://doi.org/10.1016/j.abb.2012.09.008.

Johnson, R. N., Metcalf, P. A., & Baker, J. R. (1983). Fructosamine: A new approach to the estimation of serum glycosylprotein. An index of diabetic control. *Clinica Chimica Acta, 127*(1), 87–95. https://doi.org/10.1016/0009-8981(83)90078-5.

Jud, P., & Sourij, H. (2019). Therapeutic options to reduce advanced glycation end products in patients with diabetes mellitus: A review. *Diabetes Research and Clinical Practice, 148*, 54–63. https://doi.org/10.1016/j.diabres.2018.11.016.

Karonen, M., Oraviita, M., Mueller-Harvey, I., Salminen, J.-P., & Green, R. J. (2015). Binding of an oligomeric ellagitannin series to bovine serum albumin (BSA): Analysis by isothermal titration calorimetry (ITC). *Journal of Agricultural and Food Chemistry, 63*(49), 10647–10654. https://doi.org/10.1021/acs.jafc.5b04843.

Kim, N. A., An, I. B., Lee, S. Y., Park, E.-S., & Jeong, S. H. (2012). Optimization of protein solution by a novel experimental design method using thermodynamic properties. *Archives of Pharmacal Research, 35*(9), 1609–1619. https://doi.org/10.1007/s12272-012-0912-2.

Kirchhoff, W. (1993). *EXAM, a Two-state Thermodynamic Analysis Program.* US Department of Commerce, Thermodynamics Division, National Institute of Standards and Technology, Gaithersburg, MD.

Knowler, W. C., Barrett-Connor, E., Fowler, S. E., Hamman, R. F., Lachin, J. M., Walker, E. A., & Nathan, D. M. (2002). Reduction in the incidence of type 2 diabetes with lifestyle intervention or metformin. *New England Journal of Medicine, 346*(6), 393–403. https://doi.org/10.1056/NEJMoa012512.

Kumar, A., Attri, P., & Venkatesu, P. (2010). Trehalose protects urea-induced unfolding of α-chymotrypsin. *International Journal of Biological Macromolecules, 47*(4), 540–545. https://doi.org/10.1016/j.ijbiomac.2010.07.013.

Lapolla, A., Fedele, D., Plebani, M., Garbeglio, M., Seraglia, R., D'Alpaos, M., ... Traldi, P. (1999). Direct evaluation of glycated and glyco-oxidized globins by matrix-assisted laser desorption/ionization mass spectrometry. *Rapid Communications in Mass Spectrometry, 13*, 8–14.

Lapolla, A., Traldi, P., & Fedele, D. (2005). Importance of measuring products of non-enzymatic glycation of proteins. *Clinical Biochemistry, 38*(2), 103–115. https://doi.org/10.1016/j.clinbiochem.2004.09.007.

Lee, P., & Wu, X. (2015). Review: Modifications of human serum albumin and their binding effect. *Current Pharmaceutical Design, 21*(14), 1862–1865. https://doi.org/10.2174/1381612821666150302115025.

Lewis, B. S., & Harding, J. J. (1990). The effects of aminoguanidine on the glycation (non-enzymic glycosylation) of lens proteins. *Experimental Eye Research, 50*(5), 463–467. https://doi.org/10.1016/0014-4835(90)90033-Q.

Linetsky, M., Shipova, E., Cheng, R., & Ortwerth, B. J. (2008). Glycation by ascorbic acid oxidation products leads to the aggregation of lens proteins. *Biochimica et Biophysica Acta (BBA) – Molecular Basis of Disease, 1782*(1), 22–34. https://doi.org/10.1016/j.bbadis.2007.10.003.

Linkuvienė, V., Krainer, G., Chen, W.-Y., & Matulis, D. (2016). Isothermal titration calorimetry for drug design: Precision of the enthalpy and binding constant measurements and comparison of the instruments. *Analytical Biochemistry, 515*, 61–64. https://doi.org/10.1016/j.ab.2016.10.005.

Maiti, T. K., Ghosh, K. S., & Dasgupta, S. (2006). Interaction of (−)-epigallocatechin-3-gallate with human serum albumin: Fluorescence, fourier transform infrared, circular dichroism, and docking studies. *Proteins: Structure, Function, and Bioinformatics, 64*(2), 355–362. https://doi.org/10.1002/prot.20995.

Maiti, T. K., Ghosh, K. S., Samanta, A., & Dasgupta, S. (2008). The interaction of silibinin with human serum albumin: A spectroscopic investigation. *Journal of Photochemistry and Photobiology A: Chemistry, 194*(2–3), 297–307. https://doi.org/10.1016/j.jphotochem.2007.08.028.

Malik, P., Chaudhry, N., Mittal, R., & Mukherjee, T. K. (2015). Role of receptor for advanced glycation end products in the complication and progression of various types of cancers. *Biochimica et Biophysica Acta (BBA) – General Subjects, 1850*(9), 1898–1904. https://doi.org/10.1016/j.bbagen.2015.05.020.

Marjani, N., Dareini, M., Asadzade-Lotfabad, M., Pejhan, M., Mokaberi, P., Amiri-Tehranizadeh, Z., ... Chamani, J. (2022). Evaluation of the binding effect and cytotoxicity assay of 2-ethyl-5-(4-methylphenyl) pyramido pyrazole ophthalazine trione on calf thymus DNA: Spectroscopic, calorimetric, and molecular dynamics approaches. *Luminescence: the Journal of Biological and Chemical Luminescence, 37*(2), 310–322. https://doi.org/10.1002/bio.4173.

Meloun, B., Morávek, L., & Kostka, V. (1975). Complete amino acid sequence of human serum albumin. *FEBS Letters, 58*(1–2), 134–137. https://doi.org/10.1016/0014-5793(75)80242-0.

Migliore, R., Zavalishin, M. N., Gamov, G. A., Usacheva, T. R., Sharnin, V. A., Grasso, G. I., & Sgarlata, C. (2022). Isothermal titration calorimetry investigation of the interactions between vitamin B6-derived hydrazones and bovine and human serum albumin. *Journal of Thermal Analysis and Calorimetry, 147*(9), 5483–5490. https://doi.org/10.1007/s10973-022-11200-2.

Mohamadi-Nejad, A., Moosavi-Movahedi, A. A., Hakimelahi, G. H., & Sheibani, N. (2002). Thermodynamic analysis of human serum albumin interactions with glucose: Insights into the diabetic range of glucose concentration. *The International Journal of Biochemistry & Cell Biology, 34*(9), 1115–1124. https://doi.org/10.1016/S1357-2725(02)00031-6.

Nabi, R., Alvi, S. S., Saeed, M., Ahmad, S., & Khan, M. S. (2019). Glycation and HMG-CoA reductase inhibitors: Implication in diabetes and associated complications. *Current Diabetes Reviews, 15*(3), 213–223. https://doi.org/10.2174/1573399814666180924113442.

Naik, R., & Seetharamappa, J. (2023). Elucidating the binding mechanism of an antimigraine agent with a model protein: Insights from molecular spectroscopic, calorimetric and computational approaches. *Journal of Biomolecular Structure and Dynamics, 41*(8), 3686–3701. https://doi.org/10.1080/07391102.2022.2053747.

Nakajou, K., Watanabe, H., Kragh-Hansen, U., Maruyama, T., & Otagiri, M. (2003). The effect of glycation on the structure, function and biological fate of human serum albumin as revealed by recombinant mutants. *Biochimica et Biophysica Acta (BBA) – General Subjects, 1623*(2–3), 88–97. https://doi.org/10.1016/j.bbagen.2003.08.001.

Nomi, Y., & Otsuka, Y. (2020). Isolation, identification, and proposed formation mechanism of a novel hydrophilic compound formed by Maillard reaction between pyridoxamine and pentose. *Scientific Reports, 10*(1), 1823. https://doi.org/10.1038/s41598-020-58727-8.

Nowotny, K., Jung, T., Höhn, A., Weber, D., & Grune, T. (2015). Advanced glycation end products and oxidative stress in type 2 diabetes mellitus. *Biomolecules, 5*(1), 194–222. https://doi.org/10.3390/biom5010194.

Oshitari, T. (2023). Advanced glycation end-products and diabetic neuropathy of the retina. *International Journal of Molecular Sciences, 24*(3), 2927. https://doi.org/10.3390/ijms24032927.

Ossowicz, P., Janus, E., Klebeko, J., Świątek, E., Kardaleva, P., Taneva, S., ... Guncheva, M. (2020). Modulation of the binding affinity of naproxen to bovine serum albumin by conversion of the drug into amino acid ester salts. *Journal of Molecular Liquids, 319*, 114283. https://doi.org/10.1016/j.molliq.2020.114283.

Ovung, A., Mavani, A., Ghosh, A., Chatterjee, S., Das, A., Suresh Kumar, G., ... Bhattacharyya, J. (2022). Heme protein binding of sulfonamide compounds: A correlation study by spectroscopic, calorimetric, and computational methods. *ACS Omega, 7*(6), 4932–4944. https://doi.org/10.1021/acsomega.1c05554.

Ràfols, C., Amézqueta, S., Fuguet, E., & Bosch, E. (2018). Molecular interactions between warfarin and human (HSA) or bovine (BSA) serum albumin evaluated by isothermal titration calorimetry (ITC), fluorescence spectrometry (FS) and frontal analysis capillary electrophoresis (FA/CE). *Journal of Pharmaceutical and Biomedical Analysis, 150*, 452–459. https://doi.org/10.1016/j.jpba.2017.12.008.

Rahnama, E., Mahmoodian-Moghaddam, M., Khorsand-Ahmadi, S., Saberi, M. R., & Chamani, J. (2015). Binding site identification of metformin to human serum albumin and glycated human serum albumin by spectroscopic and molecular modeling techniques: A comparison study. *Journal of Biomolecular Structure and Dynamics, 33*(3), 513–533. https://doi.org/10.1080/07391102.2014.893540.

Rebollo-Hernanz, M., Fernández-Gómez, B., Herrero, M., Aguilera, Y., Martín-Cabrejas, M. A., Uribarri, J., & del Castillo, M. D. (2019). Inhibition of the Maillard reaction by phytochemicals composing an aqueous coffee silverskin extract via a mixed mechanism of action. *Foods, 8*(10), 438. https://doi.org/10.3390/foods8100438.

Rondeau, P., Navarra, G., Cacciabaudo, F., Leone, M., Bourdon, E., & Militello, V. (2010). Thermal aggregation of glycated bovine serum albumin. *Biochimica et Biophysica Acta (BBA) – Proteins and Proteomics, 1804*(4), 789–798. https://doi.org/10.1016/j.bbapap.2009.12.003.

Sanchez-Ruiz, J. M. (1995). *Differential scanning calorimetry of proteins* (pp. 133–176). https://doi.org/10.1007/978-1-4899-1727-0_6.
Sattarahmady, N., Moosavimovahedi, A., Ahmad, F., Hakimelahi, G., Habibirezaei, M., Saboury, A., & Sheibani, N. (2007). Formation of the molten globule-like state during prolonged glycation of human serum albumin. *Biochimica et Biophysica Acta (BBA) – General Subjects, 1770*(6), 933–942. https://doi.org/10.1016/j.bbagen.2007.02.001.
Schmidt, A. M., Yan, S. D., Yan, S. F., & Stern, D. M. (2000). The biology of the receptor for advanced glycation end products and its ligands. *Biochimica et Biophysica Acta (BBA) – Molecular Cell Research, 1498*(2–3), 99–111. https://doi.org/10.1016/S0167-4889(00)00087-2.
Schmitt, A., Schmitt, J., Münch, G., & Gasic-Milencovic, J. (2005). Characterization of advanced glycation end products for biochemical studies: Side chain modifications and fluorescence characteristics. *Analytical Biochemistry, 338*(2), 201–215. https://doi.org/10.1016/j.ab.2004.12.003.
Sen, P., Fatima, S., Ahmad, B., & Khan, R. H. (2009). Interactions of thioflavin T with serum albumins: Spectroscopic analyses. *Spectrochimica Acta Part A: Molecular and Biomolecular Spectroscopy, 74*(1), 94–99. https://doi.org/10.1016/j.saa.2009.05.010.
Servetnick, D. A., Bryant, D., Wells-Knecht, K. J., & Wiesenfeld, P. L. (1996). L-Arginine inhibits in vitro nonenzymatic glycation and advanced glycosylated end product formation of human serum albumin. *Amino Acids, 11*(1), 69–81. https://doi.org/10.1007/BF00805722.
Setter, S. M., Iltz, J. L., Thams, J., & Campbell, R. K. (2003). Metformin hydrochloride in the treatment of type 2 diabetes mellitus: A clinical review with a focus on dual therapy. *Clinical Therapeutics, 25*(12), 2991–3026. https://doi.org/10.1016/S0149-2918(03)90089-0.
Shamsi, A., Al Shahwan, M., Ahamad, S., Hassan, M. I., Ahmad, F., & Islam, A. (2020). Spectroscopic, calorimetric and molecular docking insight into the interaction of Alzheimer's drug donepezil with human transferrin: Implications of Alzheimer's drug. *Journal of Biomolecular Structure and Dynamics, 38*(4), 1094–1102. https://doi.org/10.1080/07391102.2019.1595728.
Sharma, A. K., Sharma, V. R., Gupta, G. K., Ashraf, G. M., & Kamal, M. A. (2019). Advanced glycation end products (AGEs), glutathione and breast cancer: Factors, mechanism and therapeutic interventions. *Current Drug Metabolism, 20*(1), 65–71. https://doi.org/10.2174/1389200219666180912104342.
Shental-Bechor, D., & Levy, Y. (2008). Effect of glycosylation on protein folding: A close look at thermodynamic stabilization. *Proceedings of the National Academy of Sciences, 105*(24), 8256–8261. https://doi.org/10.1073/pnas.0801340105.
Sherwani, S. I., Khan, H. A., Ekhzaimy, A., Masood, A., & Sakharkar, M. K. (2016). Significance of HbA1c test in diagnosis and prognosis of diabetic patients. *Biomarker Insights, 11*. https://doi.org/10.4137/BMI.S38440.
Singh, R., Barden, A., Mori, T., & Beilin, L. (2002). Erratum to: Advanced glycation end-products: A review. *Diabetologia, 45*(2), 293. https://doi.org/10.1007/s001250100676.
Sirangelo, I., & Iannuzzi, C. (2021). Understanding the role of protein glycation in the amyloid aggregation process. *International Journal of Molecular Sciences, 22*(12), 6609. https://doi.org/10.3390/ijms22126609.
Sittiwanichai, S., Japrung, D., Mori, T., & Pongprayoon, P. (2023). Structural and dynamic alteration of glycated human serum albumin in Schiff base and Amadori adducts: A molecular simulation study. *The Journal of Physical Chemistry B*. https://doi.org/10.1021/acs.jpcb.3c02048.
Śliwińska-Hill, U., & Wiglusz, K. (2023). The effect of human serum albumin glycation on the binding of antidiabetic agent-exenatide; the multispectroscopic studies. Part II. *Biophysical Chemistry, 294*, 106948. https://doi.org/10.1016/j.bpc.2022.106948.
Sooram, B., Gupta, N., Chethireddy, V. R., Tripathi, T., & Saudagar, P. (2023). Applications of differential scanning calorimetry in studying folding and stability of proteins. *Protein Folding Dynamics and Stability*. Springer Nature Singapore, 37–60. https://doi.org/10.1007/978-981-99-2079-2_3.

Stefanowicz, P., Kijewska, M., Kluczyk, A., & Szewczuk, Z. (2010). Detection of glycation sites in proteins by high-resolution mass spectrometry combined with isotopic labeling. *Analytical Biochemistry, 400*(2), 237–243. https://doi.org/10.1016/j.ab.2010.02.011.

Su, H., & Xu, Y. (2018). Application of ITC-based characterization of thermodynamic and kinetic association of ligands with proteins in drug design. *Frontiers in Pharmacology, 9.* https://doi.org/10.3389/fphar.2018.01133.

Swamy-Mruthinti, S. (2001). Glycation decreases calmodulin binding to lens transmembrane protein, MIP. *Biochimica et Biophysica Acta (BBA) – Molecular Basis of Disease, 1536*(1), 64–72. https://doi.org/10.1016/S0925-4439(01)00031-X.

Syrový, I. (1994). Glycation of albumin: Reaction with glucose, fructose, galactose, ribose or glyceraldehyde measured using four methods. *Journal of Biochemical and Biophysical Methods, 28*(2), 115–121. https://doi.org/10.1016/0165-022X(94)90025-6.

Szkudlarek, A., Sułkowska, A., Maciążek-Jurczyk, M., Chudzik, M., & Równicka-Zubik, J. (2016). Effects of non-enzymatic glycation in human serum albumin. Spectroscopic analysis. *Spectrochimica Acta Part A: Molecular and Biomolecular Spectroscopy, 152*, 645–653. https://doi.org/10.1016/j.saa.2015.01.120.

Van Putte, L., De Schrijver, S., & Moortgat, P. (2016). The effects of advanced glycation end products (AGEs) on dermal wound healing and scar formation: A systematic review. 205951311667682 *Scars, Burns & Healing, 2.* https://doi.org/10.1177/2059513116676828.

Varshney, A., Sen, P., Ahmad, E., Rehan, M., Subbarao, N., & Khan, R. H. (2010). Ligand binding strategies of human serum albumin: How can the cargo be utilized? *Chirality, 22*(1), 77–87. https://doi.org/10.1002/chir.20709.

Wang, W.-D., Li, C., Bin, Z., Huang, Q., You, L.-J., Chen, C., ... Liu, R. H. (2020). Physicochemical properties and bioactivity of whey protein isolate-inulin conjugates obtained by Maillard reaction. *International Journal of Biological Macromolecules, 150*, 326–335. https://doi.org/10.1016/j.ijbiomac.2020.02.086.

Ward, W. H. J., & Holdgate, G. A. (2001). 7 *Isothermal titration calorimetry in drug discovery* (pp. 309–376). https://doi.org/10.1016/S0079-6468(08)70097-3.

Webster, J., Urban, C., Berbaum, K., Loske, C., Alpar, A., GÄrtner, U., ... MÜnch, G. (2005). The carbonyl scavengers aminoguanidine and tenilsetam protect against the neurotoxic effects of methylglyoxal. *Neurotoxicity Research, 7*(1–2), 95–101. https://doi.org/10.1007/BF03033780.

Westwood, M. E., & Thornalley, P. J. (1995). Molecular characteristics of methylglyoxal-modified bovine and human serum albumins. Comparison with glucose-derived advanced glycation endproduct-modified serum albumins. *Journal of Protein Chemistry, 14*(5), 359–372. https://doi.org/10.1007/BF01886793.

Wiglusz, K., Żurawska-Płaksej, E., Rorbach-Dolata, A., & Piwowar, A. (2021). How does glycation affect binding parameters of the albumin-gliclazide system in the presence of drugs commonly used in diabetes? In vitro spectroscopic study. *Molecules (Basel, Switzerland), 26*(13), 3869. https://doi.org/10.3390/molecules26133869.

Winterhalter, K. H. (1985). Non enzymatic glycosylation of proteins. *Progress in Clinical and Biological Research, 195*, 109–122.

Yamagishi, S., Maeda, S., Matsui, T., Ueda, S., Fukami, K., & Okuda, S. (2012). Role of advanced glycation end products (AGEs) and oxidative stress in vascular complications in diabetes. *Biochimica et Biophysica Acta (BBA) – General Subjects, 1820*(5), 663–671. https://doi.org/10.1016/j.bbagen.2011.03.014.

Yan, S. D., Yan, S. F., Chen, X., Fu, J., Chen, M., Kuppusamy, P., ... Stern, D. (1995). Non-enzymatically glycated tau in Alzheimer's disease induces neuronal oxidant stress resulting in cytokine gene expression and release of amyloid β-peptide. *Nature Medicine, 1*(7), 693–699. https://doi.org/10.1038/nm0795-693.

CHAPTER FIVE

Advanced glycation end products and insulin resistance in diabetic nephropathy

Kirti Parwani and Palash Mandal*
Department of Biological Sciences, P. D. Patel Institute of Applied Sciences, Charotar University of Science & Technology, Gujarat, India
*Corresponding author. e-mail address: palashmandal.bio@charusat.ac.in

Contents

1. Introduction	118
2. Advanced glycation end products	119
3. Receptors for AGEs	123
3.1 Membrane bound full length receptor for AGEs	123
3.2 Other receptors for AGEs	124
4. Role of AGEs in insulin resistance	125
4.1 Evidence from animal and in vitro studies	125
4.2 Evidences from in vivo studies	127
4.3 Evidence from studies on human subjects	128
5. AGEs and their involvement in diabetic nephropathy	129
6. Mechanisms of insulin resistance and its impact on the kidney	131
7. Inhibition of the AGE-RAGE axis: Treatments for diabetic nephropathy	134
8. Conclusion and future perspectives	138
References	139

Abstract

Insulin resistance is a central hallmark that connects the metabolic syndrome and diabetes to the resultant formation of advanced glycation end products (AGEs), which further results in the complications of diabetes, including diabetic nephropathy. Several factors play an important role as an inducer to diabetic nephropathy, and AGEs elicit their harmful effects via interacting with the receptor for AGEs Receptor for AGEs, by induction of pro-inflammatory cytokines, oxidative stress, endoplasmic reticulum stress and fibrosis in the kidney tissues leading to the loss of renal function. Insulin resistance results in the activation of other alternate pathways governed by insulin, which results in the hypertrophy of the renal cells and tissue remodeling. Apart from the glucose uptake and disposal, insulin dependent PI3K and Akt also upregulate the expression of endothelial nitric oxide synthase, that results in increasing the bioavailability of nitric oxide in the vascular endothelium, which further results in tissue fibrosis. Considering the global prevalence of diabetic nephropathy, and the impact of protein glycation, various inhibitors and treatment avenues are being developed, to prevent the progression of diabetic complications. In this chapter, we discuss the role of glycation in insulin resistance and further its impact on the kidney.

1. Introduction

Diabetes mellitus (DM) can be explained as a group of multiple abnormalities characterized by hyperglycemia either due to insulin deficiency or insulin resistance (IR) or in some cases both. The resulting hyperglycemia induces several metabolic and physiological changes including formation of advanced glycation end products (AGEs) (Vlassara & Uribarri, 2014) and IR (Yki-Järvinen & Koivisto, 1984). Diabetes is also one of the leading and most common causes of end stage renal disease (ESRD), and is associated with increased mortality and morbidity, responsible for more than 47% of new ESRD cases in the US (Caramori & Rossing, 2022).

Diabetic Nephropathy (DN) is one of the most common complications found in T2DM patients. 40% of patients with T2DM and 30% of IDDM patients develop DN (Hussain et al., 2021). DN is thought to be one of the major causes leading to chronic kidney disease (CKD) and ESRD (Alicic, Rooney, & Tuttle, 2017). In a study done incorporating 4006 T2DM patients in the United Kingdom, 28% of patients developed renal failure after the average follow up of 15 years (Retnakaran et al., 2006). In India, 34.4% prevalence of DN (Hussain, Habib, & Najmi, 2019) and a composite prevalence of 62.3% of diabetic-CKD (Dash, Agarwal, Panigrahi, Mishra, & Dash, 2018) have been reported, suggesting the higher prevalence of DN in India. As per a report of American diabetes association (ADA) (2014), more than 40% of diabetic patients are estimated to develop CKD, comprising a significant number of patients further developing ESRD requiring either dialysis or renal transplantation. IR is one of the common attributes observed in patients with moderate to severe chronic renal failure (DeFronzo & Alvestrand, 1980; Eidemak et al., 1995). There are various reports suggesting a strong association of IR and hyperinsulinemia with CKD through stage I to IV, independent of diabetes (Chen et al., 2004; Soltani, Washco, Morse, & Reisin, 2015; Yamagata et al., 2007). There are multiple biochemical and signal transduction pathways that result in IR under diabetic condition. One of the major contributors to IR is post-translational glycation of insulin, which renders it inefficient to function leading to hyperinsulinemia and hyperglycaemia (Song & Schmidt, 2012).

2. Advanced glycation end products

AGEs are a heterogenous and complex group of proteins, nucleic acid and lipids that are formed endogenously as well as during the cooking of food via the Maillard reaction. During the Maillard reaction, the reducing sugars like glucose and/or fructose non-enzymatically react with amino groups present in the proteins, lipids, and nucleic acid to form Schiff bases. Schiff base is an imine, a double bond between carbon atom of the sugar molecule and nitrogen atom of the amino acid. Amadori products are formed by the re-arrangement of the Schiff bases, where the hydrogen of the hydroxyl group in the Schiff base, bonds with the nitrogen of the C=N (imine group), resulting in the formation of a ketone, further resulting in the formation of AGEs. Apart from this, Amadori products can further breakdown via oxidation to form reactive dicarbonyls such as glyoxal, methylglyoxal (MG), which form intermediate glycation products by reacting with free amino groups of proteins. After a series of reactions like condensation, dehydration, Amadori products and intermediate glycation products, form irreversible intra- or inter-protein cross linked AGEs (Brownlee, Cerami, & Vlassara, 1988; Grandhee & Monnier, 1991). Their role in the development of various diseases like diabetes, secondary complications arising due to diabetes like diabetic nephropathy, diabetic neuropathy, cardiovascular complications has already been established, apart from being recently recognized as a role player in obesity (Gao et al., 2017). AGEs, which otherwise are found naturally in the body as a part of metabolism of various sugars, have their formation rate increased under diabetic conditions (Horie et al., 1997; Schleicher, Wagner, & Nerlich, 1997). Although the formation of AGEs was first described in 1912 (Maillard, 1912), its detailed role in pathophysiology of various diseases is still being extensively researched.

Glycolysis is the pathway which is responsible for making energy out of glucose catabolism. The glucose in the cells under normoglycemia would get metabolized via glycolysis, however, under hyperglycemic conditions, as observed in diabetics, intermediates of glycolysis get shunted into various pathways due to the inhibition of enzyme glyceraldehyde-3-phosphate dehydrogenase (GAPDH) because of hyperglycemia induced excessive superoxide radicals. These glycolytic intermediates are the prime intracellular source for the formation of three classes of dicarbonyls compounds such as glyoxal, MG, and 3-deoxyglucosone, which are the major glycating agents. GAPDH and dihydroxy acetone phosphate gets converted to MG, which further principally reacts with basic arginine groups to form argpyrimidine, a fluorescent AGE, and 5-hydro-5-methylimidazolone (Westwood & Thornalley, 1995). Lipid peroxidation in addition to glucose, can also result in the formation of carbonyl species that can interact with proteins which, further results in the formation of advanced lipoxidation end products like Nε-(carboxymethyl)lysine (CML). Also, when MG interacts with lysine in the protein, it results in the formation of Nε-(carboxyethyl) lysine (CEL), which is a homolog of CML (Miyata, Sugiyama, Suzuki, Inagi, & Kurokawa, 1999).

AGEs are also formed via alternate pathways which include polyol pathway. Under diabetic conditions, higher glucose levels lead to activation of the polyol pathway, where glucose is first converted to sorbitol by aldose reductase. The sorbitol is then acted upon by sorbitol dehydrogenase to form fructose (Gabbay, 1973; Henning, Liehr, Girndt, Ulrich, & Glomb, 2014; Singh, Bali, Singh, & Jaggi, 2014). The activation of the polyol pathway serves two advantages for the formation and accumulation of AGEs by increasing the intracellular NADH: NAD^+ ratio. Firstly, the fructose formed via the polyol pathway increases the rate of formation of intracellular AGEs as it has got 7.5-fold faster rate than glucose in forming the AGEs (Bunn & Higgins, 1981). Secondly, activation of the polyol pathway inhibits the activity of glycolytic enzymes like glyceraldehyde triphosphate dehydrogenase, thereby leading to accumulation of fructose and triose phosphates, further leading to the formation of highly reactive molecules like glyoxal, MG, 3-deoxyglucosone, which as mentioned earlier can interact and crosslink with proteins to form AGEs (Gabbay, 1973; Henning et al., 2014; Singh et al., 2014). The formation of AGEs and various physiological impacts caused by AGEs are shown in Fig. 1. Different amino acid and the type of carbohydrate reacted, results in the generation of different types of AGEs and Table 1 lists the carbohydrate sources for different AGEs and the chemical structures of various AGEs are shown in Fig. 2.

Apart from being formed endogenously, AGEs are formed during cooking via Maillard reaction as explained earlier, leading to non-enzymatic browning. It is known that cooking techniques like frying, baking,

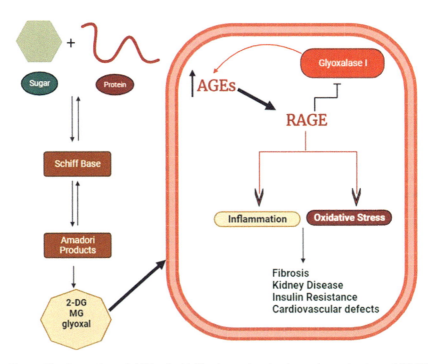

Fig. 1 The formation of AGEs via Maillard reaction leads to the activation of RAGE, which along with inhibiting the glyoxalase I that can clear AGEs, activates oxidative stress and inflammatory pathways.

Table 1 Carbohydrate sources and the corresponding AGEs.

Carbohydrate source	AGEs formed
Ribose	Pentosidine
Glucose, threose	CML
Methylglyoxal when reacts with arginine	Argypyrimidine
Methylglyoxal when reacts with lysine	Carboxyethyl lysine, MOLD
3-Deoxyglucosone when reacts with arginine	Imidazolone
3-Deoxyglucosone when reacts with lysine	DOLD
Glyoxal	GOLD

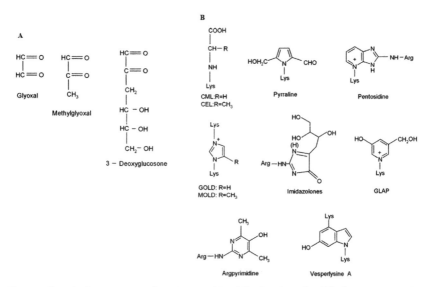

Fig. 2 Chemical structure of various AGEs (A) dicarbonyls; (B) fluorescent AGEs Pentosidine, Vesperlysine A, Argypyrimidine; and non-fluorescent AGEs like *CML*, N-carboxymethyllysine; *CEL*, N-carboxyethyllysine; *GOLD*, glyoxal-lysine dimer: *MOLD*, methylglyoxal-lysine dimer; *GLAP*, glyceraldehyde-derived pyridinium compound.

or roasting which usually happen at higher temperature, lead to increased production of AGEs in the food (Lund & Ray, 2017; Scheijen et al., 2016). Environmental factors like diet, smoking, alcohol consumption, etc. can also exert impact on the rate with which AGEs accumulate endogenously (Kumar Pasupulati, Chitra, & Reddy, 2016). Fly ash induced air-pollution has been associated with the formation of AGEs in the fibroblasts exposed to fly ash for longer time (Gursinsky et al., 2006). Factors affecting the AGE formation in the food relies much on the composition of nutrients like sugar, protein, and fat along with the techniques like temperature and duration of preparation (Lund et al., 2017; Scheijen et al., 2016). Fat rich food and the foods cooked at high temperatures are mostly rich in AGE precursor MG and AGE like CML (Koschinsky et al., 1997; McCarty, 2005). The Western diet uses fructose (Bray, Nielsen, & Popkin, 2004) and it has been known to be more reactive than glucose to form AGEs as described earlier. Also, the processed food consumed is rich in AGEs as processing of food requires treatment of food at higher temperatures for longer time, therefore leading to the formation of AGEs (Uribarri et al., 2010). These dietary AGEs when formed are absorbed by the gastrointestinal tract and play a role in accumulation of AGEs in the cells.

The understanding of the role of dietary AGEs playing a significant role in biological accumulation of AGEs came from a study in which 60% of free form of pentosidine and 2% of its peptide bound form was recovered in the urine samples of individuals fed with coffee brew and bakery products respectively (Förster, Kühne, & Henle, 2005). Also, oral intake of diet rich in AGEs in rodents led to increased levels of CML in tissues (Li et al., 2015), with higher deposition (81–320 μg CML/g dry matter) in kidneys (Tessier et al., 2016). This was also further validated by clinical trials in which higher consumption of diet rich in CML and CEL showed higher levels of their free forms in urine and plasma (Scheijen et al., 2018). Together, these studies suggest that although the endogenous AGEs are related to increased glucose levels as in case of diabetic conditions, dietary AGEs may also lead to detrimental effects as they tend to deposit in tissues.

3. Receptors for AGEs

Multiple receptors that interact with AGEs exist. Various receptors have different effects upon interacting with AGEs, some have a preventive one while some have a pathogenic effect. One of the mechanisms involved in transducing inflammatory and pathogenic effects involves the interaction of AGEs with full length Receptor for AGEs (RAGE).

3.1 Membrane bound full length receptor for AGEs

RAGE is a full length, membrane bound receptor present on numerous cell types in the body, which include peripheral immune cells, endothelial cells, smooth muscle cells, microglia, and neurons (Gao et al., 2017; Litwinoff, Hurtado Del Pozo, Ramasamy, & Schmidt, 2015; Ramasamy, Yan, & Schmidt, 2011). The other type of receptor that is endogenous secretory RAGE (esRAGE) is produced by alternative splicing and lacks the transmembrane domain. It is found majorly in the circulation. Ectodomain—shed RAGE (ecRAGE) results from the action of metalloproteases on full length RAGE (Lue, Walker, Jacobson, & Sabbagh, 2009). AGEs being the ligand for all three types of receptors, is only able to transduce the inflammatory signals via RAGE. The other receptors, endogenous secretory RAGE (esRAGE) and ecRAGE can prevent the activation of inflammatory response elicited via AGE-RAGE axis by competitively binding with AGEs, thereby preventing the binding of AGEs to RAGE (Maillard-Lefebvre et al., 2009; Yamagishi et al., 2006). The

AGE-RAGE axis primarily acts by generation of ROS and oxidative stress, mainly through the NADPH oxidase channel (Wautier et al., 2001). Oxidative stress thus further increases the formation of more AGEs. This was explained by an experiment showing that mice lacking RAGE had lower AGEs and levels of oxidative stress as compared to RAGE-expressing mice (Reiniger et al., 2010). Also, the RAGE knockout mice had lower levels of MG and AGEs in the circulation than wild type mice, although the levels of hyperglycemia were equal in both the mice (Reiniger et al., 2010). AGEs on binding to RAGE can trigger the activation of extra-cellular-signal-regulated kinase 1/2 and p38 mitogen-activated protein kinase (MAPK), rho-GTPases, Janus kinase, c-Jun N-terminal kinase and the transcription factor NF-kB (Grimm et al., 2012; Hirose, Tanikawa, Mori, Okada, & Tanaka, 2010; Li et al., 2004; Ott et al., 2014; Tanikawa, Okada, Tanikawa, & Tanaka, 2009). Further, ROS which is primarily formed due to AGE-RAGE interaction, can also induce inflammation via NF-kB activation, which further leads to upregulation of RAGE, leading to a cause-and-effect mechanism (Bongarzone, Savickas, Luzi, & Gee, 2017; Ramasamy et al., 2011).

RAGE, being a multi-ligand receptor of the immunoglobulin class of receptors, is not restricted to interact with AGEs only. It also interacts with pro-inflammatory signals obtained from high mobility group box 1 (HMGB1) (Taguchi et al., 2000), and through S100/calgranulin (Hofmann et al., 1999). They are known to upregulate the expression of matrix metalloproteinases, pro-inflammatory cytokines, and program the change in their expression in macrophages to amplify tissue damage (Schmidt et al., 2007). Hence, AGE-RAGE axis may provoke oxidative stress and inflammation; and upon recruitment of HMGB1-addressed inflammatory cells at the site of tissue stress, may exacerbate oxidative stress and trigger the formation of more AGEs.

3.2 Other receptors for AGEs

Other receptors than RAGE are found for AGEs, such as cluster of differentiation 36 (Ohgami et al., 2001), and lectin like-oxidized low-density lipoprotein 1 (Rudijanto, 2007). RAGE, type 1 and 2 macrophage scavenger receptors, oligosaccaharyl transferase–48 (AGE-R1), 80K-H phosphoprotein (AGE-R2) and galectin (AGE-R3) have been identified to be present on a wide range of cells including macrophages, endothelial cells, podocytes, monocytes, smooth muscle cells and microglia (Stitt, Bucala, & Vlassara, 1997; Thornalley, 1998). They are also reported to interact with macrophage scavenger receptors (Araki et al., 1995) or Advanced Glycation End Product Receptor (AGERs)-1–2–3 (Lai et al., 2004). These receptors are responsible for maintaining the homeostasis of AGEs by degrading the AGEs via endocytosis. AGEs can upregulate AGER1 found in almost all cells and tissues which can increase uptake and removal of AGEs (Nowotny, Jung, Höhn, Weber, & Grune, 2015). AGER1 can diminish AGE-induced oxidative stress (Cai et al., 2010 and Torreggiani et al., 2009) and can promote a sirtuin1-dependent deacetylation, thereby suppressing NF-κB (Cai et al., 2012; Uribarri et al., 2011). This receptor is mainly localized in the caveolin-rich membrane domain and, by taking up AGEs, promotes their degradation and prevents AGE-damaging effects (Stitt, Burke, Chen, McMullen, & Vlassara, 2000). AGE-R1 was the first discovered and the most expressed of these components. Its level increases in parallel with AGEs, but it is downregulated by persistently high levels of AGEs (Thornalley, 2003). AGE-R1 promotes AGE turnover by mesangial cells and contributes to counter regulation of AGE-induced inflammation (Lu et al., 2004). AGE-R2 is a membrane protein without binding activities. It is involved in stabilization of the receptor complex in combination with AGE-R3, a carbohydrate-binding protein that shows strong affinity for a wide variety of AGEs and regulates different functions, from the cell cycle to inflammation (Ribau, Hadcock, Teoh, DeReske, & Richardson, 1999; Vlassara et al., 1995). Prolonged exposure to AGEs has been associated with increased AGE-R3 expression (Iacobini et al., 2003).

4. Role of AGEs in insulin resistance

Growing evidence suggests the interesting role and association of AGE with IR, independently of diabetes mellitus. Tahara et al. (2012) reported on 322 nondiabetic Japanese subjects; serum AGE levels correlated with HOMA-IR in these subjects, and after multiple regression analysis, AGEs, along with waist circumference, glycosylated hemoglobin, and TGs, were correlated with the degree of IR.

4.1 Evidence from animal and in vitro studies

Various studies have been done on in vitro models, that suggest a probable link between AGEs and IR. Recent developments show decreased action of insulin as an impact of glycation of insulin (Boyd et al., 2000). As insulin is a very short-lived protein, with a half-life of 5–10 min, it has been reported that insulin and pro-insulin may be getting modified by glycation in the pancreas during its formation and storage. Boyd et al. (2000) prepared mono-glycated insulin which resembles the glycated insulin in vivo, that is, single glycation of phenylalanine residues in the amino terminus of the B chain of insulin. The mice under a glucose tolerance test when subjected to the modified mono-glycated insulin, showed a 20% reduced glucose lowering potency as compared to the mice treated with non-glycated insulin. Further, the mono-glycated insulin treated isolated abdominal muscles showed 20% lesser efficiency in glucose uptake, oxidation of glucose and glycogen synthesis as compared to non-glycated insulin (Boyd et al., 2000). Hunter et al. studied the IR induced by glycated insulin with the help of clamp techniques in human subjects. When mono-glycated insulin was put to test under hyper insulinemic euglycemic clamp studies, it resulted in lesser supply of exogenous glucose infusion to maintain normoglycemia and 70% more supply of glycated insulin was required to induce a similar response of control insulin-dependent uptake of glucose (Hunter et al., 2003), suggesting that glycation of insulin results in impaired insulin action.

Jia et al. (2006) showed that the arginine residue of the B chain of insulin was modified by MG, which led to reduced insulin mediated uptake of glucose in adipocytes and skeletal muscle cells compared to unmodified insulin. MG has been shown to affect insulin signaling at the molecular level by blocking tyrosine phosphorylation of insulin-receptor substrate-1 (IRS-1) and blocking PI3 Kinase activation in INS-1 cells (pancreatic β-cells) (Fiory et al., 2011). MG modification of insulin resulted in reduced insulin fibril formation and resulted in the formation of insulin native like aggregates, speculating the reduced insulin action (Oliveira et al., 2011). Apart from the direct effects of glycation on IR, there exist multiple reports that suggest the mechanisms of glycation of protein other than insulin and its impact of insulin sensitivity. TNF-α has been linked to suppress insulin activity via induction of the pro-inflammatory mechanisms involved in IR, and reports suggest that glycation of albumin results in the increased production of TNF-α, which further may result in IR (Naitoh, Kitahara, & Tsuruzoe, 2001; Miele et al., 2003). Human albumin glycation in the L6 skeletal muscle myocytes resulted in the RAGE dependent activation of PKC- α and suppression of IRS-1 action (Cassese et al., 2008).

4.2 Evidences from in vivo studies

The role of AGEs in IR has also been evaluated in animal models by studying whether inhibition of AGEs can be used as a therapy to improve insulin sensitivity. In one such study (Guo et al., 2009), Sprague-Dawley rats were treated with MG alone in drinking water to check the induction of IR. The animals in a second group received MG along with an antioxidant N-acetyl cysteine (NAC) and in a third group received MG along with TM20002, an inhibitor for AGEs. After four weeks of treatment, IR was checked by hyper insulinemic euglycemic clamp, which showed better sensitivity of insulin in the MG + NAC treated and MG + TM2002 treated groups as compared to MG alone. This shows that prevention and treatment to control AGEs can help improve IR in tissues. The same authors also showed that MG and salt co-treatment compared to MG or salt alone showed increased systolic blood pressure, increased levels of thiobarbituric acid reactive substances and albumin in urine (Guo et al., 2009). It is also important to note that in this study, AGEs derived from MG were higher in the groups treated with MG in comparison to the group which was deprived of MG. In a study done on C57BL/6 mice, diet rich in AGE MG-H1 induced IR and diabetes in non-obese animals by depleting AGE-R1 receptor and sirtuin-1 (Cai et al., 2012) both of which are known to be suppressed in chronic oxidative stress and diabetes.

In an in vivo study with KK-A(y) mice, a model often used to study obesity and type 2 diabetes, Unoki-Kubota, Yamagishi, Takeuchi, Bujo, and Saito (2010) observed that serum levels of AGEs were positively correlated with the levels of insulin. This showed an IR phenotype in the animals. To understand if glycation products were responsible for IR, they treated the animals with pyridoxamine which is an inhibitor of AGE formation. They observed that pyridoxamine was helpful in decreasing fasting insulin levels in a dose dependent manner, which improved the insulin sensitivity.

In yet another study, Kooptiwut et al. (2005), addressed the direct role of AGEs on the pancreatic β cells. They isolated islets from pancreas of two metabolically different animals DBA/2 and C57BL/6 mice. DBA/2 mice show pancreatic islet dysfunction when exposed to high glucose environment. Islets isolated from both strains were exposed to 11.1 or 40 mM of glucose in presence and absence of AGE-inhibitor, aminoguanidine. Up to 10 days, the basal insulin release was observed in islets isolated from both strains. But, as the chronic exposure to glucose continued, a decreased secretion of insulin was observed in the islets from DBA/2 mice, which in the presence of aminoguanidine was significantly higher under the same conditions. This could be due to improved islet glucokinase activity in presence of aminoguanidine. Hence, it was proven that AGEs can lead to pancreatic islet dysfunction with reduced glucokinase activity in islets. With different in vitro and in vivo findings, there is evidence that suggests that AGEs may contribute partially to the pathophysiology of IR, and hence results in obesity and type 2 diabetes mellitus.

4.3 Evidence from studies on human subjects

Several pieces of evidence suggest the existence of intriguing links between serum AGE levels and IR, even in non-diabetic human subjects. In a study of 207 healthy, non-diabetic subjects, Tan, Shiu, Wong, and Tam (2011) measured AGE levels, inflammatory markers, and homeostatic model assessment index (HOMA-IR) as a measure of IR. The serum levels of AGEs were significantly associated with IR, independently of gender. Tahara et al. (2012) also demonstrated a positive correlation of serum AGEs with IR after multiple regression analysis, as well as with glycosylated hemoglobin (HbA1C), triglycerides (TGs) and waist circumference. The authors found the trends in all 322 non-diabetic Japanese subjects irrespective of gender.

In a study of diabetic human subjects (Sarkar, Kar, Mondal, Chakraborty, & Kar, 2010), a plausible association between total carbonyl compound levels in serum with HOMA-IR levels was tested. The authors reported a significant correlation of carbonyl levels with IR. However, they did not find any significance in the role of lipid peroxidation end products with IR, conceivably suggesting a role of AGEs more than oxidative stress in having an impact on insulin sensitivity.

Several reports associate AGEs levels and its effect on IR in PCOS human subjects also, as PCOS subjects often show IR in comparison to controls. Serum levels of AGEs were higher in females with PCOS as compared to females with isolated components of the syndrome, proving the role of AGEs in ovarian dysfunction along with other metabolic abnormalities (Diamanti-Kandarakis et al., 2008). In line with this, the same group explained the role of anti-mullerian hormone (AMH), a hormone produced by granulosa cells of primary follicles in PCOS. AMH is found to be elevated in PCOS and is considered as an indicator of disturbed ovulation. The authors found a positive correlation between circulating levels of AGEs and AMH in women with PCOS, thereby suggesting a plausible link between AGEs and AMH in PCOS (Diamanti-Kandarakis et al., 2009). The effects of AGEs on the development of IR are shown in Fig. 3.

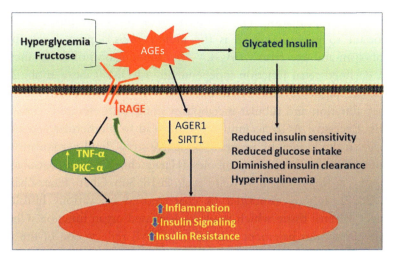

Fig. 3 Hyperglycaemia induced AGEs and its effects on insulin resistance: Activation of AGE-RAGE axis results in the activation of TNF-α and PKC-α, resulting in inflammation. Also, AGEs further upregulate RAGE by downregulating SIRT 1 and AGER1, known to improve sensitivity and AGE clearance respectively.

5. AGEs and their involvement in diabetic nephropathy

As discussed earlier, AGEs can lead to various pathophysiological effects on different organs leading to a variety of complications associated with diabetes, including DN. It is characterized as a chronic complication in diabetes, with raised albumin excretion of reduced GFR or both. Over the past years, various studies have aided in the differentiation of stages of DN in view of renal modifications.

Stage 1 marks the increase in the size of the kidney by up to 20% along with elevated renal flow. The GFR remains normal or slightly elevated without the signs of hypertension and albuminuria. DN progresses to stage 2 after two years of onset of DM, with the characteristics of thickening of basement membrane and mesangial matrix expansion, with a normalized GFR and no clinical signs. The first clinical sign of DN like microalbuminuria (30–300 mg/day) is observed in stage 3 of DN, which is manifested within 5–15 years of onset of DM. However, there are no signs of hypertension observed in patients at this stage, which is usually observed in stage 4, which also shows irreversible proteinuria and GFR below 60 ml/min/1.73 m^2. Stage 5 marks the stage of ESRD with GFR lower than 60 ml/min/1.73 m^2 and patients requiring replacement therapies like dialysis or transplantation. The involvement of AGEs with DN and CKD is important to be understood as the kidney plays a pivotal role in the metabolism and excretion of AGEs. The accumulation of AGEs increases with declined renal function independently of DM (Oberg et al., 2004). Therefore, higher levels of AGEs observed in DN are attributed to reduced kidney filtration and tubular metabolism along with elevated synthesis of AGEs. Excessive AGE accumulation can lead to glomerular and tubular damage in the kidney by inducing oxidative stress, lipid peroxidation and inflammation in the kidney tubules (Haraguchi, Kohara, Matsubayashi, Kitazawa, & Kitazawa, 2020; Kaifu et al., 2018; Sun, Chen, Hua, Zhang, & Liu, 2022). DN is associated or characterized by various histological changes in renal tissue which result in decreased organ function. Changes in the glomerulus basement membrane (GBM) reflects one of the initial signs of DN. The thickening of GBM occurs due to deposition of extracellular matrix components such as collagen and laminins. As discussed earlier, AGEs can deposit in kidneys and therefore may be one of the reasons to the change in the renal architecture. With no surprise, studies with patients and rodents have shown that AGEs crosslink with matrix proteins in the GBM (Beisswenger et al., 1995; Bohlender, Franke, Stein, & Wolf, 2005; Soulis-Liparota, Cooper, Papazoglou, Clarke, & Jerums, 1991; Bouma et al., 2003). Studies with murine models have shown that AGE-injected animals have prominent renal damage due to thickening of GBM and expansion of mesangial matrix, further confirming the detrimental role of AGEs in the development of renal damage (Nogueira, Pires, & Oliveira, 2017). Normal rodents when infused with AGEs showed widening of basement membrane, mesangial matrix expansion, and glomerulosclerosis, possibly due to increased TGF-β and/or reduced nitric oxide (NO) in the kidneys (Vlassara et al., 1994). The proximal tubules in the kidney are characterized by a good number of lysosomes and mitochondria required for protein absorption or degradation, as most of the glomerular filtrate is reabsorbed here through endocytosis. This endosomal activity of lysosomes in the tubular cells is at large felicitated by megalin, an endocytic receptor. The uptake of AGEs by the reabsorption through tubular cells leads to glucotoxicity and proximal dysfunction of the tubular lysosomes, further developing DN. This AGE-RAGE pathway results in a positive feedback loop in the proximal renal tubules, further increasing the tubular injury in DN patients with severe hyperglycaemia (Liu et al., 2022).

Further clarity on the role of AGEs in DN is discussed in reports which suggest that the accumulation of AGEs in plasma is a measure of reduced renal function (Cooper, 2001). Diabetic patients suffering from ESRD have been reported to have deposition of double the quantity of AGEs in contrast to diabetic patients without renal disease (Makita et al., 1991). The concept that AGEs cause DN is supported by the fact that inhibition of AGEs in diabetic rodents prevents albuminuria and glomerulosclerosis (Busch, Franke, Rüster, & Wolf, 2010). Non enzymatic glycation of extracellular matrix proteins like collagen and laminin renders them less reactive towards proteoglycans thereby increasing the permeability of the membrane leading to leakiness and increased permeability to albumin (Silbiger et al., 1993). This leads to loss of albumin in the urine leading to albuminuria, which is also one of the hallmarks of DN.

AGEs induce apoptosis, expression of vascular endothelial growth factor and, also stimulate the expression of MCP-1 in mesangial cells (Yamagishi et al., 2002). AGEs lead to overexpression of TGF-β, which leads to glomerulosclerosis and tubulointerstitial fibrosis in diabetic nephropathy (Yamagishi et al., 2002; Raj, Choudhury, Welbourne, & Levi, 2000). The expression of RAGE is found to be increased in podocytes of diabetic patients with DN (Tanji et al., 2000; Suzuki et al., 2006), suggesting the involvement of the AGE–RAGE axis in DN pathophysiology. AGEs upon binding to RAGE leads to the production of ROS which further stimulates the production of TGF- β, MAPK and NF-kB pathways in mesangial and tubular cells (Brownlee, 2001; Chuang, Yu, Fang, Uribarri, & He, 2007; Ha, Hwang, Park, & Lee, 2008; Yamagishi et al., 2002; Yamagishi et al., 2003) reported that podocytes which are terminally differentiated cells in kidney undergo apoptosis mediated by AGEs during the early stage of DN.

6. Mechanisms of insulin resistance and its impact on the kidney

IR is defined as a decreased responsiveness of insulin receptor signalling to insulin, which in turn results in hyperglycemia. IR plays a very important role in the disorders associated with metabolic and hemodynamic abnormalities and is also associated with higher risk of cardiovascular and CKD. Conventionally, tissue IR is understood as an underlying impairment in the response of skeletal muscle, to the metabolic and physiological actions of insulin (Muniyappa & Sowers, 2013; Shepherd & Kahn, 1999), although other organs like pancreas, adipose tissues, liver along with kidneys and heart are also insulin-dependent and insulin-sensitive. Insulin affects blood pressure by modulating the renin-angiotensin-aldosterone (RAA) system and through its effects on the sympathetic nervous system. To understand how insulin affects renal tissue and its remodelling, it is important to discuss the action of insulin. In a normoglycemic state, insulin binding to its receptor (a tyrosine-kinase receptor containing two extracellular α subunits and two trans-membrane β subunits) auto-phosphorylates the β subunits which further interact with IRS-1 and 2 and activate phosphoinositide-dependent kinase 1 via phosphoinositide-3 kinase (PI3K). This activates Akt which results in the physiological effects of insulin on glucose uptake and utilization. The disruption of this signalling pathway results in hyperinsulinemia and the excess insulin may trigger insulin-dependent growth pathways in other tissues, like the kidney (Bhandari et al., 2001; Coward et al., 2005; Nistala & Whaley-Connell, 2013).

Insulin can activate growth signalling pathways like redox sensitive serine kinase, other kinases like Rho kinase, JUN-NH2 terminal kinase, mammalian target of rapamycin/serine kinase 1 pathways, that along with growth regulate hypertrophy and fibrosis (Whaley-Connell & Sowers, 2017). Under IR, these pathways are alternatively regulated by the RAA system and the sympathetic nervous system (Oroszlán et al., 2010; Yamakawa, Tanaka, Kamei, Kadonosono, & Okuda, 2003). The activation of these pathways promotes the activation of kinases like serine kinases, which results in renal tissue remodelling by inducing growth and hypertrophy (Chen, Chen, Thomas, Kozma, & Harris, 2009; Nistala, & Whaley-Connell, 2013; Whaley-Connell et al., 2011; Whaley-Connell et al., 2012). Other than uptake and glucose disposal in normal insulinemia state, PI3K and Akt also upregulate the expression of endothelial NO synthase, that results in increasing the bioavailability of NO in the vascular endothelium (Aziz & Chaudhary, 2016; Jia, Aroor, & Sowers, 2014). Therefore, IR also leads to impaired NO dependent vascular activities, and induces tissue inflammation and fibrosis, along with concurrent reduced glucose disposal. The above findings suggest that IR regulates renal metabolic and growth pathways and influences renal plasma flow, glomerular filtration, inflammation, and fibrosis (Whaley-Connell & Sowers, 2017). One of the earlier signs observed in DN is albuminuria, that is, excessive urinary protein secretion can predict the extent and progression of the renal damage. Albuminuria results from altered permeability of the filtration barrier formed by glomerular epithelial cells and podocytes at the GBM. This renal architecture is maintained by various cytoskeleton proteins and

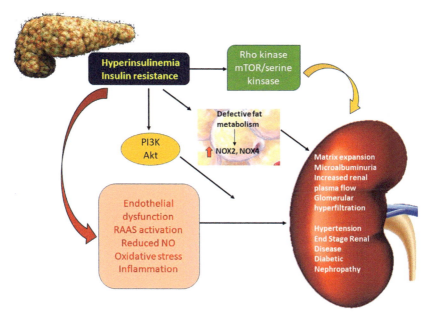

Fig. 4 The role of insulin resistance and the defective fat metabolism in the regulation of kidney functions.

other signalling proteins, thus maintaining is permeability for proper renal functions. Many in vivo and studies with human subjects have shown that diabetes associated metabolic changes alter the ultrastructure of the kidney, resulting in thickening of GBM, expansion of the mesangium, podocyte detachment, reduction in endothelial cell glycocalyx, resulting in the sclerosis of the glomerulus, that correlates with increased albuminuria (Gnudi, Coward, & Long, 2016). IR interestingly associates with the development of microalbuminuria in diabetic patients with T1DM and T2DM. IR also contributes to increased sensitivity to salt, closely associated with increased blood pressure, and albuminuria resulting in declining renal function (Vedovato et al., 2004). Fig. 4 describes how IR through its various roles impacts the kidneys, which may contribute to ESRD.

Various cells present in the kidneys are insulin sensitive with expression of functional insulin receptors, and IR in different compartments can result in different renal phenotypes. For instance, the deletion of podocyte specific insulin receptor in the glomerulus in knockout mice, developed several hallmarks of DN including albuminuria, mesangial matrix expansion, thickening of the GBM and glomerulosclerosis, in normoglycemic conditions, suggesting the role of IR independent of hyperglycaemia (Welsh et al., 2010). Deletion of the insulin receptor in tubules resulted in gluconeogenesis (Tiwari et al., 2013), while deletion in the collecting ducts resulted in lower blood pressure (Li et al., 2013), and while the deletion in the tubular epithelial cells resulted in hypertension (Tiwari et al., 2013). Of late, the role of epigenetic modulation in IR has been documented, in which fatty acid induced IR was associated with increased H3K36me2 and reduced H3K27me3 in the FOXO1 promoter region, promoting gluconeogenesis in the human urinary podocyte cell line (Kumar, Pamulapati, & Tikoo, 2016).

7. Inhibition of the AGE-RAGE axis: Treatments for diabetic nephropathy

The key aim in inhibiting the AGE-RAGE axis is strict control of hyperglycemia which would prevent the formation of AGEs and therefore would inhibit the induction of the AGE-RAGE axis. However, when such dietary intervention sometimes get difficult or ineffective, pharmacological interventions are used as an alternative strategy to reduce AGEs and inhibit RAGE. The body has endogenous mechanisms like α-ketogluteraldehyde dehydrogenase, glyoxalase, and scavengers to inhibit glycation and formation of AGEs, but hyperglycemia and activation of RAGE are reported to downregulate the production and the activity of glyoxalase I. Therefore, an urge to develop and adopt other pharmacological approaches becomes the need of the hour.

Several approaches have been explored.

(1) Antioxidants: They can prevent the generation of the free radicals due to auto-oxidation of glucose and formation of AGEs. Benfotiamine is synthesized by S-acyl derivation of thiamine and has been shown to have anti-oxidant and anti-AGE properties (Stirban et al., 2006).

(2) Amadorins: This class of Amadorins are the molecules that can prevent the conversion of Amadori products to AGEs. Vitamin B6 was identified to be the first Amadorin, which was used in the treatment of DN (Voziyan & Hudson, 2005).

(3) Dicarbonyls Scavengers: Molecules like Aminoguanidine (±)-2-isopropylidenehydrazono-4-oxo-thiazolidin-5-yl acetanilide (OPB-9195) are examples of dicarbonyls scavengers (Miyata et al., 1999). Aminoguanidine inhibits the formation of AGEs by scavenging the

precursors of AGEs. Its two crucial functional groups called hydrazine and guanidino facilitate an irreversible interaction with dicarbonyls, which further inhibit the formation of AGEs. Aminoguanidine has been shown to be effective in DN by reducing albuminuria and renal injury (Thornalley, 2003).

(4) AGE-crosslink breakers: ALT-711 (algebrium) (Peppa et al., 2006) and N-phenylthiazolium bromide (Cooper et al., 2000) could confer protection against DN and accumulation of AGEs, however the mechanism of action of the drug is not yet fully elucidated.

(5) Anti-hyperglycemic drugs: Different classes of anti-hyperglycemic drugs like biguanides (Beisswenger, Howell, Touchette, Lal, & Szwergold, 1999), sulfonylureas (Tan et al., 2007), and dipeptidyl peptidase 4 inhibitor (Sakata et al., 2013), have been tested and shown to reduce the AGEs in clinical samples. As these drugs reduce glucose levels, they can help in reversing the glycation and the formation of AGEs.

(6) Angiotensin converting enzyme (ACE) inhibitors and angiotensin receptor blockers (ARBs): ACE inhibitors like ramipril, temocaprilat, enlaprilat, captopril and perindoprilat and ARBs like valsartan, olmesartan, candesartan can prevent the ROS by metal chelation and thereby block the formation of AGEs. ARBs are also known to inhibit the expression of RAGE via peroxisome proliferator activated receptor-Υ (Yamagishi & Takeuchi, 2004).

(7) Statins: Statin like atorvastatin has been shown to prevent AGE induced RAGE expression angiotensin converting enzyme in healthy rats (Feng et al., 2011), while pravastatin prevented AGE induced tubular damage and cell death in diabetic nephropathy (Ishibashi et al., 2012).

(8) RAGE antagonist: TTP488 and low molecular weight heparin (LMWH) have been used as RAGE antagonists. LMWH has been also shown to have preventive and therapeutic effects against albuminuria and glomerulosclerosis in a DN mouse model (Myint et al., 2006). Very recently, a search for RAGE inhibitors has been undertaken, in which a RAGE aptamer was evaluated for its effects against streptozotocin-induced diabetes in rats after treatment for 4 weeks. It could successfully abrogate diabetes induced macrophage infiltration, renal dysfunction, and AGE-RAGE oxidative stress axis (Matsui et al., 2017). A specific AGE-inhibitor called FPS-ZM1 ameliorated urinary albumin excretion and renal fibrosis in AGE-treated diabetic CD1 mice (Sharma, Tupe, Wallner, & Kanwar, 2018).

(9) Decoy receptors: soluble RAGE (sRAGE) is a decoy type receptor, used as a treatment of DN. sRAGE is just an extracellular domain of membrane bound RAGE, but it lacks the transmembrane domain of the full-length RAGE receptor. Since it has the same binding domain as full-length RAGE, it can act as a decoy receptor by clearing the AGEs and prevent its binding with RAGE. In a db/db mouse model, treatment with sRAGE has been shown to decrease GBM thickening, albuminuria and glomerulosclerosis (Wendt et al., 2003).

(10) Vitamins and Polyphenols: Vitamin C and E along with combination of N-acetylcysteine with taurine and oxerutin were shown to be potential anti-glycative agents (Odetti et al., 2003). Polyphenols such as catechin have been recently identified as AGE inhibitors. In a study by Zhu et al. (2014) Catechin inhibited AGE formation and reduced the MG trapping along with expression of pro-inflammatory cytokines. A patented Chinese drug, Jiangtang decoction drug reduces accumulation of AGE and RAGE, increases phosphorylation of AKT and phosphoinositide 3-kinase (PI3K), thereby ameliorating DN by reducing the renal proinflammatory markers (Hong et al., 2017). Other class of molecules like Mangiferin, a xanthonoid improved DN in streptozotocin treated rats by decreasing the formation of AGEs, reducing oxidative stress induced MAPK and PKC signaling and inhibiting apoptotic markers (Pal, Sinha, & Sil, 2014). Resveratrol was recently found to reduce the formation of AGEs by trapping MG and forming resveratrol-MGO adducts (Shen et al., 2017). Of late, chrysin, present in herbs and bee propolis, was proven to be inhibiting AGEs induced renal fibrosis in mesangial cells in db/db mice by inhibiting the accumulation of matrix proteins in AGEs treated diabetic glomeruli (Lee et al., 2018). Recently, swertiamarin was shown to have prevented the progression of DN, in in vitro and in vivo models by preventing the RAGE/ MAPK/ TGF- β pathway. Through in silico analysis, it was also shown to block the binding of AGEs to their receptor RAGE, thereby acting as a plausible RAGE antagonist (Parwani et al., 2021).

(11) Glyoxalase 1 and Nrf-2 activators: The glyoxalase is the endogenously present AGE-detoxification enzyme which detoxifies dicarbonyls, and nrf- 2 plays a very important role in combating the oxidative stress generated by dicarbonyls and free radicals. Hence compounds that can enhance the expression of glyoxalase 1 and nrf-2 could be beneficial in preventing RAGE activation. A formulation of trans-resveratrol and

Table 2 Clinical evidences of AGE-RAGE inhibition.

Drug	Mechanism of action	Results	References
Benfotiamine	Inhibits the formation of endogenously formed AGEs	41 patients with T2DM, but without any secondary complications when administered either 900 mg/day of Benfotiamine showed significant decreased CML levels as compared to those on placebo.	Contreras, Guzman-Rosiles, Del Castillo, Gomez-Ojeda, and Garay-Sevilla (2017)
Alagebrium	AGE-breaking mechanism	57 healthy subjects above age 60 were randomly divided in 4 groups as follows: sedentary with placebo; sedentary with 200 mg/day Alagebrium; exercise with placebo and exercise with 200 mg/day Alagebrium. The subjects treated with Alagebrium showed improved function of left ventricle, which was further improved in subjects of exercise with 200 mg/day Alagebrium group.	Fujimoto et al. (2013)
Sevelamer carbonate	Inhibition of absorption of exogenously taken AGEs	In a randomised single-blinded trial, 117 patients with T2DM and stages 2 to 4 of DN, were given 1600 mg/day of sevelamer carbonate or 1200 mg/day of calcium carbonate thrice a day. Patients in sevelamer carbonate treatment group showed significantly reduced circulating and intracellular levels of AGEs like CML and MG along with improved anti-oxidant status.	Yubero-Serrano, Woodward, Poretsky, Vlassara, and Striker (2015)

hesperidin was shown to improve insulin sensitivity, reduce MG and MG derived AGEs and improve vascular functions by activating glyoxalase via nrf-2 dependent activation (Xue et al., 2016).
(12) Gut Microbiota and Prebiotics: Very recently for the first time, the role of *E. coli* strains to degrade CML into defined metabolites with the biogenic amine like carboxymethyl cadaverine as the main product has been identified, proving the role of gut microbiota in breaking down the AGEs (Hellwig et al., 2019). A study with a 12 weeks treatment of prebiotic inulin/oligofructose could prevent AGE-related outcomes in adults with pre-diabetes, by improving insulin sensitivity and reducing AGEs like MG and CML (Kellow, Coughlan, Savige, & Reid, 2014).

Apart from the above-mentioned inhibitors, several AGE inhibitors have been tested clinically for their efficacy, as mentioned in Table 2.

8. Conclusion and future perspectives

IR, and the hyperglycemia resulting from inefficient insulin action results in the generation of the AGEs, which upon interacting with their receptor RAGE elicit the inflammatory pathways that result in structural and functional defects in the kidney. The literature review from the in vitro and in vivo studies show that AGEs are also linked to the pathophysiology of IR. The IR also contributes to the disturbance in the kidney functions by promoting glomerular hypertrophy, albuminuria, increased tubular retention of sodium, hypertension, which together promote the loss of kidney functions and ESRD. Taking together the above findings, glycemic control, and the prevention of the formation of AGEs seems to be the best way to limit the progression of kidney diseases. Therefore, apart from the current therapeutic approaches discussed in this chapter, newer preventive and treatment strategies need to be designed along with getting a better insight into the cellular mechanisms, focusing on the physiology of the nephron, are needed to develop future targets for the treatment and prevention of DN. The role of AGEs and their receptors have been shown to play a pivotal role in development of IR, and since the incidences of obesity and IR are increasing at an alarming rate in younger individuals, improving the lifestyle along with development of preventive measures to augment the formation of AGEs is challenging and highly warranted.

References

Alicic, R. Z., Rooney, M. T., & Tuttle, K. R. (2017). Diabetic kidney disease: Challenges, progress, and possibilities. *Clinical Journal of the American Society of Nephrology: CJASN, 12*(12), 2032.

American Diabetes Association. (2014). Standards of medical care in diabetes—2014. Diabetes care, 37(Supplement_1), S14-S80.

Araki, N., Higashi, T., Mori, T., Shibayama, R., Kawabe, Y., Kodama, T., ... Horiuchi, S. (1995). Macrophage scavenger receptor mediates the endocytic uptake and degradation of advanced glycation end products of the Maillard reaction. *European Journal of Biochemistry/FEBS, 230*(2), 408–415.

Aziz, F., & Chaudhary, K. (2016). The triad of sleep apnea, hypertension, and chronic kidney disease: A spectrum of common pathology. *Cardiorenal Medicine, 7*(1), 74–82.

Beisswenger, P. J., Howell, S. K., Touchette, A. D., Lal, S., & Szwergold, B. S. (1999). Metformin reduces systemic methylglyoxal levels in type 2 diabetes. *Diabetes, 48*(1), 198–202.

Beisswenger, P. J., Makita, Z., Curphey, T. J., Moore, L. L., Jean, S., Brinck-Johnsen, T., & Vlassara, H. (1995). Formation of immunochemical advanced glycosylation end products precedes and correlates with early manifestations of renal and retinal disease in diabetes. *Diabetes, 44*(7), 824–829.

Bhandari, B. K., Feliers, D., Duraisamy, S., Stewart, J. L., Gingras, A. C., Abboud, H. E., ... Kasinath, B. S. (2001). Insulin regulation of protein translation repressor 4E-BP1, an eIF4E-binding protein, in renal epithelial cells. *Kidney International, 59*(3), 866–875.

Bohlender, J. M., Franke, S., Stein, G., & Wolf, G. (2005). Advanced glycation end products and the kidney. *American Journal of Physiology Renal Physiology, 289*(4), F645–F659.

Bongarzone, S., Savickas, V., Luzi, F., & Gee, A. D. (2017). Targeting the receptor for advanced glycation end products (RAGE): A medicinal chemistry perspective. *Journal of Medicinal Chemistry, 60*(17), 7213–7232.

Bouma, B., Kroon-Batenburg, L. M., Wu, Y. P., Brünjes, B., Posthuma, G., Kranenburg, O., ... Gebbink, M. F. (2003). Glycation induces formation of amyloid cross-β structure in albumin. *The Journal of Biological Chemistry, 278*(43), 41810–41819.

Boyd, A. C., Abdel-Wahab, Y. H., McKillop, A. M., McNulty, H., Barnett, C. R., O'Harte, F. P., & Flatt, P. R. (2000). Impaired ability of glycated insulin to regulate plasma glucose and stimulate glucose transport and metabolism in mouse abdominal muscle. *Biochimica Biophysica Acta General Subjects, 1523*(1), 128–134.

Bray, G. A., Nielsen, S. J., & Popkin, B. M. (2004). Consumption of high-fructose corn syrup in beverages may play a role in the epidemic of obesity. *The American Journal of Clinical Nutrition, 79*(4), 537–543.

Brownlee, M. (2001). Biochemistry and molecular cell biology of diabetic complications. *Nature, 414*(6865), 813–820.

Brownlee, M., Cerami, A., & Vlassara, H. (1988). Advanced glycosylation end products in tissue and the biochemical basis of diabetic complications. *The New England Journal of Medicine, 318*(20), 1315–1321.

Bunn, H. F., & Higgins, P. J. (1981). Reaction of monosaccharides with proteins: Possible evolutionary significance. *Science (New York, N. Y.), 213*(4504), 222–224.

Busch, M., Franke, S., Rüster, C., & Wolf, G. (2010). Advanced glycation end-products and the kidney. *European Journal of Clinical Investigation, 40*(8), 742–755.

Cai, W., Ramdas, M., Zhu, L., Chen, X., Striker, G. E., & Vlassara, H. (2012). Oral advanced glycation endproducts (AGEs) promote insulin resistance and diabetes by depleting the antioxidant defenses AGE receptor-1 and sirtuin 1. *Proceedings of the National Academy of Sciences of the United States of America, 109*(39), 15888–15893.

Cai, W., Torreggiani, M., Zhu, L., Chen, X., He, J. C., Striker, G. E., & Vlassara, H. (2010). AGER1 regulates endothelial cell NADPH oxidase-dependent oxidant stress via PKC-δ: Implications for vascular disease. *American Journal of Physiology Cell Physiology, 298*(3), C624–C634.

Caramori, M. L., & Rossing, P. (2022). Diabetic kidney disease. In *Endotext* [Internet]. MDText. com, Inc.

Cassese, A., Esposito, I., Fiory, F., Barbagallo, A. P., Paturzo, F., Mirra, P., ... Van Obberghen, E. (2008). In skeletal muscle advanced glycation end products (AGEs) inhibit insulin action and induce the formation of multimolecular complexes including the receptor for AGEs. *The Journal of Biological Chemistry, 283*(52), 36088–36099.

Chen, J. K., Chen, J., Thomas, G., Kozma, S. C., & Harris, R. C. (2009). S6 kinase 1 knockout inhibits uninephrectomy-or diabetes-induced renal hypertrophy. *American Journal of Physiology Renal Physiology, 297*(3), F585–F593.

Chen, J., Muntner, P., Hamm, L. L., Jones, D. W., Batuman, V., Fonseca, V., ... He, J. (2004). The metabolic syndrome and chronic kidney disease in US adults. *Annals of Internal Medicine, 140*(3), 167–174.

Chuang, P. Y., Yu, Q., Fang, W., Uribarri, J., & He, J. C. (2007). Advanced glycation end products induce podocyte apoptosis by activation of the FOXO4 transcription factor. *Kidney International, 72*(8), 965–976.

Contreras, C. L., Guzman-Rosiles, I., Del Castillo, D., Gomez-Ojeda, A., & Garay-Sevilla, M. E. (2017). Advanced glycation end products (AGEs) and sRAGE levels after benfotiamine treatment in diabetes mellitus type 2. *The FASEB Journal, 31*, 646-32.

Cooper, M. E. (2001). Interaction of metabolic and haemodynamic factors in mediating experimental diabetic nephropathy. *Diabetologia, 44*, 1957–1972.

Cooper, M. E., Thallas, V., Forbes, J., Scalbert, E., Sastra, S., Darby, I., & Soulis, T. (2000). The cross-link breaker, N-phenacylthiazolium bromide prevents vascular advanced glycation end-product accumulation. *Diabetologia, 43*, 660–664.

Coward, R. J., Welsh, G. I., Yang, J., Tasman, C., Lennon, R., Koziell, A., ... Mathieson, P. W. (2005). The human glomerular podocyte is a novel target for insulin action. *Diabetes, 54*(11), 3095–3102.

Dash, S. C., Agarwal, S. K., Panigrahi, A., Mishra, J., & Dash, D. (2018). Diabetes, hypertension and kidney disease combination "DHKD syndrome" is common in India. *Journal of the Association of Physicians of India, 66*(3), 30–33.

DeFronzo, R. A., & Alvestrand, A. (1980). Glucose intolerance in uremia: Site and mechanism. *The American Journal of Clinical Nutrition, 33*(7), 1438–1445.

Diamanti-Kandarakis, E., Katsikis, I., Piperi, C., Kandaraki, E., Piouka, A., Papavassiliou, A. G., & Panidis, D. (2008). Increased serum advanced glycation end-products is a distinct finding in lean women with polycystic ovary syndrome (PCOS). *Clinical Endocrinology, 69*(4), 634–641.

Diamanti-Kandarakis, E., Piouka, A., Livadas, S., Piperi, C., Katsikis, I., Papavassiliou, A. G., & Panidis, D. (2009). Anti-mullerian hormone is associated with advanced glycosylated end products in lean women with polycystic ovary syndrome. *European Journal of Endocrinology/European Federation of Endocrine Societies, 160*(5), 847–853.

Eidemak, I., Feldt-Rasmussen, B., Kanstrup, I. L., Nielsen, S. L., Schmitz, O., & Strandgaard, S. (1995). Insulin resistance and hyperinsulinaemia in mild to moderate progressive chronic renal failure and its association with aerobic work capacity. *Diabetologia, 38*, 565–572.

Feng, B., Xu, L., Wang, H., Yan, X., Xue, J., Liu, F., & Hu, J. F. (2011). Atorvastatin exerts its anti-atherosclerotic effects by targeting the receptor for advanced glycation end products. *Biochimica et Biophysica Acta Molecular Basis of Disease, 1812*(9), 1130–1137.

Fiory, F., Lombardi, A., Miele, C., Giudicelli, J., Beguinot, F., & Van Obberghen, E. (2011). Methylglyoxal impairs insulin signalling and insulin action on glucose-induced insulin secretion in the pancreatic beta cell line INS-1E. *Diabetologia, 54*, 2941–2952.

Förster, A., Kühne, Y., & Henle, T. O. (2005). Studies on absorption and elimination of dietary maillard reaction products. *Annals of the New York Academy of Sciences, 1043*(1), 474–481.

Fujimoto, N., Hastings, J. L., Carrick-Ranson, G., Shafer, K. M., Shibata, S., Bhella, P. S., ... Livingston, S. A. (2013). Cardiovascular effects of 1 year of alagebrium and endurance exercise training in healthy older individuals. *Circulation Heart Failure, 6*(6), 1155–1164.

Gabbay, K. H. (1973). The sorbitol pathway and the complications of diabetes. *The New England Journal of Medicine, 288*(16), 831–836.

Gao, Y., Bielohuby, M., Fleming, T., Grabner, G. F., Foppen, E., Bernhard, W., ... García-Cáceres, C. (2017). Dietary sugars, not lipids, drive hypothalamic inflammation. *Molecular Metabolism, 6*(8), 897–908.

Gnudi, L., Coward, R. J., & Long, D. A. (2016). Diabetic nephropathy: Perspective on novel molecular mechanisms. *Trends in Endocrinology and Metabolism: TEM, 27*(11), 820–830.

Grandhee, S. K., & Monnier, V. M. (1991). Mechanism of formation of the Maillard protein cross-link pentosidine. Glucose, fructose, and ascorbate as pentosidine precursors. *Journal of Biological Chemistry, 266*(18), 11649–11653.

Grimm, S., Ott, C., Hörlacher, M., Weber, D., Höhn, A., & Grune, T. (2012). Advanced-glycation-end-product-induced formation of immunoproteasomes: involvement of RAGE and Jak2/STAT1. *Biochemical Journal, 448*(1), 127–139.

Guo, Q., Mori, T., Jiang, Y., Hu, C., Osaki, Y., Yoneki, Y., ... Nakayama, M. (2009). Methylglyoxal contributes to the development of insulin resistance and salt sensitivity in sprague dawley rats. *Journal of Hypertension, 27*(8), 1664–1671.

Gursinsky, T., Ruhs, S., Friess, U., Diabaté, S., Krug, H. F., Silber, R. E., & Simm, A. (2006). Air pollution-associated fly ash particles induce fibrotic mechanisms in primary fibroblasts. *Biological Chemistry, 387*, 1411–1420.

Ha, H., Hwang, I. A., Park, J. H., & Lee, H. B. (2008). Role of reactive oxygen species in the pathogenesis of diabetic nephropathy. *Diabetes Research and Clinical Practice, 82*, S42–S45.

Haraguchi, R., Kohara, Y., Matsubayashi, K., Kitazawa, R., & Kitazawa, S. (2020). New insights into the pathogenesis of diabetic nephropathy: Proximal renal tubules are primary target of oxidative stress in diabetic kidney. *Acta Histochemica et Cytochemica, 53*(2), 21–31.

Hellwig, M., Auerbach, C., Müller, N., Samuel, P., Kammann, S., Beer, F., ... Henle, T. (2019). Metabolization of the advanced glycation end product N-ε-carboxymethyllysine (CML) by different probiotic E. coli strains. *Journal of Agricultural and Food Chemistry, 67*(7), 1963–1972.

Henning, C., Liehr, K., Girndt, M., Ulrich, C., & Glomb, M. A. (2014). Extending the spectrum of α-dicarbonyl compounds in vivo. *The Journal of Biological Chemistry, 289*(41), 28676–28688.

Hirose, A., Tanikawa, T., Mori, H., Okada, Y., & Tanaka, Y. (2010). Advanced glycation end products increase endothelial permeability through the RAGE/Rho signaling pathway. *FEBS Letters, 584*(1), 61–66.

Hofmann, M. A., Drury, S., Fu, C., Qu, W., Taguchi, A., Lu, Y., ... Neurath, M. F. (1999). RAGE mediates a novel proinflammatory axis: A central cell surface receptor for S100/calgranulin polypeptides. *Cell, 97*(7), 889–901.

Hong, J. N., Li, W. W., Wang, L. L., Guo, H., Jiang, Y., Gao, Y. J., Tu, P. F., & Wang, X. M. (2017). Jiangtang decoction ameliorate diabetic nephropathy through the regulation of PI3K/Akt-mediated NF-κB pathways in KK-Ay mice. *Chin Med, 12*, 1–16.

Horie, K., Miyata, T., Maeda, K., Miyata, S., Sugiyama, S., Sakai, H., ... Kurokawa, K. (1997). Immunohistochemical colocalization of glycoxidation products and lipid peroxidation products in diabetic renal glomerular lesions. Implication for glycoxidative stress in the pathogenesis of diabetic nephropathy. *The Journal of Clinical Investigation, 100*(12), 2995–3004.

Hunter, S. J., Boyd, A. C., O'Harte, F. P., McKillop, A. M., Wiggam, M. I., Mooney, M. H., ... Sheridan, B. (2003). Demonstration of glycated insulin in human diabetic plasma and decreased biological activity assessed by euglycemic-hyperinsulinemic clamp technique in humans. *Diabetes, 52*(2), 492–498.

Hussain, S., Habib, A., & Najmi, A. K. (2019). Limited knowledge of chronic kidney disease among type 2 diabetes mellitus patients in India. *International Journal of Environmental Research and Public Health, 16*(8), 1443.

Hussain, S., Jamali, M. C., Habib, A., Hussain, M. S., Akhtar, M., & Najmi, A. K. (2021). Diabetic kidney disease: An overview of prevalence, risk factors, and biomarkers. *Clinical Epidemiology and Global Health, 9*, 2–6.

Iacobini, C., Amadio, L., Oddi, G., Ricci, C., Barsotti, P., Missori, S., ... Pugliese, G. (2003). Role of galectin-3 in diabetic nephropathy. *Journal of the American Society of Nephrology: JASN, 14*(Suppl. 3), S264–S270.

Ishibashi, Y., Yamagishi, S. I., Matsui, T., Ohta, K., Tanoue, R., Takeuchi, M., ... Okuda, S. (2012). Pravastatin inhibits advanced glycation end products (AGEs)-induced proximal tubular cell apoptosis and injury by reducing receptor for AGEs (RAGE) level. *Metabolism: Clinical and Experimental, 61*(8), 1067–1072.

Jia, G., Aroor, A. R., & Sowers, J. R. (2014). Arterial stiffness: A nexus between cardiac and renal disease. *Cardiorenal Medicine, 4*(1), 60–71.

Jia, X., Olson, D. J., Ross, A. R., Wu, L., Jia, X., Olson, D. J., ... Wu, L. (2006). Structural and functional changes in human insulin induced by methylglyoxal. *The FASEB Journal, 20*(9), 1555–1557.

Kaifu, K., Ueda, S., Nakamura, N., Matsui, T., Yamada-Obara, N., Ando, R., ... Fukami, K. (2018). Advanced glycation end products evoke inflammatory reactions in proximal tubular cells via autocrine production of dipeptidyl peptidase-4. *Microvascular Research, 120*, 90–93.

Kellow, N. J., Coughlan, M. T., Savige, G. S., & Reid, C. M. (2014). Effect of dietary prebiotic supplementation on advanced glycation, insulin resistance and inflammatory biomarkers in adults with pre-diabetes: A study protocol for a double-blind placebo-controlled randomised crossover clinical trial. *BMC Endocrine Disorders, 14*(1), 1–12.

Kooptiwut, S., Kebede, M., Zraika, S., Visinoni, S., Aston-Mourney, K., Favaloro, J., ... Dunlop, M. (2005). High glucose-induced impairment in insulin secretion is associated with reduction in islet glucokinase in a mouse model of susceptibility to islet dysfunction. *Journal of Molecular Endocrinology, 35*(1), 39–48.

Koschinsky, T., He, C. J., Mitsuhashi, T., Bucala, R., Liu, C., Buenting, C., ... Vlassara, H. (1997). Orally absorbed reactive glycation products (glycotoxins): An environmental risk factor in diabetic nephropathy. *Proceedings of the National Academy of Sciences of the United States of America, 94*(12), 6474–6479.

Kumar Pasupulati, A., Chitra, P. S., & Reddy, G. B. (2016). Advanced glycation end products mediated cellular and molecular events in the pathology of diabetic nephropathy. *Biomolecular Concepts, 7*(5-6), 293–309.

Kumar, S., Pamulapati, H., & Tikoo, K. (2016). Fatty acid induced metabolic memory involves alterations in renal histone H3K36me2 and H3K27me3. *Molecular and Cellular Endocrinology, 422*, 233–242.

Lai, K. N., Leung, J. C. K., Chan, L. Y. Y., Li, F. F. K., Tang, S. C. W., Lam, M. F., ... Vlassara, H. (2004). Differential expression of receptors for advanced glycation end-products in peritoneal mesothelial cells exposed to glucose degradation products. *Clinical and Experimental Immunology, 138*(3), 466–475.

Lee, E. J., Kang, M. K., Kim, D. Y., Kim, Y. H., Oh, H., & Kang, Y. H. (2018). Chrysin inhibits advanced glycation end products-induced kidney fibrosis in renal mesangial cells and diabetic kidneys. *Nutrients, 10*(7), 882.

Li, J. H., Wang, W., Huang, X. R., Oldfield, M., Schmidt, A. M., Cooper, M. E., & Lan, H. Y. (2004). Advanced glycation end products induce tubular epithelial-myofibroblast transition through the RAGE-ERK1/2 MAP kinase signaling pathway. *The American Journal of Pathology, 164*(4), 1389–1397.

Li, L., Garikepati, R. M., Tsukerman, S., Kohan, D., Wade, J. B., Tiwari, S., & Ecelbarger, C. M. (2013). Reduced ENaC activity and blood pressure in mice with genetic knockout of the insulin receptor in the renal collecting duct. *American Journal of Physiology Renal Physiology, 304*(3), F279–F288.

Li, M., Zeng, M., He, Z., Zheng, Z., Qin, F., Tao, G., ... Chen, J. (2015). Effects of long-term exposure to free N ε-(carboxymethyl) lysine on rats fed a high-fat diet. *Journal of Agricultural and Food Chemistry, 63*(51), 10995–11001.

Litwinoff, E. M. S., Hurtado Del Pozo, C., Ramasamy, R., & Schmidt, A. M. (2015). Emerging targets for therapeutic development in diabetes and its complications: The RAGE signaling pathway. *Clinical Pharmacology and Therapeutics, 98*(2), 135–144.

Liu, B., Sun, T., Li, H., Qiu, S., Li, Y., & Zhang, D. (2022). Proximal tubular RAGE mediated the renal fibrosis in UUO model mice via upregulation of autophagy. *Cell Death & Disease, 13*(4), 399.

Lu, C., He, J. C., Cai, W., Liu, H., Zhu, L., & Vlassara, H. (2004). Advanced glycation endproduct (AGE) receptor 1 is a negative regulator of the inflammatory response to AGE in mesangial cells. *Proceedings of the National Academy of Sciences of the United States of America, 101*(32), 11767–11772.

Lue, L. F., Walker, D. G., Jacobson, S., & Sabbagh, M. (2009). Receptor for advanced glycation end products: Its role in Alzheimer's disease and other neurological diseases. *Future Neurology, 4*(2), 167–177.

Lund, M. N., & Ray, C. A. (2017). Control of Maillard reactions in foods: Strategies and chemical mechanisms. *Journal of Agricultural and Food Chemistry, 65*(23), 4537–4552.

Maillard, L. C. (1912). Action of amino acids on sugars. Formation of melanoidins in a methodical way. *Compte-Rendu de l'Academie des Sciences, 154*, 66–68.

Maillard-Lefebvre, H., Boulanger, E., Daroux, M., Gaxatte, C., Hudson, B. I., & Lambert, M. (2009). Soluble receptor for advanced glycation end products: A new biomarker in diagnosis and prognosis of chronic inflammatory diseases. *Rheumatology, 48*(10), 1190–1196.

Makita, Z., Radoff, S., Rayfield, E. J., Yang, Z., Skolnik, E., Delaney, V., ... Vlassara, H. (1991). Advanced glycosylation end products in patients with diabetic nephropathy. *The New England Journal of Medicine, 325*(12), 836–842.

Matsui, T., Higashimoto, Y., Nishino, Y., Nakamura, N., Fukami, K., & Yamagishi, S. I. (2017). RAGE-aptamer blocks the development and progression of experimental diabetic nephropathy. *Diabetes, 66*(6), 1683–1695.

McCarty, M. F. (2005). The low-AGE content of low-fat vegan diets could benefit diabetics–though concurrent taurine supplementation may be needed to minimize endogenous AGE production. *Medical Hypotheses, 64*(2), 394–398.

Miele, C., Riboulet, A., Maitan, M. A., Romano, C., Formisano, P., Giudicelli, J., ... Van Obberghen, E. (2003). Human glycated albumin affects glucose metabolism in L6 skeletal muscle cells by impairing insulin-induced insulin receptor substrate (IRS) signaling through a protein kinase Cα-mediated mechanism. *The Journal of Biological Chemistry, 278*(48), 47376–47387.

Miyata, T., Ishikawa, S., Asahi, K., Inagi, R., Suzuki, D., Horie, K., ... Kurokawa, K. (1999). 2-Isopropylidenehydrazono-4-oxo-thiazolidin-5-ylacetanilide (OPB-9195) treatment inhibits the development of intimal thickening after balloon injury of rat carotid artery: Role of glycoxidation and lipoxidation reactions in vascular tissue damage. *FEBS Letters, 445*(1), 202–206.

Miyata, T., Sugiyama, S., Suzuki, D., Inagi, R., & Kurokawa, K. (1999). Increased carbonyl modification by lipids and carbohydrates in diabetic nephropathy. *Kidney International, 56*, S54–S56.

Muniyappa, R., & Sowers, J. R. (2013). Role of insulin resistance in endothelial dysfunction. *Reviews in Endocrine & Metabolic Disorders, 14*, 5–12.

Myint, K. M., Yamamoto, Y., Doi, T., Kato, I., Harashima, A., Yonekura, H., ... Hashimoto, N. (2006). RAGE control of diabetic nephropathy in a mouse model: Effects of RAGE gene disruption and administration of low–molecular weight heparin. *Diabetes, 55*(9), 2510–2522.

Naitoh, T., Kitahara, M., & Tsuruzoe, N. (2001). Tumor necrosis factor-α is induced through phorbol ester-and glycated human albumin-dependent pathway in THP-1 cells. *Cellular Signalling, 13*(5), 331–334.

Nistala, R., & Whaley-Connell, A. (2013). Resistance to insulin and kidney disease in the cardiorenal metabolic syndrome; role for angiotensin II. *Molecular and Cellular Endocrinology, 378*(1-2), 53–58.

Nogueira, A., Pires, M. J., & Oliveira, P. A. (2017). Pathophysiological mechanisms of renal fibrosis: A review of animal models and therapeutic strategies. *In Vivo (Athens, Greece), 31*(1), 1–22.

Nowotny, K., Jung, T., Höhn, A., Weber, D., & Grune, T. (2015). Advanced glycation end products and oxidative stress in type 2 diabetes mellitus. *Biomolecules, 5*(1), 194–222.

Oberg, B. P., McMenamin, E., Lucas, F. L., McMonagle, E., Morrow, J., Ikizler, T. A. L. P., & Himmelfarb, J. (2004). Increased prevalence of oxidant stress and inflammation in patients with moderate to severe chronic kidney disease. *Kidney International, 65*(3), 1009–1016.

Odetti, P., Pesce, C., Traverso, N., Menini, S., Maineri, E. P., Cosso, L., Valentini, S., Patriarca, S., Cottalasso, D., Marinari, U. M., & Pronzato, M. A. (2003). Comparative trial of N-acetyl-cysteine, taurine, and oxerutin on skin and kidney damage in long-term experimental diabetes. *Diabetes, 52*(2), 499–505.

Ohgami, N., Nagai, R., Ikemoto, M., Arai, H., Kuniyasu, A., Horiuchi, S., & Nakayama, H. (2001). Cd36, a member of the class b scavenger receptor family, as a receptor for advanced glycation end products. *The Journal of Biological Chemistry, 276*(5), 3195–3202.

Oliveira, L. M., Lages, A., Gomes, R. A., Neves, H., Família, C., Coelho, A. V., & Quintas, A. (2011). Insulin glycation by methylglyoxal results in native-like aggregation and inhibition of fibril formation. *BMC Biochemistry, 12*, 1–13.

Oroszlán, M., Bieri, M., Ligeti, N., Farkas, A., Meier, B., Marti, H. P., & Mohacsi, P. (2010). Sirolimus and everolimus reduce albumin endocytosis in proximal tubule cells via an angiotensin II-dependent pathway. *Transplant Immunology, 23*(3), 125–132.

Ott, C., Jacobs, K., Haucke, E., Santos, A. N., Grune, T., & Simm, A. (2014). Role of advanced glycation end products in cellular signaling. *Redox Biology, 2*, 411–429.

Pal, P. B., Sinha, K., & Sil, P. C. (2014). Mangiferin attenuates diabetic nephropathy by inhibiting oxidative stress mediated signaling cascade, TNFα related and mitochondrial dependent apoptotic pathways in streptozotocin-induced diabetic rats. *PLoS One, 9*(9), e107220.

Parwani, K., Patel, F., Bhagwat, P., Dilip, H., Patel, D., Thiruvenkatam, V., & Mandal, P. (2021). Swertiamarin mitigates nephropathy in high-fat diet/streptozotocin-induced diabetic rats by inhibiting the formation of advanced glycation end products. *Archives of Physiology and Biochemistry*, 1–19.

Peppa, M., Brem, H., Cai, W., Zhang, J. G., Basgen, J., Li, Z., ... Uribarri, J. (2006). Prevention and reversal of diabetic nephropathy in db/db mice treated with alagebrium (ALT-711). *American Journal of Nephrology, 26*(5), 430–436.

Raj, D. S., Choudhury, D., Welbourne, T. C., & Levi, M. (2000). Advanced glycation end products: A Nephrologist's perspective. *American Journal of Kidney Diseases: The Official Journal of the National Kidney Foundation, 35*(3), 365–380.

Ramasamy, R., Yan, S. F., & Schmidt, A. M. (2011). Receptor for AGE (RAGE): Signaling mechanisms in the pathogenesis of diabetes and its complications. *Annals of the New York Academy of Sciences, 1243*(1), 88–102.

Reiniger, N., Lau, K., McCalla, D., Eby, B., Cheng, B., Lu, Y., ... Rosario, R. (2010). Deletion of the receptor for advanced glycation end products reduces glomerulosclerosis and preserves renal function in the diabetic OVE26 mouse. *Diabetes, 59*(8), 2043–2054.

Retnakaran, R., Cull, C. A., Thorne, K. I., Adler, A. I., & Holman, R. R. UKPDS Study Group. (2006). Risk factors for renal dysfunction in type 2 diabetes: UK Prospective Diabetes Study 74. *Diabetes, 55*(6), 1832–1839.

Ribau, J. C. O., Hadcock, S. J., Teoh, K., DeReske, M., & Richardson, M. (1999). Endothelial adhesion molecule expression is enhanced in the aorta and internal mammary artery of diabetic patients. *The Journal of Surgical Research, 85*(2), 225–233.

Rudijanto, A. (2007). The expression and down stream effect of lectin like-oxidized low density lipoprotein 1 (LOX-1) in hyperglycemic state. *Acta Medica Indonesiana, 39*(1), 36–43.

Sakata, K., Hayakawa, M., Yano, Y., Tamaki, N., Yokota, N., Eto, T., ... Sagara, S. (2013). Efficacy of alogliptin, a dipeptidyl peptidase-4 inhibitor, on glucose parameters, the activity of the advanced glycation end product (AGE)–receptor for AGE (RAGE) axis and albuminuria in Japanese type 2 diabetes. *Diabetes/Metabolism Research and Reviews, 29*(8), 624–630.

Sarkar, P., Kar, K., Mondal, M. C., Chakraborty, I., & Kar, M. (2010). Elevated level of carbonyl compounds correlates with insulin resistance in type 2 diabetes. *Annals of the Academy of Medicine of Singapore, 39*(12), 909.

Scheijen, J. L., Clevers, E., Engelen, L., Dagnelie, P. C., Brouns, F., Stehouwer, C. D., & Schalkwijk, C. G. (2016). Analysis of advanced glycation endproducts in selected food items by ultra-performance liquid chromatography tandem mass spectrometry: Presentation of a dietary AGE database. *Food Chemistry, 190*, 1145–1150.

Scheijen, J. L., Hanssen, N. M., Van Greevenbroek, M. M., Van der Kallen, C. J., Feskens, E. J., Stehouwer, C. D., & Schalkwijk, C. G. (2018). Dietary intake of advanced glycation endproducts is associated with higher levels of advanced glycation endproducts in plasma and urine: The CODAM study. *Clinical Nutrition (Edinburgh), 37*(3), 919–925.

Schleicher, E. D., Wagner, E., & Nerlich, A. G. (1997). Increased accumulation of the glycoxidation product N (epsilon)-(carboxymethyl) lysine in human tissues in diabetes and aging. *The Journal of Clinical Investigation, 99*(3), 457–468.

Schmidt, A. M., Clynes, R., Moser, B., Yan, S. F., Ramasamy, R., & Herold, K. (2007). Receptor for AGE (RAGE): Weaving tangled webs within the inflammatory response. *Current Molecular Medicine, 7*(8), 743–751.

Sharma, I., Tupe, R. S., Wallner, A. K., & Kanwar, Y. S. (2018). Contribution of myo-inositol oxygenase in AGE: RAGE-mediated renal tubulointerstitial injury in the context of diabetic nephropathy. *American Journal of Physiology Renal Physiology, 314*(1), F107–F121.

Shen, Y., Xu, Z., & Sheng, Z. (2017). Ability of resveratrol to inhibit advanced glycation end product formation and carbohydrate-hydrolyzing enzyme activity, and to conjugate methylglyoxal. *Food Chem, 216*, 153–160.

Shepherd, P. R., & Kahn, B. B. (1999). Glucose transporters and insulin action—Implications for insulin resistance and diabetes mellitus. *The New England Journal of Medicine, 341*(4), 248–257.

Silbiger, S., Crowley, S., Shan, Z., Brownlee, M., Satriano, J., & Schlondorff, D. (1993). Nonenzymatic glycation of mesangial matrix and prolonged exposure of mesangial matrix to elevated glucose reduces collagen synthesis and proteoglycan charge. *Kidney International, 43*(4), 853–864.

Singh, V. P., Bali, A., Singh, N., & Jaggi, A. S. (2014). Advanced glycation end products and diabetic complications. *Korean Journal of Physiology & Pharmacology, 18*(1), 1–14.

Soltani, Z., Washco, V., Morse, S., & Reisin, E. (2015). The impacts of obesity on the cardiovascular and renal systems: Cascade of events and therapeutic approaches. *Current Hypertension Reports, 17*(2), 7.

Song, F., & Schmidt, A. M. (2012). Glycation and insulin resistance: Novel mechanisms and unique targets? *Arteriosclerosis, Thrombosis, and Vascular Biology, 32*(8), 1760–1765.

Soulis-Liparota, T., Cooper, M., Papazoglou, D., Clarke, B., & Jerums, G. (1991). Retardation by aminoguanidine of development of albuminuria, mesangial expansion, and tissue fluorescence in streptozocin-induced diabetic rat. *Diabetes, 40*(10), 1328–1334.

Stirban, A., Negrean, M., Stratmann, B., Gawlowski, T., Horstmann, T., Gotting, C. H., ... Vlassara, H. (2006). Benfotiamine prevents macro-and microvascular endothelial dysfunction and oxidative stress following a meal rich in advanced glycation end products in individuals with type 2 diabetes. *Diabetes Care, 29*(9), 2064–2071.

Stitt, A. W., Bucala, R., & Vlassara, H. (1997). Atherogenesis and advanced glycation: Promotion, progression, and prevention. *Annals of the New York Academy of Sciences, 811*(1), 115–129.

Stitt, A. W., Burke, G. A., Chen, F., McMullen, C. B. T., & Vlassara, H. (2000). Advanced glycation end product receptor interactions on microvascular cells occur within caveolin-rich membrane domains. *The FASEB Journal, 14*(15), 2390–2392.

Sun, H., Chen, J., Hua, Y., Zhang, Y., & Liu, Z. (2022). New insights into the role of empagliflozin on diabetic renal tubular lipid accumulation. *Diabetology & Metabolic Syndrome, 14*(1), 1–12.

Suzuki, D., Toyoda, M., Yamamoto, N., Miyauchi, M., Katoh, M., Kimura, M., ... Yagame, M. (2006). Relationship between the expression of advanced glycation end-products (AGE) and the receptor for AGE (RAGE) mRNA in diabetic nephropathy. *Internal Medicine (Tokyo, Japan), 45*(7), 435–441.

Taguchi, A., Blood, D. C., del Toro, G., Canet, A., Lee, D. C., Qu, W., ... Hofmann, M. A. (2000). Blockade of RAGE–amphoterin signalling suppresses tumour growth and metastases. *Nature, 405*(6784), 354–360.

Tahara, N., Yamagishi, S. I., Matsui, T., Takeuchi, M., Nitta, Y., Kodama, N., ... Imaizumi, T. (2012). Serum levels of advanced glycation end products (AGEs) are independent correlates of insulin resistance in nondiabetic subjects. *Cardiovascular Therapeutics, 30*(1), 42–48.

Tan, K. C., Shiu, S. W., Wong, Y., & Tam, X. (2011). Serum advanced glycation end products (AGEs) are associated with insulin resistance. *Diabetes/Metabolism Research and Reviews, 27*(5), 488–492.

Tan, K. C. B., Chow, W. S., Tso, A. W. K., Xu, A., Tse, H. F., Hoo, R. L. C., ... Lam, K. S. L. (2007). Thiazolidinedione increases serum soluble receptor for advanced glycation end-products in type 2 diabetes. *Diabetologia, 50*, 1819–1825.

Tanikawa, T., Okada, Y., Tanikawa, R., & Tanaka, Y. (2009). Advanced glycation end products induce calcification of vascular smooth muscle cells through RAGE/p38 MAPK. *Journal of Vascular Research, 46*(6), 572–580.

Tanji, N., Markowitz, G. S., Fu, C., Kislinger, T., Taguchi, A., Pischetsrieder, M., ... D'agati, V. D. (2000). Expression of advanced glycation end products and their cellular receptor RAGE in diabetic nephropathy and nondiabetic renal disease. *Journal of the American Society of Nephrology: JASN, 11*(9), 1656–1666.

Tessier, F. J., Niquet-Léridon, C., Jacolot, P., Jouquand, C., Genin, M., Schmidt, A. M., ... Boulanger, E. (2016). Quantitative assessment of organ distribution of dietary protein-bound 13C-labeled Nε-carboxymethyllysine after a chronic oral exposure in mice. *Molecular Nutrition & Food Research, 60*(11), 2446–2456.

Thornalley, P. J. (1998). Cell activation by glycated proteins. AGE receptors, receptor recognition factors and functional classification of AGEs. *Cellular and Molecular Biology, 44*(7), 1013–1023.

Thornalley, P. J. (2003). Glyoxalase I–structure, function and a critical role in the enzymatic defence against glycation. *Biochemical Society Transactions, 31*(6), 1343–1348.

Thornalley, P. J. (2003). Use of aminoguanidine (Pimagedine) to prevent the formation of advanced glycation endproducts. *Archives of Biochemistry and Biophysics, 419*(1), 31–40.

Tiwari, S., Singh, R. S., Li, L., Tsukerman, S., Godbole, M., Pandey, G., & Ecelbarger, C. M. (2013). Deletion of the insulin receptor in the proximal tubule promotes hyperglycemia. *Journal of the American Society of Nephrology: JASN, 24*(8), 1209.

Torreggiani, M., Liu, H., Wu, J., Zheng, F., Cai, W., Striker, G., & Vlassara, H. (2009). Advanced glycation end product receptor-1 transgenic mice are resistant to inflammation, oxidative stress, and post-injury intimal hyperplasia. *The American Journal of Pathology, 175*(4), 1722–1732.

Unoki-Kubota, H., Yamagishi, S. I., Takeuchi, M., Bujo, H., & Saito, Y. (2010). Pyridoxamine, an inhibitor of advanced glycation end product (AGE) formation ameliorates insulin resistance in obese, type 2 diabetic mice. *Protein and Peptide Letters, 17*(9), 1177–1181.

Uribarri, J., Cai, W., Ramdas, M., Goodman, S., Pyzik, R., Chen, X., ... Vlassara, H. (2011). Restriction of advanced glycation end products improves insulin resistance in human type 2 diabetes: Potential role of AGER1 and SIRT1. *Diabetes Care, 34*(7), 1610–1616.

Uribarri, J., Woodruff, S., Goodman, S., Cai, W., Chen, X. U. E., Pyzik, R., ... Vlassara, H. (2010). Advanced glycation end products in foods and a practical guide to their reduction in the diet. *Journal of the American Dietetic Association, 110*(6), 911–916.

Vedovato, M., Lepore, G., Coracina, A., Dodesini, A. R., Jori, E., Tiengo, A., ... Trevisan, R. (2004). Effect of sodium intake on blood pressure and albuminuria in Type 2 diabetic patients: The role of insulin resistance. *Diabetologia, 47*, 300–303.

Vlassara, H., & Uribarri, J. (2014). Advanced glycation end products (AGE) and diabetes: Cause, effect, or both? *Current Diabetes Reports, 14*, 1–10.

Vlassara, H., Li, Y. M., Imani, F., Wojciechowicz, D., Yang, Z., Liu, F. T., & Cerami, A. (1995). Identification of galectin-3 as a high-affinity binding protein for advanced glycation end products (AGE): A new member of the AGE-receptor complex. *Molecular Medicine (Cambridge, Mass.), 1*, 634–646.

Vlassara, H., Striker, L. J., Teichberg, S., Fuh, H., Li, Y. M., & Steffes, M. (1994). Advanced glycation end products induce glomerular sclerosis and albuminuria in normal rats. *Proceedings of the National Academy of Sciences of the United States of America, 91*(24), 11704–11708.

Voziyan, P. A., & Hudson, B. G. (2005). Pyridoxamine: The many virtues of a maillard reaction inhibitor. *Annals of the New York Academy of Sciences, 1043*(1), 807–816.

Wautier, M. P., Chappey, O., Corda, S., Stern, D. M., Schmidt, A. M., & Wautier, J. L. (2001). Activation of NADPH oxidase by AGE links oxidant stress to altered gene expression via RAGE. *American Journal of Physiology. Endocrinology and Metabolism, 280*(5), E685–E694.

Welsh, G. I., Hale, L. J., Eremina, V., Jeansson, M., Maezawa, Y., Lennon, R., ... Caunt, C. J. (2010). Insulin signaling to the glomerular podocyte is critical for normal kidney function. *Cell Metabolism, 12*(4), 329–340.

Wendt, T. M., Tanji, N., Guo, J., Kislinger, T. R., Qu, W., Lu, Y., ... Stein, G. (2003). RAGE drives the development of glomerulosclerosis and implicates podocyte activation in the pathogenesis of diabetic nephropathy. *The American Journal of Pathology, 162*(4), 1123–1137.

Westwood, M. E., & Thornalley, P. J. (1995). Molecular characteristics of methylglyoxal-modified bovine and human serum albumins. Comparison with glucose-derived advanced glycation endproduct-modified serum albumins. *Journal of Protein Chemistry, 14*(5), 359–372.

Whaley-Connell, A. T., Habibi, J., Nistala, R., DeMarco, V. G., Pulakat, L., Hayden, M. R., ... Sowers, J. R. (2012). Mineralocorticoid receptor-dependent proximal tubule injury is mediated by a redox-sensitive mTOR/S6K1 pathway. *American Journal of Nephrology, 35*(1), 90–100.

Whaley-Connell, A., & Sowers, J. R. (2017). Insulin resistance in kidney disease: Is there a distinct role separate from that of diabetes or obesity. *Cardiorenal Medicine, 8*(1), 41–49.

Whaley-Connell, A., Habibi, J., Panfili, Z., Hayden, M. R., Bagree, S., Nistala, R., ... Ferrario, C. M. (2011). Angiotensin II activation of mTOR results in tubulointerstitial fibrosis through loss of N-cadherin. *American Journal of Nephrology, 34*(2), 115–125.

Xue, M., Weickert, M. O., Qureshi, S., Kandala, N. B., Anwar, A., Waldron, M., ... Rabbani, N. (2016). Improved glycemic control and vascular function in overweight and obese subjects by glyoxalase 1 inducer formulation. *Diabetes, 65*(8), 2282–2294.

Yamagata, K., Ishida, K., Sairenchi, T., Takahashi, H., Ohba, S., Shiigai, T., ... Koyama, A. (2007). Risk factors for chronic kidney disease in a community-based population: A 10-year follow-up study. *Kidney International, 71*(2), 159–166.

Yamagishi, S. I., Adachi, H., Nakamura, K., Matsui, T., Jinnouchi, Y., Takenaka, K., ... Shigeto, Y. (2006). Positive association between serum levels of advanced glycation end products and the soluble form of receptor for advanced glycation end products in nondiabetic subjects. *Metabolism: Clinical and Experimental, 55*(9), 1227–1231.

Yamagishi, S. I., Inagaki, Y., Okamoto, T., Amano, S., Koga, K., Takeuchi, M., & Makita, Z. (2002). Advanced glycation end product-induced apoptosis and overexpression of vascular endothelial growth factor and monocyte chemoattractant protein-1 in human-cultured mesangial cells. *The Journal of Biological Chemistry, 277*(23), 20309–20315.

Yamagishi, S. I., Inagaki, Y., Okamoto, T., Amano, S., Koga, K., & Takeuchi, M. (2003). Advanced glycation end products inhibit de novo protein synthesis and induce TGF-β overexpression in proximal tubular cells. *Kidney International, 63*(2), 464–473.

Yamagishi, S., & Takeuchi, M. (2004). Nifedipine inhibits gene expression of receptor for advanced glycation end products (RAGE) in endothelial cells by suppressing reactive oxygen species generation. *Drugs Under Experimental and Clinical Research, 30*(4), 169–175.

Yamakawa, T., Tanaka, S. I., Kamei, J., Kadonosono, K., & Okuda, K. (2003). Phosphatidylinositol 3-kinase in angiotensin II-induced hypertrophy of vascular smooth muscle cells. *European Journal of Pharmacology, 478*(1), 39–46.

Yki-Järvinen, H., & Koivisto, V. A. (1984). Insulin resistance in type I diabetes: Prevalence, pathogenesis and therapeutic approaches. *Annals of Clinical Research, 16*(2), 74–83.

Yubero-Serrano, E. M., Woodward, M., Poretsky, L., Vlassara, H., & Striker, G. E. (2015). Effects of sevelamer carbonate on advanced glycation end products and antioxidant/pro-oxidant status in patients with diabetic kidney disease. *Clinical Journal of the American Society of Nephrology: CJASN, 10*(5), 759.

Zhu, D., Wang, L., Zhou, Q., Yan, S., Li, Z., Sheng, J., & Zhang, W. (2014). (+)-Catechin ameliorates diabetic nephropathy by trapping methylglyoxal in type 2 diabetic mice. *Molecular Nutrition & Food Research, 58*(12), 2249–2260.

CHAPTER SIX

The Maillard reactions: Pathways, consequences, and control

Delia B. Rodriguez-Amaya* and Jaime Amaya-Farfan
School of Food Engineering, University of Campinas, Campinas, SP, Brazil
*Corresponding author. e-mail address: dramaya2012@gmail.com

Contents

1. Introduction	150
2. Stages of the Maillard reactions	150
2.1 Reactions of the initial stage	152
2.2 Reactions of the intermediate stage	154
2.3 Reactions of the final stage	164
3. Influencing factors	167
4. Consequences of the Maillard reactions	170
4.1 Impact on food quality	170
4.2 Impact on human health	170
5. Strategies for controlling the Maillard reactions	173
6. Final considerations	174
References	175

Abstract

The century old Maillard reactions continue to draw the interest of researchers in the fields of Food Science and Technology, and Health and Medical Sciences. This chapter seeks to simplify and update this highly complicated, multifaceted topic. The simple nucleophilic attack of an amine onto a carbonyl group gives rise to a series of parallel and subsequent reactions, occurring simultaneously, resulting into a vast array of low and high mass compounds. Recent research has focused on: (1) the formation and transformation of α-dicarbonyl compounds, highly reactive intermediates which are essential in the development of the desired color and flavor of foods, but also lead to the production of the detrimental advanced glycation end products (AGEs); (2) elucidation of the structures of melanoidins in different foods and their beneficial effects on human health; and (3) harmful effects of AGEs on human health. Considering that MRs have both positive and negative consequences, their control to accentuate the former and to mitigate the latter, is also being conscientiously investigated with the use of modern techniques and technology.

1. Introduction

Non-enzymatic browning occurs in foods through different routes: oxidative degradation of ascorbic acid, oxidation of phenolic compounds, polymerization of the cleavage products of anthocyanins, caramelization, and the Maillard reactions (MRs). First described in 1912 by Louis-Camille Maillard, the latter is the most investigated. Widely observed in food systems and termed MR in Food Science and Technology, this process also occurs in the human body and is known as protein glycation in Health and Medical Sciences.

The MRs occur during food processing and meal preparation (cooking, frying, grilling, and baking), contributing to food sensory characteristics (color, aroma, taste, and texture). Maillard reaction products (MRPs) are consumed everyday worldwide through foods such as dairy products, bakery products, breakfast cereals, honey, soy sauce, coffee, beer, chocolate, grilled meat, dehydrated vegetables, and processed fruits.

Although generally called Maillard reaction, this browning actually consists of an extremely complex series of reactions and interactions, occurring simultaneously and sequentially, producing a vast array of low-mass and high-mass compounds. Thus, in this chapter, this browning is referred to as the Maillard reactions (MRs).

2. Stages of the Maillard reactions

Hodges's description of the basic pathways of the MRs in 1953 was the first coherent and comprehensive organization of the many reactions already reported at the time to be part of the Maillard browning. The Hodge (1953) scheme (Fig. 1) divided the MRs into three major stages:
- The initial (or early) stage consists of sugar-amine condensation and the Amadori rearrangement. The products are colorless, without absorption in the ultraviolet region.
- The intermediate (or advanced) stage is made up of sugar dehydration, sugar fragmentation, and amino acid degradation (Strecker degradation). The products are colorless or pale yellow, with strong absorption in the ultraviolet region.
- The final stage is composed of aldol condensation, aldehyde-amine polymerization, formation of heterocyclic nitrogen compounds, and production of polymers/co-polymers (melanoidins). The products are highly colored (brown).

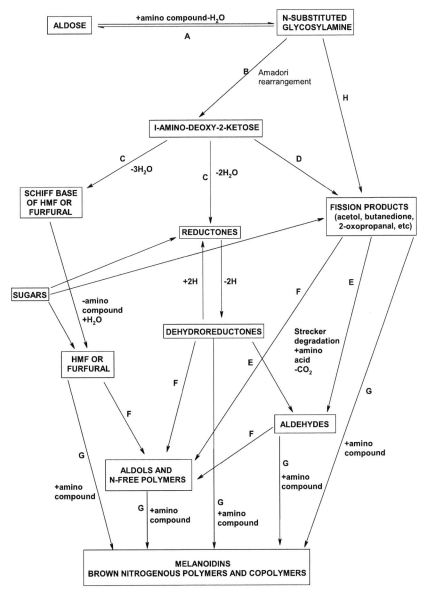

Fig. 1 The Hodge Scheme as presented by Nursten (2005). A, Sugar-amine condensation; B, Amadori rearrangement; C, sugar dehydration; D, sugar fragmentation; E, amino acid degradation; F, aldol condensation; G, aldehyde-amine condensation; H, free radical reactions. Reaction H was added to the scheme to incorporate the more recently discovered free radical breakdown of Maillard intermediates. *Reproduced with permission from Nursten, H., 2005. Maillard reaction: chemistry, biochemistry and implications. London: Royal Society of Chemistry.*

2.1 Reactions of the initial stage
2.1.1 Sugar-amine condensation
In the well-defined initial stage, the MRs commence with the nucleophilic attack of a free electron pair on the nitrogen of an amino group of an amine, amino acid, peptide or protein, on the electron deficient carbonyl group of a reducing sugar (Fig. 2A).

The ε-amino group of lysine, and to a lesser extent, the thiol, guanidyl, imidazole, and indole groups of cysteine, arginine, histidine, and tryptophan,

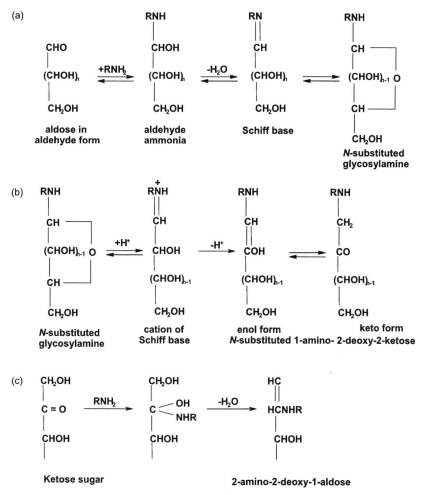

Fig. 2 Reactions of the initial stage: (a) sugar-amine condensation), (b) Amadori rearrangement, and (c) Heyns rearrangement.

respectively, are the reactive groups in amino acids. Aside from these side chains, the α-amino groups of amino acids in proteins/peptides can also participate in the reaction. Glucose and fructose are the most important reacting sugars; ribose is the major reactant in meat. Among the disaccharides, lactose of dairy products and maltose in cereal products (e.g., malt) are important browning precursors. Not being a reducing sugar, sucrose by itself is not a reactant, but it can be easily cleaved into glucose and fructose, as in heating, thereby becoming a participant. Sugars bound as glycosides (e.g., in glycoproteins, glycolipids, and heteroglycosides) are less reactive, but the aglycone can also be released during heating, forming free reducing saccharides (Sikorski, Pokorny, & Damodaran, 2008).

Elimination of water follows the amino-carbonyl reaction, forming the unstable Schiff base, which cyclizes to an N-substituted aldosylamine if the sugar is an aldose (Fig. 2A). If a ketose (e.g., fructose) is involved, a ketosylamine is formed.

2.1.2 Amadori and Heyns rearrangement

Amadori rearrangement takes place with the aldosylamine yielding 1-amino-1-deoxy-2-ketose, which is called the Amadori compound (Fig. 2B). This rearrangement occurs spontaneously even at 25 °C. A ketosylamine undergoes the Heyns rearrangement to form 2-amino-2-deoxy-aldose (Fig. 2C).

Both Amadori and Heyns rearrangements begin with protonation, which causes opening of the hemiacetal ring (Davidek & Davidek, 2004). Loss of a hydrogen proton leads, via enaminol, to an aminodeoxyketose from an aldosylamine or to an aminodeoxyaldose from a ketosylamine.

The aldosylamine can react with a second molecule of aldose and form a dialdosylamine, which rearranges to a diketosamine. For example, di-D-glucosylglycine yields di-D-fructosylglycine (Davidek & Davidek, 2004).

Through the years, Amadori compounds had been encountered in various heated and stored foods such as malt, soy sauce, roasted cocoa, dehydrated fruits, dehydrated vegetables and other vegetable products, malts and beers (e.g., Hashiba, 1978; Heinzler & Eichner, 1991; Reuter & Eichner, 1989; Sanz, del Castillo, Corzo, & Olano, 2001; Wittmann & Eichner, 1989; Yu, Zhang, & Zhang, 2018). The major Amadori compound was found to be fructosylglutamate in dried tomatoes, fructosylproline in dried apricots (Davidek, Kraehenbuehl, Devaud, Robert, & Blank, 2005), and lactulosyllysine in milk (Aalaei, Rayner, & Sjöholm, 2019).

The Schiff base can undergo several reactions without passing through the Amadori rearrangement as enumerated by Davidek and Davidek (2004):
- transamination, converting the amino acid into the corresponding α-keto acid and the sugar into an α-amino alcohol derivative (Yaylayan, 2003).
- decarboxylation, resulting in an intermediate that hydrolyzes to form a Strecker aldehyde and unsaturated amino alditol (Cremer, Vollenbroeker, & Eichner, 2000), which can be transformed to 3-hydroxy-2-methylpyridine.
- mainly under alkaline conditions, the Namiki pathway, involving the cleavage of the sugar moiety of the Schiff base to a C2 fragment, glycolaldehyde imine in the enol form, which on dimerization and oxidation forms a 1,4-dialkylpyrazinium radical cation (Hayashi & Namiki, 1980; Namiki, 1988).

2.2 Reactions of the intermediate stage

Much less defined and remarkably more complicated, the intermediate stage (now often called the advanced stage) encompasses many parallel and sequential reactions, occurring simultaneously, involving variable pathways including dehydration, cyclization, retroaldolization, enolization, oxidation, fragmentation, acid hydrolysis, isomerization, rearrangement, free radical reaction, and further condensation. Varying with existing conditions, numerous compounds with differing molecular weights and compositions are formed, including heterocyclic, carboxylic, and aliphatic compounds, such as reductones, dehydroreductones, fission products, pyrazines, furans, pyrroles, and aldols (Locas & Yaylayan, 2010). Aside from the decomposition of the Amadori and Heyns compounds, degradation of the amino acids and the sugars takes place independently.

2.2.1 Sugar dehydration

The Amadori compounds undergo enolization, which occurs in two ways (Fig. 3) (Davidek & Davidek, 2004; Nursten, 2005). At pH7 or below, a 1,2-enolization (sometimes called 3-deoxyosone pathway) occurs with the formation of 3-deoxyosones and furfural from pentoses or hydroxymethylfurfural from hexoses as main products. At pH > 7, the degradation of the Amadori compound involves mainly 2,3-enolization (the 1-deoxyosone pathway), forming 1-deoxyosones, reductones (e.g., 4-hydroxy-5-methyl-2,3-dihydrofuran-3-one) and an array of fission products (e.g., acetol, pyruvaldehyde,

The Maillard reactions

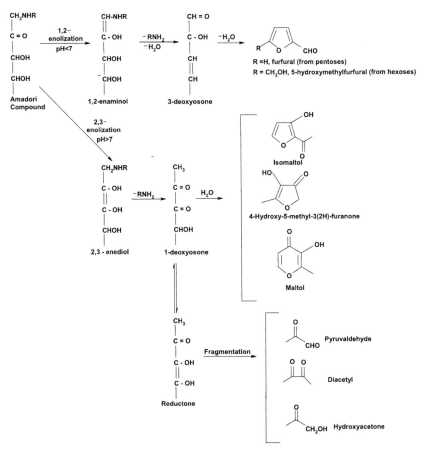

Fig. 3 Deamination, dehydration, and fragmentation of the Amadori compound. Based on Amaya-Farfan and Rodriguez-Amaya (2021), Nursten (2005), and Resconi et al. (2013).

and diacetyl), which are highly reactive and react further (Martins, Jongen, & Van Boekel, 2001). Carbonyl groups can condense with free amino groups, resulting in the incorporation of nitrogen into the reaction products.

2.2.2 Sugar fragmentation

Multiple fragmentation of the sugar moiety constitutes branch points and establishes many parallel reaction pathways, adding to the enormous complexity of the MRs. These fragmentations occur principally by retroaldolization and oxidative fission; hydrolysis, and dehydration are also involved.

Examples of sugar fragmentation products are (Nursten, 2005): glycolaldehyde, glyceraldehyde, 2-oxopropanal, acetol, dihydroxyacetone, acetoin, butanedione, ethanol, aldol, propanal, pyruvic acid, levulinic acid, saccharinic acid, lactic acid, acetic acid, formic acid, and formaldehyde. Glyoxal can be obtained by oxidation of glycolaldehyde and 2-oxopropanal, by retroaldolization of 1- and 3-deoxyglucosone, or by hydrolysis of diacetylformoin, which can also give rise to butanedione by reduction, dehydration, and hydrolysis. 2,3-Pentanedione can be produced from butanedione by aldol reaction with formaldehyde, dehydration, and reduction or by aldol condensation of hydroxyacetone and acetaldehyde, followed by dehydration.

2.2.3 Strecker degradation

Rizzi (1999) recognized three roles for the Strecker degradation in the formation of flavor compounds in processed foods:
- the major pathway for conversion of amino acids into structurally related aldehydes of significant flavor value.
- a relatively low energy route for mobilizing amino acids' nitrogen and sulfur to form ammonia, hydrogen sulfide, and many flavor-significant S/N/O-containing heterocyclic compounds.
- a reduction mechanism for conversion of dicarbonyls into acyloin, thereby opening the way for still more diverse flavor compound formation.

In this degradation, α-dicarbonyl (α-DC) or vinylogous dicarbonyl compounds react with amino acids, producing aldehydes and α-aminoketones (Fig. 4A). The amino acid is decarboxylated and deaminated to form an aldehyde (called Strecker aldehyde), containing one carbon less than the original amino acid. The transfer of the amino group to the dicarbonyl forms aminoketone, which is an important intermediate in the formation of several classes of heterocyclic compounds, including pyrazines (Fig. 4B) oxazoles, and thiozoles.

Examples of aldehydes formed from amino acids by the Strecker degradation are methional from methionine, phenylacetaldehyde from phenylalanine, methylpropanol from valine, 3-methylbutanal from leucine, and 2-methylbutanal from isoleucine. Strecker aldehydes of certain amino acids, particularly methionine, phenylalanine, and leucine, have very low odor thresholds and contribute to the aroma of thermally processed foods (Hofmann, Münch, & Schieberle, 2000). Strecker aldehydes identified in different milk products are 3-methylbutanal in pasteurized and sterilized

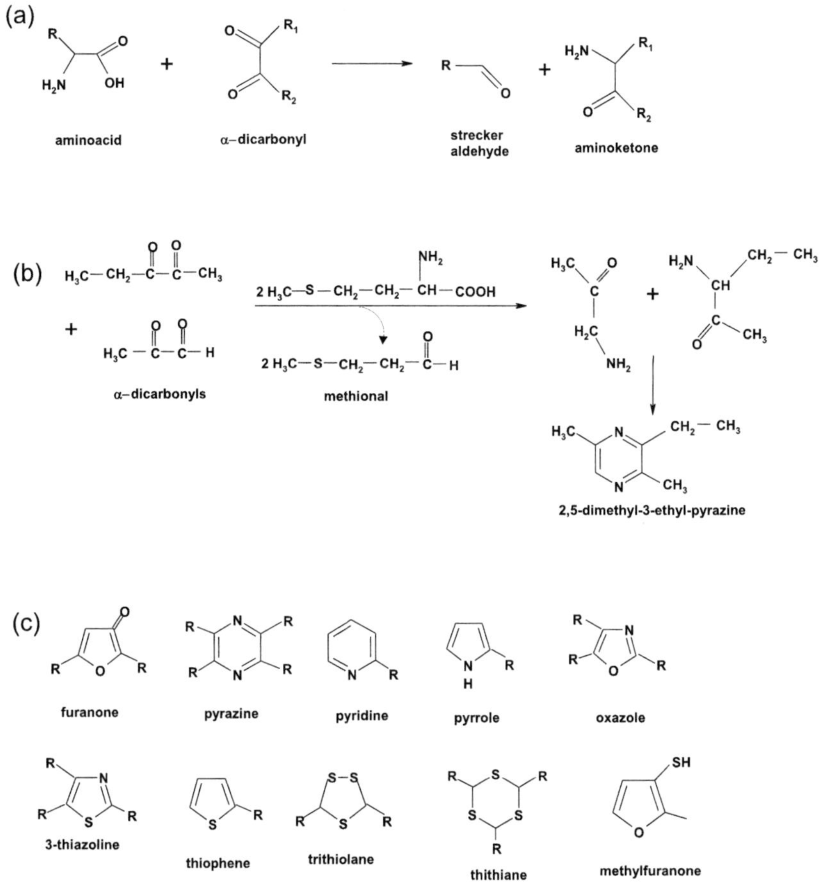

Fig. 4 (a) Strecker degradation, (b) formation of an alkyl pyrazine through the Strecker degradation, (c) heterocyclic compounds formed by the Maillard reactions. Based on Lindsay (1996), Resconi et al. (2013), and Yaylayan (2003).

milk, 2-methylbutanal and isobutanal in UHT milk, and 2-methylbutanal and isobutanal in milk powder (O'Brien, 2009).

The term "Strecker degradation" originally referred only to reactions induced by carbonyl compounds (Schönberg & Moubacher, 1952). It now refers to all types of oxidative deamination of amino acids. Rizzi (2008) cited Strecker and Strecker-like reactions between amino acids and lipid oxidation products, and terpenes or o-quinones, as potential sources of novel flavor compounds. Moreover, reactions in which Strecker aldehydes are formed directly from Amadori compounds, instead of passing through amino acid/carbonyl reactions, can occur.

The production of Strecker aldehydes and α-amino carbonyl compounds, critical intermediates in the generation of aroma compounds during the Maillard reactions, is considered a major role of the Strecker degradation (Yaylayan, 2003). However, they can also be produced independently of each other by pathways other than that of Strecker degradation: directly from Amadori rearrangement products, from amino acids through oxidative decarboxylation and thermal reactions, and through α-DC assisted oxidative decarboxylation.

2.2.4 Formation and transformation of reactive intermediates

The Amadori rearrangement and the Strecker degradation control the balance among critically important intermediates (i.e., α-DC, α-hydroxycarbonyl, 2-amino carbonyls and 2-(amino acid)-carbonyl compounds), during the MRs and therefore regulate the aromagenic versus chromogenic pathways (Yaylayan, 2003). The importance of the Strecker degradation in aromagenesis is manifested in its concentration of 2-amino carbonyl species, depletion of α-DCs, both oxidatively and non-oxidatively, and generation of Strecker aldehydes. On the other hand, the Amadori rearrangement can contribute to aromagenesis through the production of α-DCs, oxidatively and non-oxidatively, and through the formation of 2-amino carbonyls in the presence of ammonia.

Recent research has focused on the α-DCs (Fig. 5), which have drawn considerable attention because of their role in the formation of food flavor and color as well as AGEs (advanced glycation end products) (Liu et al., 2023; Starowicz & Zieliński, 2019). These compounds are highly reactive towards nucleophilic groups. It can react with terminal and ε-amino, imidazole, indolyl, mercapto, and guanidine functional groups, leading to the crosslinking of proteins and the generation of AGEs.

α-DCs are mainly produced through the MRs, but are also generated through caramelization and lipid peroxidation, or enzymatically by microorganisms in fermented foods (Yan, Wu, & Xue, 2023). In the MRs, α-DCs are formed through fragmentation of Amadori or Heyns products; they may also be produced through sugar degradation.

Amadori compounds can decompose to α-DCs of various chain lengths, such as 3-deoxyhexos-2-ulose, 1-deoxy-2,3-hexodiulose, 2-oxopropanal, butane-2,3-dione, glyoxal. Deoxyosones further give rise to carbonyl compounds, such as methylglyoxal and hydroxyacetaldehyde (Cui et al., 2021). Oxidation of hydroxyacetaldehyde results in glyoxal formation.

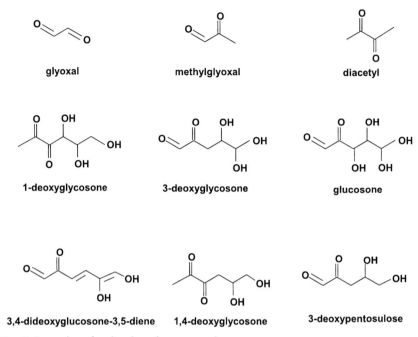

Fig. 5 Examples of α-dicarbonyl compounds.

The very reactive α-DCs then initiate a series of further reactions that finally result in the formation of volatiles and brown pigments. α-DCs such as glyoxal, methylglyoxal, and diacetyl are known precursors of heterocyclic flavor compounds (Feng, Zhou, Wang, Wang, & Xia, 2021). Moreover, the α-DCs are starting materials for polymerization reactions which lead to formation of carbohydrate-based melanoidins (Kroh, Fiedler, & Wagner, 2008).

The three major C_6-α-DCs glucosone, 1-deoxyglucosone, and 3-deoxyglucosone (3-DG), examined under Maillard conditions, formed different heterocyclic products, such as pyranones, furanones, furans, and the corresponding N-heterocycles in the presence of amino components (Haase, Kanzler, Hildebrandt, & Kroh, 2017).

It is well known that the *α*-DCs can cause the Strecker degradation of amino acids, forming aldehydes that are significant to food flavor. For example, the Strecker degradation of leucine generates 3-methylbutanal with chocolate odors. The Strecker degradation of cysteine releases H_2S, mercaptoacetaldehyde, acetaldehyde, formaldehyde, and NH_3, which react with the carbonyl compounds, especially the *α*-DCs, generating the sulfur-containing meaty flavors, such as thiols, thiophenes, and thiazoles

(Hou et al., 2017). The α-DCs participate in the formation of the heterocyclic compounds such as pyrazines (Fig. 4B), pyranones, furans, furanones, and pyrroles, and the sulfur-containing compounds such as thiols, thiophenes, and thiazoles that are important to food flavor (Hellwig & Henle, 2014; Starowicz & Zieliński, 2019).

A multitude of Maillard DCs, ranging from the intact carbon backbone down to C3 and C2 fragments, had been detected in several carbohydrate systems (e.g., in glucose, maltose, or ascorbic acid reactions). Smuda and Glomb (2013) comprehensively reviewed the five major mechanisms reported in the literature for dicarbonyl decomposition: (1) retro-aldol fragmentation, (2) hydrolytic α-DC cleavage, (3) oxidative α-DC cleavage, (4) hydrolytic β-DC cleavage, and (5) amine-induced β-DC cleavage.

The formation of colored products under Maillard conditions was shown for active methylene compounds such as 3-deoxyglucosone (Bruhns, Kanzler, Degenhardt, Koch, & Kroh, 2019), methylglyoxal (Kanzler, Wustrack, & Rohn, 2021), and norfuraneol (Kanzler & Haase, 2020). More importantly, the highly reactive α-DCs may react with lysine and arginine side chains of proteins to yield peptide-bound amino acid derivatives, the AGEs.

Maasen et al. (2021) cited studies that determined the DC concentrations in foods and drinks (e.g., honey, coffee, soft drinks, beer, wine, and soy sauce), especially in sugar-rich or fermented products. The DCs most often included are methylglyoxal, glyoxal, 3-deoxyglucosone, 3-deoxygalactosone, glucosone, diacetyl, and 3,4-dideoxyglucosone-3-ene. An extensive study quantified methylglyoxal, 3-deoxyglucosone, and 3-deoxygalactosone in 173 foods and drinks. The highest concentrations were found in balsamic vinegar, honey, and candy (Degen, Hellwig, & Henle, 2012). 3-Deoxyglucosone was the most predominant; methylglyoxal was of minor importance quantitatively in all foods studied, except for honey.

Maasen et al. (2021) generated a comprehensive dietary database of the three major DCs (methylglyoxal, glyoxal, and 3-deoxyglucosone) in 223 foods and drinks commonly consumed in the Western diet, covering all main food groups, e.g. vegetables, fruit, cheese, fish, meat, milk, and tea. The amount of DCs was mainly related to that of carbohydrates/simple sugars and the type of sugar used. Heat treatment increased the DC concentrations. Of the three DCs measured, 3-deoxyglucosone was the most abundant in most foods and drinks.

Davídek, Robert, Devaud, Vera, and Blank (2006) and Robert, Vera, Kervella, Davidek, and Blank (2005) provided evidence for direct formation of acids from C_6-α-DCs by an oxidative mechanism and incorporation of ^{17}OH group into the carboxylic group. At least two reaction mechanisms could lead to carboxylic acids: a hydrolytic β-DC cleavage as the major pathway and an alternative minor pathway via oxidative α-DC cleavage induced by oxidizing species (Davidek et al., 2006).

2.2.5 Generation of aroma/flavor compounds

In the 1970s to the 1990s, volatile compounds generated by the MRs were extensively studied. By the 1990s, more than 2500 volatiles, characteristic of thermally processed foods (e.g., cooked meat, coffee, potato chips, and baked goods) had been identified (Mottram, 1994; Reineccius, 1990). Numerous heterocyclic compounds have been encountered such as furanones, pyrazines, pyrroles, oxazoles, 3-thiazolines, trithiolanes, thithianes and methylfuranones (Fig. 4C). Other volatiles produced by the MRs are esters, acids, ketones, aldehydes, alcohols, alkenes, alkanes, amines, and mercaptans.

Nursten (2005) classified the volatile products of the Maillard reactions into three groups:
- simple sugar dehydration/fragmentation products, encompassing furans (e.g., hydroxymethylfurfural), pyrones (e.g., maltol), cyclopentenes, (e.g., methyl cyclopentenolone), carbonyls, and acids (e.g., acetic acid);
- simple amino acid degradation products, such as the Strecker degradation aldehydes (e.g., methional);
- volatiles produced by further interactions, such as pyrroles, pyridines, imidazoles, pyrazines, oxazoles, and thiazoles.

Jousse, Jongen, Agterof, Russell, and Braat (2006), on the other hand, categorized the volatile compounds into four classes: pyrroles and other nitrogen-containing heterocyclic compounds, furans and other oxygen containing heterocyclic compounds, carbonyls, and pyrazines.

Degradation of ARPs generates numerous aroma compounds, including aldehydes, ketones, furans, pyrazines, thiophenes, thiazoles, oxazoles, disulfides. As desirable taste enhancer and flavor precursor, it is suggested that ARPs have great prospect of application as substitutes of monosodium glutamate.

Cui et al. (2021) cited various transformations of ARPs into aroma compounds. Oxidation of hydroxyacetaldehyde results in glyoxal formation. The aldehyde generated from pyruvic acid can react with hydroxyacetaldehyde to form butane-2,3-dione. Dicarbonyls, through the Strecker degradation of amino acids, form Strecker aldehydes and α-amino ketone that can generate alkylpyrazines (Fig. 4). These alkylpyrazines give nutty, roasted, and baked cereal aroma. The Strecker degradation of cysteine forms mercaptoacetaldehyde, acetaldehyde and hydrogen sulfide as the primary degradation products. γ-Hydroxyl glutamic acid firstly eliminates a formic acid, and then undergoes the Strecker degradation to yield malonaldehyde (Zhao et al., 2018). In addition to Strecker degradation products, the products of deoxyosone dehydration are also important aroma compounds. Furaneol has an intense caramel-like odor. Furans, furanones, and pyranones impart sweet, burnt, and caramel-like aroma. The Maillard reaction of cysteine has been widely used for meat flavor generation (Zhai et al., 2019).

2.2.6 Formation of advanced glycation end products

Although AGEs can be formed in other pathways (e.g., autoxidation of glucose, peroxidation of lipids, polyol pathway), they are produced mainly by the MRs (Liu et al., 2023; Phuong-Nguyen, McNeill, Aston-Mourney, & Rivera, 2023). Reactive intermediates of the MRs, particularly α-DCs, react with free amino acids, peptides, and proteins, forming low-molecular-weight AGEs or high-molecular-weight AGEs.

Although all free amino acids can form AGEs, lysine and arginine are the major contributing amino acids to AGEs formation (Fig. 6) (Poulsen et al., 2013). Most of the characterized AGEs are derived from modifications of lysine, such as N-ε-(carboxymethyl)lysine (CML), N-ε-(carboxyethyl)lysine (CEL), pyrraline, methylglyoxal–lysine dimer (MOLD), glyoxal–lysine dimer (GOLD), and pentosidine. MOLD, GOLD, and pentosidine are cross-linked compounds derived from two lysine residues (MOLD and GOLD) or from one lysine and one arginine residue (pentosidine). Arginine can also be modified to methylglyoxal-derived hydroimidazolinone isomers (MG-H$_n$).

2.2.7 Free radical breakdown of Maillard intermediates

Reaction H has been inserted into the Hodge scheme (Fig. 1) to represent the later discovered free-radical breakdown of Maillard intermediates.

Rizzi (2003) reviewed studies from a historical perspective and concluded that free radical intermediates played an important role in the MRs. More work was recommended to establish the structures of these highly reactive radicals and to elucidate their involvement in Maillard chemistry.

Towards the end of the 1970s and the beginning of the 1980s MR researchers started to seriously consider the participation of free radicals in

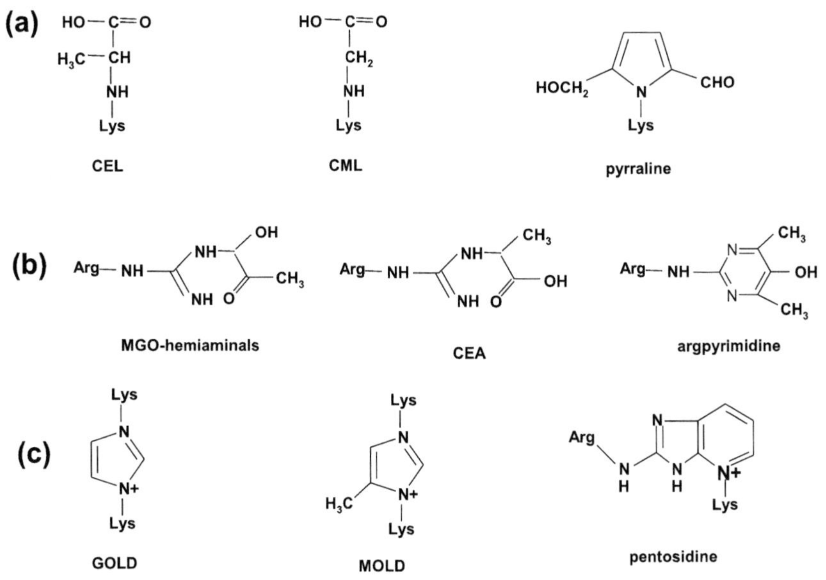

Fig. 6 Examples of advanced glycation end products. N-ε-(carboxyethyl)lysine (CEL), N-ε-(carboxymethyl)lysine (CML), N-7-(1-carboxyethyl)arginine (CEA), (glyoxal–lysine dimer (GOLD), methylglyoxal–lysine dimer (MOLD).

Maillard browning (Rizzi, 2003). Hayashi and Namiki (1980, 1981) reported the formation of free radicals at the beginning of the intermediate stage by the reduction of N,N-dialkylpyrazinium, which can be formed by the condensation of two-carbon enaminols. Namiki and Hayashi (1983) proposed the existence of a new pathway to browning in the MRs, involving sugar fragmentation and free radical formation prior to the Amadori rearrangement.

The radicals were identified as pyrazinium radical cations in model systems and in wheat bread crust and roasted cocoa and coffee (Hofmann, Bors, & Stettmaier, 1999a, 1999b), which could be formed by reaction of an amine with either α-DCs or carbohydrates (Bin, Peterson, & Elias, 2012). EPR spectroscopy of melanoidins isolated from heated aqueous solutions of bovine serum albumin and glycolaldehyde and from wheat bread crust, roasted cocoa, and coffee beans revealed the presence of 1,4-bis (5-amino-5-carboxy-1-pentyl)pyrazinium radical cation (CROSSPY) (Hofmann et al., 1999b). These findings led Hofmann, Bors, and Stettmaier (2002) to propose that the protein bound CROSSPY has a fundamental role in melanoidin formation.

Thermally generated free radicals were detected in bread, the crust exhibiting 2–5 times higher free radical concentration than the middle of the bread (Yordanov & Mladenova, 2004). Because of the relatively low (150–220 °C) temperature of the thermal treatment, the studied free radicals could be assumed to appear in the course of the MRs and not at the point that the material is carbonized.

Highly reactive intermediary radicals were detected when lysine and glucose were heated at intermediate water activity at pH7.0 and 8.0 (Yin et al., 2013). Generation of the radicals was favored by more alkaline conditions (pH8.0) and stimulated by presence of the transition metal ion Fe^{2+}.

2.3 Reactions of the final stage

At the final stage of the Maillard reactions, aldol condensation and formation of heterocyclic nitrogenous compounds occur with simultaneous polymerization of reactive intermediates and products (e.g., enaminol products, low-molecular-weight sugar analogues, unsaturated carbonyl products) to form the high-molecular-weight melanoidins. The formation of intermediates and their products differs greatly by reactant and reaction conditions, and an enormous variety of compounds are formed, resulting in brown polymers of remarkable complexity and diversity.

The active α-DCs and heterocyclic aldehydes formed from ARPs are considered vital precursors for melanoidin generation (Bork, Haase, Rohn, & Kanzler, 2022; Cui et al., 2021). ARPs therefore play an important role for melanoidin formation as building-block reservoirs.

2.3.1 Aldol condensation and the Michael reaction

Aldol and Michael-type reactions are crucial for the formation of melanoidins. Aldehydes are formed by sugar dehydration, sugar fragmentation, and Strecker degradation in the intermediate stage and they can react with each other by aldol condensation (Kanzler & Haase, 2020; Nursten, 2005). Additional carbonyl compounds derived from the oxidation of lipids may also participate. Moreover, Kanzler and Haase (2020) had shown a mechanism by which the sugar fragment norfuraneol could undergo aldolization and condense with other carbonyls of amino acid origin, or from fragmented sugars, and yield adducts with an increased number of conjugated double bonds, which can be further extended if the reaction is complemented by Michael additions.

Norfuraneol incubated with methylglyoxal or diacetyl at elevated temperatures demonstrated that aldol reactions led to the formation of heterogeneous carbohydrate-based oligomers, possible contributors to the browning of the reaction mixtures (Bork et al., 2022). Redox reactions were also identified as important participants, resulting in an increasing number of double bonds in the detected reaction products.

2.3.2 Aldehyde–amine condensation

The final stage is characterized by the polymerization of low-mass carbonyls with amines binding simultaneously to α-terminal, ε- and other amino groups of different polypeptide chains. Polymerization and copolymerization reactions lead to the formation of brown-colored, heterogenous, high-molecular-weight, highly cross-linked, nitrogen-containing polymers known by the generic name of melanoidins. Heterocyclic ring systems such as pyridines, pyrazines, pyrroles, and imidazoles have also been shown to be involved (Nursten, 2005).

2.3.3 Formation of melanoidins

Three theories have been raised for the formation of melanoidins (Echavarría, Pagán, & Ibarz, 2012; Moreira, Nunes, Domingues, & Coimbra, 2012; Wang, Qian, & Yao, 2011):

- high-mass melanoidins are formed by the polymerization of low-mass reaction intermediates (via polycondensation), such as furans, pyrroles, pyrrolopyrroles, and/or their derivatives, formed in the later stages of the MRs (Hayase, Usui, & Watanabe, 2006; Tressl, Wondrak, Krüger, & Rewicki, 1998);
- high-mass melanoidins are derived from cross-linking colored low-mass MRPs to proteins via reactive side chains of lysine, arginine, or cysteine (Hofmann, 1998a,1998b);
- the melanoidin skeleton is built primarily from sugar degradation products formed in the early stages of the Maillard reactions branched with amino compounds, such as amino acids (Cämmerer, Jalyschko, & Kroh, 2002; Kroh et al., 2008).

Melanoidins have also been grouped into the following structural categories (Fogliano & Morales, 2011):

- low molecular weight-colored compounds crosslinked by free amino groups of lysine or arginine in proteins;
- furan and/or pyroles units forming repeating units of melanoidins by polycondensation reactions;
- skeletons that are primarily made up of sugar degradation products that are formed in the early stages of the MRs, polymerized and interconnected by amino compounds.

- skeletons that are primarily made up as proteins connected with MRPs (i.e., melanoproteins).

Melanoidins are chemically varied, negatively charged, furan ring-containing, nitrogen-containing brown polymers. They may contain carbonyl, carboxyl, amine, amide, pyrrole, indole, azomethine, ester, ether, methyl, hydroxyl, or phenolic groups. Aside from the diversity of sources, characterization of melanoidins is rendered difficult by the degree of processing that generates these pigments.

Although melanoidins were initially thought to be solely high-mass compounds, later work has shown that there is also a low-mass fraction (Bekedam et al., 2008b; Ćosović, Vojvodić, Bošković, Plavšić, & Lee, 2010; Glösl et al., 2004). Low-molecular-weight (LMW) melanoidins isolated from coffee brew had similar features as the high-molecular-weight (HMW) coffee melanoidins (Bekedam et al., 2008b). They contained 3% nitrogen, indicating that amino acids or proteins were at some point incorporated. Glucose was the main sugar and phenolic groups were also found (Bekedam, Schols, van Boekel, & Smit, 2008c).

Melanoidins produced in foods are predominantly HMW compounds (Wang et al., 2011). Considering that the molecular weight of melanoidins is highly dependent on the heating intensity and HMW melanoidins are produced at longer times, Wang et al. (2011) suggested that LMW melanoidins might be formed at the initial stages of the MRs, subsequently polymerizing or cross-linking with other MRPs to eventually result in HMW melanoidins during the later stages.

Melanoidins are produced during heat processing of a wide range of food products such as coffee, bread, honey, milk, malt, cooked meat, dark beer, processed tomato, balsamic vinegar, wine, and soy sauce. Melanoidins from coffee have been the most investigated.

Thermal degradation of melanoidins isolated from bread crust, coffee, and tomato sauce generated important flavor volatiles, particularly furans, carbonyl compounds, 1,3-dioxolanes, pyrroles, pyrazines, pyridines, thiophenes, and phenols (Adams, Borrelli, Fogliano, & De Kimpe, 2005). Melanoidin-rich fractions extracted from coffee, barley coffee, dark beer, and traditional balsamic vinegar (Tagliazucchi & Verzelloni, 2014) were formed mainly of carbohydrates, phenolic compounds, and proteins. Glucose was the most abundant sugar incorporated into the melanoidins.

Many studies on different coffee products have now confirmed that the major components of coffee melanoidins are polysaccharides, proteins, and phenolic compounds (Bekedam, De Laat, Schols, van Boekel, & Smit, 2007; Bekedam, Loots, Schols, van Boekel, & Smit, 2008a,2008b; Gniechwitz et al., 2008; Moreira et al., 2012; Nunes & Coimbra, 2007). Arabinogalactans in the form of arabinogalactan-proteins were found in green and roasted coffee beans as well as in green and roasted coffee brews, including instant coffee (Capek, Matulová, Navarini, & Suggi-Liverani, 2010; Gniechwitz et al., 2008; Matulová, Capek, Kaneko, Navarini, & Liverani, 2011). Rhamnoarabinosyl and rhamnoarabinoarabinosyl side chains were reported by Nunes, Reis, Silva, Domingues, and Coimbra (2008) as structural features of coffee arabinogalactan-proteins.

No difference was observed in the amino acid composition among the HMW coffee melanoidin fractions (Bekedam, Schols, van Boekel, & Smit, 2006). Alanine, aspartic acid/asparagine, glutamicacid/glutamine, and glycine were the amino acids most frequently encountered in all fractions, whereas histidine, lysine, methionine, and tyrosine were the least abundant. Arginine was consistently absent (Bekedam et al., 2006; Nunes & Coimbra, 2007). The presence of hydroxyproline in melanoidin fractions containing also arabinose and galactose residues is considered as further evidence for the existence of arabinogalactan-proteins complexed to coffee melanoidin structures.

A comparative study of the chemical composition of HMW melanoidins isolated from instant soluble coffee and instant soluble barley was undertaken by Antonietti et al. (2022). The barley melanoidins were almost exclusively composed of glucose (76%) whereas the coffee mealanoidins were composed mostly by arabinogalactans (~41%) and lower amounts of galactomannan (14%). Melanoidins from the latter presented a significantly higher content of phenolic compounds.

3. Influencing factors

The rate, course, and extent of the MRs depend on multiple factors such as: the type and concentration of reducing sugars and of the amino compounds; the physicochemical parameters, such as the pH and water activity (a_w); the presence of metal ions and pro-Maillard or anti-Maillard reation compounds, such as phenolics, sulfites, free radicals and carbonyl compounds from lipid peroxidation, dehydroascorbic acid; presence of pro- and antioxidants; and processing method and conditions, especially the temperature and duration of the treatment (Abd El-Salam & El-Shibiny, 2018; Ames, 1990; Feng et al., 2021; Fu, Zhang, Soladoye, & Aluko, 2020; Lund & Ray, 2017; Perez-Locas & Yaylayan, 2010; Shaheen, Shorbagi, Lotenzo, & Farag, 2022; Sikorski et al., 2008).

Type and concentration of the reactants: The reactivity of the carbonyl group follows the following order (Nursten, 1981; Sikorski et al., 2008):
- Aldoses are more reactive than ketoses.
- Trioses react more rapidly than tetroses, tetroses more rapidly than pentoses, which are more reactive than hexoses.
- Monosaccharides are more reactive than di- or polysaccharides.
- α-DC derivatives react more readily than aldehydes, which are more reactive than ketones; reducing sugars are less reactive, followed by oxo-acids.
- Acyclic form of sugar reacts more readily than the cyclic form.

Reactivity tends to decrease with increasing molecular weight because of steric hindrance (O'Mahony, Drapala, Mulcahy, & Mulvihill, 2017). HuiLing, ZhiQiang, Chun, ZhiNi, and YongQuan (2010) found the reaction activity of sugars to be in the order of xylose > galactose > glucose > fructose. As expected, no reaction occurred when sucrose was used.

Primary amines are more reactive than secondary amines, while tertiary amines are inactive (Poulsen et al., 2013). In ribose-amino acid Maillard model systems, the order of reactivity after heating for 10 h at 100 °C was: ribose-lysine > -cysteine > -isoleucine ≈-glycine (Hemmler et al., 2018). The amino acid precursors were responsible for the molecular characteristics of the products while the sugar precursors drove the reaction rates.

The ratio of the sugars and the amino acids has significant influence on the formation of the brown pigments, with the sugars having a promotional effect while the amino acids seem to have a preventive role (O'Brien, 2009).

The effect of the initial content of sugars, total proteins, free and total lysine on the formation of MRPs was investigated by Žilić, Aktağ, Dodig, and Gökmen (2021) in sweet cookies made of wholegrain flour. 3-Deoxyglucosone was the predominant α-DC in all cereal cookies. The highest content of AGE was detected in cookies made from hulless oat flour rich in proteins and total lysine. In addition, a high correlation between protein-bound Maillard reaction products in the cookies and the total proteins and the total lysine content in the flours was found.

Temperature and reaction time: Temperature is widely considered as the most important influencing factor. The MRs are highly sensitive to temperature, but they also take place at room temperature although at a very slow rate.

That the rate of Maillard browning increases with rising temperature was reported by Maillard (1912), later confirmed by many authors. The reactions are slow at low temperatures, but the rate doubles with every 10 °C increase in temperature between 40 and 70 °C.

Interestingly, thermal treatment affects not only the rate but also the flavoring characteristic. High temperature (>100 °C) increased the formation of meaty aroma generated by xylose and chicken peptide through a Maillard reaction system while lower temperature and longer heating tended to generate a broth-like taste (i.e., umami and kokumi) (Liu, Liu, He, Song, & Chen, 2015).

pH. The pH has considerable effect on the rate and direction of the MRs. The reaction rate is maximum close to neutrality (pH6–7). In acid medium (pH lower than the pK_a value of the amino group), the amine groups in the amino acids are predominantly protonated, eliminating their nucleophilicity, thus preventing their reaction with carbonyl groups (O'Mahony et al., 2017; van Boekel, 2001). On the other hand, protonation of the carbonyl group increases the reactivity. Consequently, the reaction rate reaches a maximum in a slightly acidic medium in the reaction with amine while this occurs in a slightly basic medium in the case of amino acids. Basicity also increases the open-chair form of hexose sugars, rendering them more available for interaction.

As previously stated, sucrose can only participate in the MRs when the glycosidic bond is hydrolyzed, and the reducing monosaccharides constituents are released. This is facilitated at a low pH and high moisture content.

The pathway towards melanoidin formation is also affected by pH. At acidic pH, the Amadori compounds will degrade and polymerize via 1,2-eneaminol formation, producing furfurals., Alkaline conditions favor the 2,3-enediol pathway, generating reductones, dicarbonyls, aldehydes, furaneols, pyrones, and other fission products (Ames, 1990; Nursten,1981).

Water content and activity: The MRs occur easily when the moisture content is about 30%; it is maximal at aw in the range of 0.60–0.85 (O'Brien & Morrisey, 1989). A maximum is reached because the reactions occur less readily at high aw since the reactants are diluted, whereas at lower aw, although the reactants become more concentrated, their mobility is limited (Ames, 1990; Van Boekel, 2001).

Metal ions: Cu^{2+}, Fe^{2+} and Zn^{2+}, can accelerate the MRs (Ramonaityté, Keršienė, Adams, Tehrani, & De Kimpe, 2009). Both Fe^{3+} and Fe^{2+} accelerated the reaction, but the reaction rate by Fe^{3+} was faster than that by Fe^{2+} (HuiLing et al., 2010). Ca^{2+}, Mg^{2+} Sn^{2+} and Mn^{2+} inhibited the reactions, K^+ had little effect, and Na^+ had no effect. Phosphates and citrates increase the rate of the reaction.

4. Consequences of the Maillard reactions
4.1 Impact on food quality

The MRs bring both beneficial and negative effects on foods:
- browning, which is desirable in baking and roasting, but not in foods which have weak color or a color of their own;
- production of the volatile compounds that constitute the much-appreciated aromas/flavors of cooked, baked, roasted, fried, and fermented foods (e.g., bread, coffee, cocoa, roasted meat, and dark beer) but also generation of off-flavors in ultrahigh-temperature processed milk, pilsner beers, dehydrated foods, grilled meat or fish, and overheated tea or coffee.
- modification of protein functionality in foods, such as solubility, emulsifying foaming, gelation, and textural properties.
- production of reductones and antioxidants, which can contribute to the stabilization of foods against oxidative deterioration.

While the influence of the MRs on food color and flavor has drawn much attention, the connection between MRPs and textural properties has not been focused on (Starowicz & Zieliński, 2019). Textural properties have been mostly investigated in meat products, the change in the texture being estimated as the difference in linkage formations between proteins and polysaccharides.

Kukuminato, Koyama, and Kosekia (2021) investigated the antibacterial potential of various combinations of reducing sugars and amino acids. Their findings suggested that melanoidins generated from xylose with phenylalanine and/or proline could be used as potential novel alternative food preservatives to control pathogenic bacteria.

MRPs (e.g., Amadori compounds, reactive reductones, premelanoidins, and melanoidins) possess excellent antioxidant ability in many food products through chelation of metal ions, breakdown of radical chains and hydrogen peroxide, and scavenging of reactive oxygen species (Nooshkam, Varidi, & Bashash, 2019).

4.2 Impact on human health

Two effects of the MRs have long been of concern in nutrition. First, the binding of the ε-amino group of lysine residues, which occurs even during storage at room temperature, renders this indispensable amino acid unavailable for the human body. This situation is aggravated by the fact that lysine is a limiting amino acid in many foods. Second, the interaction of MRPs with nutritionally essential metals (Ca, Mg, Cu, Zn, Fe) prevents mineral absorption and metabolism.

Thermal treatments are accompanied by protein unfolding, which may increase digestibility by increasing the availability of cleavage sites (Joubran, Moscovici, & Lesmes, 2015). On the other hand, digestibility may be lowered when proteins are modified by the MRs (Sanwar Gilani, Wu Xiao, & Cockell, 2012; Sanz, Corzo-Martinez, Rastall, Olano, & Moreno, 2007; Seiquer et al., 2006). This may be explained by the modification of enzymatic cleavage sites for intestinal proteases, which may reduce proteolysis (Chevalier, Chobert, Genot, & Haertlé, 2001; Joubran et al., 2015). In addition, MR-induced protein aggregates may protect proteins during digestion. More important, however, is the change that can be induced in the innumerable blood proteins and hormons, and membrane-bound receptors and transporters which can lose functionality after modification.

More recently, researchers have focused on the health benefits of melanoidins. Health promoting biological activities have been attributed to melanoidins: antioxidative, antimicrobial, anticariogenic, anti-inflammatory, antihypertensive, antiallergenic, antimutagenic, chemopreventive, phase I (carcinogen-activating enzymes) and II (detoxifying enzymes) enzyme-modulating, tumor growth-inhibiting, *Helicobacter pylori* suppressing, prebiotic, and gut microbiota regulating (ALjahdali, Gadonna-Widehem, Anton, & Carbonero, 2020; Borelli & Fogliano, 2005; Hiramoto et al., 2004; Langner & Rzeski, 2014; Morales, Somoza, & Fogliano, 2012; Moreira et al., 2012; Mu, Wang, & Kitts, 2016; Nooshkam, Varidi, & Verma, 2020; Shaheen et al., 2022; Wang et al., 2011). The antimicrobial properties refer to both the melanoidin's capacity to improve the shelf-life of food (antimicrobial effect on spoilage microbiota) and to boost food safety (antimicrobial effect on pathogens) (Diaz-Morales, Ortega-Heras, Diez-Maté, Gonzalez-SanJose, & Muñiz, 2022).

During food processing and storage, the MRs can produce substances with documented toxic or carcinogenic effects, such as acrylamide, heterocyclic amines, furan (ALjahdali & Carbonero, 2019; Anese & Suman, 2013; Mogol & Gökmen, 2016; Rifai & Saleh, 2020). α-DCs are considered precursors to these hazardous substances (Yan et al., 2023). The α-DCs themselves may have detrimental effects on human health. In foods α-DCs can react with proteins that have amino and sulphydryl groups to further generate MRPs (Amoroso, Maga, & Daglia, 2013; Liu et al., 2023). Therefore, high content of α-DCs in foods would have a negative impact on the digestibility and nutritional value of protein. In vitro results showed that during gastroduodenal digestion, α-DCs react with digestive enzymes to produce carbonylated proteins (Amoroso et al., 2013).

In recent years the main concern has been the AGEs, which are implicated in the pathogenesis of chronic diseases, such as diabetes, cardiovascular diseases, chronic kidney disease, osteoporosis, osteoarthritis, cancer, Alzheimer's disease and other neurodegenerative diseases, as well as retinal aging, endothelial dysfunction, obesity, and loss of immunity, discussed in numerous reviews (e.g., Cai et al., 2014; Ge et al., 2022; Geng et al., 2023; Glenn & Stitt, 2009; Hellwig & Henle, 2014; Indyk, Bronowicka-Szydełko, Gamian, & Kuzan, 2021; Jiang, et al., 2022; Koska, Gerstein, Beisswenger, & Reaven, 2022; Monnier & Taniguchi, 2016; Nass et al., 2007; Poulsen et al., 2013; Robert & Labat-Robert, 2014; Robert, Labat-Robert, & Robert, 2010; Vlassara & Uribarri, 2014). AGEs may be involved in the development of chronic inflammation by acting as inflammatory components and affecting the gut microbiome (Toda, Hellwig, Henle, & Vieths, 2019). Phuong-Nguyen et al. (2023) summarized the current evidence of the association between a high-AGE diet and poor health outcomes, focusing on the relation between dietary AGEs and alterations in the gastrointestinal structure, modifications in enteric neurons, and microbiota reshaping.

AGEs are classified as endogenous and exogenous AGEs. Endogenous AGEs are produced by the glycosylation of proteins and sugars in the human body. Exogenous AGEs are ingested or inhaled from the outside; most of which are brought into the body through diet and are known as dietary AGEs (O'Brien, 2009).

The dietary AGEs are largely formed at the intermediate and final stages of the MRs. There is a significant positive correlation between AGEs content in blood and that of dietary AGEs among healthy people, indicating that dietary AGEs are main sources of accumulated AGEs in the human body (Liu et al., 2023).

Phuong-Nguyen et al. (2023) provided evidence of the deleterious effects of AGEs on the gastrointestinal tract, markedly altering the gut structure, leading to increased intestinal permeability and reduced expression of enteric neurons, and reshaping the microbiota composition.

Clarke, Dordevic, Tan, Ryan, and Coughlan (2016) examined 12 dietary AGE interventions with a total of 293 participants. The evidence presented indicated that a high AGE diet might contribute to risk factors associated with chronic disease, such as inflammation and oxidative stress. However, due to a lack of high-quality randomized trials, more research was deemed necessary for chronic kidney disease.

A large multinational prospective cohort study across 20 anatomical cancer sites, does not support the hypothesis that dietary AGEs contribute to a higher cancer incidence (Córdova et al., 2022). The association between the intake of three well-characterized dietary AGEs, (CML, CEL and MG-H1) and the risk of overall and site-specific cancers in the European Prospective Investigation into Cancer and Nutrition (EPIC) cohort was examined. A higher intake of dietary AGEs was not associated with an increased risk of overall cancer and most cancer types studied. A non-linear weak positive association was observed between higher intakes of MG-H1 and the risk of prostate cancer. In contrast, a strong inverse association was observed with the risk of laryngeal cancer.

5. Strategies for controlling the Maillard reactions

Considering that the MRs provoke both desirable and undesirable effects, the need for control is evident. The complexity of this process and the many influencing factors make this task extremely challenging.

Optimization of reaction variables has been among the first measures recommended (BeMiller & Huber, 2008; Nursten, 2005), including:
- removing one of the reactants;
- decreasing the temperature and time of heat treatment;
- adjusting the pH
- adjusting the water content or water activity;
- minimizing the presence of metal ions;
- using chemical inhibitors, such as SO_2 or bisulphites.

Jaeger et al. (2010), on the other hand, advocated the use of recent advances in food processing, such as:
- improved thermal technologies, such as ohmic heating, that allow direct heating of the product to overcome the heat transfer limitations of conventional thermal processing, thereby reducing thermal exposure during food preservation;
- nonthermal technologies (e.g., high hydrostatic pressure and pulsed electric fields), can extend the shelf-life of food products without the application of heat;

- removal of Maillard reaction substrates in raw food materials by the application of pulsed electric field cell disintegration and extraction as well as enzymatic conversion.

Lund and Ray (2017) contended that the development of efficient strategies to control the MRs in foods requires an understanding of reaction mechanisms and how reaction conditions affected the MPs. The following approaches were presented:
- addition of functional ingredients (e.g., plant polyphenols and vitamins);
- enzymatic inhibition of MRs (e.g. using fructosamine oxidase, fructoseamine kinase, and carbohydrate oxidases);
- inhibition of MRs by alternative processing such as ohmic heating, pulsed electric field, high-pressure processing, and encapsulation of metal ions.

The first approach aims at targeting: (1) reactive sites of MRs (modification of reducing sugars, blocking or modifying amines), (2) intermediates of MRs (trapping of α-DCs, scavenging of Maillard-derived radicals, action of pyridoxamine on Amadori products); and (3) MRPs (trapping of Strecker aldehydes, trapping of acrylamide).

Liu et al. (2023) classified the current means of controlling AGEs into three categories: (1) using AGE-breakers, (2) inhibiting AGEs receptors, and (3) inhibiting AGEs formation. Polyphenols as a group of natural inhibitors have anti-glycosylation properties making them suitable to be used as AGEs and α-DC inhibitors in food systems. Liu et al. (2023) tabulated 24 studies on polyphenol compounds extracted from natural plants with AGEs inhibiting potentials.

Strategies for reducing α-DC levels in foods have also been sought. These strategies include the use of new processing technologies, formula modification, and supplementation with α-DC scavengers (e.g., phenolic compounds) (Yan et al., 2023).

6. Final considerations

Over a century after their discovery, the MRs continue to fascinate and challenge researchers around the world. As greater knowledge unfolds on a certain aspect, the need for more clarifications in another facet of this extremely complicated process becomes evident. So, what changes with time are the features that are focused on. In the last decade, the highly reactive intermediates, characteristics of the melanoidins of different food products, formation of AGEs, and health effects of melanoidins and AGEs have taken center stage. There can be no doubt that the MRs will continue to be investigated for years to come, the focus changing to what are considered of great importance at the moment.

References

Aalaei, K., Rayner, M., & Sjöholm, I. (2019). Chemical methods and techniques to monitor early Maillard reaction in milk products: A review. *Critical Reviews in Food Science and Nutrition, 59*, 1829–1839.

Abd El-Salam, M. H., & El-Shibiny, S. (2018). Glycation of whey proteins: Technological and nutritional implications. *International Journal of Biological Macromolecules, 112*, 83–92.

Adams, A., Borrelli, R. C., Fogliano, V., & De Kimpe, N. (2005). Thermal degradation studies of food melanoidins. *Journal of Agricultutal and Food Chemistry, 53*, 4136–4142.

ALjahdali, N., & Carbonero, F. (2019). Impact of Maillard reaction products on nutrition and health: Current knowledge and need to understand their fate in the human digestive system. *Critical Reviews in Food Science and Nutrition, 59*, 474–487.

ALjahdali, N., Gadonna-Widehem, P., Anton, P. M., & Carbonero, F. (2020). Gut microbiota modulation by dietary barley malt melanoidins. *Nutrients, 12*, 241.

Amaya-Farfan, J., & Rodriguez-Amaya, D. B. (2021). The Maillard reactions. In J. Amaya-Farfan, & D. B. Rodriguez-Amaya (Eds.). *Chemical changes during processing and storage of foods. Implications for food quality and human health* (pp. 215–263). London: Academic Press.

Ames, J. M. (1990). Control of the Maillard reaction in food systems. *Trends in Food Science & Technology, 1*, 150–154.

Amoroso, A., Maga, G., & Daglia, M. (2013). Cytotoxicity of α-dicarbonyl compounds submitted to in vitro simulated digestion process. *Food Chemistry, 140*, 654–659.

Anese, M., & Suman, M. (2013). Mitigation strategies of furan and 5-hydroxymethylfurfural in food. *Food Research International, 51*, 257–264.

Antonietti, S., Silva, A. M., Simões, C., Almeida, D., Félix, L. M., Papetti, A., & Nunes, F. M. (2022). Chemical composition and potential biological activity of melanoidins from instant soluble coffee and instant soluble barley: A comparative study. *Frontiers in Nutrition, 9*, 825584.

Bekedam, E. K., De Laat, M. P. F. C., Schols, H. A., van Boekel, M. A. J. S., & Smit, G. (2007). Arabinogalactan proteins are incorporated in negatively charged coffee brew melanoidins. *Journal of Agricultural and Food Chemistry, 55*, 761–768.

Bekedam, E. K., Loots, M. J., Schols, H. A., van Boekel, M. A. J. S., & Smit, G. (2008a). Roasting effects on formation mechanisms of coffee brew melanoidins. *Journal of Agricultural and Food Chemistry, 56*, 7138–7145.

Bekedam, E. K., Roos, E., Henk, A., Schols, H. A., van Boekel, M. A. J. S., & Smit, G. (2008b). Low molecular weight melanoidins in coffee brew. *Journal of Agricultural and Food Chemistry, 56*, 4060–4067.

Bekedam, E. K., Schols, H. A., van Boekel, M. A. J. S., & Smit, G. (2006). High molecular weight melanoidins from coffee brew. *Journal of Agricultural and Food Chemistry, 54*, 7658–7666.

Bekedam, E. K., Schols, H. A., van Boekel, M. A. J. S., & Smit, G. (2008c). Incorporation of chlorogenic acids in coffee brew melanoidins. *Journal of Agricultural and Food Chemistry, 56*, 2055–2063.

BeMiller, J. N., & Huber, K. C. (2008). Carbohydrates. In S. Damodaran, K. L. Parkin, & O. R. Fennema (Eds.). *Fennema's food chemistry* (pp. 83–154)(fourth ed.). Boca Raton: CRC Press.

Bin, Q., Peterson, D. G., & Elias, R. J. (2012). Influence of phenolic compounds on the mechanisms of pyrazinium radical generation in the Maillard reaction. *Journal of Agricultural and Food Chemistry, 60*, 5482–5490.

Bork, L. V., Haase, P. T., Rohn, S., & Kanzler, C. (2022). Formation of melanoidins—Aldol reactions of heterocyclic and short-chain Maillard intermediates. *Food Chemistry, 380*, 131852.

Borrelli, R. C., & Fogliano, V. (2005). Bread crust melanoidins as potential prebiotic ingredients. *Molecular Nutrition & Food Research, 49*, 673–678.

Bruhns, P., Kanzler, C., Degenhardt, A. G., Koch, T. J., & Kroh, L. W. (2019). Basic structure of melanoidins formed in the Maillard reaction of 3-deoxyglucosone and γ-aminobutyric acid. *Journal of Agricultural and Food Chemistry, 67*, 5197–5203.

Cai, W., Uribarri, J., Zhu, L., Chen, X., Swamy, S., Zhao, Z., ... Vlassara, H. (2014). Oral glycotoxins are a modifiable cause of dementia and the metabolic syndrome in mice and humans. *Proceedings of the National Academy of Sciences of the United States of America, 111*, 4940–4945.

Cämmerer, B., Jalyschko, W., & Kroh, L. W. (2002). Intact carbohydrate structures as part of the melanoidin skeleton. *Journal of Agricultural and Food Chemistry, 50*, 2083–2087.

Capek, P., Matulová, M., Navarini, L., & Suggi-Liverani, F. (2010). Structural features of an arabinogalactan-protein isolated from instant coffee powder of *Coffea arabica* beans. *Carbohydrate Polymers, 80*, 180–185.

Chevalier, F., Chobert, J.-M., Genot, C., & Haertlé, T. (2001). Scavenging of free radicals, antimicrobial, and cytotoxic activities of the Maillard reaction products of α-lactoglobulin glycated with several sugars. *Journal of Agricultural and Food Chemistry, 49*, 5031–5038.

Clarke, R. E., Dordevic, A. L., Tan, S. M., Ryan, L., & Coughlan, M. T. (2016). Dietary advanced glycation end products and risk factors for chronic disease: A systematic review of randomised controlled trials. *Nutrients, 8*, 125.

Córdova, R., Mayén, A. L., Knaze, V., Aglago, E. K., Schalkwijk, C., Wagner, K. H., ... Freisling, H. (2022). Dietary intake of advanced glycation end products (AGEs) and cancer risk across more than 20 anatomical sites: A multinational cohort study. *Cancer Communications, 42*, 1041–1045.

Ćosović, B., Vojvodić, V., Bošković, N., Plavšić, M., & Lee, C. (2010). Characterization of natural and synthetic humic substances (melanoidins) by chemical composition and adsorption measurements. *Organic Geochemistry, 41*, 200–205.

Cremer, D. R., Vollenbroeker, M., & Eichner, K. (2000). Investigation of the formation of Strecker aldehydes from the reaction of Amadori rearrangement products with α-amino acids in low moisture model system. *European Food Research and Technology, 211*, 400–403.

Cui, H., Yu, J., Zhai, Y., Feng, L., Chen, P., Hayat, K., ... Ho, C.-T. (2021). Formation and fate of Amadori rearrangement products in Maillard reaction. *Trends in Food Science and Technology, 115*, 391–408.

Davidek, T., & Davidek, J. (2004). Chemistry of the Maillard reaction in foods. In P. Tomasik (Ed.). *Chemical and functional properties of food saccharides* Boca Baton: CRC Press (Chapter 18).

Davidek, T., Kraehenbuehl, K., Devaud, S., Robert, F., & Blank, I. (2005). Analysis of Amadori compounds by high-performance cation exchange chromatography coupled to tandem mass spectrometry. *Analytical Chemistry, 77*, 140–147.

Davídek, T., Robert, F., Devaud, S., Vera, F. A., & Blank, I. (2006). Sugar fragmentation in the Maillard reaction cascade: Formation of short-chain carboxylic acids by a new oxidative alpha-dicarbonyl cleavage pathway. *Journal of Agricultural and Food Chemistry, 54*, 6677–6684.

Degen, J., Hellwig, M., & Henle, T. (2012). 1,2-Dicarbonyl compounds in commonly consumed foods. *Journal of Agricultural and Food Chemistry, 60*, 7071–7079.

Diaz-Morales, N., Ortega-Heras, M., Diez-Maté, A. M., Gonzalez-SanJose, M. L., & Muñiz, P. (2022). Antimicrobial properties and volatile profile of bread and biscuits melanoidins. *Food Chemistry, 373*, 131648 (Part B).

Echavarría, A. P., Pagán, J., & Ibarz, A. (2012). Melanoidins formed by Maillard reaction in food and their biological activity. *Food Engineering Reviews, 4*, 203–223.

Feng, T., Zhou, Y., Wang, X., Wang, X., & Xia, S. (2021). α-Dicarbonyl compounds related to antimicrobial and antioxidant activity of Maillard reaction products derived from xylose, cysteine and corn peptide hydrolysate. *Food Bioscience, 41*, 100951.

Fogliano, V., & Morales, F. J. (2011). Estimation of dietary intake of melanoidins from coffee and bread. *Food and Function, 2*, 117–123.

Fu, Y., Zhang, Y., Soladoye, O. P., & Aluko, R. E. (2020). Maillard reaction products derived from food protein-derived peptides: Insights into flavor and bioactivity. *Critical Reviews in Food Science and Nutrition, 60*, 3429–3442.

Ge, W., Jie, J., Yao, J., Li, W., Cheng, Y., & Lu, W. (2022). Advanced glycation end products promote osteoporosis by inducing ferroptosis in osteoblasts. *Molecular Medicine Reports, 25*, 140.

Geng, Y., Mou, Y., Xie, Y., Ji, J., Chen, F., Liao, X., ... Ma, L. (2023). Dietary advanced glycation end products: An emerging concern for processed foods. *Food Reviews International*. https://doi.org/10.1080/87559129.2023.2169867.

Glenn, J. V., & Stitt, A. W. (2009). The role of advanced glycation end products in retinal ageing and disease. *Biochimica et Biophysica Acta, 1790*, 1109–1116.

Glösl, S., Wagner, K. H., Draxler, A., Kaniak, M., Lichtenecker, S., Sonnleitner, A., ... Elmadfa, I. (2004). Genotoxicity and mutagenicity of melanoidins isolated from a roasted glucose-glycine model in human lymphocyte cultures, intestinal Caco-2 cells and in the *Salmonella typhimurium* strains TA98 and TA102 applying the AMES test. *Food and Chemical Toxicology, 42*, 1487–1495.

Gniechwitz, D., Reichardt, N., Meiss, E., Ralph, J., Steinhart, H., Blaut, M., & Bunzel, M. (2008). Characterization and fermentability of an ethanol soluble high molecular weight coffee fraction. *Journal of Agricultural and Food Chemistry, 56*, 5960–5969.

Haase, P. T., Kanzler, C., Hildebrandt, J., & Kroh, L. W. (2017). Browning potential of C_6-α-dicarbonyl compounds under Maillard conditions. *Journal of Agricultural and Food Chemstry, 65*, 1924–1931.

Hashiba, H. (1978). Isolation and identification of Amadori compounds from soy sauce. *Agricultural and Biological Chemistry, 42*, 763–768.

Hayase, F., Usui, T., & Watanabe, H. (2006). Chemistry and some biological effects of model melanoidins and pigments as Maillard intermediates. *Molecular Nutrition & Food Research, 50*, 1171–1179.

Hayashi, T., & Namiki, M. (1980). Formation of two-carbon sugar fragment at an early stage of the browning reaction of sugar with amine. *Agricultural and Biological Chemistry, 44*, 2575–2580.

Hayashi, T., & Namiki, M. (1981). On the mechanism of free radical formation during browning reaction of sugars with amino compounds. *Agricultural and Biological Chemistry, 45*, 933–939.

Heinzler, M., & Eichner, K. (1991). Behaviour of Amadori compounds during cocoa processing. Part I. Formation and decomposition. *Zeitschrift für Lebensmittel-Untersuchung und Forschung, 192*, 24–29.

Hellwig, M., & Henle, T. (2014). Baking, ageing, diabetes: A short history of the Maillard reaction. *Angewandte Chemie International Edition, 53*, 10316–10329.

Hemmler, D., Roullier-Gall, C., Marshall, J. W., Rychlik, M., Taylor, A. J., & Schmitt-Kopplin, P. (2018). Insights into the Chemistry of Non-Enzymatic Browning Reactions in Different Ribose-Amino Acid Model Systems. *Scientific Reports, 8*, 16879.

Hiramoto, S., Itoh, K., Shizuuchi, S., Kawachi, Y., Morishita, Y., Nagase, M., ... Kimura, N. (2004). Melanoidin, a food protein-derived advanced Maillard reaction product, suppresses *Helicobacter pylori* in vitro and in vivo. *Helicobacter, 9*, 429–435.

Hodge, J. E. (1953). Chemistry of browning reactions in model systems. *Journal of Agricultutal and Food Chemistry, 1*, 928–943.

Hofmann, T. (1998a). Studies on the relationship between molecular weight and the color potency of fractions obtained by thermal treatment of glucose/amino acid and glucose/protein solutions by using ultracentrifugation and color dilution techniques. *Journal of Agricultural and Food Chemistry, 46*, 3891–3895.

Hofmann, T. (1998b). Studies on melanoidin-type colorants generated from the Maillard reaction of protein-bound lysine and furan-2-carboxaldehydes chemical characterisation of a red coloured domain. *Zeitschrift für Lebensmitteluntersuchung und -Forschung A, 206*, 251–258.

Hofmann, T., Bors, W., & Stettmaier, K. (1999a). Studies on radical intermediates in the early stage of the nonenzymatic browning reaction of carbohydrates and amino acids. *Journal of Agricultural and Food Chemistry, 47*, 379–390.

Hofmann, T., Bors, W., & Stettmaier, K. (1999b). Radical-assisted melanoidin formation during thermal processing of foods as well as under physiological conditions. *Journal of Agricultural and Food Chemistry, 47*, 391–396.

Hofmann, T., Bors, W., & Stettmaier, K. (2002). CROSSPY: A radical intermediate of melanoidin formation in roasted coffee. In M. J. Morello, F. Shahidi, & C.-T. Ho (Eds.). *Free radicals in foods: Chemistry, nutrition, and health effects* (pp. 49–68). ACS Symposium Series, vol. 807, Chapter 4.

Hofmann, T., Münch, P., & Schieberle, P. (2000). Quantitative model studies on the formation of aroma active aldehydes and acids by Strecker-type reactions. *Journal of Agricultural and Food Chemistry, 48*, 434–440.

Hou, L., Xie, J., Zhao, J., Zhao, M., Fan, M., Xiao, Q., ... Chen, F. (2017). Roles of different initial Maillard intermediates and pathways in meat flavor formation for cysteine-xylose-glycine model reaction systems. *Food Chemistry, 232*, 135–144.

HuiLing, W., ZhiQiang, W., Chun, H., ZhiNi, P., & YongQuan, C. (2010). Factors affecting the Maillard reaction. *Modern Food Science and Technology, 26*, 441–444.

Indyk, D., Bronowicka-Szydełko, A., Gamian, A., & Kuzan, A. (2021). Advanced glycation end products and their receptors in serum of patients with type 2 diabetes. *Scientific Reports, 11*, 13264.

Jaeger, H., Janositz, A., & Knorr, D. (2010). The Maillard reaction and its control during food processing. The potential of emerging technologies. *Pathologie Biologie, 58*, 207–213.

Jiang, T., Zhang, Y., Dai, F., Liu, C., Hu, H., & Zhang, Q. (2022). Advanced glycation end products and diabetes and other metabolic indicators. *Diabetology & Metabolic Syndrome, 14*, 104.

Joubran, Y., Moscovici, A., & Lesmes, U. (2015). Antioxidant activity of bovine alpha lactalbumin Maillard products and evaluation of their in vitro gastro-duodenal digestive proteolysis. *Food Function, 6*, 1229–1240.

Jousse, F., Jongen, T., Agterof, W., Russell, S., & Braat, P. (2006). Simplified kinetic scheme of flavor formation by the Maillard reaction. *Journal of Food Science, 67*, 2534–2542.

Kanzler, C., & Haase, P. T. (2020). Melanoidins formed by heterocyclic Maillard reaction intermediates via aldol reaction and Michael addition. *Journal of Agricultural and Food Chemistry, 68*, 332–339 2020.

Kanzler, C., Wustrack, F., & Rohn, S. (2021). High-resolution mass spectrometry analysis of melanoidins and their precursors formed in a model study of the Maillard reaction of methylglyoxal with l-alanine or l-lysine. *Journal of Agricultural and Food Chemistry, 69*, 11960–11970.

Koska, J., Gerstein, H. C., Beisswenger, P. J., & Reaven, P. D. (2022). Advanced glycation end products predict loss of renal function and high-risk chronic kidney disease in type 2 diabetes. *Diabetes Care, 45*, 684–691.

Kroh, L. W., Fiedler, T., & Wagner, J. (2008). α-Dicarbonyl compounds—Key intermediates for the formation of carbohydrate-based melanoidins. *Annals of the New York Academy of Sciences, 26*, 210–215.

Kukuminato, S., Koyama, K., & Kosekia, S. (2021). Antibacterial properties of melanoidins produced from various combinations of Maillard reaction against pathogenic bacteria. *Microbiology Spectrum, 9* e01142-21.

Langner, E., & Rzeski, W. (2014). Biological properties of melanoidins: A review. *International Journal of Food Properties, 17*, 344–353.

Lindsay, R. C. (1996). Food chemistry. In O. R. Fennema (Ed.). (third ed.). New York: Marcel Dekker, Inc Flavor.

Liu, J., Liu, M., He, C., Song, H., & Chen, F. (2015). Effect of thermal treatment on the flavor generation from Maillard reaction of xylose and chicken peptide. *LWT—Food Science and Technology, 64*, 316–325.

Liu, Y., Lu, L., Yuan, S., Guo, Y., Yao, W., Zhou, W., & Yu, H. (2023). Formation of advanced glycation end-products and α-dicarbonyl compounds through Maillard reaction: Solutions from natural polyphenols. *Journal of Food Composition and Analysis, 120*, 105350.

Locas, C. P., & Yaylayan, V. A. (2010). The Maillard reaction and food quality deterioration. In L. H. Skibsted, J. Risbo, & M. L. Andersen (Eds.). *Chemical deterioration and physical instability of food and beverages* (pp. 70–94). Oxford: Woodhead Publishing Limited.

Lund, M. N., & Ray, C. A. (2017). Control of Maillard reaction in foods: Strategies and chemical mechanisms. *Journal of Agricultural and Food Chemistry, 65*, 4537–4552.

Maasen, K., Scheijen, J. L. J. M., Opperhuizen, A., Stehouwer, C. D. A., Van Greevenbroek, M. M., & Schalkwijk, C. G. (2021). Quantification of dicarbonyl compounds in commonly consumed foods and drinks; presentation of a food composition database for dicarbonyls. *Food Chemistry, 339*, 128063.

Maillard, L. C. (1912). Action des acides aminés sur les sucres: Formation des mélanoidines por voie méthodique. *Comptes rendus de l'Académie des Sciences, 154*, 66–68.

Martins, S. I. F. S., Jongen, W. M. F., & Van Boekel, M. A. J. S. (2001). A review of Maillard reaction in food and implications to kinetic modelling. *Trends in Food Science and Technology, 11*, 364–373.

Matulová, M., Capek, P., Kaneko, S., Navarini, L., & Liverani, F. S. (2011). Structure of arabinogalactan oligosaccharides derived from arabinogalactan-protein of *Coffea arabica* instant coffee powder. *Carbohydrate Research, 346*, 1029–1036.

Mogol, B. A., & Gökmen, V. (2016). Thermal process contaminants: Acrylamide, chloropropanols and furan. *Current Opinion in Food Science, 7*, 86–92.

Monnier, V. M., & Taniguchi, N. (2016). Advanced glycation in diabetes, aging and age-related diseases: Conclusions. *Glycoconjugate Journal, 33*, 691–692.

Morales, F. J., Somoza, V., & Fogliano, V. (2012). Physiological relevance of dietary melanoidins. *Amino Acids, 42*, 1097–1109.

Moreira, A. S., Nunes, F. M., Domingues, M. R., & Coimbra, M. A. (2012). Coffee melanoidins: Structures, mechanisms of formation and potential health impacts. *Food & Function, 3*, 903–915.

Mottram, D. S. (1994). Some aspects of the chemistry of meat flavour. In F. Shahidi (Ed.). *The flavour of meat and meat products* (pp. 210–230). Berlin: Springer.

Mu, K., Wang, S., & Kitts, D. D. (2016). Evidence to indicate that Maillard reation products can provide selective antimicrobial activity. *Integrative Food, Nutrition and Metabolism, 3*, 330–335.

Namiki, M. (1988). Chemistry of Maillard reaction: Recent studies on the browning reaction mechanism and the development of antioxidants and mutagens. *Advances in Food Research, 38*, 115–184.
Namiki, M., & Hayashi, T. (1983). A new mechanism of the Maillard reaction involving sugar fragmentation and free radical formation. *ACS Symposium Series, 215*, 21–46.
Nass, N., Bartling, B., Navarette Santos, A., Scheubel, R. J., Börgermann, J., Silber, R. E., & Simm, A. (2007). Advanced glycation end products, diabetes and ageing. *Zeitschrift für Gerontologie und Geriatrie, 40*, 349–356.
Nooshkam, M., Varidi, M., & Bashash, M. (2019). The Maillard reaction products as foodborn antioxidant and antibrowning agents in model and real food systems. *Food Chemistry, 275*, 644–660.
Nooshkam, M., Varidi, M., & Verma, D. K. (2020). Functional and biological properties of Maillard conjugates and their potential application in medical and food: A review. *Food Research International, 131* Article 109003.
Nunes, F. M., & Coimbra, M. A. (2007). Melanoidins from coffee infusions. Fractionation, chemical characterization, and effect of the degree of roast. *Journal of Agricultural and Food Chemistry, 55*, 3967–3977.
Nunes, F. M., Reis, A., Silva, A. M. S., Domingues, M. R. M., & Coimbra, M. A. (2008). Rhamnoarabinosyl and rhamnoarabinoarabinosyl side chains as structural features of coffee arabinogalactan. *Phytochemistry, 69*, 1573–1585.
Nursten, H. E. (1981). Recent developments in studies of the Maillard reaction. *Food Chemistry, 6*, 263–277.
Nursten, H. (2005). *The Maillard reaction: Chemistry, biochemistry and implications.* Cambridge: The Royal Society of Chemistry.
O'Brien, J. (2009). Non-enzymatic degradation pathways of lactose and their significance in dairy products. In P. McSweeney, & P. Fox (Eds.). *Advanced Dairy Chemistry* (pp. 231–294). New York: Springer.
O'Mahony, J. A., Drapala, K. P., Mulcahy, E. M., & Mulvihill, D. M. (2017). Controlled glycation of milk proteins and peptides: Functional properties. *International Dairy Journal, 67*, 16–34.
Perez-Locas, C., & Yaylayan, V. A. (2010). The Maillard reaction and food quality deterioration. In L. H. Skibsted, J. Risbo, & M. L. Andersen (Eds.). *Chemical deterioration and physical instability of food and beverages* (pp. 70–94). Woodhead Publishing.
Phuong-Nguyen, K., McNeill, B. A., Aston-Mourney, K., & Rivera, L. R. (2023). Advanced glycation end-products and their effects on gut health. *Nutrients, 15*, 405.
Poulsen, M. W., Hedegaard, R. V., Andersen, J. M., de Courten, B., Bügel, S., Nielsen, J., ... Dragsted, L. O. (2013). Advanced glycation endproducts in food and their effects on health. *Food and Chemical Toxicology, 60*, 10–37.
Ramonaitytė, D. T., Keršienė, M., Adams, A., Tehrani, K. A., & De Kimpe, N. (2009). The interaction of metal ions with Maillard reaction products in a lactose–glycine model system. *Food Research International, 42*, 331–336.
Reineccius, G. A. (1990). The influence of Maillard reaction on the sensory properties of foods. In P. A. Finot, H. U. Aeschbacher, R. F. Hurrel, & R. Liardon (Eds.). *The Maillard reaction in food processing, human nutrition and physiology* (pp. 157–170). Basel: Birkhäuser.
Resconi, V. C., Escudero, A., & Campo, M. M. (2013). The development of aromas in ruminant meat. *Molecules (Basel, Switzerland), 18*, 6748–6781.
Reutter, M., & Eichner, K. (1989). Separation and determination of Amadori compounds by HPLC and post-column reaction. *Zeitschrift für Lebensmittel-Untersuchung und Forschung, 188*, 28–35.
Rifai, L., & Saleh, F. A. (2020). A review on acrylamide in food: Occurrence, toxicity, and mitigation strategies. *International Journal of Toxicology, 39*, 93–102.

Rizzi, G. P. (1999). The Strecker degradation and its contribution to food flavor. In R. Teranishi, E. L. Wick, & I. Hornstein (Eds.). *Flavor chemistry. Thirty years of progress* (pp. 335–343). New York: Kluwer Academic/Plenum Publishers.

Rizzi, G. P. (2003). Free radicals in the Maillard reaction. *Food Reviews International, 19*, 375–395.

Rizzi, G. P. (2008). The Strecker degradation of amino acids: Newer avenues for flavor formation. *Food Reviews International, 24*, 416–435.

Robert, L., & Labat-Robert, J. (2014). Role of the Maillard reaction in aging and age-related diseases. Studies at the cellular-molecular level. *Clinical Chemistry and Laboratory Medicine, 52*, 5–10.

Robert, L., Labat-Robert, J., & Robert, A.-M. (2010). The Maillard reaction. From nutritional problems to preventive medicine. *Pathologie Biologie, 58*, 200–206.

Robert, F., Vera, F. A., Kervella, F., Davidek, T., & Blank, I. (2005). Elucidation of chemical pathways in the Maillard reaction by ^{17}O NMR spectroscopy. *Annals of the New York Academy of Sciences, 1043*, 63–72.

Sanwar Gilani, G., Wu Xiao, C. W., & Cockell, K. A. (2012). Impact of antinutritional factors in food proteins on the digestibility of proteins and the bioavailability of amino acids and on protein quality. *British Journal of Nutrition, 108*, S315–S332.

Sanz, M. L., Corzo-Martinez, M., Rastall, R. A., Olano, A., & Moreno, F. J. (2007). Characterization and in vitro digestibility of bovine β-lactoglobulin glycated with galactooligosaccharides. *Journal of Agricultural and Food Chemistry, 55*, 7916–7925.

Sanz, M. L., del Castillo, M. D., Corzo, N., & Olano, A. (2001). Formation of Amadori compounds in dehydrated fruits. *Journal of Agricultural and Food Chemistry, 49*, 5228–5231.

Schönberg, A., & Moubacher, R. (1952). The Strecker degradation of α-amino acids. *Chemical Reviews, 50*, 261–277.

Seiquer, I., Díaz-Alguacil, J., Delgado-Andrade, C., Lopez-Frias, M., Hoyos, A. M., Galdo, G., & Navarro, M. P. (2006). Diets rich in Maillard reaction products affect protein digestibility in adolescent males aged 11–14 y. *American Journal of Clinical Nutrition, 83*, 1082–1088.

Shaheen, S., Shorbagi, M., Lotenzo, J. M., & Farag, M. A. (2022). Dissecting dietary melanoidins: Formation mechanisms, gut interactions and functional properties. *Critical Reviews in Food Science and Nutrition, 62*, 8954–8971.

Sikorski, Z. E., Pokorny, J., & Damodaran, S. (2008). Physical and chemical interactions of components in food systems. In S. Damodaran, K. L. Parkin, & O. R. Fennema (Eds.). *Fennema's food chemistry* (pp. 849–883). (fourth ed.). Boca Raton: CRC Press.

Smuda, M., & Glomb, M. A. (2013). Fragmentation pathways during Maillard-induced carbohydrate degradation. *Journal of Agricultural and Food Chemistry, 61*, 10198–10208.

Starowicz, M., & Zieliński, H. (2019). How Maillard reaction influences sensorial properties (color, flavor and texture) of food products? *Food Reviews International, 35*, 707–725.

Tagliazucchi, D., & Verzelloni, E. (2014). Relationship between the chemical composition and the biological activities of food melanoidins. *Food Science and Biotechnology, 23*, 561–568.

Toda, M., Hellwig, M., Henle, T., & Vieths, S. (2019). Influence of the Maillard reaction on the allergenicity of food proteins and the development of allergic inflammation. *Current Allergy and Asthma Reports, 19*, 4.

Tressl, R., Wondrak, G. T., Krüger, R.-P., & Rewicki, D. (1998). New melanoidin-like Maillard polymers from 2-deoxypentoses. *Journal of Agricultural and Food Chemistry, 46*, 104–110.

Van Boekel, M. A. J. S. (2001). Kinetic aspects of the Maillard reaction: A critical review. *Die Nahrung, 45*, 150–159.

Vlassara, H., & Uribarri, J. (2014). Advanced glycation end products (AGE) and diabetes: Cause, effect, or both? *Current Diabetes Reports, 14*, 453.

Wang, H.-Y., Qian, H., & Yao, W.-R. (2011). Melanoidins produced by the Maillard reaction: Structure and biological activity. *Food Chemistry, 128,* 573–584.

Wittmann, R., & Eichner, K. (1989). Detection of Maillard products in malts beers, and brewing couleurs. *Zeitschrift für Lebensmittel-Untersuchung und –Forschung, 188,* 212–220.

Yan, S., Wu, L., & Xue, X. (2023). α-Dicarbonyl compounds in food products: Comprehensively understanding their occurrence, analysis, and control. *Comprehensive Reviews in Food Science and Food Safety, 22,* 1387–1417.

Yaylayan, V. A. (2003). Recent advances in the chemistry of Strecker degradation and Amadori rearrangement: Implications to aroma and color formation. *Food Science and Technology Research, 9,* 1–6.

Yordanov, N. D., & Mladenova, R. (2004). EPR study of free radicals in bread. *Spectrochimica Acta, Part A: Molecular and Biomolecular Spectroscopy, 60,* 1395–1400.

Yu, J., Zhang, S., & Zhang, L. (2018). Evaluation of the extent of initial Maillard reaction during cooking some vegetables by direct measurement of the Amadori compounds. *Journal of the Science of Food and Agriculture, 98,* 190–197.

Zhai, Y., Cui, H., Hayat, K., Hussain, S., Tahir, M. U., Yu, J., ... Ho, C.-T. (2019). Interaction of added L-cysteine with 2-threityl-thiazolidine-4-carboxylic acid derived from the xylose-cysteine system affecting its Maillard browning. *Journal of Agricultural and Food Chemistry, 67,* 8632–8640.

Zhao, J., Wang, T., Xie, J., Xiao, Q., Cheng, J., Chen, F., ... Sun, B. (2018). Formation mechanism of aroma compounds in a glutathione-glucose reaction with fat or oxidized fat. *Food Chemistry, 270,* 436–444.

Žilić, S., Aktağ, I. G., Dodig, D., & Gökmen, V. (2021). Investigations on the formation of Maillard reaction products in sweet cookies made of different cereals. *Food Research International, 144,* 110352.

CHAPTER SEVEN

Structural changes in hemoglobin and glycation

Amanda Luise Alves Nascimento, Ari Souza Guimarães, Tauane dos Santos Rocha, Marilia Oliveira Fonseca Goulart, Jadriane de Almeida Xavier*, and Josué Carinhanha Caldas Santos*

Federal University of Alagoas, Institute of Chemistry and Biotechnology, Campus A. C. Simões, Maceió, Alagoas, Brazil
*Corresponding authors. e-mail address: jadrianexavier@iqb.ufal.br; josue@iqb.ufal.br

Contents

1. Introduction	184
2. Glycation of hemoglobin	187
2.1 Relationship between glycated hemoglobin and carbonyl stress	190
2.2 The use of HbA1c in plasma glucose monitoring	193
2.3 Glycation vs. fructation	194
3. Hb glycation: *in vitro* protocols evaluation	195
4. Structural changes in Hb after glycation	199
5. Techniques and methods for assessing structural and conformational changes in glycated Hb	200
5.1 UV–vis absorption spectroscopy	201
5.2 Molecular fluorescence spectroscopy	206
5.3 Mass spectrometry	209
5.4 Circular dichroism	210
5.5 X-ray diffraction	210
5.6 Dynamic light scattering	211
5.7 Microscopy	211
5.8 Nuclear magnetic resonance	212
5.9 Fourier transform infrared spectroscopy	212
5.10 Raman spectroscopy	213
5.11 Miscellaneous techniques	213
6. Trends	215
7. Conclusions	217
Acknowledgements	217
References	217

Vitamins and Hormones, Volume 125
ISSN 0083-6729, https://doi.org/10.1016/bs.vh.2024.02.001
Copyright © 2024 Elsevier Inc. All rights are reserved, including those for text and data mining, AI training, and similar technologies.

Abstract

Hemoglobin (Hb) is a hemeprotein found inside erythrocytes and is crucial in transporting oxygen and carbon dioxide in our bodies. In erythrocytes (Ery), the main energy source is glucose metabolized through glycolysis. However, a fraction of Hb can undergo glycation, in which a free amine group from the protein spontaneously binds to the carbonyl of glucose in the bloodstream, resulting in the formation of glycated hemoglobin (HbA1c), widely used as a marker for diabetes. Glycation leads to structural and conformational changes, compromising the function of proteins, and is intensified in the event of hyperglycemia. The main changes in Hb include structural alterations to the heme group, compromising its main function (oxygen transport). In addition, amyloid aggregates can form, which are strongly related to diabetic complications and neurodegenerative diseases. Therefore, this chapter discusses in vitro protocols for producing glycated Hb, as well as the main techniques and biophysical assays used to assess changes in the protein's structure before and after the glycation process. This more complete understanding of the effects of glycation on Hb is fundamental for understanding the complications associated with hyperglycemia and for developing more effective prevention and treatment strategies.

1. Introduction

Hemoglobin (Hb) is a globular protein from the group of hemeproteins and represents most of the protein content of red blood cells, also called erythrocytes (Ery), corresponding to around 95% of the average volume of these cells (Yoshida, Prudent, & D'alessandro, 2019). Its structure consists of a tetramer, and each of the four subunits contains a heme group (Fig. 1) (Kettisen & Bulow, 2021; Longo, Piolatto, Ferrero, & Piga, 2021). Hb, through the heme group, performs its primary function, the transportation of O_2 from the lungs to the body's tissues and organs, as well as assisting in the transport of CO_2 from the tissues back to the lungs to remove this gas from the body (Ahmed, Ghatge, & Safo, 2020). Hb is also involved in several other functions, including the regulation of oxygen-dependent NO metabolism, maintenance of serum pH, and the ability to participate in body heat exchange during the oxygenation/deoxygenation cycle (Kosmachevskaya & Topunov, 2018).

There are different isoforms of Hb, which differ in subunit composition and expression, varying during human development. The most common type in healthy adults (95–98%) is hemoglobin A (HbA), composed of two α subunits and two β subunits. Hemoglobin A2 (HbA2) is found in a smaller proportion (2–3%), comprised of two α subunits and two δ subunits, and has a lower affinity for O_2 than HbA. The main Hb isoform in

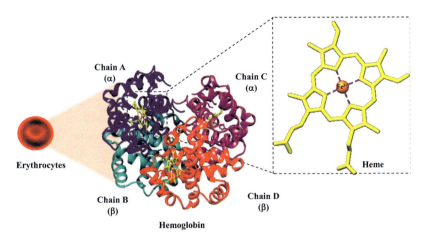

Fig. 1 Tetrameric structure of hemoglobin (PDB ID: 1A3N) present in erythrocytes, indicating the four subunits (α and β) and their respective heme groups.

fetuses and newborns is fetal hemoglobin (HbF), comprised of two α subunits and two γ subunits. HbF has a greater affinity for oxygen than HbA, facilitating the transfer of O_2 from the mother's blood to the fetus during pregnancy (Guo et al., 2019; Kaufman et al., 2023). Mutations in the β-globin gene can lead to the formation of HbA variants, resulting in different forms of hemoglobin. Among the most common variants are hemoglobin S (HbS), which is responsible for sickle cell anemia, and hemoglobin C (HbC), which can cause mild hemolytic anemia (Ding et al., 2018).

A fraction of HbA can undergo covalent modification through a spontaneous chemical reaction due to adding a free amine group from the protein to the carbonyl of reducing sugars. This process occurs through an irreversible non-enzymatic reaction called glycation, leading to the formation of the total fraction of glycated hemoglobin (HbA1) and various subtypes such as HbA1a, HbA1b and HbA1c (Heo et al., 2021), with HbA1c being defined as the most stable adduct of this reaction (Ding et al., 2018).

The first record of HbA1c was in 1955 when Kunkel and Wallenius observed a new Hb, in small quantities in adult human blood, but with electrophoretic properties different from HbA (Kunkel & Wallenius, 1955). In 1957, hemoglobin fractions that eluted faster than HbA were identified (Allen et al., 1958). About 10 years later, already termed HbA1c, glycated hemoglobin was defined as the fraction of HbA modified by the addition of a hexose to the *N*-terminal group of the β-chain and unique both as a glycoprotein and in the mode of sugar binding to the protein (Bookchin & Gallop, 1968). It also was observed that HbA1c was elevated in the erythrocytes of diabetic patients (Rahbar et al., 1969), and for the first time, the relationship between HbA1c, average glycemia, and chronic complications in diabetic patients was described (Trivelli et al., 1971), while Koenig et al. (1976) observed a correlation between HbA1c and average blood glucose levels. In 1993, the American Association for Clinical Chemistry (AACC) established a subcommittee to set an international standard for the measurement of HbA1c (Little et al., 2001). The American Diabetes Association (ADA) (2010) recommended HbA1c levels for diagnosing diabetes and pre-diabetes. Considering the impact of diabetes as a public health problem worldwide, in 2021, the World Health Organization (WHO) (2023) launched the Global Diabetes Compact to promote improvements in the prevention and control of the disease, especially in low- and middle-income countries. In addition, in 2022, the World Health Assembly endorsed five global targets for diabetes coverage and treatment by 2030 (WHO, 2023). Fig. 2 illustrates a timeline of the scientific evolution of the clinical use of HbA1c in diabetes from its discovery to the present day.

The process of Hb glycation leads to structural and, consequently, conformational changes in the protein's native structure, compromising its function, such as oxygen transportation by the heme group. In diabetic patients, Hb glycation is even more evident due to hyperglycemia and is

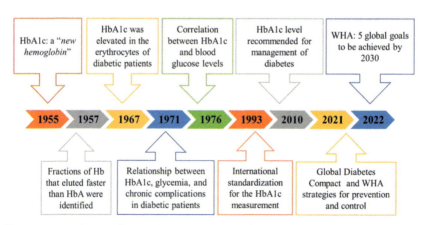

Fig. 2 Historical scientific developments in the clinical use of glycated hemoglobin (HbA1c).

associated with various complications. In this context, it is important to evaluate the consequences of glycation on Hb by exploring different experimental strategies for monitoring these changes, which could provide valuable insights into understanding the pathophysiology of the disease and help develop more effective and targeted therapeutic strategies.

This chapter discusses the main concepts involved in the process of glycation and carbonyl stress related to the formation of HbA1c and its use as a marker for diabetes. This process is closely related to the conformational changes that occur in the protein, so topics such as in vitro protocols for Hb glycation, structural changes in glycated Hb, and the main biophysical assays for assessing changes in protein structure are covered.

2. Glycation of hemoglobin

Glycation, also known as the Maillard reaction, is defined as a sequence of non-enzymatic reactions that begins with the addition of a nucleophilic group from a biomolecule to the carbonyl of reducing sugar, such as glucose and fructose, generating the so-called advanced glycation end products (AGEs) (Torres et al., 2018). This reaction also leads to the formation of reactive carbonyl species (RCS), such as α, and β-dicarbonyl compounds, which, when reacting with the amino group of the amino acid side chain, lead to the formation of AGEs (Thornalley, 2005). However, RCS are not only formed by the classic glycation reaction pathway but can be generated by various oxidation reactions in the body. Thus, although the Maillard reaction is well established as a trigger for the formation of AGEs, all the reactions that lead to the formation of RCS in the body can contribute to their formation. The condition characterized by the steady state of RCS, which leads to the accumulation of AGEs, is called carbonyl stress. RCS, especially α,β-dicarbonyls such as methylglyoxal (MG), have attracted particular attention due to their reactivity, which is around 20,000 times greater than that of fructose or glucose and can react with biomolecules to directly generate the respective AGEs (Kosmachevskaya et al., 2021; Thornalley, 2005).

In recent years, numerous AGEs have been identified in vivo and in vitro, which can be classified into different groups based on their chemical structures and ability to emit fluorescence (Perrone et al., 2020; Popovaet al., 2010) (Fig. 3), among which are: (i) fluorescent AGEs with cross-links: pentosidine (Sirangelo & Iannuzzi, 2021), vespertysine A (Tessier et al., 1999),

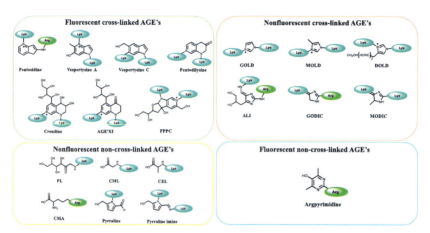

Fig. 3 Examples of AGEs according to their classification considering the intrinsic fluorescence and the presence or absence of cross-linking in the compound structures.

vespertysine C, pentodilysine (Popova et al., 2010), crossline (Sirangelo & Iannuzzi, 2021), AGE'XI, FPPC (Popova, et al., 2010); (ii) non-fluorescent AGEs with cross-links: glyoxal-lysine dimer (GOLD) (Klaus et al., 2018), methylglyoxal-lysine dimer (MOLD) (Ahmed et al., 1997; Nemet et al., 2006), DOLD (Perrone et al., 2020), arginine-lysine imidazole cross-links (ALI) (Popova et al., 2010; Sirangelo & Iannuzzi, 2021), glyoxal imidazo-limine lysine-arginine (GODIC) (Klaus et al., 2018), methylglyoxal-lysine dimer (MODIC) (Nemet et al., 2006); (iii) non-fluorescent AGEs without cross-links: N-fructosyl-lysine (FL) (Sirangelo & Iannuzzi, 2021), N-ε-(carboxymethyl)lysine (CML), N-(1-carboxyethyl)lysine (CEL) (Ahmed et al., 1997; Sirangelo & Iannuzzi, 2021), N-ϖ-carboxymethyl-arginine (CMA) (Odani et al., 2001), Pyrraline (Liang et al., 2019; Perrone et al., 2020), Pyrraline imine (Popova et al., 2010) and (iv) fluorescent AGEs without cross-links: Argpyrimidine (Nemet et al., 2006).

Despite the structural variability of AGEs, there is a common characteristic, such as the ability to form stable covalent bonds with proteins, lipids, and nucleotides (Miller et al., 2003), resulting in the structural modification of these compounds, which can lead to functional alterations in proteins and lipids, contributing to affecting the normal function of cells and tissues. Thus, monitoring the fluorescence emission of systems containing proteins and reducing sugars is one of the analytical strategies used to assess the degree of AGE formation, as well as the inhibition of their formation by different compounds in vitro studies.

The formation of HbA1c occurs mainly due to the exposure of Hb to blood glucose. During this reaction, the free aldehyde of glucose condenses with the N-amino group of the terminal valine in one of the β chains of Hb (β-Val-1), forming an intermediate cabinolamine, which undergoes dehydration producing an unstable Schiff base, which by Amadori rearrangement, leads to the formation of more stable compounds, the ketoamines or fructosamines, also known as Amadori products, such as HbA1c (Fig. 4), and consequently act as precursors of AGEs (González-Viveros et al., 2021; Kosmachevskaya et al., 2021).

Protein glycation alters secondary and tertiary structure, leading to a loss of function or abnormality (Bakhti et al., 2007; Huang et al., 2016; Kazemi et al., 2018). Sen et al., (2005) showed that HbA1c undergoes structural changes and functional modifications, exhibiting more surface Trp residues and higher thermal stability than normal Hb and resulting in the conversion of α-helix to β-sheet structures. Additionally, interactions in HbA1c are usually weaker than in Hb, thus leading to the release of Fe(II), which can catalyze the formation of radicals and increase the oxidative stress associated with diabetes (Bose & Chakraborti, 2008).

Hb glycation process alters the physical properties of erythrocytes and can interfere with the basic function of cell-related oxygen and carbon dioxide transportation (Samanta, 2021; Shin et al., 2008; Watala et al., 1996). Under normal conditions, red blood cells have a characteristic biconcave disk shape, which gives them greater flexibility. This shape helps them to deform and pass through narrow capillaries, promoting blood circulation (Diez-Silva et al., 2010; Elgsaeter & Mikkelsen, 1991). The cell membrane in these globules mainly comprises lipids and proteins, such as spectrin, which provides elasticity, promoting deformation and flexibility (Barbarino et al., 2021; Mariana, 2023; Nash & Gratzer, 1993). Therefore, glycation of Hb can cause a loss of flexibility and deformity in red blood cells,

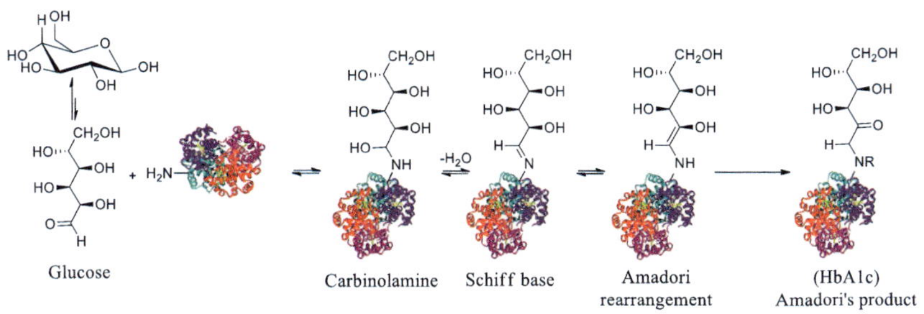

Fig. 4 Simplified representation of the glycated hemoglobin (HbA1c) formation process.

making them more rigid and less malleable, mainly due to changes in the structure of red blood cell membrane proteins, affecting blood flow and impairing the deformability of erythrocytes (Kosmachevskaya et al., 2021; Shin et al., 2007; Singh & Shin, 2009). Besides, glycation of Hb can alter its ability to bind/release oxygen, thus, consequently impacting the erythrocyte's ability to transport O_2 to the tissues (Elliott, 2008; Negre-Salvayre et al., 2009).

The glycation process of Hb depends on the pK_a of amino groups' and neighboring residues' charge and steric effects. In general, the pK_a values of the α-amino groups of the N-terminal residues, especially that of βVal-1 ($pK_a \sim 7$), are lower than those of the ε-amino groups of the Lys residues (between 10 and 11) in Hb (Shapiro et al., 1979). Thus, after forming the Amadori product, glucose can still bind to N-terminal groups of valine residues, conferring an additive negative charge to the respective glycation products (Chen et al., 2022). Thus, as Hb and HbA1c have different isoelectric points, this effect is exploited in separation techniques, such as capillary electrophoresis and cation exchange chromatography ((Chen et al., 2022; Quan, 2008). Additionally, using mass spectrometry, it has been shown that other Hb sites can also be glycated (Heo et al., 2021; Muralidharan et al., 2020), in addition to β-Val-1, α-Lys-61 (Heo et al., 2021; Rabbani & Thornalley, 2021), and β-Lys-66 (Rabbani & Thornalley, 2021) are possible glycation sites (Heo et al., 2021). This process occurs symmetrically between the α and β globin, with a preference for the unmodified globin (Muralidharan et al., 2020). Structural changes resulting from glycation lead to variations in the conformation of Hb, compromising its main functions and triggering deleterious physiological processes to human health.

2.1 Relationship between glycated hemoglobin and carbonyl stress

In order to perform their vital function, the Ery needs to maintain an efficient and productive metabolism. In this context, glucose metabolism is fundamental in the Ery since it is the main source of energy for these cells, which have a unique characteristic: they have no nucleus or mitochondria, making them exclusively dependent on glycolysis to generate energy (Sae-lee et al., 2022; Wijk & Solinge, 2005). The glucose transporter GLUT1, expressed in Ery, is responsible for glucose uptake and transportation to the glycolytic pathway (Kosmachevskaya et al., 2021). In Ery, the intracellular concentration of methylglyoxal (MG) depends on glucose metabolism in the glycolytic pathway (Fig. 5). Triose-phosphate isomerase

Structural changes in hemoglobin and glycation

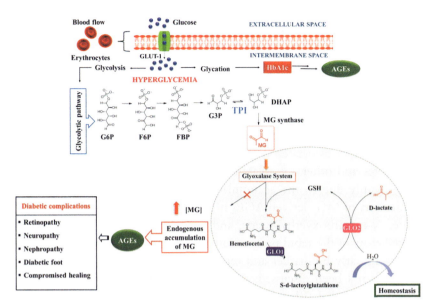

Fig. 5 Glycation reaction process and carbonyl stress in diabetes and its complications. The type 1 glucose transporter (GLUT-1) is present in erythrocytes and favors the passage of glucose from the extracellular environment, i.e., the bloodstream, to the intracellular environment. In hyperglycemia, where there is an increase in the amount of glucose available in the blood, the entry of glucose into the cell via GLUT-1 is also increased. Inside the cell, excess glucose can serve as a substrate for the glycation of hemoglobin, resulting in the formation of HbA1c and the direct formation of advanced glycation end products (AGEs). In addition, the availability of glucose molecules influences glycolytic activity within the cell. The glycolytic pathway involves a series of reactions, including glucose 6-phosphate (G6P), fructose 6-phosphate (F6P), fructose 1,6-bisphosphate (FBP), glyceraldehyde 3-phosphate (3GP). The enzyme triosephosphate isomerase (TPI) plays an important role in the balance between G3P and dihydroxyacetone phosphate (DHAP). When this balance is not maintained, and an excess of DHAP occurs, it can be degraded to methylglyoxal (MG) by MG synthase. Therefore, alterations in TPI function can lead to the endogenous accumulation of MG, favoring the formation of AGEs. After MG formation, a detoxification system called the glyoxalase system comes into play. In turn, this system is made up of two enzymes: glyoxalase I (GLO1) and II (GLO2), and GSH, which act to convert MG into less reactive molecules, such as D-lactate, to mitigate the formation of AGEs. The action of the glyoxalase system leads to organic homeostasis. On the other hand, when these enzymes cannot adequately perform their detoxification functions, there is an increase in their endogenous concentration and, consequently, a greater formation of AGEs. This condition is directly related to the main complications of diabetes *mellitus* (DM), such as retinopathy, neuropathy, nephropathy, diabetic foot, and impaired healing.

is the enzyme that adjusts the balance between glyceraldehyde-3-phosphate (G3P) and dihydroxyacetone phosphate (DHAP), so a decrease in its activity leads to the accumulation of DHAP, which is spontaneously degraded, leading to the formation of MG, thus contributing to the endogenous accumulation of this carbonyl species (Kosmachevskaya et al., 2021).

In diabetic patients, the concentration of MG can be up to 6 times higher, which is associated with the development of vascular complications of diabetes and other age-related diseases, such as neurodegenerative diseases, mainly due to the accumulation of AGEs (Barinova et al., 2023). Biomarkers of MG-derived AGEs in plasma, such as N-ε-(carboxymethyl) lysine, N-ε-(carboxyethyl)lysine, and hydroimidazolones, have been used as predictors of the rapid progression of diabetic nephropathy. The role of MG as an early biomarker and monitoring of long-term metabolic complications has been discussed; therefore, assays to quantify MG in blood plasma are needed (Bhat et al., 2019). MG can promote rapid glycation through interaction with cellular lipids, deoxyribonucleic acid (DNA), and extracellular proteins, mainly with lysine residues. therefore, when MG levels are above the normal physiological level, as in hyperglycemia, cellular dysfunction, genotoxic effects, and negative health consequences can occur (Saeed et al., 2020).

On the other hand, a group of enzymes is capable of playing an important role in the detoxification of MG and other RCS formed in the glycation reaction. This group is called the glyoxalase system (GLO), which is composed of two enzymes, glyoxalase I (GLO1) and glyoxalase II (GLO2), together with reduced glutathione (GSH), which catalyze the conversion of reactive acyclic α-oxoaldehydes into their corresponding α-hydroxy acids and is considered the primary mechanism for limiting the formation of AGEs (Aragonès et al., 2021). Fig. 5 shows the glycation reaction, process, and carbonyl stress in diabetes and its complications.

Considering that AGE levels depend on blood glucose concentration, several studies have investigated the role of GLO1 in diabetes. Studies in mice and rats have shown that overexpression of GLO1 could delay the onset of diabetes complications, such as retinopathy and nephropathy (Berner et al., 2012; Brouwers et al., 2014; Giacco et al., 2014). *In vitro*, a lower expression of GLO1 in cells treated with high glucose concentrations was associated with the accumulation of cellular MG (Yao & Broenlee, 2010). Red blood cells from diabetic rats showed higher levels of GLO1 compared to those without the disease, i.e., without excessive exposure to glucose, and this evidence is related to studies carried out on humans, indicating a possible compensatory response to carbonyl stress (Aragonès et al., 2021; Atkins & Thornally, 1989; McLellan & Thornalley, 1989). Studies have shown increased plasma MG and D-lactate in humans, both in fasting hyperglycemia and the postprandial period. Together, these data suggest that carbonyl stress is a determining factor in the onset of the micro- and macrovascular complications of diabetes (Maessen et al., 2015; Rabbani & Thornalley, 2019).

2.2 The use of HbA1c in plasma glucose monitoring

HbA1c is currently considered an important marker for screening, diagnosing, and monitoring individual glycemic control, reflecting the average blood glucose over the last three months, corresponding to the half-life of Ery, so its measurement is useful for indicating hyperglycemia. In addition, the clinical use of this marker reflects the progression of microvascular lesions in diabetes *mellitus* (DM) and is widely used in this condition (Guo et al., 2019; Hempe & Hsia, 2022; Muralidharan et al., 2020).

According to the ADA (2022), HbA1c should be used as a parameter for screening and diagnosing type 2 DM, with the values determined as the cut-off point for pre-diabetes: 5.7–6.4%, and for diabetes: ≥6.5% of total Hb content, in the absence of unequivocal hyperglycemia. Therefore, two HbA1c tests with abnormal results in the same or different samples are recommended for diagnosis. However, it is worth noting that for individuals with altered red cell turnover, such as hemolytic or iron deficiency anemia, HbA1c should not be used for diagnostic purposes.

HbA1c is not commonly used as a diagnostic parameter for gestational diabetes mellitus (GDM) or to assess the effects on the fetus in the context of GDM. Although there is no standardization of the classification and diagnostic criteria for GDM, health organizations such as the ADA and the WHO recommend adopting guidelines (Mendonça et al., 2022). Thus, the guideline is to diagnose GDM between the 24th and 28th week of pregnancy by analyzing the pregnant woman's glycemic parameters using a fasting blood glucose test and an oral glucose tolerance test. HbA1c values, however, can be important in early pregnancy to rule out the presence of pre-existing diabetes (Mendonça et al., 2022; Metzger et al., 2010). A systematic review based on a meta-analysis aimed at assessing diagnostic accuracy concluded that the assessment of GDM based on HbA1c values results in few cases of false positives; however, it can lead to high false negatives (Renz et al., 2019). One reason could be that the HbA1c content does not provide real-time information on glucose levels, an essential aspect of the accurate diagnosis of GDM.

Although HbA1c levels are directly determined by serum glucose and Hb concentrations, studies have shown that other factors can interfere with them, such as race, age, diet, medication use, inflammatory processes, and diseases such as obesity and asthma. Therefore, the individual variability of this marker should be carefully monitored (Guo et al., 2019; Hempe & Hsia, 2022; Muralidharan et al., 2020). A systematic review with meta-analysis conducted to evidence about the variability of HbA1c during several consultations of the same individual with vascular complications and death in individuals with type 2 diabetes concluded that a high variability of this marker is related to greater vascular complications and even death in people affected by the disease, emphasizing the importance of care in preventing adverse outcomes in the public studied (Qu et al., 2022).

In this context, considering the possible individual variability of HbA1c and the limitations of its clinical use, Hempe et al., (2002) proposed a mathematical model called the Hemoglobin Glycation Index (HGI), aimed at quantifying measurements and predicting the HbA1c value, which has since been widely used. In addition, the glycation gap (GGap) and glycated albumin (GA) are also increasingly employed in clinical practice. It is worth noting that the lifespan of erythrocytes does not influence GA and is, therefore, an excellent alternative indicator for assessing glycemic control. Furthermore, it is recognized that the higher the levels of HGI, GGap, and GA, the greater the possibility of risk associated with diabetes complications and cardiovascular diseases. Therefore, the use of these indices is important for individualizing the treatment of diabetic individuals, as well as for preventing their complications (Wang et al., 2022; Xing et al., 2023).

2.3 Glycation vs. fructation

When discussing the glycation process and its health implications, it is essential to better understand some definitions. Although some authors use the term fructation to refer specifically to the reaction between the amino group of a biomolecule and the carbonyl group of the open form of fructose (reducing sugar), the term glycation itself has a broader scope. It can occur between any nucleophilic group of a biomolecule and the carbonyl of reducing sugar (glucose or fructose, for example) or even a RCS. Thus, the terminology fructation is usually used to specify that a reaction with fructose carried out the glycation of that protein.

Fructose has been widely used in vitro protein glycation experiments. It was chosen because it is the most common monosaccharide in the human diet and is more reactive than glucose. Thus, the generation of AGEs occurs in a shorter time, resulting in faster structural changes in the glycated protein. Although aldehydes are more electrophilic than ketones, fructose (ketose) is in more significant proportion in the open form in solution than glucose (aldose). This is due to the stability of the pyranosidic form of glucose, making it more likely to be in the cyclic form, which is not reactive. As a result, fructose is ~7.5 times more reactive than glucose (Torres et al., 2018; Vistoli et al., 2013).

In vitro glycation studies allow glycation to be mimicked under simulated physiological conditions and play a crucial role in investigating this reaction's effects on proteins and discovering potential inhibitors. However, in addition to monitoring the formation/inhibition of fluorescent AGEs, it is also essential to carry out assays that can corroborate the assessment of the extent of glycation. During the glycation reaction, the protein's free amine group is compromised, so monitoring amine groups before and after the reaction is an important parameter; greater blocking of amine groups has shown a positive correlation with the higher formation of AGEs. Since glycation leads to changes in the native structure of the protein, assays that evaluate structural changes in the protein are essential for a more complete evaluation. In addition, RCS assays are also necessary, as they lead to much faster formation of AGEs due to their high reactivity. Thus, compounds that can capture RCS are essential for inhibiting the glycation process since these species are formed not only by the classical route.

3. Hb glycation: *in vitro* protocols evaluation

Glycated Hb is an important biochemical marker for DM. Thus, models to mimic this process have been evaluated, as well as inhibition of Hb glycation in vitro, to understand the formation of AGES, the consequences on the native structure of the protein, and, in this way, to evaluate possible glycation inhibitors.

Antiglycants are substances that can potentially reduce and prevent the damage caused by glycation. Thus, different compounds of synthetic or natural origin can act as inhibitors at any stage of AGE formation, either by reducing or removing glycation intermediates and end products

(Nagaiet al., 2014). The mechanism of action of the inhibitor varies according to the structure and conditions of the environment, so there are a variety of mechanisms, and they can act as sugar competitors, protein competitors, RAGE blockers, eliminating reactive oxygen species and reactive carbonyl species, chelating transition metals and breaking covalent cross-links (Abbas et al., 2016; Feng et al., 2023; Song et al., 2021). Among the best-known and used in vitro assays as an anti-glycating standard, highlight aminoguanidine (pimagedine), which, although it has been the subject of phase III clinical trials, has had its use suspended due to adverse side effects. Metformin is used to treat type 2 DM, which, like aminoguanidine, reacts with dicarbonyls such as MG. Another relevant substance is acetylsalicylic acid, as it can transfer an acetyl group to amino acid residues, such as lysine, preventing the glycation reaction from occurring with glucose and contributing to the reduction of AGEs (Torres et al., 2018). Several possible inhibitors, natural or synthetic, are being studied concerning the inhibition of AGEs and the structural changes resulting from glycation.

Generally, Hb is glycated in vitro in a system containing the protein, phosphate buffer, and reducing sugar or RCS and incubated at 37 °C for weeks, and the progression of glycation is assessed during this period using spectroscopic techniques. However, as we can see (Table 1), there is still no standardization regarding the experimental conditions for the in vitro glycation process of Hb. There is a wide variation in the experimental systems, especially concerning the Hb, reducing sugars, and RCS ratio. These variations directly impact the extent of protein glycation, resulting in the inability to compare results between studies. Incubation time, for example, varies greatly and is directly related to the extent of glycation. Therefore, the lack of standardization makes it difficult to coherently compare results in experimental trials aimed at evaluating substances capable of attenuating and/or inhibiting glycation, as well as in the process of assessing the structural and conformational variations of the protein during the glycation process simulating physiological conditions. Analyzing and monitoring the structural changes resulting from the glycation process is vital to understanding the mechanisms underlying this biochemical reaction and its biological implications. Specific techniques/assays that enable precise and detailed analysis of the extent of glycation, especially in the presence of possible inhibitors, are needed to investigate these alterations. There are several approaches for this purpose, each with advantages and limitations.

Table 1 Experimental protocols used for Hb glycation in vitro. Experimental conditions

[Hb] (µM)	Carbonyl compound or RCS	Ratio [carbonyl compound]/[Hb]	Buffer	pH	T (°C)	Incubation (days)	References
77.5	F or ribose (200 mM)	2600	Phosphate (20 mM)	7.2	25	5	Hasan & Naeem (2020)
155	G (40 mM)	258	Phosphate (50 mM)	7.4	37	35	Sahebi & Divsalar (2016)
31	MG or GO or GA or GC (10 mM)	323	Phosphate (100 mM)	7.4	37	28	Lee et al., (2021)
155	G and F (40 mM)	258	Phosphate (50 mM)	7.4	37	35	Kazemi et al., (2019)
100	Acrylamide (1 mM)	10	Phosphate (100 mM)	7.4	37	2	Favinha et al. (2020)
100	MG (0.10 mM)	1	Phosphate (50 mM)	6.6	25	3	Bose et al. (2013)
100	F (0–0.16 mM)	0–2	Phosphate (50 mM)	6.6	22	5	Bose & Chakraborti, (2008)

(continued)

Table 1 Experimental protocols used for Hb glycation in vitro. (cont'd)
Experimental conditions

[Hb] (μM)	Carbonyl compound or RCS	Ratio [carbonyl compound]/[Hb]	Buffer	pH	T (°C)	Incubation (days)	References
2000	F (50 mM)	25	Phosphate (50 mM)	7.4	37	5	Sattarahmady et al. (2014)
1860	G (111 mM)	60	Only in water	—	37	5	Bae & Lee, (2004)
1000	G or F (200 mM)	200	Phosphate (100 mM)	8.0	25	0.083	Ioannou & Varotsis, (2017)
15	F (30 mM)	2000	Phosphate (50 mM)	7.4	37	20	Ghazanfari-Sarabi et al. (2019)
15	F (30 mM)	2000	Phosphate (50 mM)	7.4	37	20	Bakhti et al. (2007)

Legend: RCS (reactive carbonyl species); F (fructose); G (glucose); GO (glyoxal); MG (methylglyoxal); GA (DL-glyceraldehyde); GC (glycolaldehyde dimer).

4. Structural changes in Hb after glycation

The main chemical changes that occur in Hb result from reactions or processes that can modify the structure and function of the protein. Among the most relevant reactions/processes are: (i) oxidation: Hb can undergo oxidation, leading to the formation of disulfide bonds between the polypeptide chains and the inability to transport oxygen from the lungs to the tissues and remove carbon dioxide from the tissues to the lungs (intrinsic enzymatic activity) (Buehler & Alayash, 2007; Ramos et al., 2021); (ii) denaturation: exposure of Hb to extreme conditions of pH, temperature and pressure can lead to denaturation of the protein, resulting in loss of its biological function (Kristinsson & Hultin, 2004; Rieder, 1970); (iii) glycation: Hb can react with reducing sugars, such as glucose, leading to the formation of advanced glycation end products (AGEs). Glycation alters the structure of Hb, reducing its solubility and triggering a loss of activity, thus preventing it from binding to oxygen and consequently reducing Hb's ability to transport the gas to the tissues (Kazemi et al., 2018; Kazemi et al., 2019).

Glycation leads to one of the most significant chemical changes in Hb, resulting in the formation of cross-links between the polypeptide chains, thus increasing the rigidity of the protein and reducing its ability to bind and transport oxygen (Sirangelo & Iannuzzi, 2021). Along these lines, the glycation process of Hb can affect its structure in various ways. In general, the main structural changes to Hb after glycation are: (i) alteration of the overall electrical charge: the reaction of sugar molecules with Hb can alter the protein's electrical charge, which affects its solubility and interactions with other molecules (Abdalla et al., 2022); (ii) chemical modification of subunits: glycation can affect the individual subunits of Hb, resulting in changes in the conformation and function of the protein (Kazemi et al., 2019); (iii) formation of aggregates: glycation can lead to the formation of Hb aggregates, which can interfere with the normal function of the cell and lead to complications in patients with diabetes (Hasan & Naeem, 2020; Rahmanifar & Miroliaei, 2020).

Since the glycation process affects the structure of proteins, it plays a fundamental role in forming amyloid aggregates, both in vivo and in vitro. The formation of amyloid structures is characterized as a process in which abnormally folded proteins aggregate into poorly soluble fibrils devoid of normal biological activity. Both processes affect the normal functions performed by the protein, enabling adverse effects to occur (Stefani & Dobson, 2003). Thus, the impact of glycation on the process of protein amyloid aggregation has been widely studied, including the role of AGEs in the process of β-amyloid peptide (Aβ) formation associated with Alzheimer's disease (Li et al., 2013; Tsoi et al., 2023), as well as the association of Hb glycation modifications on amyloidogenic properties and their role in the accumulation of additional amyloid during the progression of diabetes (Sirangelo & Iannuzzi, 2021).

Conformational and functional changes in Hb because of the glycation process can also trigger changes in the heme group, which the Soret band can assess. In the molecular absorption spectrum of Hb, the Soret band is one of the most intense and reflects the absorption of electromagnetic radiation in the 400–450 nm range, attributed to the absorption of light by the heme group present in the protein (Moulick et al., 2007). Usually, the most common electronic transition in the Soret band is $\pi \rightarrow \pi^*$. The Fe(II) ion of the metalloporphyrin ring (heme group) accepts the free electrons of the N atoms of the pyrrole rings, while the electrons of the metal ion are donated to the porphyrin molecule, which allows a flow of electrons within the delocalized π system (Zheng et al., 2008).

Thus, the structural alterations of Hb can be monitored directly or indirectly. Various techniques and assays have been developed and used to evaluate the conformation of the protein, including molecular fluorescence, circular dichroism (CD), UV–vis molecular absorption, electrochemical techniques (Li et al., 2009; Zhao et al., 2011), FTIR, Raman spectroscopy (Barman et al., 2012) and mass spectroscopy (Muralidharan et al., 2020), among others, which can provide information on the folding/unfolding of proteins as a result of the glycation process due to the structural changes that have occurred.

5. Techniques and methods for assessing structural and conformational changes in glycated Hb

The structural or conformational changes resulting from the glycation process of Hb, whether through the formation of aggregates (amyloid fibrils), alterations in the subunits (Soret bands, or the formation of AGEs), as well as other structural changes that occur in Hb, altering its natural functions, can be assessed using different techniques. Thus, the interaction between electromagnetic radiation and matter, such as UV–vis absorption spectroscopy and molecular fluorescence, Raman spectroscopy, circular dichroism, microscopy, and X-ray diffraction, among others, are used to obtain direct and indirect information regarding changes in the structure of Hb resulting from the glycation process (Table 2) (Dodero & Messina, 2013; Pignataro et al., 2020). In this context, there are direct biophysical methods that allow the assessment of structural and conformational changes in Hb after glycation. In contrast, indirect methods do not directly assess glycated Hb or its conformational change but rather the effects of these changes due to functional assessment or changes in physical properties (Perrone et al., 2020; Sarmah & Roy, 2022; Zaman et al., 2019).

5.1 UV–vis absorption spectroscopy

UV–visible absorption spectroscopy (UV–vis) makes it possible to measure the absorption of electromagnetic radiation by the sample at different wavelengths, so the absorption profile and intensity of Hb can be used to assess the extent of the glycation process and conformational changes in the protein (Pignataro et al., 2020). The UV–vis spectrum of Hb shows the Soret band between 410 and 422 nm and two Q bands (~440 and 573 nm). Information about the secondary structure of the heme group can be defined from the evaluation of the Soret band (Heydari et al., 2022; Zheng et al., 2008). Usually, the glycation process leads to a hypochromic effect and a ~10 nm blue shift in the Soret band, in addition to the disappearance of the Q bands concerning the native Hb, indicating a change in the conformation of the Hb, which may be due to factors such as the presence of free radicals generated by glycation, which oxidize the iron in the heme, influencing the function performed by the protein (Sahebi & Divsalar, 2016).

Conformational changes and the formation of AGEs due to glycation in Hb can be assessed using colorimetric methods. In these, colorimetric probes or specific reagents are utilized to measure various parameters related to structural changes in Hb due to glycation. Amadori's product (ketoamine), considered an early AGE, can be determined using the NBT (nitroblue tetrazolium) assay. NBT is a light-yellow dye that, when in contact with ketamine, is reduced, forming a blue-colored chromophore known as formazan (525 nm) (Esackimuthu & Saraswathi, 2021). The TNBSA (2,4,6-trinitrobenzenesulfonic acid) assay is used to quantify the modification of lysine residues in the glycation process. This method is based on the reaction of TNBSA with the primary amine groups present on the lysine residues, forming N-trinitrophenylamines, a chromophore that can be measured between 335 and 340 nm (Cayot & Tainturier, 1997; Esackimuthu & Sarasathi, 2021). The dinitrophenylhydrazine (DNPH)

Table 2 Techniques for assessing conformational changes in glycated hemoglobin to obtain direct and/or indirect information.

Technique	Information Direct	Information Indirect	Advantage	Disadvantages	Conformational change assessed	References
UV–vis absorption spectroscopy	X		Low cost; simple; fast; non-destructive; versatile; instrumentation easily available	Low sensitivity, interference from species that absorb in the same spectral region evaluated	Amyloid fibrils, Soret band, protein tertiary structure	Moulick et al. (2007), Oliveira & Domingues (2018), Saeed & Abolaban (2021), Wang et al. (2017).
Molecular fluorescence spectroscopy	X	X	Fast, simple, non-destructive, and enables multiple applications	Requires specific probes to determine non-fluorescent AGEs; interference from species emitting in the same spectral region	Fluorescent and non-fluorescent AGEs, protein tertiary structure; amyloid fibrils	Perrone et al. (2020), Pignataro et al. (2020), Sen et al. (2005), Wang et al. (2017), Zaman et al. (2019)
Mass spectroscopy	X		High sensitivity, reproducibility, and accurate results	Complexity and availability of instruments, restricted dynamic range, destructive technique.	Formation of AGEs, fragmentation profile, secondary and tertiary structure of proteins	Cerqueira et al. (2020), Chen et al. (2022), Pundir & Chawla (2014),

Circular dichroism	×	Fast; small sample size; non-destructive	Limited application, low sensitivity in the visible region, organic solvent spectral cutoff	Fibrillar aggregation, Soret band, protein secondary and tertiary structure	Bose & Chakraborti (2008), Kumagai et al. (2017) Wang et al. (2017)
X-ray diffraction	×	High-resolution, non-destructive	Requires regular crystals, causes radiation damage, limited instrumental availability	Secondary and tertiary protein structure; amyloid fibrils	Carugo & Carugo (2005), Maveyraud & Mourey (2020), Mertens & Svergun (2010), Ogawa et al. (2019), Zaman et al. (2019).
Dynamic light scattering (DLS)	×	Wide applicability, fast, small amount of sample required, non-destructive	Limited analysis range, need to approximate spherical shape; low resolution.	Amyloid fibrils, monitoring the formation of protein aggregates	Moulick et al. (2007), Naftaly et al. (2021) Oliveira & Domingues (2018), Uskoković (2012), Zaman et al. (2019)

(*continued*)

Table 2 Techniques for assessing conformational changes in glycated hemoglobin to obtain direct and/or indirect information. (cont'd)

Technique	Information Direct	Information Indirect	Advantage	Disadvantages	Conformational change assessed	References
Microscopy	×		High resolution; high sensitivity; non-destructive, imaging	Sample preparation, vacuum conditions, limitation of structural information, and size	Formation of protein and fibril aggregates	Agronskaia et al. (2008), Bouma et al. (2003), Dybas et al. (2022), Iram et al. (2013), Shimasaki et al. (2017)
Nuclear magnetic resonance (NMR) spectroscopy	×	×	High resolution; versatile; non-destructive	Instrumental complexity, sample preparation (deuterated solvents), broadening of spectral lines	Heme group; secondary and tertiary structure; fibril structure	Cheng, Li, Li, Lu, & Wang (2000), Cho et al. (2022), Khajehpour, et al. (2006), Loquet et al. (2018), Zuppi et al. (2002)

Fourier transform infrared spectroscopy (FTIR)	×	Non-destructive; easy sample preparation	Low sensitivity, water content (spectral interference)	Secondary structure, protein fibrils	Dodero & Messina (2013), Khajehpour et al. (2006), Oberg & Fink (1998), Saeed & Abolaban (2021)
Raman spectroscopy	×	High reproducibility, non-destructive, applicability to biological cells and tissues	Presence of spurious background signals; weak Raman scattering	Secondary and tertiary structure of proteins	Barman et al. (2012), Barth & Zscherp (2002), Noothalapati et al. (2021), Wang et al. (2017)

assay measures carbonyl group content, a key biomarker to produce reactive oxygen species (ROS). After the reaction of dinitrophenylhydrazine with the glycated protein, the chromophore formed can be quantified at 370 nm (Rahmanifar & Miroliaei, 2020). Hb can catalyze the degradation of H_2O_2. Thus, to determine the peroxidase-like activity of Hb, the o-dianisidine method is used, which forms a chromophore that can be measured at 450 nm (Esackimuthu & Sarasathi, 2021). Congo red (CR) is a symmetrical sulfonated azo dye with a hydrophobic center consisting of a biphenyl group spaced between negatively charged sulfonate groups (Fig. 6) and is used to monitor the formation of protein aggregates. The complex (native CR-Hb) has an absorption maximum of ~500 nm, indicative of the interaction of CR with Hb with distorted and asymmetric conformations. However, when CR binds to amyloid fibrils, there is a bathochromic shift of around 30 nm (Hasan & Naeem, 2020).

5.2 Molecular fluorescence spectroscopy

Steady-state molecular fluorescence spectroscopy makes it possible to measure the emission of radiation from a sample when it is excited at a specific wavelength. After glycation, the fluorescent properties of Hb can be altered due to changes in its three-dimensional structure, for example, the oxidation process of the protein due to exposure to reactive species generated during glycation, which can increase the fluorescence emission of Hb (Rahmanifar & Miroliaei, 2020; Sattarahmady et al., 2007). Molecular fluorescence analysis of glycated Hb can be carried out in a relatively simple, fast, and non-invasive manner. However, the results should be evaluated cautiously, considering other biophysical techniques (Sarmah & Roy, 2022) due to the possibility of the influence of different fluorophores in complex systems.

The intrinsic fluorescence of Hb is a sensitive index of changes in the protein's conformation. Hb contains six Trp and ten Tyr residues, which are restricted to hydrophobic regions closer to the heme group and are the main amino acid residues responsible for the macromolecule's intrinsic fluorescence emission, with Trp β37 being the most pronounced (Iram et al., 2013; Silva et al., 2020). Hb has a maximum fluorescence emission at ~340 nm (λ_{ex} = 280 or 295 nm); however, with the formation of AGEs that present autofluorescence (Fig. 3), the changes caused by the process of formation of these products in Hb can be recorded by the increase in fluorescence intensity between 440 and 480 nm (λ_{ex} = 334, 330, 365 and

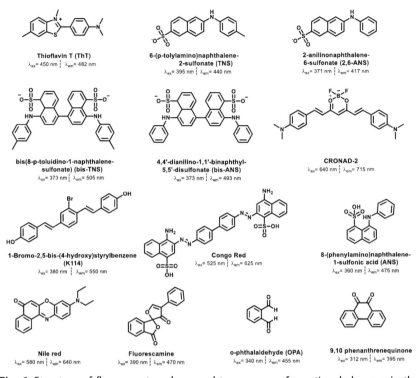

Fig. 6 Structure of fluorescent probes used to assess conformational changes in the structure of glycated Hb.

370 nm, among others depending on the AGE) (Iram et al., 2013; Perrone et al., 2020). This assay has a dual function, as it can be used to assess the formation of AGEs and their inhibition when the assays are carried out in the presence of compounds with these characteristics to reduce the extent of glycation. This way, Hb is glycated in the presence and absence of the possible inhibitor evaluated. Usually a pure compound (synthetic or natural) or natural extract, and inhibition is verified by the lower formation of AGEs, characterized by a decrease in fluorescence intensity compared to the control system, in which Hb is glycated in the absence of the inhibitor. However, one of the main limitations of this assay is that it only measures the formation of fluorescent AGEs, although the reaction can also produce non-fluorescent AGEs, which would then be measured. Another limitation can occur when the possible inhibitor being evaluated emits fluorescence in the same spectral region as the AGEs, causing overlap and thus making it difficult to perform a quantitative assessment of the process.

Generally, the formation of glycation products (λ_{em} = 450 nm) can also be assessed by resonance energy transfer (FRET) from Trp to AGEs (Iram et al., 2013). On the other hand, some AGEs do not show intrinsic fluorescence (Fig. 3), requiring fluorescent probes to assess conformational changes in the protein. Some fluorescent probes, therefore, can bind to proteins through covalent bonds and/or non-covalent interactions, thus providing an additional possibility to study conformational changes or structural changes in proteins (Hawe et al., 2008; Wang et al., 2017). Thioflavin T (ThT) is one of the most widely used fluorescent probes to investigate the fibrillar and aggregated states of Hb due to its high sensitivity and selectivity; however, even though it can determine mature fibrillar states, ThT cannot detect intermediate states or amorphous structures of the protein (Kazemi et al., 2019; Dodero & Messina, 2013). Therefore, is necessary the development and use of probes for these purposes, such as N-arylaminonaphthalene sulfonate (NAS) derivatives: TNS, 2,6-NA, bis-TNS, bis-ANS (Celej et al., 2008), besides CRONAD-2 (Ran et al., 2009), K114, and Congo red (Dodero & Messina, 2013; Iram et al., 2013) (Fig. 6). In addition, other probes such as 1-anilinonaphthalene-8-sulfonic acid (ANS) (Silva et al., 2020) and Nile red can provide information on the hydrophobicity of glycated Hb by interacting with hydrophobic amino acid residues exposed on the surface of the protein . Furthermore, the amount of arginine in glycated Hb can be determined using the fluorescent probe 9,10-phenanthrenoquinone (Fig. 6); the method involves the reaction of the probe with arginine residues present in the Hb, resulting in the formation of a fluorescent product, with the intensity of the fluorescence being proportional to the amount of the amino acid present in the sample (Esackimuthu & Sarasathi, 2021). Other assays can be used to measure the extent of glycation, evaluate possible inhibitors, and assess the formation of AGEs. The determination of free amine groups, which are affected by the glycation reaction and, therefore, glycated proteins, have a reduction in these groups. Fluorescamine and o-phthalaldehyde (OPA) (Fig. 6) are fluorogenic probes widely used to determine these groups. Fluorescamine can react with amines, guanidines, alcohols, and thiols, but only the reaction with primary amines can give rise to fluorescent products (Derayea & Samir, 2020). For this reason, fluorescamine is widely used to assess the amount of free amino groups in proteins, including glycated Hb. Fluorescamine reacts directly and selectively with primary amines in an alkaline medium (pH 8.5), producing a fluorescent product (λ_{ex} = 390 nm/λ_{em} = 470 nm). OPA is more soluble and stable in aqueous buffers (λ_{ex} = 340 nm/λ_{em} = 455 nm), as well as being around 5–10 times more sensitive than fluorescamine (Benson & Hare, 1975). However, while fluorescamine reacts directly with the amine functional group, OPA requires the addition of a thiol compound to form the fluorophore. In both methods, the fluorescence intensity is proportional to the amount of free amino groups in the protein, allowing quantification in glycated Hb treated with the anti-glycating compound. The results are generally expressed as a percentage (%) of the free amino group, where the emission from the control system is 100%.

5.3 Mass spectrometry

Mass spectrometry (MS) is one of the most sensitive and versatile techniques for studying proteins, enabling identification and quantification, and is therefore useful for assessing conformational changes in the three-dimensional structure of Hb after glycation, as well as investigating the molecular mass and fragmentation profile (α and β chains) of glycated Hb compared to non-glycated Hb, whose tetramer has three charge states (16^+, 17^+, 18^+) in the mass spectrum (Lapolla et al., 2001; Muralidharan et al., 2020; Roberts et al., 2001). Due to the clinical importance of HbA1c, different MS-based techniques are used to allow direct assessment of conformational changes in glycated Hb, such as electrospray ionization mass spectrometry (ESI-MS) and matrix-assisted laser desorption ionization mass spectrometry (MALDI-MS), which allow assessment of the composition and conformational changes of proteins without fragmenting (Cerqueira et al., 2020; Hattan et al., 2016; Heo et al., 2021; Muralidharan et al., 2020; Yadav & Mandal, 2022; Zhang et al., 2001). Collision-induced dissociation mass spectrometry (CID-MS) makes it possible to assess the secondary and tertiary structure of HbA1c after enzymatic digestion processes (Heo et al., 2021; Shin et al., 2021), information on molecular fragments can also be obtained by time-of-flight secondary ion mass spectrometry (TOF-SIMS) which makes it possible to assess the secondary structure and detect the heme group (Martínez-Negro et al., 2021; Tahoun et al., 2022). In addition, MALDI-TOF MS provides information on the molecular mass of Hb compared to HbA1c (Siddiqui et al., 2018). The identification of the different glycation sites by liquid chromatography coupled with tandem mass spectrometry (LC-MS/MS) is an essential step in understanding the effects of glycation on protein function (Heo et al., 2021; Ioannou & Varotsis, 2017; Mou et al., 2022; Rubino et al., 2009), in addition, heme adducts from the α, β and $β_{Gly}$ chains can be monitored (Muralidharan et al., 2020).

5.4 Circular dichroism

Circular dichroism (CD) is a spectroscopic technique that allows the structural conformation of proteins and nucleic acids to be analyzed. In the evaluation of glycated Hb, CD can be used to assess the protein's conformational change after glycation. CD measures the difference between circularly polarized electromagnetic radiation absorption when the magnetic field is applied clockwise and counterclockwise (García-Etxarri & Dionne, 2013; Yao & Liu, 2018). CD measurement can provide information on the secondary structure of proteins, such as α-helix and β-sheet content, and the tertiary structure (Iram et al., 2013; Pignataro et al., 2020). UV CD spectra are very sensitive to variations in protein structure and are especially useful for determining changes in the secondary structure of these macromolecules. Native Hb typically shows absorption at approximately 208 and 222 nm, bands indicative of the α-helicoidal structure, with 208 nm being associated with the α-helix transition (π → π*), while 222 nm can be related to the α-helix and coiled-coil transition (n → π*) (Heydari et al., 2022; Kryscio et al., 2012). Glycation of Hb results in cross-linked β-intermolecular sheets, common in AGEs. On the other hand, the near-UV CD spectra of proteins provide information on aromatic side chains and disulfide bonds. Absorption near 280–290 nm in the CD spectrum is attributed to aromatic residues, which give Hb a characteristic profile (Hasan & Naeem, 2020; Li, Nagai, & Nagai, 2000). This suggests that native Hb has a well-defined tertiary structure. However, the glycation process results in a decrease in the signal at 280 nm, indicating a loss of the tertiary structure. This results in the exposure of some non-polar internal groups to the solvent, thus making the surface of the protein more hydrophobic compared to the native state. Regarding the secondary structure, a high similarity was observed between Hb and HbA1c, indicating that the α-helix chains and β-sheets were not significantly affected by the glycation process (Iram et al., 2013; Rabbani et al., 2017; Śmiga et al., 2021).

5.5 X-ray diffraction

X-ray diffraction is a structural analysis technique that provides information on the 3D structure of crystallized proteins due to the interaction of the absorbed X-rays with the crystalline structure of the sample (Carugo & Carugo, 2005; Maveyraud & Mourey, 2020; Wang et al., 2017). For this reason, this technique can be used to evaluate the conformational changes of Hb, monitoring the domains affected during glycation due to changes in its crystal structure, as well as the structural dynamics and dispersion profile of the proteins (Heo et al., 2021; Zhang et al., 2008). This technique is considered the standard for identifying amyloid fibrils. It enables the study of atomic details related to the reflection profile of the crossed β-structure, which allows inferences to be made about fibril conformations (Luo et al., 2018; Zaman et al., 2019). The meridional reflection at around 4.8 Å is associated with the distance between the β ribbons, while an equatorial reflection in the 6–12 Å range corresponds to the distance between β sheets; this behavior is observed when the fibril samples are aligned, however, when not aligned two ring reflections appear at the same distances (Cao & Mezzenga, 2019).

5.6 Dynamic light scattering

Dynamic light scattering (DLS) measures the size of particles solvated in solution, including macromolecules such as proteins, and assesses the hydrodynamic radius of native and glycated Hb (Naftaly et al., 2021). Changes in the globular structure of hydrodynamic radii can infer conformational changes due to the formation of protein aggregates since these generally affect the size of protein molecules, possibly due to the expansion of HbA1c domains due to conformational changes (Cao & Mezzenga, 2019; Jha & Venkatesu, 2016; Siddiqui et al., 2018).

5.7 Microscopy

Transmission and scanning electron microscopy are imaging techniques used to evaluate the structure and organization of proteins, glycation, and other processes that can cause structural changes (Badar et al., 2020). Transmission electron microscopy (TEM) allows the visualization of internal structures of a sample containing proteins, obtaining information such as aggregate size and morphology. In the analysis of glycated Hb, the formation of protein aggregates induced by glycation can be assessed, which can be visualized (Hasan & Naeem, 2020; Khan et al., 2021; Naftaly et al., 2021). On the other hand, scanning electron microscopy (SEM) can be used to visualize the surface of glycated Hb, assessing the formation of protein aggregates and amyloid fibrils (Liu et al., 2023). In this technique, electrons are directed at the surface of the sample, making it possible to map and obtain high-resolution three-dimensional images of the surface (Iram et al., 2013; Tahoun et al., 2022). Atomic force microscopy (AFM) provides a sub-nanometric view of the morphology of aggregates, which is useful for investigating the fibrils and three-dimensional structures of biological samples by obtaining structural aspects such as torsion and dimensions (Loyola-Leyva et al., 2021; Ruggeri et al., 2019; Zaman et al., 2019). These microscopic techniques are also applied to provide structural details on the length scale of amyloid fibrils, usually of 2–8 protofilaments, each approximately 2–7 nm in diameter (Cao & Mezzenga, 2019).

5.8 Nuclear magnetic resonance

Nuclear magnetic resonance (NMR) is a non-invasive technique that uses external magnetic fields and radiofrequency pulses to excite and detect the resonance of atomic nuclei in a sample (Bothwell & Griffin, 2011; Cho et al., 2022). This technique can be used to assess the three-dimensional structure of the protein, detect conformational changes induced by glycation and determine the structure of amyloid fibrils (solid-state NMR) (Haris & Severcan, 1999; Loquet et al., 2018; Tycko, 2011; Wang et al., 2017; Zaman et al., 2019). In the evaluation of Hb, NMR spectra before and after glycation are compared, and the differences in the spectra can provide information on changes in the structure of the protein, such as the glycation profile on the lysine residues and conformational changes arising from changes in the heme group, as well as in the secondary and tertiary structure, besides the identifying the AGEs produced during glycation (Cho et al., 2022; Khodarahmi et al., 2012; Neglia et al., 1985; Polykretis et al., 2020; Tahoun et al., 2022).

5.9 Fourier transform infrared spectroscopy

Fourier transform infrared spectroscopy (FTIR) is a technique used to detect conformational changes in the secondary structure of Hb (Wang et al., 2017). FTIR can identify variations in the molecular structure of glycated Hb by detecting the presence of new infrared absorption bands formed due to glycation (Kazemi et al., 2018). These bands indicate changes in the chemical bonds in Hb, such as the C=O and –NH bonds, associated with the stretching and vibration of the carbonyl and amide group (Barth & Zscherp, 2002). Changes in the FTIR absorption bands in the 1600–1700 cm^{-1} region may indicate changes in the secondary and, consequently, three-dimensional structure of Hb after the glycation process (Iram et al., 2013; Militello et al., 2004).

5.10 Raman spectroscopy

Raman spectroscopy measures the inelastic scattering of electromagnetic radiation by a sample, making it possible to observe molecular vibrational modes and obtain information on structure and conformation (Noothalapati et al., 2021). Raman spectroscopy makes it possible to evaluate conformational changes in the secondary and tertiary structure of proteins, such as Hb, due to the glycation process, assessed through spectral changes, which make it possible to differentiate Hb from HbA1c (Barman et al., 2012). Thus, a shift in Raman bands is observed in glycated Hb compared to normal protein, resulting from the presence of covalent bonds; in addition, the intensity and width of the bands can provide information on protein conformation (Dybas et al., 2022; Kuhar et al., 2021). The six characteristic spectral bands of Hb: band I (1340–1390 cm^{-1}), band II (1470–1505 cm^{-1}), band III (1535–1575 cm^{-1}), band IV (1550–1590 cm^{-1}), band V (1605–1645 cm^{-1}) and band VI (1560–1600 cm^{-1}), these high frequency regions (1200–1700 cm^{-1}) reflect the vibration pattern of the heme group and are a sensitive band for structural changes. A shift in spectral signals can be produced when glucose binds to Hb, with a characteristic band occurring $\sim$1106 cm^{-1} related to the δ(C–O–C) stretching bond (González-Viveros et al., 2021; Ma et al., 2019; Tahoun et al., 2022).

In summary, spectroscopic techniques can be sensitive and non-destructive and generally require little material for analysis (Mallya et al., 2013). Biophysical assays are important tools for evaluating the glycation and conformational change of Hb and understanding how these processes affect the structure and function of the protein (Lapolla et al., 2001). HbA1c analysis methods can be further divided into two categories: methods based on structural differences, which are associated with the biophysical techniques, and methods based on charge differences, exploiting separation techniques as high-resolution ion-exchange liquid chromatography and capillary electrophoresis, and electrochemical techniques (Neelofar & Ahmad, 2017; Sherwani et al., 2016).

5.11 Miscellaneous techniques

High-performance liquid chromatography (HPLC) is widely used to quantify glycated Hb in blood. Thus, chromatographic methods are currently the reference for HbA1c analysis (Alzahrani et al., 2023; Trouiller-Gerfaux et al., 2019), such as cation exchange high-performance liquid chromatography, which separates and quantifies the different forms of Hb according to their electrical charge and size, since HbA1c has a different isoelectric point from non-glycated Hb, as HbA1c has a greater number of negative charges compared to native Hb, facilitating the elution of the glycated species (Chen et al., 2022; Ioannou & Varotsis, 2017; Sharma et al., 2020; Yadav & Mandal, 2022). Thus, this separation technique can be coupled with different detectors to obtain information on conformational changes (Ioannou & Varotsis, 2019).

Electrophoresis separates and visualizes the different protein isoforms, including Hb, according to their electrical charge and size. HbA1c has a distinct electrical charge than non-glycated Hb due to the presence of sugar groups attached to it, which allows it to be separated and visualized by gel electrophoresis (Chen et al., 2022). This electrophoretic separation makes it possible to monitor changes in the amount of glycated and normal Hb (Antonello et al., 2022). Analysis using sodium dodecyl sulfate-polyacrylamide gel electrophoresis (SDS-PAGE) confirms that the migration of HbA1c is slower compared to native Hb due to the increase in molecular mass resulting from modifications to the protein structure (Śmiga et al., 2021). In capillary electrophoresis, Hb variants can be separated according to surface charges so that the elution profile can vary depending on the nature of the change in the surface charge of the tetramer or glycated form; however, this variation can provide an inadequate estimate of HbA1c (Chen et al., 2022; Yadav & Mandal, 2022).

Electrochemical techniques are practical and useful for monitoring structural changes in Hb, as they provide data on the electrochemical behavior of species, usually associated with the development of rapid analysis biosensors. One of the advantages of these techniques is real-time monitoring, in addition to continuous measurements (Hussain et al., 2017; Magar et al., 2021; Teymourian et al., 2020) however, they have limitations regarding time, sample quantity and low selectivity (Abdalla et al., 2022; Duanghathaipornsuk et al., 2021; Eissa et al., 2019; Molazemhosseini et al., 2016; Moon et al., 2017). Electrical impedance spectroscopy (EIS) is a non-invasive, simple-to-instrument technique that uses the measurement of the electrical resistance of a sample to evaluate properties such as cell permeability and cell adhesion and can thus assess changes in blood cell function resulting from Hb glycation (Abdalla et al., 2022; Moon et al., 2017). EIS evaluates changes in behaviors that occur as a result of glycation (Bohli et al., 2017; Jain & Chauhan, 2017), as well as changes in the electrical properties of HbA1c, since by analyzing the impedance spectrum it is possible to obtain information about changes in electronic and structural properties in glycated Hb (Bertok et al., 2019; Boonyasit et al., 2016; Duanghathaipornsuk et al., 2021; Park et al., 2008; Yazdanpanah et al., 2015). Cyclic voltammetry (CV) is an electrochemical technique that can be used to obtain information on the kinetics, redox behavior, and electron transfer properties of the protein and be correlated with conformational changes induced by glycation since it allows the simultaneous detection of the oxidation potentials of HbA1c and native Hb (Bohli et al., 2017; Noviana et al., 2022; Eissa et al., 2019; Tiwari et al., 2022). As Hb undergoes structural modifications during glycation and its electrochemical behavior can be affected as the surface charges change, this process is monitored by CV, which allows the detection of these changes due to changes in redox potentials (Jain & Chauhan, 2017; Yadav & Mandal, 2022).

Finally, differential scanning calorimetry (DSC) is a thermal analysis technique used to determine the stability of biological macromolecules in the presence of molecules that induce conformational changes. The thermal energy released or absorbed is measured during a chemical reaction or a conformational change in the protein, so this technique is used to assess the effects of glycation on the stability of Hb (Siddiqui et al., 2018). In the evaluation of HbA1c, a DSC can be used to determine the enthalpy related to the thermal denaturation of the protein, thus evaluating the thermal stability of glycated Hb compared to non-glycated Hb (Bose & Chakraborti, 2008; Ishtikhar et al., 2021). This technique makes it possible to measure phase transitions, conformational changes, the thermal stability of a protein, and other biological changes.

Thus, instrumental methods and techniques are useful for evaluating changes in the structure and properties of Hb because of structural and conformational changes during the glycation process. Each technique provides different information, depending on its principles, advantages, and limitations. Therefore, in general, using different instrumental strategies is recommended to obtain a complete characterization since no universal technique exists.

6. Trends

Analytical techniques have been employed to investigate conformational changes in Hb and HbA1c. Among these techniques, sensors play a crucial role in detecting and quantifying these changes when coupled with different techniques (Han et al., 2017; Sharma et al., 2020). Portable devices/sensors based on electrochemical detection, for example, are common and have wide applicability in environmental monitoring, the food industry, and clinical diagnostics due to several advantages such as low cost, fast response, and simple manufacture (Bertok et al., 2019; Magar at al., 2021; Zhang, Li, Yang, Chang, & Simone, 2020). These sensors use the electrochemical properties of HbA1c to monitor changes in redox behavior and electron transfer characteristics and thus evaluate conformational changes (Duanghathaipornsuk et al., 2021; Yadav & Mandal, 2022). Modified electrodes can be used in association with nanomaterials to produce electrochemical sensors for reliable assessment of Hb and HbA1c levels in blood samples, enabling a reliable approach to sensitive point-of-care testing (POCT) (Ponsanti et al., 2022; Zhan et al., 2022; Zhang et al., 2020).

Also noteworthy are molecular fluorescence-based optical sensors, based on the development of fluorescent probes designed to bind to glycated regions of HbA1c selectively, exhibiting changes in fluorescence intensity or emission wavelength after interaction with the target molecule (Li et al., 2018; Li et al., 2022). As previously discussed, a limitation regarding absorption or emission spectroscopic techniques using probes is associated with the spectral overlap with (i) HbA1c, (ii) AGEs with autofluorescence, or (iii) in the evaluation of inhibitors. Thus, there is a need to develop new methods to overcome these limitations. Li et al., (2022) synthesized glutathione (GSH) functionalized Au-Ag nanoparticles (GSH-AuAg NPs) that showed fluorescence attributed to aggregation-induced emission, making it possible to determine Hb. Guo et al. (2022) developed a NIR fluorescent probe (CBP) activated from the reaction with H_2O_2, which was highly sensitive to the increase in viscosity with an increase in fluorescence at 635 nm and exhibited high selectivity for Aβ1–42 aggregates, increasing fluorescence intensity. In addition, this fluorescent probe also showed applicability for imaging β-amyloid fibrils. Ghadami et al., (2023) synthesized ten benzylideneindandione derivatives as fluorescent probes for amyloid fibrils, showing high binding affinity with these systems, allowing the conformational changes undergone by Hb to be monitored in real-time, with high selectivity and multimodal detection, since they can combine their fluorescent and colorimetric properties (Fakayode et al., 2023; Mohammadinejad et al., 2022; Wang et al., 2020).

7. Conclusions

The structural changes in Hb resulting from glycation represent a critical aspect in understanding the pathogenesis of glycemic disorders since this non-enzymatic reaction leads to the formation of AGEs and amyloid fibrils that can affect the conformation and functionality of the protein, including changes in the heme group environment, disruption of tertiary and quaternary structures and changes in the accessibility of active sites. These changes can be monitored using different techniques. This enables the detection, quantification, investigation, and characterization of glycated species, offering an understanding of the effects related to glycation and the structural dynamics of Hb since these can significantly interfere with the oxygen transport capacity, stability, and physiological role Hb plays. Understanding the structural changes in Hb resulting from glycation is necessary to develop diagnostic tools and therapeutic strategies to monitor glycemic disorders effectively.

Acknowledgements

The authors thank the Federal University of Alagoas (UFAL) for the infrastructure, the Coordenação de Aperfeiçoamento de Pessoal de Nível Superior - Brazil (CAPES, Finance Code 001), Conselho Nacional de Desenvolvimento Científico e Tecnológico (CNPq), Fundação de Amparo à Pesquisa do Estado de Alagoas (FAPEAL), and Financiadora de Estudos e Projetos (Finep). Finally, JCCS and MOFG are recipients of research fellowships from CNPq.

References

Abbas, G., Al-Harrasi, A., Hussain, H., Hussain, J., Rashid, R., & Choudhary, M. (2016). Antiglycation therapy: Discovery of promising antiglycation agents for managing diabetic complications. *Pharmaceutical Biology, 54*(2), 198–206.

Abdalla, S., Farsaci, F., Tellone, E., Shirbeeny, W., Hassan, A. M., Bahabri, F., & Kandil, S. (2022). Hemoglobin glycation increases the electric charges on red blood cells: Effects of dielectric polarization. *Materials Chemistry and Physics, 276*, 125348.

ADA. (2010). Standards of medical care in diabetes-2010. *Diabetes Care, 33*(1), S11–S61.

Agronskaia, A. V., Valentijn, J. A., Van Driel, L. F., Schneijdenberg, C. T., Humbel, B. M., En Henegouwen, P. M. V. B., & Gerritsen, H. C. (2008). Integrated fluorescence and transmission electron microscopy. *Journal of Structural Biology, 164*(2), 183–189.

Ahmed, M. H., Ghatge, M. S., & Safo, M. K. (2020). Hemoglobin: Structure, function and allostery. *Subcellular Biochemistry, 94*, 345–382.

Ahmed, M. U., Frye, E. B., Degenhardt, T. P., Thorpe, S. R., & Baynes, J. W. (1997). N^ε-(Carboxyethyl)lysine, a product of the chemical modification of proteins by methylglyoxal, increases with age in human lens proteins. *The Biochemical Journal, 324*, 565–570.

Allen, D. W., Schroeder, W. A., & Balog, J. Z. (1958). Observations on the chromatographic heterogeneity of normal adult and fetal human hemoglobin: A study of the effects of crystallization and chromatography on the heterogeneity and isoleucine content. *Journal of the American Chemical Society, 80*, 1628–1634.

Alzahrani, B. A., Salamatullah, H. K., Alsharm, F. S., Baljoon, J. M., Abukhodair, A. O., Ahmed, M. E., & Radi, S. (2023). The effect of different types of anemia on HbA1c levels in non-diabetics. *BMC Endocrine Disorders, 23*(1), 24.

Antonello, G., Lo Monaco, C., Napoli, P., Solimando, D., Curcio, C., Barberio, G., & Nigra, M. (2022). Two co-inherited hemoglobin variants revealed by capillary electrophoresis during quantification of glycated hemoglobin. *Clinical Chemistry and Laboratory Medicine (CCLM), 60*(6), 886–890.

Aragonès, G., Rowan, S., Francisco, S. G., Whitcomb, E. A., Yang, W., Perini-Villanueva, G., ... Bejarano, E. (2021). The glyoxalase system in age-related diseases: Nutritional intervention as anti-ageing strategy. *Cells, 10*(8), 1852.

Atkins, T. W., & Thornally, P. J. (1989). Erythrocyte glyoxalase activity in genetically obese (ob/ob) and streptozotocin diabetic mice. *Diabetes Res, 11*(3), 125–129.

Badar, A., Arif, Z., Qais, F. A., Islam, S. N., & Alam, K. (2020). Carbamylation of human serum albumin generates high-molecular weight aggregates: Fine characterization by multi-spectroscopic methods and electron microscopy. *International Journal of Biological Macromolecules, 164*, 2380–2388.

Bae, J. W., & Lee, M. H. (2004). Effect and putative mechanism of action of ginseng on the formation of glycated hemoglobin in vitro. *Journal of Ethnopharmacology, 91*(1), 137–140.

Bakhti, M., Habibi-Rezaei, M., Moosavi-Movahedi, A. A., & Khazaei, M. R. (2007). Consequential alterations in haemoglobin structure upon glycation with fructose: Prevention by acetylsalicylic acid. *Journal of Biochemistry, 141*(6), 827–833.

Barbarino, F., Wäschenbach, L., Cavalho-Lemos, V., Dillenberger, M., Becker, K., Gohlke, H., & Cortese-Krott, M. M. (2021). Targeting spectrin redox switches to regulate the mechanoproperties of red blood cells. *Biological Chemistry, 402*(3), 317–331.

Barinova, K. V., Serebryakova, M. V., Melnikova, A. K., Medvedeva, M. V., Muronetz, V. I., & Schmalhausen, E. V. (2023). Mechanism of inactivation of glyceraldehyde-3-phosphate dehydrogenase in the presence of methylglyoxal. *Archives of Biochemistry and Biophysics, 733*, 109485.

Barman, I., Dingari, N. C., Kang, J. W., Horowitz, G. L., Dasari, R. R., & Feld, M. S. (2012). Raman spectroscopy-based sensitive and specific detection of glycated hemoglobin. *Analytical Chemistry, 84*(5), 2474–2482.

Barth, A., & Zscherp, C. (2002). What vibrations tell about proteins. *Quarterly Reviews of Biophysics, 35*(4), 369–430.

Benson, J. R., & Hare, P. E. (1975). o-Phthalaldehyde: Fluorogenic detection of primary amines in the picomole range. Comparison with fluorescamine and ninhydrin. *Proceedings of the National Academy of Sciences, 72*(2), 619–622.

Berner, A. K., Brouwers, O., Pringle, R., Klaassen, I., Colhoun, L., McVicar, C., ... Stitt, A. W. (2012). Protection against methylglyoxal-derived AGEs by regulation of glyoxalase 1 prevents retinal neuroglial and vasodegenerative pathology. *Diabetologia, 55*(3), 845–854.

Bertok, T., Lorencova, L., Chocholova, E., Jane, E., Vikartovska, A., Kasak, P., & Tkac, J. (2019). Electrochemical impedance spectroscopy based biosensors: Mechanistic principles, analytical examples and challenges towards commercialization for assays of protein cancer biomarkers. *ChemElectroChem, 6*(4), 989–1003.

Bhat, L. R., Vedantham, S., Krishnan, U. M., & Rayappan, J. B. B. (2019). Methylglyoxal – An emerging biomarker for diabetes mellitus diagnosis and its detection methods. *Biosensors and Bioelectronics, 133*, 107–124.

Bohli, N., Chammem, H., Meilhac, O., Mora, L., & Abdelghani, A. (2017). Electrochemical impedance spectroscopy on interdigitated gold microelectrodes for glycosylated human serum albumin characterization. *IEEE Transactions on Nanobioscience, 16*(8), 676–681.

Bookchin, R., & Gallop, P. (1968). Structure of hemoglobin AIc: Nature of the N-terminal β chain blocking group. *Biochemical and Biophysical Research Communications, 32*(1), 86–93.

Boonyasit, Y., Laiwattanapaisal, W., Chailapakul, O., Emneus, J., & Heiskanen, A. R. (2016). Boronate-modified interdigitated electrode array for selective impedance-based sensing of glycated hemoglobin. *Analytical Chemistry, 88*(19), 9582–9589.

Bose, T., Bhattacherjee, A., Banerjee, S., & Chakraborti, A. S. (2013). Methylglyoxal-induced modifications of hemoglobin: Structural and functional characteristics. *Archives of Biochemistry and Biophysics, 529*(2), 99–104.

Bose, T., & Chakraborti, A. S. (2008). Fructose-induced structural and functional modifications of hemoglobin: Implication for oxidative stress in diabetes mellitus. *Biochimica et Biophysica Acta (BBA)—General Subjects, 1780*(5), 800–808.

Bothwell, J. H., & Griffin, J. L. (2011). An introduction to biological nuclear magnetic resonance spectroscopy. *Biological Reviews, 86*(2), 493–510.

Bouma, B., Kroon-Batenburg, L. M., Wu, Y. P., Brünjes, B., Posthuma, G., Kranenburg, O., & Gebbink, M. F. (2003). Glycation induces formation of amyloid cross-β structure in albumin. *Journal of Biological Chemistry, 278*(43), 41810–41819.

Brouwers, O., Niessen, P. M. G., Miyata, T., Østergaard, J. A., Flyvbjerg, A., Peutz-Kootstra, C. J., ... Schalkwijk, C. G. (2014). Glyoxalase-1 overexpression reduces endothelial dysfunction and attenuates early renal impairment in a rat model of diabetes. *Diabetologia, 57*(1), 224–235.

Buehler, P. W., & Alayash, A. I. (2007). Oxidation of hemoglobin: Mechanisms of control in vitro and in vivo. *Transfusion Alternatives in Transfusion Medicine, 9*(4), 204–212.

Cao, Y., & Mezzenga, R. (2019). Food protein amyloid fibrils: Origin, structure, formation, characterization, applications and health implications. *Advances in Colloid and Interface Science, 269*, 334–356.

Carugo, O., & Carugo, K. D. (2005). When X-rays modify the protein structure: Radiation damage at work. *Trends in Biochemical Sciences, 30*(4), 213–219.

Cayot, P., & Tainturier, G. (1997). The quantification of protein amino groups by the trinitrobenzenesulfonic acid method: A reexamination. *Analytical Biochemistry, 249* 1884-200.

Celej, M. S., Jares-Erijman, E. A., & Jovin, T. M. (2008). Fluorescent N-arylamino-naphthalene sulfonate probes for amyloid aggregation of alpha-synuclein. *Biophysical Journal, 94*, 4867–4879.

Cerqueira, L. B., Fachi, M. M., Kawagushi, W. H., Pontes, F. L. D., de Campos, M. L., & Pontarolo, R. (2020). New validated method for quantification of glycated hemoglobin by LC-QToF-MS: Is HRMS able to quantify clinical samples? *Journal of the American Society for Mass Spectrometry, 31*(6), 1172–1179.

Chen, Z., Shao, L., Jiang, M., Ba, X., Ma, B., & Zhou, T. (2022). Interpretation of HbA1c lies at the intersection of analytical methodology, clinical biochemistry and hematology. *Experimental and Therapeutic Medicine, 24*(6), 1–11.

Cheng, Y., Li, Y., Li, R., Lu, J., & Wang, K. (2000). Orally administered cerium chloride induces the conformational changes of rat hemoglobin, the hydrolysis of 2, 3-DPG and the oxidation of heme-Fe (II), leading to changes of oxygen affinity. *Chemico-Biological Interactions, 125*(3), 191–208.

Cho, S., Duong, V. A., Mok, J. H., Joo, M., Park, J. M., & Lee, H. (2022). Enrichment and analysis of glycated proteins. *Reviews in Analytical Chemistry, 41*(1), 83–97.

Derayea, S. M., & Samir, E. (2020). A review on the use of fluorescamine as versatile and convenient analytical probe. *Microchemical Journal, 156*, 104835.

Diez-Silva, M., Dao, M., Han, J., Lim, C. T., & Suresh, S. (2010). Shape and biomechanical characteristics of human red blood cells in health and disease. *MRS Bulletin, 35*(5), 382–388.

Ding, L., Xu, Y., Liu, S., Bi, Y., & Xu, Y. (2018). Hemoglobin A1c and diagnosis of diabetes. *Journal of Diabetes, 10*(5), 365–372.

Dodero, V. I., & Messina, P. V. (2013). Analyzing the solution state of protein structure, interactions, and ligands by spectroscopic methods. *Proteins in Solution and at Interfaces: Methods and Applications in Biotechnology and Materials Science*, 73–98.

Duanghathaipornsuk, S., Reaver, N. G., Cameron, B. D., & Kim, D. S. (2021). Adsorption kinetics of glycated hemoglobin on aptamer microarrays with antifouling surface modification. *Langmuir: The ACS Journal of Surfaces and Colloids, 37*(15), 4647–4657.

Dybas, J., Alcicek, F. C., Wajda, A., Kaczmarska, M., Zimna, A., Bulat, K., & Marzec, K. M. (2022). Trends in biomedical analysis of red blood cells–Raman spectroscopy against other spectroscopic, microscopic and classical techniques. *TrAC Trends in Analytical Chemistry, 146*, 116481.

Eissa, S., Almusharraf, A. Y., & Zourob, M. (2019). A comparison of the performance of voltammetric aptasensors for glycated haemoglobin on different carbon nanomaterials-modified screen printed electrodes. *Materials Science and Engineering: C, 101*, 423–430.

Elgsaeter, A., & Mikkelsen, A. (1991). Shapes and shape changes in vitro in normal red blood cells. *Biochimica et Biophysica Acta (BBA)-Reviews on Biomembranes, 1071*(3), 273–290.

Elliott, S. (2008). Erythropoiesis-stimulating agents and other methods to enhance oxygen transport. *British Journal of Pharmacology, 154*(3), 529–541.

Esackimuthu, P., & Saraswathi, N. T. (2021). Non enzymatic covalent modification by glycolysis end product converts hemoglobin into its oxidative stress potency state. *Biochemical and Biophysical Research Communications, 534*, 387–394.

Fakayode, S. O., Lisse, C., Medawala, W., Brady, P. N., Bwambok, D. K., Anum, D., & Grant, C. (2023). Fluorescent chemical sensors: Applications in analytical, environmental, forensic, pharmaceutical, biological, and biomedical sample measurement, and clinical diagnosis. *Applied Spectroscopy Reviews*, 1–89.

Feng, N., Feng, Y., Tan, J., Zhou, C., Xu, J., Chen, Y., ... Wu, Q. (2023). Inhibition of advance glycation end products formation, gastrointestinal digestion, absorption and toxicity: A comprehensive review. *International Journal of Biological Macromolecules, 249*, 125814.

Favinha, A. G., Barreiro, D. S., Martins, J. N., O'Toole, P., & Pauleta, S. R. (2020). Acrylamide-hemoglobin adduct: A spectroscopic study. *Spectrochimica Acta Part A: Molecular and Biomolecular Spectroscopy, 241*, 118644.

García-Etxarri, A., & Dionne, J. A. (2013). Surface-enhanced circular dichroism spectroscopy mediated by nonchiral nanoantennas. *Physical Review B, 87*(23), 235409.

Ghadami, S. A., Ahadi-Amandi, S., Khodarahmi, R., Ghanbari, S., & Adibi, H. (2023). Synthesis of benzylidene-indandione derivatives as quantification of amyloid fibrils. *Biophysical Chemistry, 296*, 106982.

Ghazanfari-Sarabi, S., Habibi-Rezaei, M., Eshraghi-Naeeni, R., & Moosavi-Movahedi, A. A. (2019). Prevention of haemoglobin glycation by acetylsalicylic acid (ASA): A new view on old mechanism. *PLoS One, 14*(4), e0214725.

Giacco, F., Du, X., D'Agati, V. D., Milne, R., Sui, G., Geoffrion, M., & Brownlee, M. (2014). Knockdown of glyoxalase 1 mimics diabetic nephropathy in nondiabetic mice. *Diabetes, 63*(1), 291–299.

González-Viveros, N., Castro-Ramos, J., Gómez-Gil, P., & Cerecedo-Núñez, H. H. (2021). Characterization of glycated hemoglobin based on Raman spectroscopy and artificial neural networks. *Spectrochimica Acta—Part A: Molecular and Biomolecular Spectroscopy, 247*.

Guo, Y., Leng, H., Wang, Y., Shi, W.-J., Zhang, L., & Yan, J. (2022). A novel H_2O_2-activated NIR fluorescent probe for imaging β-amyloid fibrils and mitochondrial viscosity. *Dyes and Pigments, 206*, 110665.

Guo, W., Zhou, Q., Jia, Y., & Xu, J. (2019). Increased levels of glycated hemoglobin A1c and iron deficiency anemia: A review. *Medical Science Monitor, 25*, 8371–8378.

Han, Y. D., Kim, K. R., Park, Y. M., Song, S. Y., Yang, Y. J., Lee, K., & Yoon, H. C. (2017). Boronate-functionalized hydrogel as a novel biosensing interface for the glycated hemoglobin A1c (HbA1c) based on the competitive binding with signaling glycoprotein. *Materials Science and Engineering: C, 77*, 1160–1169.

Haris, P. I., & Severcan, F. (1999). Caracterização espectroscópica FTIR da estrutura de proteínas em meios aquosos e não aquosos. *Journal of Molecular Catalysis B: Enzymatic, 7*(1-4), 207–221.

Hasan, S., & Naeem, A. (2020). Consequence of macromolecular crowding on aggregation propensity and structural stability of haemoglobin under glycating conditions. *International Journal of Biological Macromolecules, 162*, 1044–1053.

Hattan, S. J., Parker, K. C., Vestal, M. L., Yang, J. Y., Herold, D. A., & Duncan, M. W. (2016). Analysis and quantitation of glycated hemoglobin by matrix assisted laser desorption/ionization time of flight mass spectrometry. *Journal of the American Society for Mass Spectrometry, 27*(3), 532–541.

Hawe, A., Sutter, M., & Jiskoot, W. (2008). Extrinsic Fluorescent Dyes as Tools for Protein Characterization. *Pharm. Res. 25*(7), 1487–1499.

Hempe, J. M., Gomez, R., McCarter, R. J., & Chalew, S. A. (2002). High and low hemoglobin glycation phenotypes in type 1 diabetes. *Journal of Diabetes and its Complications, 16*(5), 313–320.

Hempe, J. M., & Hsia, D. S. (2022). Variation in the hemoglobin glycation index. *Journal of Diabetes and its Complications, 36*(7), 108223.

Heo, C. E., Kim, M., Son, M. K., Hyun, D. G., Heo, S. W., & Kim, H. I. (2021). Ion mobility mass spectrometry analysis of oxygen affinity-associated structural changes in hemoglobin. *Journal of the American Society for Mass Spectrometry, 32*(10), 2528–2535.

Heydari, M., Khatibi, A., & Zarrabi, M. (2022). The effect of formetanate hydrochloride on the glycated human hemoglobin. *Iranian Journal of Chemistry and Chemical Engineering, 41*(5), 1804–1811.

Heydari, M., Khatibi, A., & Zarrabi, M. (2022). The Effect of Formetanate Hydrochloride on the Glycated Human Hemoglobin. *Iranian Journal of Chemistry and Chemical Engineering, 41*(5), 1804–1811.

Huang, Y. T., Liao, H. F., Wang, S. L., & Lin, S. Y. (2016). Glycation and secondary conformational changes of human serum albumin: Study of the FTIR spectroscopic curve-fitting technique. *Aims Biophysics, 3*(2), 247–260.

Hussain, K. K., Moon, J. M., Park, D. S., & Shim, Y. B. (2017). Electrochemical detection of hemoglobin: A review. *Electroanalysis, 29*(10), 2190–2199.

Ioannou, A., & Varotsis, C. (2017). Modifications of hemoglobin and myoglobin by Maillard reaction products (MRPs). *PLoS One, 12*(11), e0188095.

Ioannou, A., & Varotsis, C. (2019). *Reaction of hemoglobin with the schiff base intermediate of the glucose/asparagine reaction: Formation of a hemichrome. Polyphenols in plants.* Academic Press, 317–325.

Iram, A., Alam, T., Khan, J. M., Khan, T. A., Khan, R. H., & Naeem, A. (2013). Molten globule of hemoglobin proceeds into aggregates and advanced glycated end products. *PLoS One, 8*(8) e72075.

Ishtikhar, M., Siddiqui, Z., Ahmad, A., Ashraf, J. M., Arshad, M., Doctor, N., & Khan, R. H. (2021). Phytochemical thymoquinone prevents hemoglobin glycoxidation and protofibrils formation: A biophysical aspect. *International Journal of Biological Macromolecules, 190*, 508–519.

Jain, U., & Chauhan, N. (2017). Glycated hemoglobin detection with electrochemical sensing amplified by gold nanoparticles embedded N-doped graphene nanosheet. *Biosensors and Bioelectronics, 89*, 578–584.

Jha, I., & Venkatesu, P. (2016). Unprecedented improvement in the stability of hemoglobin in the presence of promising green solvent 1-allyl-3-methylimidazolium chloride. *ACS Sustainable Chemistry & Engineering, 4*(2), 413–421.

Kaufman, D. P., Khattar, J., & Lappin, S. L. (2023). *Physiology, fetal hemoglobin*. Treasure Island: StatPearls Publishing.

Kazemi, F., Divsalar, A., & Saboury, A. A. (2018). Structural analysis of the interaction between free, glycated and fructated hemoglobin with propolis nanoparticles: A spectroscopic study. *International Journal of Biological Macromolecules, 109*, 1329–1337.

Kazemi, F., Divsalar, A., Saboury, A. A., & Seyedarabi, A. (2019). Propolis nanoparticles prevent structural changes in human hemoglobin during glycation and fructation. *Colloids and Surfaces B: Biointerfaces, 177*, 188–195.

Kettisen, K., & Bülow, L. (2021). Introducing negatively charged residues on the surface of fetal hemoglobin improves yields in *Escherichia coli*. *Frontiers in Bioengineering and Biotechnology, 9*.

Khajehpour, M., Dashnau, J. L., & Vanderkooi, J. M. (2006). Infrared spectroscopy used to evaluate glycosylation of proteins. *Analytical Biochemistry, 348*(1), 40–48.

Khan, M. S., Tabrez, S., Al-Okail, M. S., Shaik, G. M., Bhat, S. A., Rehman, T. M., & AlAjmi, M. F. (2021). Non-enzymatic glycation of protein induces cancer cell proliferation and its inhibition by quercetin: Spectroscopic, cytotoxicity and molecular docking studies. *Journal of Biomolecular Structure and Dynamics, 39*(3), 777–786.

Khodarahmi, R., Karimi, S. A., Kooshk, M. R. A., Ghadami, S. A., Ghobadi, S., & Amani, M. (2012). Comparative spectroscopic studies on drug binding characteristics and protein surface hydrophobicity of native and modified forms of bovine serum albumin: Possible relevance to change in protein structure/function upon non-enzymatic glycation. *Spectrochimica Acta Part A: Molecular and Biomolecular Spectroscopy, 89*, 177–186.

Klaus, A., Rau, R., & Glomb, M. A. (2018). Modification and crosslinking of proteins by glycolaldehyde and glyoxal: A model system. *Journal of Agricultural and Food Chemistry, 66*(41), 10835–10843.

Koenig, R. J., Peterson, C. M., Jones, R. L., Saudek, C., Lehrman, M., & Cerami, A. (1976). Correlation of glucose regulation and hemoglobin A1c in diabetes mellitus. *The New England Journal of Medicine, 295*(8), 417–420.

Kosmachevskaya, O. V., Novikova, N. N., & Topunov, A. F. (2021). Carbonyl stress in red blood cells and hemoglobin. *Antioxidants, 10*(2), 1–23.

Kosmachevskaya, O. V., & Topunov, A. F. (2018). Alternate and additional functions of erythrocyte hemoglobin. *Biochemistry (Moscow), 83*(12–13), 1575–1593.

Kristinsson, H. G., & Hultin, H. O. (2004). Changes in trout hemoglobin conformations and solubility after exposure to acid and alkali pH. *Journal of Agricultural and Food Chemistry, 52*(11), 3633–3643.

Kryscio, D. R., Fleming, M. Q., & Peppas, N. A. (2012). Conformational studies of common protein templates in macromolecularly imprinted polymers. *Biomedical Microdevices, 14*, 679–687.

Kuhar, N., Sil, S., & Umapathy, S. (2021). Potential of Raman spectroscopic techniques to study proteins. *Spectrochimica Acta Part A: Molecular and Biomolecular Spectroscopy, 258*, 119712.

Kumagai, P. S., DeMarco, R., & Lopes, J. L. (2017). Advantages of synchrotron radiation circular dichroism spectroscopy to study intrinsically disordered proteins. *European Biophysics Journal, 46*, 599–606.

Kunkel, H. G., & Wallenius, G. (1955). New hemoglobin in normal adult blood. *Science (New York, N. Y.), 122*(3163), 288.

Lapolla, A., Fedele, D., & Traldi, P. (2001). Diabetes and mass spectrometry. *Diabetes/Metabolism Research and Reviews, 17*(2), 99–112.

Lee, J. H., Samsuzzaman, Md, Park, M. G., Park, S. J., & Kim, S. Y. (2021). Methylglyoxal-derived hemoglobin advanced glycation end products induce apoptosis and oxidative stress in human umbilical vein endothelial cells. *International Journal of Biological Macromolecules, 187*, 409–421.

Li, D., Fang, C., Li, H., & Tu, Y. (2022). Fluorescence/electrochemiluminescence approach for instant detection of glycated hemoglobin index. *Analytical Biochemistry, 659*, 114958.

Li, D., Yan, J., Fang, C., & Tu, Y. (2022). Label-free detection of hemoglobin using GSH-AuAg NPs as fluorescent probe by dual quenching mechanism. *Sensors and Actuators: B. Chemical, 355*, 131291.

Li, R., Nagai, Y., & Nagai, M. (2000). Changes of tyrosine and tryptophan residues in human hemoglobin by oxygen binding: Near-and far-UV circular dichroism of isolated chains and recombined hemoglobin. *Journal of Inorganic Biochemistry, 82*(1-4), 93–101.

Li, X. H., Du, L. L., Cheng, X. S., Jiang, X., Zhang, Y., Lv, B. L., & Zhou, X. W. (2013). Glycation exacerbates the neuronal toxicity of β-amyloid. *Cell Death & Disease, 4*(6), e673.

Li, X., Zheng, W., Zhang, L., Yu, P., Lin, Y., Su, L., & Mao, L. (2009). Effective electrochemical method for investigation of hemoglobin unfolding based on the redox property of heme groups at glassy carbon electrodes. *Analytical Chemistry, 81*(20), 8557–8563.

Li, Z., Yuan, H., Yuan, W., Su, Q., & Li, F. (2018). Upconversion nanoprobes for biodetections. *Coordination Chemistry Reviews, 354*, 155–168.

Liang, Z., Chen, X., Yang, Z., Lai, Y., Yang, Y., Lei, C., & Zeng, Y. (2019). Pyrraline formation modulated by sodium chloride and controlled by encapsulation with different coating materials in the maillard reaction. *Biomolecules, 9*, 721.

Little, R. R., Rohlfing, C. L., Wiedmeyer, H.-M., Myers, G. L., Sacks, D. B., & Goldstein, D. E. (2001). The national glycohemoglobin standardization program: A five-year progress report. *Clinical Chemistry, 47*(11), 1985–1992.

Liu, L., Guo, J., Wang, Z., Duan, X., Zhang, X., Yu, Y., & Wu, T. (2023). Effect of interaction between catechin and glycated porcine hemoglobin on its antioxidant functional and structural properties. *LWT*, 114868.

Longo, F., Piolatto, A., Ferrero, G. B., & Piga, A. (2021). Ineffective erythropoiesis in β-Thalassaemia: Key steps and therapeutic options by drugs. *International Journal of Molecular Sciences, 22*(13).

Loquet, A., El Mammeri, N., Stanek, J., Berbon, M., Bardiaux, B., Pintacuda, G., & Habenstein, B. (2018). 3D structure determination of amyloid fibrils using solid-state NMR spectroscopy. *Methods (San Diego, Calif.), 138*, 26–38.

Loyola-Leyva, A., Loyola-Rodríguez, J. P., Terán-Figueroa, Y., Camacho-Lopez, S., González, F. J., & Barquera, S. (2021). Application of atomic force microscopy to assess erythrocytes morphology in early stages of diabetes. A pilot study. *Micron (Oxford, England: 1993), 141*, 102982.

Luo, F., Gui, X., Zhou, H., Gu, J., Li, Y., Liu, X., & Liu, C. (2018). Atomic structures of FUS LC domain segments reveal bases for reversible amyloid fibril formation. *Nature Structural & Molecular Biology, 25*(4), 341–346.

Ma, G., Chen, H., Zhang, Q., Ma, J., Yu, Q., Han, L., & Song, R. (2019). Protective characterization of low dose sodium nitrite on yak meat myoglobin in a hydroxy radical oxidation environment: Fourier Transform Infrared spectroscopy and laser Micro-Raman spectroscopy. *LWT, 116*, 108556.

Maessen, D. E., Hanssen, N. M., Scheijen, J. L., van der Kallen, C. J., van Greevenbroek, M. M., Stehouwer, C. D., & Schalkwijk, C. G. (2015). Post–glucose load plasma α-dicarbonyl concentrations are increased in individuals with impaired glucose metabolism and type 2 diabetes: The CODAM study. *Diabetes Care, 38*(5), 913–920.

Magar, H. S., Hassan, R. Y., & Mulchandani, A. (2021). Espectroscopia de impedância eletroquímica (EIS): Princípios, construção e aplicações de biosensoriamento. *Sensores, 21*(19), 6578.

Mallya, M., Shenoy, R., Kodyalamoole, G., Biswas, M., Karumathil, J., & Kamath, S. (2013). Absorption spectroscopy for the estimation of glycated hemoglobin (HbA1c) for the diagnosis and management of diabetes mellitus: A pilot study. *Photomedicine and Laser Surgery, 31*(5), 219–224.

Mariana, M. D. (2023). Alteration of some red blood cell components in diabetes mellitus. *Editorial Board, 1*(101), 17–26.

Martínez-Negro, M., González-Rubio, G., Aicart, E., Landfester, K., Guerrero-Martínez, A., & Junquera, E. (2021). Insights into colloidal nanoparticle-protein corona interactions for nanomedicine applications. *Advances in Colloid and Interface Science, 289*, 102366.

Maveyraud, L., & Mourey, L. (2020). Protein X-ray crystallography and drug discovery. *Molecules (Basel, Switzerland), 25*(5), 1030.

McLellan, A. C., & Thornalley, P. J. (1989). Glyoxalase activity in human red blood cells fractioned by age. *Mechanisms of Ageing and Development, 48*(1), 63–71.

Mendonça, E., Fragoso, M., Oliveira, J., Xavier, J., Goulart, M., & Oliveira, A. (2022). Gestational diabetes mellitus: The crosslink among inflammation, nitroxidative stress, intestinal microbiota and alternative therapies. *Antioxidants, 11*(1), 129.

Mertens, H. D., & Svergun, D. I. (2010). Structural characterization of proteins and complexes using small-angle X-ray solution scattering. *Journal of Structural Biology, 172*(1), 128–141.

Metzger, B., Gabbe, S., Persson, B., Buchanan, T., Catalano, P., Damm, P., ... Schmidt, M. (2010). International association of diabetes and pregnancy study groups recommendations on the diagnosis and classification of hyperglycemia in pregnancy. *Diabetes Care, 33*(3), 676–682.

Militello, V., Casarino, C., Emanuele, A., Giostra, A., Pullara, F., & Leone, M. (2004). Aggregation kinetics of bovine serum albumin studied by FTIR spectroscopy and light scattering. *Biophysical Chemistry, 107*(2), 175–187.

Miller, A. G., Meade, S. J., & Gerrard, J. A. (2003). New insights into protein crosslinking via the Maillard reaction: Structural requirements, the effect on enzyme function, and predicted efficacy of crosslinking inhibitors as anti-ageing therapeutics. *Bioorganic & Medicinal Chemistry, 11*(6), 843–852.

Mohammadinejad, A., Heydari, M., Kazemi Oskuee, R., & Rezayi, M. (2022). A critical systematic review of developing aptasensors for diagnosis and detection of diabetes biomarkers. *Critical Reviews in Analytical Chemistry, 52*(8), 1795–1817.

Molazemhosseini, A., Magagnin, L., Vena, P., & Liu, C. C. (2016). Single-use disposable electrochemical label-free immunosensor for detection of glycated hemoglobin (HbA1c) using differential pulse voltammetry (DPV). *Sensors, 16*(7), 1024.

Moon, J. M., Kim, D. M., Kim, M. H., Han, J. Y., Jung, D. K., & Shim, Y. B. (2017). A disposable amperometric dual-sensor for the detection of hemoglobin and glycated hemoglobin in a finger prick blood sample. *Biosensors and Bioelectronics, 91*, 128–135.

Mou, L., Hu, P., Cao, X., Chen, Y., Xu, Y., He, T., & He, R. (2022). Comparison of bovine serum albumin glycation by ribose and fructose in vitro and in vivo. *Biochimica Et Biophysica Acta (BBA)-Molecular Basis of Disease, 1868*(1), 166283.

Moulick, R. G., Bhattacharya, J., Roy, S., Basak, S., & Dasgupta, A. K. (2007). Compensatory secondary structure alterations in protein glycation. *Biochimica et Biophysica Acta (BBA)-Proteins and Proteomics, 1774*(2), 233–242.

Muralidharan, M., Bhat, V., & Mandal, A. K. (2020). Structural analysis of glycated human hemoglobin using native mass spectrometry. *The FEBS Journal, 287*(6), 1247–1254.

Naftaly, A., Izgilov, R., Omari, E., & Benayahu, D. (2021). Revealing advanced glycation end products associated structural changes in serum albumin. *ACS Biomaterials Science & Engineering, 7*(7), 3179–3189.

Nagai, R., Shirakawa, J., Ohno, R., Moroishi, N., & Nagai, M. (2014). Inhibition of AGEs formation by natural products. *Amino Acids, 46*(2), 261–266.

Nash, G. B., & Gratzer, W. B. (1993). Structural determinants of the rigidity of the red cell membrane. *Biorheology, 30*(5–6), 397–407.

Neelofar, K., & Ahmad, J. (2017). Uma visão geral da glicação in vitro e in vivo da albumina: um potencial marcador de doença no diabetes mellitus. *Glycoconjugate Journal, 34*, 575–584.

Neglia, C. I., Cohen, H. J., Garber, A. R., Thorpe, S. R., & Baynes, J. W. (1985). Characterization of glycated proteins by ^{13}C NMR spectroscopy. Identification of specific sites of protein modification by glucose. *Journal of Biological Chemistry, 260*(9), 5406–5410.

Negre-Salvayre, A., Salvayre, R., Augé, N., Pamplona, R., & Portero-Otin, M. (2009). Hyperglycemia and glycation in diabetic complications. *Antioxidants & Redox Signaling, 11*(12), 3071–3109.

Nemet, I., Varga-Defterdarović, L., & Turk, Z. (2006). Methylglyoxal in food and living organisms. *Molecular Nutrition and Food Research, 50*(12), 1105–1117.

Noothalapati, H., Iwasaki, K., & Yamamoto, T. (2021). Non-invasive diagnosis of colorectal cancer by Raman spectroscopy: Recent developments in liquid biopsy and endoscopy approaches. *Spectrochimica Acta Part A: Molecular and Biomolecular Spectroscopy, 258*, 119818.

Noviana, E., Siswanto, S., & Budi Hastuti, A. A. (2022). Advances in nanomaterial-based biosensors for determination of glycated hemoglobin. *Current Topics in Medicinal Chemistry, 22*(27), 2261–2281.

Oberg, K. A., & Fink, A. L. (1998). A new attenuated total reflectance Fourier transform infrared spectroscopy method for the study of proteins in solution. *Analytical Biochemistry, 256*(1), 92–106.

Odani, H., Iijima, K., Nakata, M., Miyata, S., Kusunoki, H., Yasuda, Y., ... Fujimoto, D. (2001). Identification of Nv-carboxymethylarginine, a new advanced glycation end-product in serum proteins of diabetic patients: Possibility of a new marker of aging and diabetes. *Biochemical and Biophysical Research Communications, 285*, 1232–1236.

Ogawa, N., Kimura, T., Umehara, F., Katayama, Y., Nagai, G., Suzuki, K., & Ichimura, M. (2019). Creation of haemoglobin A1c direct oxidase from fructosyl peptide oxidase by combined structure-based site specific mutagenesis and random mutagenesis. *Scientific Reports, 9*(1), 1–13.

Oliveira, C., & Domingues, L. (2018). Guidelines to reach high-quality purified recombinant proteins. *Applied Microbiology and Biotechnology, 102*, 81–92.

Park, J. Y., Chang, B. Y., Nam, H., & Park, S. M. (2008). Selective electrochemical sensing of glycated hemoglobin (HbA1c) on thiophene-3-boronic acid self-assembled monolayer covered gold electrodes. *Analytical Chemistry, 80*(21), 8035–8044.

Perrone, A., Giovino, A., Benny, J., & Martinelli, F. (2020). Advanced glycation end products (AGEs): Biochemistry, signaling, analytical methods, and epigenetic effects. *Oxidative Medicine and Cellular Longevity, 2020*, 1–18.

Pignataro, M. F., Herrera, M. G., & Dodero, V. I. (2020). Evaluation of peptide/protein self-assembly and aggregation by spectroscopic methods. *Molecules (Basel, Switzerland), 25*(20), 4854.

Polykretis, P., Luchinat, E., Boscaro, F., & Banci, L. (2020). Methylglyoxal interaction with superoxide dismutase 1. *Redox biology, 30*, 101421.

Ponsanti, K., Ngernyuang, N., Tangnorawich, B., Na-Bangchang, K., Boonprasert, K., Tasanarong, A., & Pechyen, C. (2022). A novel electrochemical-biosensor microchip based on MWCNTs/AuNPs for detection of glycated hemoglobin (HbA1c) in diabetes patients. *Journal of the Electrochemical Society, 169*(3), 037520.

Popova, E. A., Mironova, R. S., & Odjakova, M. K. (2010). Non-enzymatic glycosylation and deglycating enzymes. *Biotechnology & Biotechnological Equipment, 24*(3), 1928–1935.

Pundir, C. S., & Chawla, S. (2014). Determination of glycated hemoglobin with special emphasis on biosensing methods. *Analytical Biochemistry, 444*, 47–56.

Qu, F., Shi, Q., Wang, Y., Shen, Y., Zhou, K., Pearson, E. R., & Li, S. (2022). Visit-to-visit glycated hemoglobin A1c variability in adults with type 2 diabetes: A systematic review and meta-analysis. *Chinese Medical Journal, 135*(19), 2294–2300.

Quan, C., et al. (2008). A study in glycation of a therapeutic recombinant humanized monoclonal antibody: Where it is, how it got there, and how it affects charge-based behavior. *Analytical Biochemistry, 373*(2), 179–191.

Rabbani, G., Baig, M. H., Jan, A. T., Lee, E. J., Khan, M. V., Zaman, M., & Choi, I. (2017). Binding of erucic acid with human serum albumin using a spectroscopic and molecular docking study. *International Journal of Biological Macromolecules, 105*, 1572–1580.

Rabbani, N., & Thornalley, P. J. (2019). Glyoxalase 1 modulation in obesity and diabetes. *Antioxidants & Redox Signaling, 30*(3), 354–374.

Rabbani, N., & Thornalley, P. J. (2021). Protein glycation—Biomarkers of metabolic dysfunction and early-stage decline in health in the era of precision medicine. *Redox Biology, 42*.

Rahbar, S., Blumenfeld, O., & Ranney, H. M. (1969). Studies of an unusual hemoglobin in patients with diabetes mellitus. *Biochemical and Biophysical Research Communications, 36*(5), 838–843.

Rahmanifar, E., & Miroliaei, M. (2020). Differential effect of biophenols on attenuation of AGE-induced hemoglobin aggregation. *International Journal of Biological Macromolecules, 151*, 797–805.

Ramos, L. D., Mantovani, M. C., Sartori, A., Dutra, F., Stevani, C. V., & Bechara, E. J. (2021). Aerobic co-oxidation of hemoglobin and aminoacetone, a putative source of methylglyoxal. *Free Radical Biology and Medicine, 166*, 178–186.

Ran, C., Xu, X., Raymond, S. B., Ferrara, B. J., Neal, K., Bacskai, B. J., ... Moore, A. (2009). Design, synthesis, and testing of difluoroboron-derivatized curcumins as near-infrared probes for in vivo detection of amyloid-β deposits. *Journal American Chemical Society, 131*(42), 15257–15261.

Renz, P. B., Chume, F. C., Timm, J. R., Pimentel, A. L., & Camargo, J. L. (2019). Diagnostic accuracy of glycated hemoglobin for gestational diabetes mellitus: a systematic review and meta-analysis. *Clinical Chemistry and Laboratory Medicine (CCLM), 57*(10), 1435–1449.

Rieder, R. F. (1970). Hemoglobin stability: Observations on the denaturation of normal and abnormal hemoglobins by oxidant dyes, heat, and alkali. *The Journal of Clinical Investigation, 49*(12), 2369–2376.

Roberts, N. B., Amara, A. B., Morris, M., & Green, B. N. (2001). Long-term evaluation of electrospray ionization mass spectrometric analysis of glycated hemoglobin. *Clinical Chemistry, 47*(2), 316–321.

Rubino, F. M., Pitton, M., Di Fabio, D., & Colombi, A. (2009). Toward an "omic" physiopathology of reactive chemicals: Thirty years of mass spectrometric study of the protein adducts with endogenous and xenobiotic compounds. *Mass Spectrometry Reviews, 28*(5), 725–784.

Ruggeri, F. S., Šneideris, T., Vendruscolo, M., & Knowles, T. P. (2019). Atomic force microscopy for single molecule characterisation of protein aggregation. *Archives of Biochemistry and Biophysics, 664*, 134–148.

Saeed, A., & Abolaban, F. (2021). Spectroscopic study of the effect of low dose fast neutrons on the hemoglobin structure. *Spectrochimica Acta Part A: Molecular and Biomolecular Spectroscopy, 261*, 120082.

Saeed, M., Kausar, M. A., Singh, R., Siddiqui, A. J., & Akhter, A. (2020). The role of glyoxalase in glycation and carbonyl stress induced metabolic disorders. *Current Protein & Peptide Science, 21*(9), 846–859.

Sae-Lee, W., McCafferty, C. L., Verbeke, E. J., Havugimana, P. C., Papoulas, O., McWhite, C. D., ... Marcotte, E. M. (2022). The protein organization of a red blood cell. *Cell Reports, 40*(3), 111103.

Sahebi, U., & Divsalar, A. (2016). Synergistic and inhibitory effects of propolis and aspirin on structural changes of human hemoglobin resulting from glycation: An in vitro study. *Journal of the Iranian Chemical Society, 13*(11), 2001–2011.

Samanta, S. (2021). Glycated hemoglobin and subsequent risk of microvascular and macrovascular complications. *Indian Journal of Medical Sciences, 73*(2), 230–238.

Sarmah, S., & Roy, A. S. (2022). A review on prevention of glycation of proteins: Potential therapeutic substances to mitigate the severity of diabetes complications. *International Journal of Biological Macromolecules, 195*, 565–588.

Sattarahmady, N., Heli, H., Moosavi-Movahedi, A. A., & Karimian, K. (2014). Deferiprone: Structural and functional modulating agent of hemoglobin fructation. *Molecular Biology Reports, 41*(3), 1723–1729.

Sattarahmady, N., Moosavi-Movahedi, A. A., Ahmad, F., Hakimelahi, G. H., Habibi-Rezaei, M., Saboury, A. A., & Sheibani, N. (2007). Formation of the molten globule-like state during prolonged glycation of human serum albumin. *Biochimica et Biophysica Acta (BBA)-General Subjects, 1770*(6), 933–942.

Sen, S., Kar, M., Roy, A., & Chakraborti, A. S. (2005). Effect of nonenzymatic glycation on functional and structural properties of hemoglobin. *Biophysical Chemistry, 113*(3), 289–298.

Shapiro, R., McManus, M., Garrick, L., McDonald, M. J., & Bunn, H. F. (1979). Nonenzymatic glycosylation of human hemoglobin at multiple sites. *Metabolism, 28*(4), 427–430.

Sharma, P., Panchal, A., Yadav, N., & Narang, J. (2020). Analytical techniques for the detection of glycated haemoglobin underlining the sensors. *International Journal of Biological Macromolecules, 155*, 685–696.

Sherwani, S. I., Khan, H. A., Ekhzaimy, A., Masood, A., & Sakharkar, M. K. (2016). Significance of HbA1c test in diagnosis and prognosis of diabetic patients. *Biomarker Insights, 11* BMI-S38440.

Siddiqui, Z., Ishtikhar, M., & Ahmad, S. (2018). d-Ribose induced glycoxidative insult to hemoglobin protein: An approach to spot its structural perturbations. *International Journal of Biological Macromolecules, 112*, 134–147.

Shimasaki, T., Yoshida, H., Kamitori, S., & Sode, K. (2017). X-ray structures of fructosyl peptide oxidases revealing residues responsible for gating oxygen access in the oxidative half reaction. *Scientific Reports, 7*(1), 2790.

Shin, A., Connolly, S., Little, R., & Kabytaev, K. (2021). Quantitation of glycated albumin by isotope dilution mass spectrometry. *Clinica Chimica Acta, 521*, 215–222.

Shin, S., Ku, Y., Babu, N., & Singh, M. (2007). Erythrocyte deformability and its variation in diabetes mellitus. *Indian Journal of Experimental Biology, 45*, 121–128.

Shin, S., Ku, Y. H., Suh, J. S., & Singh, M. (2008). Rheological characteristics of erythrocytes incubated in glucose media. *Clinical Hemorheology and Microcirculation, 38*(3), 153–161.

Silva, M. M., De Araújo Dantas, M. D., Da Silva Filho, R. C., Dos Santos Sales, M. V., De Almeida Xavier, J., Leite, A. C. R., ... Santos, J. C. C. (2020). Toxicity of thimerosal in biological systems: Conformational changes in human hemoglobin, decrease of oxygen binding capacity, increase of protein glycation and amyloid's formation. *International Journal of Biological Macromolecules, 154*, 661–671.

Singh, M., & Shin, S. (2009). Changes in erythrocyte aggregation and deformability in diabetes mellitus: A brief review. *Indian Journal of Experimental Biology, 47*, 7–15.

Sirangelo, I., & Iannuzzi, C. (2021). Understanding the role of protein glycation in the amyloid aggregation process. *International Journal of Molecular Sciences, 22*(12), 6609.

Śmiga, M., Smalley, J. W., Ślęzak, P., Brown, J. L., Siemińska, K., Jenkins, R. E., & Olczak, T. (2021). Glycation of host proteins increases pathogenic potential of Porphyromonas gingivalis. *International Journal of Molecular Sciences, 22*(21), 12084.

Song, Q., Liu, J., Dong, L., Wang, X., & Zhang, X. (2021). Novel advances in inhibiting advanced glycation end product formation using natural compounds. *Biomedicine & Pharmacotherapy, 140*, 111750.

Stefani, M., & Dobson, C. M. (2003). Protein aggregation and aggregate toxicity: New insights into protein folding, misfolding diseases and biological evolution. *Journal of Molecular Medicine, 81*, 678–699.

Tahoun, M., Engeser, M., Namasivayam, V., Sander, P. M., & Müller, C. E. (2022). Chemistry and analysis of organic compounds in dinosaurs. *Biology, 11*(5), 670.

Tessier, F., Obrenovich, M., & Monnier, V. M. (1999). Structure and mechanism of formation of human lens fluorophore LM-1. *Journal of Biological Chemistry, 274*, 20796–20804 30, 23.

Teymourian, H., Barfidokht, A., & Wang, J. (2020). Electrochemical glucose sensors in diabetes management: An updated review (2010–2020). *Chemical Society Reviews, 49*(21), 7671–7709.

Thornalley, P. J. (2005). Dicarbonyl intermediates in the Maillard reaction. *Annals of the New York Academy of Sciences, 1043*, 111–117.

Tiwari, N., Chatterjee, S., Kaswan, K., Chung, J. H., Fan, K. P., & Lin, Z. H. (2022). Recent advancements in sampling, power management strategies and development in applications for non-invasive wearable electrochemical sensors. *Journal of Electroanalytical Chemistry*116064.

Torres, N. M. P., De, O., Xavier, J., De, A., Goulart, M. O. F., Alves, R. B., & Freitas, R. P. de (2018). The chemistry of advanced glycation end-products. *Revista Virtual de Química*, 375–392.

Trivelli, L. A., Ranney, H. M., & Lai, H.-T. (1971). Hemoglobin components in patients with diabetes mellitus. *New England Journal of Medicine, 284*(7), 353–357.

Trouiller-Gerfaux, P., Podglajen, E., Hulo, S., Richeval, C., Allorge, D., Garat, A., & Dauchet, L. (2019). The association between blood cadmium and glycated haemoglobin among never-, former, and current smokers: A cross-sectional study in France. *Environmental Research, 178*, 108673.

Tsoi, P. S., Quan, M. D., Ferreon, J. C., & Ferreon, A. C. M. (2023). Aggregation of disordered proteins associated with neurodegeneration. *International Journal of Molecular Sciences, 24*(4), 3380.

Tycko, R. (2011). Solid-state NMR studies of amyloid fibril structure. *Annual Review of Physical Chemistry, 62*, 279–299.

Uskoković, V. (2012). Dynamic light scattering based microelectrophoresis: Main prospects and limitations. *Journal of Dispersion Science and Technology, 33*(12), 1762–1786.

Vistoli, G., De Maddis, D., Cipak, A., Zarkovic, N., Carini, M., & Aldini, G. (2013). Advanced glycoxidation and lipoxidation end products (AGEs and ALEs): An overview of their mechanisms of formation. *Free Radical Research, 47*(sup1), 3–27.

Wang, C., Wang, X., Chan, H. N., Liu, G., Wang, Z., Li, H. W., & Wong, M. S. (2020). Amyloid-β oligomer-targeted gadolinium-based NIR/MR dual-modal theranostic nanoprobe for Alzheimer's disease. *Advanced Functional Materials, 30*(16), 1909529.

Wang, K., Sun, D. W., Pu, H., & Wei, Q. (2017). Principles and applications of spectroscopic techniques for evaluating food protein conformational changes: A review. *Bioorganic & Technology (Elmsford, N. Y.), 67*, 207–219.

Wang, S., Gu, L., Chen, J., Jiang, Q., Sun, J., Wang, H., & Wang, L. (2022). Association of hemoglobin glycation index and glycation gap with cardiovascular disease among US adults. *Diabetes Research and Clinical Practice, 190*, 109990.

Watala, C., Golański, J., Witas, H., Gurbiel, R., Gwoździński, K., & Trojanowski, Z. (1996). The effects of in vivo and in vitro non-enzymatic glycosylation and glycoxidation on physico-chemical properties of haemoglobin in control and diabetic patients. *The International Journal of Biochemistry & Cell Biology, 28*(12), 1393–1403.

Wijk, R., & Solinge, W. W. (2005). The energy-less red blood cell is lost: Erythrocyte enzyme abnormalities of glycolysis. *Blood, 106*(13), 4034–4042.

World Health Organization (WHO). (2023). *Diabetes*. Available from: https://www.who.int/news-room/fact-sheets/detail/diabetes. Accessed 30.04.2023.

Xing, Y., Zhen, Y., Yang, L., Huo, L., & Ma, H. (2023). Association between hemoglobin glycation index and non-alcoholic fatty liver disease. *Frontiers in Endocrinology, 14*.

Yadav, N., & Mandal, A. K. (2022). Interference of hemoglobin variants in HbA1c quantification. *Clinica Chimica Acta, 539*, 55–65.

Yao, D., & Brownlee, M. (2010). Hyperglycemia-induced reactive oxygen species increase expression of the receptor for advanced glycation end products (RAGE) and RAGE ligands. *Diabetes, 59*(1), 249–255.

Yao, K., & Liu, Y. (2018). Enhancing circular dichroism by chiral hotspots in silicon nanocube dimers. *Nanoscale, 10*(18), 8779–8786.

Yazdanpanah, S., Rabiee, M., Tahriri, M., Abdolrahim, M., & Tayebi, L. (2015). Glycated hemoglobin-detection methods based on electrochemical biosensors. *TrAC Trends in Analytical Chemistry, 72*, 53–67.

Yoshida, T., Prudent, M., & D'alessandro, A. (2019). Red blood cell storage lesion: Causes and potential clinical consequences. *Blood Transfusion = Trasfusione Del Sangue, 17*(1), 27–52.

Zaman, M., Khan, A. N., Zakariya, S. M., & Khan, R. H. (2019). Protein misfolding, aggregation and mechanism of amyloid cytotoxicity: An overview and therapeutic strategies to inhibit aggregation. *International Journal of Biological Macromolecules, 134*, 1022–1037.

Zhan, Z., Li, Y., Zhao, Y., Zhang, H., Wang, Z., Fu, B., & Li, W. J. (2022). A review of electrochemical sensors for the detection of glycated hemoglobin. *Biosensors, 12*(4), 221.

Zhang, H., Li, D., Yang, Y., Chang, H., & Simone, G. (2020). On-resonance islands of Ag-nanowires sense the level of glycated hemoglobin for diabetes diagnosis. *Sensors and Actuators B: Chemical, 321*, 128451.

Zhang, X., Medzihradszky, K. F., Cunningham, J., Lee, P. D., Rognerud, C. L., Ou, C. N., & Witkowska, H. E. (2001). Characterization of glycated hemoglobin in diabetic patients: Usefulness of electrospray mass spectrometry in monitoring the extent and distribution of glycation. *Journal of Chromatography B: Biomedical Sciences and Applications, 759*(1), 1–15.

Zhang, Y., Bhatt, V. S., Sun, G., Wang, P. G., & Palmer, A. F. (2008). Site-selective glycosylation of hemoglobin on Cys β93. *Bioconjugate Chemistry, 19*(11), 2221–2230.

Zhao, X., Mai, Z., Dai, Z., & Zou, X. (2011). Direct probing of the folding/unfolding event of bovine hemoglobin at montmorillonite clay modified electrode by adsorptive-transfer voltammetry. *Talanta, 84*(1), 148–154.

Zheng, W., Shan, N., Yu, L., & Wang, X. (2008). UV–visible, fluorescence and EPR properties of porphyrins and metalloporphyrins. *Dyes and Pigments, 77*, 153–157.

Zuppi, C., Messana, I., Tapanainen, P., Knip, M., Vincenzoni, F., Giardina, B., & Nuutinen, M. (2002). Proton nuclear magnetic resonance spectral profiles of urine from children and adolescents with type 1 diabetes. *Clinical Chemistry, 48*(4), 660–662.

CHAPTER EIGHT

Attenuation of albumin glycation and oxidative stress by minerals and vitamins: An *in vitro* perspective of dual-purpose therapy

Ashwini Dinkar Jagdale[a], Rahul Shivaji Patil[b], and Rashmi Santosh Tupe[a,*]

[a]Symbiosis School of Biological Sciences (SSBS), Symbiosis International (Deemed University) (SIU), Pune, Maharashtra, India
[b]Vascular Biology Center, Medical College of Georgia, Augusta University, Augusta, GA, United States
*Corresponding author. e-mail address: rashmi.tupe@ssbs.edu.in

Contents

1. Introduction	232
2. Materials and methods	234
2.1 Materials	234
3. Methods	235
3.1 *In vitro* glycation of albumin	235
3.2 Estimation of glycation markers	235
3.3 Estimation of structural modification	236
3.4 Effect of the dose-dependent response of minerals on BSA glycation	237
3.5 Effects of selected minerals, vitamins, and glycated albumin on HEK-293 cells	237
3.6 Cell treatment	237
3.7 Statistical analysis	239
4. Results and discussion	239
4.1 Effect of minerals on BSA glycation	239
4.2 Hyperchromic effect	239
4.3 Dose dependant effect of minerals on glycation markers	241
4.4 Effect of fat-soluble vitamins on BSA glycation	241
4.5 Effect of water-soluble vitamins on BSA glycation	242
4.6 Effect of vitamins and minerals on cytotoxicity and oxidative stress in HEK-293 cells	244
5. Conclusion and future directions	246
Acknowledgments	246
References	247

Abstract

Nonenzymatic glycation of proteins is accelerated in the context of elevated blood sugar levels in diabetes. Vitamin and mineral deficiencies are strongly linked to the onset and progression of diabetes. The antiglycation ability of various water- and fat-soluble vitamins, along with trace minerals like molybdenum (Mo), manganese (Mn), magnesium (Mg), chromium, etc., have been screened using Bovine Serum Albumin (BSA) as *in vitro* model. BSA was incubated with methylglyoxal (MGO) at 37 °C for 48 h, along with minerals and vitamins separately, along with controls and aminoguanidine (AG) as a standard to compare the efficacy of the minerals and vitamins. Further, their effects on renal cells' (HEK-293) antioxidant potential were examined. Antiglycation potential is measured by monitoring protein glycation markers, structural and functional modifications. Some minerals, Mo, Mn, and Mg, demonstrated comparable inhibition of protein-bound carbonyl content and ß-amyloid aggregation at maximal physiological concentrations. Mo and Mg protected the thiol group and free amino acids and preserved the antioxidant potential. Vitamin E, D, B1 and B3 revealed significant glycation inhibition and improved antioxidant potential in HEK-293 cells as assessed by estimating lipid peroxidation, SOD and glyoxalase activity. These results emphasize the glycation inhibitory potential of vitamins and minerals, indicating the use of these micronutrients in the prospect of the therapeutic outlook for diabetes management.

1. Introduction

Diabetes mellitus (DM) is a metabolic disorder that causes persistently elevated blood sugar levels over a prolonged period. Type II DM also referred to as non-insulin-dependent diabetes mellitus (NIDDM), develops when β-cells lose their responsiveness to insulin. The factors contributing to its pathogenesis include a diet high in fat and calories, a sedentary lifestyle, and obesity (Tuomilehto et al., 2001). The frequency of DM is rising rapidly worldwide, a prevalent disease. According to a report by (Magliano & Boyko, 2021) over 537 million people were estimated to be suffering from this disease. If this trend continues, there will be 693 million diabetics worldwide by the year 2045 (Cho et al., 2018). People with DM are at high risk of developing macrovascular and microvascular complications, affecting the heart, blood vessels, kidney, cornea of eyes and nerves (Harding et al., 2019). These complications associated with DM are primarily due to the increased concentration of advanced glycation end products (AGEs), oxidative stress, and activation of different cellular signaling pathways (Bangar et al., 2022; Fan & Monnier, 2021; Taguchi, & Fukami, 2023).

Methylglyoxal and other glycating agents such as glucose, fructose, galactose, and lactose interact non-enzymatically with free amino groups of proteins to form AGEs, or Maillard reaction products, which modify the protein's structure and function (Singh et al., 2014). As albumin is the most abundant protein of the plasma, making up 50% of the plasma proteins, it becomes an ultimate target of the glycating agents. Bovine serum albumin (BSA) is a globular protein of 585 amino acid residues and a homolog of human serum albumin (HSA). BSA has a high (76%) sequence homology to HSA, and both have been the subject of extensive *in vitro* glycation research (Sadowska-Bartosz et al., 2015). The link between diabetes and trace elements has been found in numerous investigations (Khan & Awan, 2014; Sanjeevi et al., 2018), and their concentrations are reported to be lowered in patients with DM (Siddiqui et al., 2014). Several reports manifest the pivotal role of vitamins and minerals in insulin resistance, and their deficiency may lead to adverse effects and escalation of Type II diabetes progression (Dubey et al., 2020; Vinson & Howard, 1996). Due to the antioxidant potential and role in glucose metabolism, various vitamin and mineral supplements are admissible in managing DM to prevent secondary complications (Martini et al., 2010).

Molybdenum (Mo) has insulin-mimetic activity, and Molybdate partly mimics insulin-promoted metabolic effects in *Drosophila melanogaster* (Bohdana et al., 2014). Molybdate treatment for diabetic rats has enhanced lipid and glucose metabolism (Ozcelikay et al., 1996). Zeng et al. (2008) reported that tetrathiomolybdate partially improves hyperglycemia in streptozotocin (STZ)-induced diabetic rats and in the db/db mice model. Tetrathiomolybdate was also used for treating fibrotic, inflammatory, and autoimmune diseases, including the non-obese diabetic mouse model (Brewer et al., 2006). Sodium molybdate protects the cell's inherent antioxidant system by lowering lipid peroxidation in alloxan-induced diabetic rats (Panneerselvam & Govindasamy, 2004). Diabetes and renal failure are more common among people with manganese (Mn) deficiencies, according to studies on the general population. Koh et al. (2014) studied the association of blood Mn levels in chronic diseases such as diabetes, hypertension, ischemic heart disease, and renal dysfunction in the Korean general population.

Vitamin E, alpha-tocopherol, is a compelling fat-soluble vitamin with potent antioxidant and anti-inflammatory effects. Li et al. investigated the impact of vitamin E supplementation in diabetic people on LDL oxidative susceptibility, glycation, and AGEs alterations in vitro (Li et al., 1996). Several studies demonstrated that vitamin E inhibits glycation of haemoglobin, which manifests as a diagnostic biomarker of diabetes, in STZ-induced diabetic rats (Je et al., 2001) by inhibiting malondialdehyde (MDA) formation and lipid peroxidation (Minamiyama et al., 2008). A meta-analysis study of randomized controlled trials reported that patients with Type II diabetes who use vitamin C supplements may experience better glycemic control and a decline in blood pressure (Mason et al., 2021). A previous study also reported that vitamin C at 10 mM concentration showed significant glycation inhibition and reduced oxidative stress by lowering advanced oxidation protein products (AOPP) and AGEs formation in the glycated BSA model *in vitro* (Grzebyk & Piwowar, 2016). Research suggests that vitamin B6, an antioxidant, can help to prevent the formation of AGEs. Additionally, studies have demonstrated that it may lower the risk of diabetes as well as diabetic complications (Mascolo & Vernì, 2020). A thorough comparison and analysis of the antiglycation potential of these vitamins and minerals have yet to be conducted.

The generation of AGEs in diabetes is linked to increased oxidative stress and inflammation, which can lead to secondary complications associated with high blood sugar levels. Using vitamins and minerals as dietary supplements to combat or at least control the deleterious effects of protein glycation is an additive regimen for managing Type II DM. Based on the evidence presented, this study has focused on exploring the potential of trace elements, fat-soluble vitamins, and water-soluble vitamins as antiglycation agents.

2. Materials and methods
2.1 Materials

Bovine serum albumin (BSA, Fraction V, A2153), sodium azide, Methyl glyoxal (MGO), Aminoguanidine, 2, 4-dinitrophenylhydrazine (DNPH), 5, 5-dithiobis (2-nitrobenzoic acid) (DTNB), Nitroblue tetrazolium chloride (NBT), and Congo red were obtained from Sigma (St. Louis, MO, USA). Thiobarbituric acid, 3-(4, 5-dimethylthiazol-2-yl)-2, 5-diphenyltetrazolium bromide (MTT), Fetal Bovine Serum (FBS) and p-Benzoquinone were acquired from Himedia, India. The Human Embryonic Kidney cell line (HEK-293) was purchased from National Centre for Cell Sciences, Pune, India. All other reagents were of analytical grades, such as minerals—Manganese (Mn), Selenium (Se), Magnesium (Mg), Mo, Chromium (Cr), Sodium (Na), Potassium (K), Copper (Cu), Iodine (I), Calcium (Ca), Iron (Fe) and Vitamins.

3. Methods
3.1 *In vitro* glycation of albumin

BSA was glycated using the procedure outlined by McPherson et al. with few changes (McPherson et al., 1988). Glycated samples were prepared by incubating BSA (0.37 mM) with methylglyoxal (MGO-10 mM) prepared in phosphate buffer saline (PBS) (200 μM, pH 7.4) including 0.02% sodium azide, along with vitamins and minerals, and incubated at 37 °C for 48 h. The physiological concentrations of minerals and vitamins used were as Mn (36.6 ng/mL), Se (120 ng/mL), Mg (24.3 μg/mL), Mo (5.4 ng/mL), Cr (1.4 ng/mL), Na (3.31 mg/mL), K (167 μg/mL), Cu (1.2 μg/mL), I (50 ng/mL), Ca (104 μg/mL), Fe (1.34 μg/mL), Vit E (10.6 μg/mL), Vit D (50 μg/mL), Vit A (324 ng/mL), Vit K (3.2 ng/mL), Vit C (8.97 μg/mL), Vit B1 (21 ng/mL), Vit B2 (32 ng/mL), Vit B3 (17 ng/mL), Vit B5 (1.65 μg/mL), Vit B7 (0.404 ng/mL), Vit B9 (0.094 ng/mL) and Vit B12 (0.611 ng/mL) (Diem & Lentner, 1970). Standard inhibitor AG (BSA+MGO+AG), negative control (native BSA), and positive control (glycated BSA- BSA+MGO) groups were incubated under similar conditions. After incubation, samples were filter sterilized by passing through 0.22 μm membrane filters under aseptic conditions during the entire process. After incubation, unbound MGO was eliminated by dialyzing against PBS, and the dialysate was utilized for additional investigation. The protein concentration was determined by the Lowry method using BSA as a standard (Lowry et al., 1951).

3.2 Estimation of glycation markers
3.2.1 Fructosamine content

Fructosamine content was analysed using the (NBT) assay, which Baker et al. (1985) described. In carbonate buffer (10 mM, pH 10.35), 0.75 mM of NBT solution was prepared. NBT solution, 800 μL, was incubated with 40 μL of test samples at 37 °C for 30 min, and at 530 nm, absorbance was measured (Genesys 10 S UV-Visible, Thermo Scientific, USA). A typical 1-deoxy-1-moepholinofructose curve was used to quantify the fructosamine concentration, which was then represented in nM/mg of protein. The percentage inhibition with minerals and vitamins was determined, taking the positive control as 100%.

3.2.2 Protein carbonyl groups

The method published by Tupe and Agte (2010) measured the carbonyl group in glycated samples. In 2.5 M HCl, 10 mM DNPH was prepared.

The glycated sample (500 μL) was first incubated with DNPH solution at room temperature for 1 h and then precipitated with TCA (1 mL, 20%). A 1:1 v/v mixture of ethanol and ethyl acetate was used to wash the precipitate and the precipitated protein was solubilized in 6 M urea (1 mL), and absorbance was taken at 365 nm. Using the DNPH molar extinction coefficient of 21 mM^{-1} cm^{-1}, the protein carbonyl group content was estimated and expressed in mM/mg of protein. The percent inhibition with minerals was determined, taking the positive control as 100%.

3.3 Estimation of structural modification

3.3.1 β-amyloid aggregates

The previously established method (Nirwal et al.,2021) was used to measure the aggregation of glycated proteins. A solution of Congo red (100 μM) was made in PBS (pH 7.4) with ethanol (10%, v/v). Congo red solution was incubated with the glycated sample (100 μL) for 20 min at room temperature, and absorbance was measured at 530 nm. The percent inhibition with minerals and vitamins was determined, taking the positive control as 100%.

3.3.2 Thiol group estimation

The method described by Ellman (1959) was used to assess the free thiol groups in samples treated with glycated albumin, vitamins, and minerals. The glycated sample (250 μL) was incubated with (DTNB) (0.5 mM, 750 μL) for 15 min. Absorbance was measured at 410 nm. Free thiol content was calculated using the molar extinction coefficient of 5, 5′- Dithiobis (2-nitrobenzoic acid) (ε 410 nm = 13.6 mM^{-1} cm^{-1}). The percent protection of the free thiol groups was calculated by taking negative control as 100%.

3.3.3 Free amino group

The free amino group in the treated sample was assessed according to the method given by Aćimović et al. (2011). A glycated sample (100 μL) and an equal volume of PBS were mixed with 40 μL of 0.1 M p-Benzoquinone prepared in DMSO and incubated at 37 °C for 30 min, and absorbance was measured at 480 nm. The concentration of the free amino group was estimated using a standard curve of alanine (10–90 mM) and expressed in mM/mg protein.

3.3.4 Hyperchromicity shift

A UV-Visible spectrophotometer (Thermo Fisher Scientific, Waltham, MA, USA) was used to record the absorption spectra of native and glycated albumin in the 200–400 nm range (Baraka-Vidot et al., 2014). To study changes in the native and glycated albumin sample in the presence of minerals Mn, Mg, and Mo at different concentrations.

3.4 Effect of the dose-dependent response of minerals on BSA glycation

Various concentrations (Mn- 4.7, 18.3, 36.6, and 54.9 ng/mL and Mo- 0.5, 3.6, 5.4, 10.8 ng/mL) of selected minerals were checked for their potential to inhibit BSA glycation by measuring their effect on fructosamine content, a protein carbonyl group, β-amyloid aggregation, and thiol group in dose dependant aspect.

3.5 Effects of selected minerals, vitamins, and glycated albumin on HEK-293 cells

Human Embryonic Kidney cells (HEK-293) were purchased from National Centre for Cell Sciences, NCCS, Pune, India. Cells were cultured in Dulbecco's modified Eagle's medium (DMEM) from Himedia (India), supplemented with 10% FBS from Gibco (Canada). Cells were then incubated at 37 °C in a CO_2 incubator with 5% CO_2 (CO28IR-10, New Brunswick Scientific, USA). Treatment of minerals and vitamins was given to HEK-293 cells for 24 h at 37 °C.

3.5.1 Cell viability assay

The percentage of decreased 3-(4, 5-dimethylthiazol-2-yl)-2, 5-diphenyltetrazolium bromide serve as an indirect indicator of cell viability (Nabi et al., 2019). Using a hemocytometer, HEK-293 cells were seeded into a 96-well plate at a final cell count of 20,000 cells per well. After confluency, a medium (0.01% FBS) containing glycated samples (100 μL), minerals, and vitamins was added to the cells. After 24 h incubation, the medium was withdrawn, MTT (5 mg/mL, 150 μL/well) was added, and the plate was then incubated in a CO_2 incubator for 4 h in the dark. Lysis solution (100 μL, 50% dimethylformamide, 20% sodium dodecyl sulfate, pH 7.4) was added to the cells to dissolve the formazan crystals and incubated for 1 h. After that, the MTT reduction level was evaluated by detecting its absorbance at 570 nm with a microplate reader (680, Bio-Rad, USA).

3.6 Cell treatment

In 25 cm^2 culture flasks, HEK293 cells were seeded with 5 mL of DMEM containing 10% FBS and incubated at 37 °C in a CO_2 incubator. After 90% confluency, the cells were treated with glycated albumin, minerals, and vitamin separately (2 mL) for 24 h in optiMEM media. After the cells had been incubated, they were trypsinized, and the trypsinized cell pellet was then suspended in 500 μL of lysis buffer (20 mM tris, 100 mM sodium chloride, 1 mM EDTA, and 10% Triton X-100) and centrifuged for 15 min using a 4000 rpm at 4 °C. The supernatant was used for the analysis of superoxide dismutase (SOD), glyoxalase enzymes, and lipid peroxidation (LPO) (Gaikwad et al., 2022).

3.6.1 Superoxide dismutase (SOD)
The activity of SOD was estimated by the methodology of Kono et al. (2002). The assay mixture consisted of 962 μL of carbonate buffer, 15 μL of NBT, 3 μL of Triton X-100, 10 μL of EDTA, 10 μL hydroxylamine hydrochloride and 10 μL of cell lysate (equivalent to 100 μg of protein). The increase in absorbance was recorded at 540 nm. Monitoring the rate of inhibition of NBT reduction will provide a calculation for the activity of SOD. The enzyme concentration that results in a half-maximal inhibition of NBT reduction is known as one unit.

3.6.2 Glyoxalase I activity
The glyoxalase enzyme activity was estimated according to the method described by Arai et al. (2014) with some modifications. The assay mixture consisted of 20 μL cell lysate and 180 μL of hemithioacetal solution (1.5 mM MGO + 2 mM GSH in 50 mM sodium phosphate buffer pH 6.6). The reaction mixture was kept at 37°C for 5 min. Then enzyme activity was determined by measuring the formation of S-D-lactoylglutathione from hemithioacetal at 240 nm and expressed as μM of S-D-lactoylglutathione formed/min/mg of protein.

3.6.3 Lipid peroxidation
MDA was assessed according to the method described by Linden et al. (2008). TCA (1 mL, 10%) was added to the treated cell lysate (200 μL), then centrifuged at 2000 rpm for 15 min. The supernatant (1 mL) was transferred to a different tube containing thiobarbituric acid (2 mL, 0.67% in 0.25 M HCl) and incubated in a boiling water bath for 25 min and then in ice-cold water for 10 min. Utilizing a molar extinction coefficient of 153,000 $M^{-1}cm^{-1}$ and an absorbance measurement at 535 nm, the amount of MDA was determined. The final results were represented as nM of MDA per mg of protein.

3.7 Statistical analysis

All data were represented as the mean ± SD of four (n = 4) independent experiments. Statistical analysis was performed using one-way ANOVA. The significance of the results was determined by comparison with positive control glycated samples using Student's t-test. All samples were compared with glycated BSA (unless not mentioned), where the level of significance was indicated as *($p < 0.05$), **($p < 0.01$).

4. Results and discussion
4.1 Effect of minerals on BSA glycation

Early glycation products like carbonyls and fructosamine were significantly inhibited by Mo (5.4 ng/mL), Mn (36.6 ng/mL), and Mg (24.3 μg/mL) with $p < 0.05$ (Fig. 1A and B). Similarly, all the minerals except Mg and Se inhibited the formation of protein cross-links by preventing amyloid formation; however, Mo and Cr showed prominent inhibition, as depicted in Fig. 1C ($p < 0.05$). A free thiol group is a strong nucleophile that embodies albumin's anti-oxidant capability. Its level was significantly reduced in glycated BSA (5.06 nM/mg protein) compared to native BSA (12.84 nM/mg protein). Mo and Mn have prevented the oxidation of free thiol groups remarkably (Fig. 1D). Mn deficiency is also associated with increased oxidative stress as this is an essential cofactor of the mitochondrial enzyme SOD, and this is relatable to the present results where Mn prevented oxidation of free thiol group of BSA. In the current study, Mn and Mo revealed significant antiglycation activity regarding fructosamine, protein-bound carbonyl, and thiol group modification.

4.2 Hyperchromic effect

As evident in Fig. 2, the native structure of albumin is disrupted upon glycation, which is reflected in the hyperchromic shift of the absorbance of protein at 280 nm. Glycated albumin exhibits greater absorption between 280 and 340 nm than native albumin due to structural modification and aromatic amino acid exposure (Adeshara & Tupe, 2016). Mg and Mn conferred significant protection against albumin structural alteration. Mg is a cofactor necessary for glucose uptake into cells and for the metabolism of carbohydrates participates in insulin's cellular action (Lopez-Ridaura et al., 2004). Although there is no direct evidence of the role of Mn in glycation inhibition, hypomagnesemia is a common trait in the development of DM,

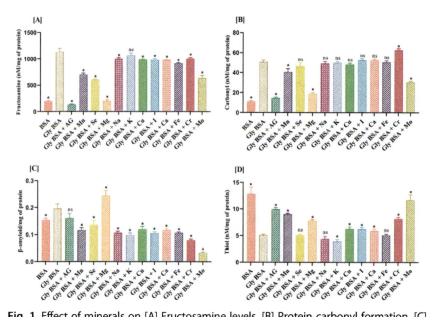

Fig. 1 Effect of minerals on [A] Fructosamine levels, [B] Protein carbonyl formation, [C] β-amyloid aggregation, and [D] Free thiol levels upon *in vitro* glycation of albumin. Each observation represents the mean ± SD (n = 4). P-value *p < 0.05 (*- against glycated BSA).

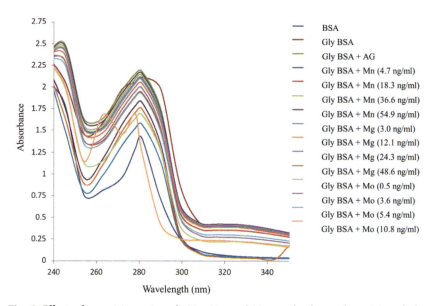

Fig. 2 Effect of promising minerals Mn, Mg and Mo on the hyperchromicity of glycated albumin. Each observation represents the mean ± SD (n = 4).

and some reports indicated that magnesium intake lowers the incidence of Type II DM by reducing insulin resistance (Hata et al., 2013). BSA includes many metal binding sites; each has a unique binding affinity value based on the metal. MD simulation studies revealed that Mg^{2+} could interact with the carboxylate groups on Asp and Glu residues of BSA during heat treatment, acting as a disulfide bridge to stabilize the protein's secondary and tertiary structures and preventing BSA aggregation (Liu et al., 2020). This could contribute to Mg's ability to avoid structural changes in BSA.

4.3 Dose dependant effect of minerals on glycation markers

By assessing different glycation markers, Mn and Mo were screened for their dose-dependent effect on glycation inhibition. A clear trend was observed showing a dose-dependent decrease in protein-bound carbonyl content and β-amyloid formation by Mn treatment compared to glycated BSA (Fig. 3C). This result is similar to the study reported by Tarwadi and Agte (2011) where along with other micronutrients Mn at 0.03, 0.06, and 1 ppm concentrations demonstrated a significant reduction in AGEs fluorescence and protein carbonyls in glycated BSA *in vitro*. Glycation causes globular proteins to unfold and refold by forming β-cross structures, which condense into β-amyloid aggregation formation (Bouma et al., 2003). The microenvironment of lysine and arginine residues is changed by the covalent binding of glycation agents, resulting in polypeptide unfolding, which is mechanically stressed by AGEs-bridged cross-links. In the present study, Mn exhibited significant inhibition of β-amyloid formation in a dose-dependent manner.

4.4 Effect of fat-soluble vitamins on BSA glycation

Among fat-soluble vitamins, all the vitamins inhibited the formation of fructosamine and β-amyloid, comparable to the standard inhibitor AG (Fig. 4A and C). However, none had shown significant inhibition of protein carbonyl content (Fig. 4B).

As evident in Fig. 4D, vitamins E and D significantly protected the free thiol group of albumin from oxidation. The finding corroborated a related report where vitamin E *in vitro* showed potent inhibition of AGEs formation from glycated HSA (Schleicher et al., 1997). Jain et al. (1991) reported that vitamin E reduces erythrocyte haemoglobin glycation by inhibiting the MDA formation. For a period of two months, administering a daily dose of vitamin E to diabetic patients has been shown to noticeably lower their fasting HbA1c levels (Ceriello et al., 1991). In contrast, in another study, vitamin E supplementation did not reduce glycation in intracellular or plasma proteins (Reaven et al., 1995).

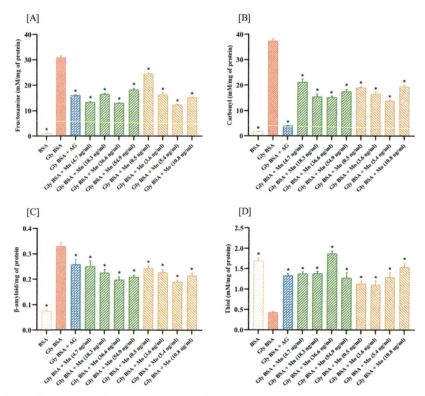

Fig. 3 Dose dependent effect of Mn and Mo on glycation markers [A] Fructosamine, [B] Carbonyl content, [C] β-amyloid aggregation, [D] Free thiol level. Each observation represents the mean ± SD (n = 4). P-value *p < 0.05 (*- against glycated BSA).

Cholecalciferol and calcitriol, a constituent of vitamin D supplements at concentrations of 0.1 μM and 0.1 nM, cause structural and functional modifications. According to a docking study, calcitriol and cholecalciferol can form noncovalent interactions with HSA domains 2 and 3, with domain 2 having a higher binding affinity and domain 2 also being one of the significant glycation sites of HSA. Vitamin D interaction with HSA impairs the protein's ability to bind glucose moieties to avoid HSA glycation (Iqbal et al., 2016).

4.5 Effect of water-soluble vitamins on BSA glycation

Water-soluble vitamins, including B1, B3, B6, B7, B9, and B12, significantly influence DM and its associated complications (Deshmukh et al., 2020). The early glycation product fructosamine formation was significantly inhibited by the vitamins B7, B1, and B9 (Fig. 5A). Even though vitamins did not considerably reduce carbonyl content (Fig. 5B), vitamin B1, B12, and C

Attenuation of albumin glycation and oxidative stress by minerals and vitamins 243

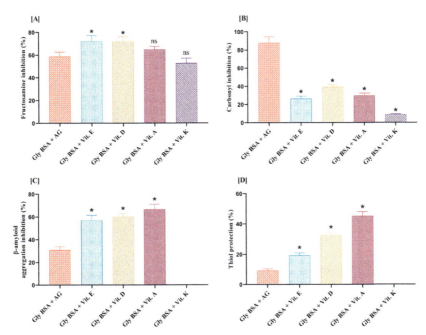

Fig. 4 Effect of fat-soluble vitamins on [A] Fructosamine inhibition, [B] Carbonyl inhibition, [C] β-amyloid aggregation, and [D] Free thiol levels upon *in vitro* BSA glycation. Each observation represents the mean ± SD (n = 4). P-value *p < 0.05 (*- against glycated BSA + AG).

inhibited the formation of β-amyloid aggregates significantly (Fig. 5C). The thiol group was significantly protected by vitamin B1, B3, and B9 (Fig. 5D).

Vitamin B1 at 500 µM concentration reduces 54.06% fructosamine content as compared to glycated HSA *in vitro* and vitamin B1 treatment, 15 mg/kg/day for 30 days reduced lipid peroxidation and exert a beneficial effect in terms of various enzymatic activities in alloxan-induced diabetic rats (Abdullah et al., 2021). In 1996, Vinson and Howard noted that vitamin C effectively inhibited the glycation of BSA and serum proteins in human subjects. Vitamin C and glucose share structural similarity, and at physiologic pH, vitamin C exists in the form of the ascorbate anion, which is a molecule with two ionizable –OH groups, and due to the presence of carbonyl groups, it reacts strongly with an amino group of proteins inhibiting glycation (Dakhale et al., 2011). It's worth noting that vitamin B3 has been found to effectively decrease HSA glycation, preserve the protein's secondary structure, and shield against DNA damage. Moreover, it binds to HSA at Sudlow's site I in a natural, spontaneous waywith a binding constant ranging around 10^4 M^{-1} (Abdullah et al., 2017).

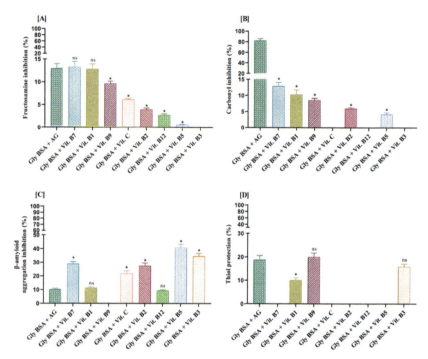

Fig. 5 Effect of water-soluble vitamins on [A] Fructosamine inhibition, [B] Protein carbonyl inhibition, [C] β-amyloid aggregation inhibition, and [D] Free thiol protection upon *in vitro* glycation of albumin. Each observation represents the mean ± SD (n = 4). P-value *p < 0.05 (*- against glycated BSA + AG).

4.6 Effect of vitamins and minerals on cytotoxicity and oxidative stress in HEK-293 cells

To understand how vitamins and minerals protect the body, renal cells were exposed to glycated samples for analysis. It was noted that vitamin B1, B3, and mineral Mo had protected the cells from glycation-induced cytotoxicity (Fig. 6A). Minerals like Mn and Mo, as well as the vitamins A, D, B1, and B3, were able to considerably lower oxidative stress by maintaining the activity of the antioxidant enzyme SOD (Fig. 6B). It has been reported that vitamin D protects against oxidative and inflammatory damage to the retinal epithelial and endothelial cells (Hernandez et al., 2021). Vitamin D, Mn, and Mo significantly reduce lipid peroxidation in HEK-293 cells as compared to glycated BSA (Fig. 6C). Cellular glyoxalase enzyme activity was improved in cells treated with the minerals Mn and Mo as compared to a positive control (Fig. 6D).

Studies have suggested that inhaling Mn significantly reduced the expression of the genes for the inflammatory proteins such as amyloid

Attenuation of albumin glycation and oxidative stress by minerals and vitamins 245

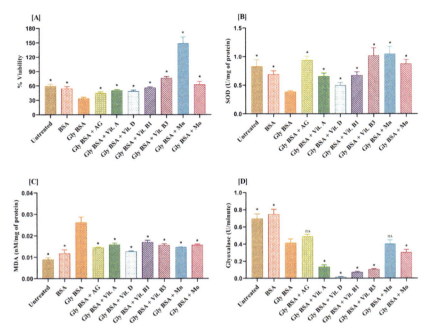

Fig. 6 Effect of minerals and vitamins on [A] Cell viability, [B] SOD activity, [C] Lipid peroxidation, and [D] Glyoxalase enzyme activity of HEK-293 cells. Each observation represents the mean ± SD (n = 4). P-value *p < 0.05. (*- against glycated BSA).

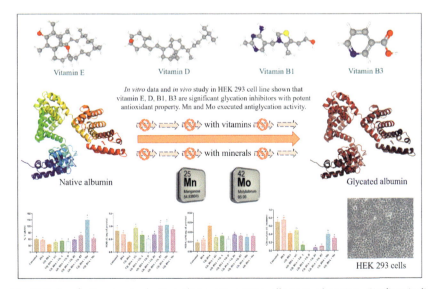

Fig. 7 Role of vitamins and minerals in preventing albumin glycation. Finding indicates that vitamin E, D, B1, B2 and minerals like Mn and Mo exhibited significant antiglycation and antioxidant potential.

precursor protein, cyclooxygenase-2, and transforming growth factor-β (HaMai et al., 2006). Kunzler et al. (2017) investigated the cytotoxic effect of vitamin A on the human SH-SY5Y neuroblastoma cell line, and it was reported that concentrations higher than 7 μM dramatically decreased cell viability; also vitamin A treatment at concentrations of 5, 7, 10, and 20 μM increased the rate of DCFH oxidation showing significant depletion of reactive oxygen species formation. Vitamin D treatment has not changed GLO-1 gene expression in diabetic rats (Derakhshanian et al., 2019).

5. Conclusion and future directions

Our study aimed to identify minerals and vitamins that possess both antiglycation and antioxidant potential. Among the 23 screened, Mn, Mo, and vitamins E, D, E, B1, and B3 demonstrated effective inhibition of albumin glycation. We specifically chose these vitamins and minerals to explore their protective abilities against cell toxicity induced by glycated albumin in HEK-293 cells (Fig. 7). Vitamin B3, and Mn were particularly promising due to their potent antioxidant activity, which increased the SOD activity. Different studies have demonstrated that a deficiency in certain micronutrients, including Mg, Mn, Fe, and vitamin D, can contribute to the onset of diabetes. Our study focuses on the role of these essential vitamins and minerals in preventing glycation in terms of inhibition of fructosamine, protein carbonyl groups, β-amyloid aggregates, and free thiol groups.

Supplementation of these micronutrients could serve as a dual-purpose therapy for treating DM. This research highlights the potential of natural, inexpensive, and non-toxic micronutrients for preventing and treating diabetic complications. However, further research is required to accurately identify the specific molecular mechanisms that minerals and vitamins utilize to prevent protein glycation.

Acknowledgments

Dr. Ashwini D. Jagdale would like to acknowledge the Indian Council of Medical Research (ICMR)—Department of Health Research for providing a grant under the Woman Scientist programme [File No. R.12013/09/2022-HR], and Dr. Rashmi S. Tupe acknowledges the support of the Ministry of Ayurveda, Yoga, and Naturopathy, Unani, Siddha, and Homeopathy (AYUSH), Government of India [Grant no: Z. 28015/103/2012-HPC (EMR)-AYUSH].

References

Abdullah, K. M., Qais, F. A., Ahmad, I., & Naseem, I. (2017). Inhibitory effect of vitamin B3 against glycation and reactive oxygen species production in HSA: An in vitro approach. *Archives of Biochemistry and Biophysics, 627*, 21–29. https://doi.org/10.1016/j.abb.2017.06.009.

Abdullah, K. M., Arefeen, A., Shamsi, A., Alhumaydhi, F. A., & Naseem, I. (2021). Insight into the in vitro antiglycation and in vivo antidiabetic effects of thiamine: Implications of vitamin B1 in controlling diabetes. *ACS Omega, 6*(19), 12605–12614. https://doi.org/10.1021/acsomega.1c00631.

Aćimović, J. M., Jovanović, V. B., Veselinović, M. R., Srećković, V. D., & Mandić, L. M. (2011). Method for monitoring the protein amino group changes during carbonylation. *Clinical Biochemistry, 44*(12), 994–999. https://doi.org/10.1016/j.clinbiochem.2011.05.019.

Adeshara, K., & Tupe, R. (2016). Antiglycation and cell protective actions of metformin and glipizide in erythrocytes and monocytes. *Molecular Biology Reports, 43*(3), 195–205. https://doi.org/10.1007/s11033-016-3947-5.

Arai, M., Nihonmatsu-Kikuchi, N., Itokawa, M., Rabbani, N., & Thornalley, P. J. (2014). Measurement of glyoxalase activities. *Biochemical Society Transactions, 42*(2), 491–494. https://doi.org/10.1042/BST20140010.

Baker, J. R., Metcalf, P. A., Johnson, R. N., Newman, D., & Rietz, P. (1985). Use of protein-based standards in automated colorimetric determinations of fructosamine in serum. *Clinical Chemistry, 31*(9), 1550–1554.

Bangar, N. S., Gvalani, A., Ahmad, S., Khan, M. S., & Tupe, R. S. (2022). Understanding the role of glycation in the pathology of various non-communicable diseases along with novel therapeutic strategies. *Glycobiology, 32*(12), 1068–1088. https://doi.org/10.1093/glycob/cwac060.

Baraka-Vidot, J., Navarra, G., Leone, M., Bourdon, E., Militello, V., & Rondeau, P. (2014). Deciphering metal-induced oxidative damages on glycated albumin structure and function. *Biochimica et Biophysica Acta, 1840*(6), 1712–1724. https://doi.org/10.1016/j.bbagen.2013.12.017.

Bouma, B., Kroon-Batenburg, L. M., Wu, Y. P., Brünjes, B., Posthuma, G., Kranenburg, O., ... Gebbink, M. F. (2003). Glycation induces formation of amyloid cross-beta structure in albumin. *The Journal of Biological Chemistry, 278*(43), 41810–41819. https://doi.org/10.1074/jbc.M303925200.

Brewer, G. J., Dick, R., Zeng, C., & Hou, G. (2006). The use of tetrathiomolybdate in treating fibrotic, inflammatory, and autoimmune diseases, including the non-obese diabetic mouse model. *Journal of Inorganic Biochemistry, 100*(5-6), 927–930. https://doi.org/10.1016/j.jinorgbio.2005.10.007.

Ceriello, A., Giugliano, D., Quatraro, A., Donzella, C., Dipalo, G., & Lefebvre, P. J. (1991). Vitamin E reduction of protein glycosylation in diabetes. New prospect for prevention of diabetic complications? *Diabetes Care, 14*(1), 68–72 https://doi.org/10.2337/diacare.14.1.68.

Cho, N. H., Shaw, J. E., Karuranga, S., Huang, Y., da Rocha Fernandes, J. D., Ohlrogge, A. W., & Malanda, B. (2018). IDF Diabetes Atlas: Global estimates of diabetes prevalence for 2017 and projections for 2045. *Diabetes Research and Clinical Practice, 138*, 271–281. https://doi.org/10.1016/j.diabres.2018.02.023.

Dakhale, G. N., Chaudhari, H. V., & Shrivastava, M. (2011). Supplementation of vitamin C reduces blood glucose and improves glycosylated hemoglobin in type 2 diabetes mellitus: A randomized, double-blind study. *Advances in Pharmacological Sciences*195271. https://doi.org/10.1155/2011/195271.

Derakhshanian, H., Djalali, M., Mohammad Hassan, M. H., Alvandi, E., Eshraghian, M. R., Mirshafiey, A., ... Djazayery, A. (2019). Vitamin D suppresses cellular pathways of diabetes complication in liver. *Iranian Journal of Basic Medical Sciences, 22*(6), 690–694. https://doi.org/10.22038/ijbms.2019.36054.8584.

Deshmukh, S. V., Prabhakar, B., & Kulkarni, Y. A. (2020). Water soluble vitamins and their role in diabetes and its complications. *Current Diabetes Reviews, 16*(7), 649–656. https://doi.org/10.2174/1573399815666190916114040.

Diem, K., & Lentner, C. (1970). *Documenta Geigy: Scientific tables* (7th ed.). Basel, Switzerland: CIBA-Geigy Limited, 667.

Dubey, P., Thakur, V., & Chattopadhyay, M. (2020). Role of minerals and trace elements in diabetes and insulin resistance. *Nutrients, 12*(6), 1864. https://doi.org/10.3390/nu12061864.

Ellman, G. L. (1959). Tissue sulfhydryl groups. *Archives of Biochemistry and Biophysics, 82*(1), 70–77. https://doi.org/10.1016/0003-9861(59)90090-6.

Fan, X., & Monnier, V. M. (2021). Protein posttranslational modification (PTM) by glycation: Role in lens aging and age-related cataractogenesis. *Experimental Eye Research, 210*, 108705. https://doi.org/10.1016/j.exer.2021.108705.

Gaikwad, D. D., Bangar, N. S., Apte, M. M., Gvalani, A., & Tupe, R. S. (2022). Mineralocorticoid interaction with glycated albumin downregulates NRF—2 signaling pathway in renal cells: Insights into diabetic nephropathy. *International Journal of Biological Macromolecules, 220*, 837–851. https://doi.org/10.1016/j.ijbiomac.2022.08.095.

Grzebyk, E., & Piwowar, A. (2016). Inhibitory actions of selected natural substances on formation of advanced glycation endproducts and advanced oxidation protein products. *BMC Complementary and Alternative Medicine, 16*(1), 381. https://doi.org/10.1186/s12906-016-1353-0.

HaMai, D., Rinderknecht, A. L., Guo-Sharman, K., Kleinman, M. T., & Bondy, S. C. (2006). Decreased expression of inflammation-related genes following inhalation exposure to manganese. *Neurotoxicology, 27*(3), 395–401. https://doi.org/10.1016/j.neuro.2005.11.004.

Harding, J. L., Pavkov, M. E., Magliano, D. J., Shaw, J. E., & Gregg, E. W. (2019). Global trends in diabetes complications: A review of current evidence. *Diabetologia, 62*(1), 3–16. https://doi.org/10.1007/s00125-018-4711-2.

Hata, A., Doi, Y., Ninomiya, T., Mukai, N., Hirakawa, Y., Hata, J., ... Kiyohara, Y. (2013). Magnesium intake decreases Type 2 diabetes risk through the improvement of insulin resistance and inflammation: the Hisayama Study. *Diabetic Medicine: A Journal of the British Diabetic Association, 30*(12), 1487–1494. https://doi.org/10.1111/dme.12250.

Hernandez, M., Recalde, S., González-Zamora, J., Bilbao-Malavé, V., Sáenz de Viteri, M., Bezunartea, J., ... García-Layana, A. (2021). Anti-inflammatory and anti-oxidative synergistic effect of vitamin D and nutritional complex on retinal pigment epithelial and endothelial cell lines against age-related macular degeneration. *Nutrients, 13*(5), 1423. https://doi.org/10.3390/nu13051423.

Iqbal, S., Alam, M. M., & Naseem, I. (2016). Vitamin D prevents glycation of proteins: An in vitro study. *FEBS Letters, 590*(16), 2725–2736. https://doi.org/10.1002/1873-3468.12278.

Jain, S. K., Levine, S. N., Duett, J., & Hollier, B. (1991). Reduced vitamin E and increased lipofuscin products in erythrocytes of diabetic rats. *Diabetes, 40*(10), 1241–1244. https://doi.org/10.2337/diab.40.10.1241.

Je, H. D., Shin, C. Y., Park, H. S., Huh, I. H., & Sohn, U. D. (2001). The comparison of vitamin C and vitamin E on the protein oxidation of diabetic rats. *Journal of Autonomic Pharmacology, 21*(5–6), 231–236. https://doi.org/10.1046/j.1365-2680.2001.00226.x.

Khan, A. R., & Awan, F. R. (2014). Metals in the pathogenesis of type 2 diabetes. *Journal of Diabetes and Metabolic Disorders, 13*(1), 16. https://doi.org/10.1186/2251-6581-13-16.

Koh, E. S., Kim, S. J., Yoon, H. E., Chung, J. H., Park, C. W., Chang, Y. S., & Shin, S. J. (2014). Association of blood manganese level with diabetes and renal dysfunction: A cross-sectional study of the Korean general population. *BMC Endocrine Disorders, 14*, 24. https://doi.org/10.1186/1472-6823-14-24.

Kono, Y., Okada, S., Tazawa, Y., Kanzaki, S., Mura, T., Ueta, E., ... Otsuka, Y. (2002). Response of anti-oxidant enzymes mRNA in the neonatal rat liver exposed to 1,2,3,4-tetrachlorodibenzo-p-dioxin via lactation. *Pediatrics International: Official Journal of the Japan Pediatric Society, 44*(5), 481–487. https://doi.org/10.1046/j.1442-200x.2002.01608.x.

Kunzler, A., Kolling, E. A., Da Silva, J. D. Jr, Gasparotto, J., de Bittencourt Pasquali, M. A., Moreira, J. C. F., & Gelain, D. P. (2017). Retinol (Vitamin A) increases α-synuclein, β-amyloid peptide, tau phosphorylation and RAGE content in human SH-SY5Y neuronal cell line. *Neurochemical Research, 42*(10), 2788–2797. https://doi.org/10.1007/s11064-017-2292-y.

Li, D., Devaraj, S., Fuller, C., Bucala, R., & Jialal, I. (1996). Effect of alpha-tocopherol on LDL oxidation and glycation: In vitro and in vivo studies. *Journal of Lipid Research, 37*(9), 1978–1986.

Linden, A., Gülden, M., Martin, H. J., Maser, E., & Seibert, H. (2008). Peroxide-induced cell death and lipid peroxidation in C6 glioma cells. *Toxicology In Vitro: An International Journal Published in Association with BIBRA, 22*(5), 1371–1376. https://doi.org/10.1016/j.tiv.2008.02.003.

Liu, X., Zhang, W., Liu, J., Pearce, R., Zhang, Y., Zhang, K., ... Liu, B. (2020). Mg^{2+} inhibits heat-induced aggregation of BSA: The mechanism and its binding site. *Food Hydrocolloids, 101*, 105450. https://doi.org/10.1016/j.foodhyd.2019.105450.

Lopez-Ridaura, R., Willett, W. C., Rimm, E. B., Liu, S., Stampfer, M. J., Manson, J. E., & Hu, F. B. (2004). Magnesium intake and risk of type 2 diabetes in men and women. *Diabetes Care, 27*(1), 134–140. https://doi.org/10.2337/diacare.27.1.134.

Lowry, O. H., Rosebrough, N. J., Farr, A. L., & Randall, R. J. (1951). Protein measurement with the Folin phenol reagent. *The Journal of Biological Chemistry, 193*(1), 265–275.

Magliano, D. J., Boyko, E. J., & IDF Diabetes Atlas 10th edition scientific committee . (2021). *IDF DIABETES ATLAS*. (10th ed.). International Diabetes Federation.

Martini, L. A., Catania, A. S., & Ferreira, S. R. (2010). Role of vitamins and minerals in prevention and management of type 2 diabetes mellitus. *Nutrition Reviews, 68*(6), 341–354. https://doi.org/10.1111/j.1753-4887.2010.00296.x.

Mascolo, E., & Vernì, F. (2020). Vitamin B6 and diabetes: Relationship and molecular mechanisms. *International Journal of Molecular Sciences, 21*(10), 3669. https://doi.org/10.3390/ijms21103669.

Mason, S. A., Keske, M. A., & Wadley, G. D. (2021). Effects of vitamin C supplementation on glycemic control and cardiovascular risk factors in people with type 2 diabetes: A GRADE-assessed systematic review and meta-analysis of randomized controlled trials. *Diabetes Care, 44*(2), 618–630. https://doi.org/10.2337/dc20-1893.

McPherson, J. D., Shilton, B. H., & Walton, D. J. (1988). Role of fructose in glycation and cross-linking of proteins. *Biochemistry, 27*(6), 1901–1907. https://doi.org/10.1021/bi00406a016.

Minamiyama, Y., Takemura, S., Bito, Y., Shinkawa, Y., Tsukioka, T., Nakahira, A., ... Okada, S. (2008). Supplementation of α-tocopherol improves cardiovascular risk factors via the insulin signalling pathway and reduction of mitochondrial reactive oxygen species in type II diabetic rats. *Free Radical Research, 42*(3), 261–271. https://doi.org/10.1080/10715760801898820.

Nabi, R., Alvi, S. S., Shah, A., Chaturvedi, C. P., Iqbal, D., Ahmad, S., & Khan, M. S. (2019). Modulatory role of HMG-CoA reductase inhibitors and ezetimibe on LDL-AGEs-induced ROS generation and RAGE-associated signalling in HEK-293 Cells. *Life Sciences, 235*, 116823. https://doi.org/10.1016/j.lfs.2019.116823.

Nirwal, S., Bharathi, V., & Patel, B. (2021). Amyloid-like aggregation of bovine serum albumin at physiological temperature induced by cross-seeding effect of HEWL amyloid aggregates. *Biophysical Chemistry, 278*(1–12), 106678. https://doi.org/10.1016/j.bpc.2021.106678.

Ozcelikay, A. T., Becker, D. J., Ongemba, L. N., Pottier, A. M., Henquin, J. C., & Brichard, S. M. (1996). Improvement of glucose and lipid metabolism in diabetic rats treated with molybdate. *The American Journal of Physiology, 270*(2 Pt 1), E344–E352. https://doi.org/10.1152/ajpendo.1996.270.2.E344.

Panneerselvam, S. R., & Govindasamy, S. (2004). Effect of sodium molybdate on the status of lipids, lipid peroxidation and antioxidant systems in alloxan-induced diabetic rats. *Clinica Chimica Acta; International Journal of Clinical Chemistry, 345*(1-2), 93–98. https://doi.org/10.1016/j.cccn.2004.03.005.

Reaven, P. D., Herold, D. A., Barnett, J., & Edelman, S. (1995). Effects of Vitamin E on susceptibility of low-density lipoprotein and low-density lipoprotein subfractions to oxidation and on protein glycation in NIDDM. *Diabetes Care, 18*(6), 807–816. https://doi.org/10.2337/diacare.18.6.807.

Rovenko, B. M, Perkhulyn, N. V., Lushchak, O. V., Storey, J. M., Storey, K. B., & Lushchak, V. I. (2014). Molybdate partly mimics insulin-promoted metabolic effects in Drosophila melanogaster. *Comparative Biochemistry and Physiology Part C: Toxicology & Pharmacology, 165*, 76–82. https://doi.org/10.1016/j.cbpc.2014.06.002.

Sadowska-Bartosz, I., Stefaniuk, I., Galiniak, S., & Bartosz, G. (2015). Glycation of bovine serum albumin by ascorbate in vitro: Possible contribution of the ascorbyl radical? *Redox Biology, 6*, 93–99. https://doi.org/10.1016/j.redox.2015.06.017.

Sanjeevi, N., Freeland-Graves, J., Beretvas, S. N., & Sachdev, P. K. (2018). Trace element status in type 2 diabetes: A meta-analysis. *Journal of Clinical and Diagnostic Research, 12*(5), OE01–OE08. https://doi.org/10.7860/JCDR/2018/35026.11541.

Schleicher, E. D., Wagner, E., & Nerlich, A. G. (1997). Increased accumulation of the glycoxidation product N(epsilon)-(carboxymethyl) lysine in human tissues in diabetes and aging. *The Journal of Clinical Investigation, 99*(3), 457–468. https://doi.org/10.1172/JCI119180.

Siddiqui, K., Bawazeer, N., & Joy, S. S. (2014). Variation in macro and trace elements in progression of type 2 diabetes. *The Scientific World Journal, 2014*, 461591. https://doi.org/10.1155/2014/461591.

Singh, V. P., Bali, A., Singh, N., & Jaggi, A. S. (2014). Advanced glycation end products and diabetic complications. *The Korean Journal of Physiology & Pharmacology: Official Journal of the Korean Physiological Society and the Korean Society of Pharmacology, 18*(1), 1–14. https://doi.org/10.4196/kjpp.2014.18.1.1.

Taguchi, K., & Fukami, K. (2023). RAGE signaling regulates the progression of diabetic complications. *Frontiers in Pharmacology, 14*, 1128872. https://doi.org/10.3389/fphar.2023.1128872.

Tarwadi, K. V., & Agte, V. V. (2011). Effect of micronutrients on methylglyoxal-mediated in vitro glycation of albumin. *Biological Trace Element Research, 143*(2), 717–725. https://doi.org/10.1007/s12011-010-8915-7.

Tuomilehto, J., Lindström, J., Eriksson, J. G., Valle, T. T., Hämäläinen, H., Ilanne-Parikka, P., ... Uusitupa, M. Finnish Diabetes Prevention Study Group. (2001). Prevention of type 2 diabetes mellitus by changes in lifestyle among subjects with impaired glucose tolerance. *The New England Journal of Medicine, 344*(18), 1343–1350. https://doi.org/10.1056/NEJM200105033441801.

Tupe, R. S., & Agte, V. V. (2010). Role of zinc along with ascorbic acid and folic acid during long-term in vitro albumin glycation. *The British Journal of Nutrition, 103*(3), 370–377. https://doi.org/10.1017/S0007114509991929.

Vinson, J. A., & Howard, T. B. (1996). Inhibition of protein glycation and advanced glycation end products by ascorbic acid and other vitamins and nutrients. *The Journal of Nutritional Biochemistry, 7*(12), 659–663. https://doi.org/10.1016/S0955-2863(96)00128-3.

Zeng, C., Hou, G., Dick, R., & Brewer, G. J. (2008). Tetrathiomolybdate is partially protective against hyperglycemia in rodent models of diabetes. *Experimental Biology and Medicine (Maywood, N. J.), 233*(8), 1021–1025. https://doi.org/10.3181/0801-RM-10.

CHAPTER NINE

Non-enzymatic glycation and diabetic kidney disease

Anil K. Pasupulati[a,*], Veerababu Nagati[a], Atreya S.V. Paturi[a], and G. Bhanuprakash Reddy[b,*]

[a]Department of Biochemistry, University of Hyderabad, Hyderabad, India
[b]Department of Biochemistry, ICMR–National Institute of Nutrition, Hyderabad, India
*Corresponding authors. e-mail address: anilkumar@uohyd.ac.in; geereddy@yahoo.com

Contents

1. Kidney architecture	252
2. Kidney dysfunction in diabetes	254
3. Altered glucose homeostasis	256
4. Non-enzymatic glycation (NEG)- a post-translational modification	258
4.1 The chemistry of NEG	259
4.2 AGEs	259
4.3 Diversity in AGEs formation	262
5. Fate of AGEs	262
6. Pathology of AGEs in DKD	264
6.1 AGEs contribute to ROS, inflammation and cell death	264
6.2 AGEs and renin-angiotensin system (RAS)	265
6.3 AGEs and reactivation of developmental pathways	267
6.4 AGEs and phenotypic switch of renal cells	268
6.5 AGEs and complement activation	270
7. Strategies to prevent AGEs formation and DKD	272
7.1 Hypoglycemic agents	273
7.2 RAGE inhibitors	273
7.3 AGE-inhibitors	274
7.4 AGE-breakers	274
7.5 Dietary/nutritional interventions	275
8. Summary and future perspective	276
Acknowledgments	278
References	278

Abstract

Chronic diabetes leads to various complications including diabetic kidney disease (DKD). DKD is a major microvascular complication and the leading cause of morbidity and mortality in diabetic patients. Varying degrees of proteinuria and reduced glomerular filtration rate are the cardinal clinical manifestations of DKD that eventually progress into end-stage renal disease. Histopathologically, DKD is characterized by renal hypertrophy, mesangial expansion, podocyte injury, glomerulosclerosis, and tubulointerstitial fibrosis, ultimately leading to renal replacement therapy. Amongst the many mechanisms, hyperglycemia contributes to the pathogenesis of DKD via a mechanism known as non-enzymatic glycation (NEG). NEG is the irreversible conjugation of reducing sugars onto a free amino group of proteins by a series of events, resulting in the formation of initial Schiff's base and an Amadori product and to a variety of advanced glycation end products (AGEs). AGEs interact with cognate receptors and evoke aberrant signaling cascades that execute adverse events such as oxidative stress, inflammation, phenotypic switch, complement activation, and cell death in different kidney cells. Elevated levels of AGEs and their receptors were associated with clinical and morphological manifestations of DKD. In this chapter, we discussed the mechanism of AGEs accumulation, AGEs-induced cellular and molecular events in the kidney and their impact on the pathogenesis of DKD. We have also reflected upon the possible options to curtail the AGEs accumulation and approaches to prevent AGEs mediated adverse renal outcomes.

1. Kidney architecture

One of the most intricate and vital organs in the human body is the kidneys. By controlling electrolytes, acid-base and fluid balance, as well as removing toxic byproducts of the metabolism, the kidneys help maintain homeostasis (Bello-Reuss & Reuss, 1983). Further, kidneys help to retain the large biomolecules in the blood by preventing their loss into the urine. The kidneys synthesize several local hormones or autocrine/paracrine regulators, such as erythropoietin, renin-angiotensin system (RAS) components, vitamin D3, prostaglandins, adrenomedullin, and endothelins (Mukoyama & Nakao, 2005). Anatomically, the kidney is divided into the renal papilla, renal pelvis, medulla, and outer cortex. The medulla is further indistinctly divided into the inner medulla, the inner stripe of the outer medulla, and the outer stripe of the outer medulla. The vasculature of the kidney is ideally adapted to increase blood flow to the energy-demanding cortex. Each part of the kidney consists of defined segments of the nephron, the functional unit of the kidney (Fig. 1). Nephron is composed of a glomerulus and renal tubule. Glomeruli are the filtering portion of the kidney and have a more complex structure comprising capillaries, parietal epithelium, visceral epithelium, and mesangial cells. Based on the anatomical and functional characteristics, the renal tubule is divided into the proximal convoluted tubule, Henle's loop, and distal convoluted tubule. Though the renal tubule is involved in selective reabsorption and determines the final composition of urine, the glomerulus provides permselectivity and

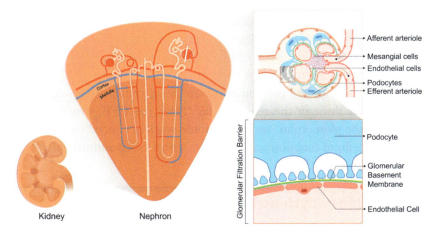

Fig. 1 Architecture of the kidney. The nephron, the premier functional unit, is distributed in both cortex and medullary regions of the kidney. Glomerulus and tubule are two regions of the nephron. The glomerular filtration barrier that performs renal filtration comprises the endothelium of blood vessels, basement membrane, and podocytes.

aids in blood filtration. The glomeruli filters ~180 g of glucose and 50 g of amino acids daily, and the proximal tubule of the nephron reabsorbs 99.8% of the glomerular filtrate (Curthoys & Moe, 2014).

The glomerulus, the kidney's filtering unit, is a specialized bundle of capillaries without being surrounded by the interstitium. However, ultrafiltration is accomplished by the three components of the glomerular filtration barrier (GFB) working in a concert. The proximal component of the GFB is a fenestrated capillary endothelium characterized by individual fenestra on the order of 70–100 nm in diameter (Alallam, Choukaife, Seyam, Lim, & Alfatama, 2023). The second component of GFB is the glomerular basement membrane (GBM), a complex mesh of extracellular proteins, including type IV collagen, laminins, fibronectins, and proteoglycans. The distal component of the GFB is visceral epithelial cells, known as podocytes. These specialized cells possess primary processes that ultimately branch into regularly spaced foot processes that enwrap and provide epithelial coverage to the glomerular capillaries. Inter-digitating podocyte foot processes help create a filtration slit diaphragm (SD) and support to help sustain the integrity of the free-standing capillary loops. SD serves as a size and charge-selective barrier, preventing plasma proteins from filtering into the urine, thus curbing protein loss. Podocytes provide growth factors to maintain endothelial cells and supply components of the basement membrane. Therefore, podocytes receive greater attention due to their distinct function and prominent location in glomerular biology. Mesangial cells, another significant cell type of glomerulus, contributes to the integrity of the glomerular tuft and the dynamic nature of filtration. Together, the glomerulus helps the formation of the primary filtrate that enters a space delimited by glomerular epithelial cells before modification during transit through the renal tubule. Thus, the GFB, structural components of the glomerulus, and renal tubule work in concert to achieve the nephron's definite function and the urine's final composition.

2. Kidney dysfunction in diabetes

Diabetic Kidney Disease (DKD) refers to specific pathologic structural and functional changes in the kidneys of patients with type 1 and type 2 diabetes mellitus (T1/T2DM). DKD is clinically characterized by persistent albuminuria and a progressive decline in kidney function. DKD is the leading cause of chronic kidney disease (CKD) and leads to end-stage kidney disease (ESKD) (Oshima et al., 2021). It should be noted that 30% to 40% of patients with DM develop DKD (Bakris, 2019). Likely, people with T1 DM of five or more years duration with albuminuria develop DKD, whereas the prevalence can range widely in T2DM. With the onset of DM, kidney weight and size increase by 15%, which remains even after a progressive reduction in kidney function. Hypertrophy of both glomerular and tubular components, thickening of basement membrane in both these compartments, progressive accumulation of extracellular matrix proteins and the resultant expansion of the mesangium, and podocyte injury are the prominent early manifestations of DKD. The presence of nodular lesions in the kidney of DM patients was first described by Kimmelstiel and Wilson in 1936. The nodules are typically acellular and positive by a periodic acid-Schiff stain. Although these nodules are pathognomonic for DKD, they are reported in 10–50% of biopsies from patients with T1 or T2DM (Mise et al., 2017). Kidneys of DKD patients are often presented with diffuse glomerular lesions characterized by diffuse mesangial matrix expansion. Arteriolar lesions in afferent and efferent vessels are also prominent in kidneys presented with DKD. Over time, hyaline material replaces the entire vessel wall structure, which is highly specific for DM. As DKD progresses, kidneys undergo several pathological changes, including proliferative changes and tubular atrophy, ultimately resulting in interstitial fibrosis and glomerulosclerosis. More than 50% of structural aberrations with glomerulosclerosis, tubular dysfunction and tubulointerstitial fibrosis impair kidney function (Tan et al., 2020).

The majority of the kidney cells, including podocytes, mesangial cells, and tubulointerstitial cells, are afflicted with noxious stimuli prevalent in the DM milieu. Partial or severe injury or dysfunction in one of these cell types extends to all kidney cells and affects the renal outcome. The clinical diagnosis of kidney function is based on proteinuria, estimated glomerular filtration rate (eGFR), or both. Albuminuria is the most crucial predictor of declining kidney function and the development of DKD. The importance of assessing proteinuria as a diagnostic requirement, as well as a measure of disease severity and risk of progression, has been supported by numerous studies over the years (Chen, Knicely, & Grams, 2019; Haider & Aslam, 2023; Wu et al., 2014). DKD, with a continuous decline in kidney function, progresses towards CKD and renal replacement therapy or death, as indicated by an increased blood creatinine concentration or a lowering eGFR. According to KDIGO (Kidney Disease: Improving Global Outcomes) guidelines, urinary albumin and eGFR help diagnose and predict the progression of DKD (KDIGO, 2020, 2021) (Fig. 2). American Diabetes Association (ADA) stated that albuminuria levels were explained by the albumin-to-creatine ratio (ACR). An average ACR level is recommended to be < 30 mg/g; if ACR 30 mg/g-299 mg/g is

	Stage 1	Stage 2	Stage 3	Stage 4	Stage 5
UACR (mg/g/24 h)	<30 mg/g Silent stage		30-299 mg/g Microalbuminuria	>300 mg/g Macroalbuminuria	Decreasing Uremic, ESKD
GFR (ml/mim/ 1.73 m²)	>90 Slight increase	89-60	59-30	<29-15	<15
Renal Features	Hyper-filtration Hypertrophy of kidney	Non-specific GBM thickening & expansion of interstitium	Distinct thickening of tubular and GBM & Marked mesangial expansion, Arteriolar hyalinosis	Pronounced nodular, diffuse inter-capillary sclerosis & reduced cellularity	Advanced, diffuse, & nodular sclerosis, reduced cellularity, encroachment of the mesangium on capillary lumina
Chronology following early diagnosis	Pre-diabetes or onset of diabetes	First 5 years	6-15 yrs	15-25 yrs	25-30 yrs

Fig. 2 Map representing various stages of DKD progression along with renal manifestations based on the decline in kidney function as evidenced by UACR and GFR.

moderate microalbuminuria, whereas ≥ 300 mg/g is macro-albuminuria, high risk of kidney failure. In similar lines, GFR < 60 mL/min/1.73 m^2 indicates reduced kidney function and kidney failure by a GFR < 15 mL/min/1.73 m^2 (Pavkov, Collins, Coresh, & Nelson, 2018). Although DKD can be diagnosed clinically, kidney biopsy is considered a gold standard test for diagnosis and prognosis, which is usually only carried out when a different renal pathology is suspected. Notably, despite having biopsy-proven DKD, about 25% of patients with T2DM and impaired kidney function show little to no proteinuria, according to the recent data (Santoro et al., 2021). The pathophysiology leading to the development of DKD and subsequent ESKD is the result of the diabetic milieu, which directly or indirectly affects kidney biology vis-a-vis its function.

3. Altered glucose homeostasis

Glucose homeostasis, the tight regulation of blood glucose levels (5.5 mM), is crucial to maintain fully functional biological processes, organs, and overall normal health. Glucose homeostasis is maintained by several endocrine mediators which regulate glucose metabolism predominantly in the liver, skeletal muscle, and adipose tissue. Insulin lowers blood glucose, whereas glucagon, somatostatin, cortisol, growth hormone, and adrenaline antagonize insulin action and contribute to an increase in blood glucose. Either absolute deficiency of insulin (T1DM) or relative deficiency of insulin or its activity (T2DM) is presented with a decline in glycolytic flux, one of the main reasons for hyperglycemia. Under normal conditions, glucose is converted to glucose-6-phosphate immediately after being taken up by insulin-sensitive cells. Decreased insulin/glucagon ratio in diabetes manifests in the stimulation of glucose-6-phosphatase activity and decreased glucokinase activity in liver. The phosphorylation rate of glucose is 30% lower in the diabetic liver when compared with a healthy liver. Dephosphorylation of glucose-6-phosphate and impaired phosphorylation of glucose to glucose-6-phosphate manifest in elevated glucose in circulation. In diabetes, liver cells show a 60% reduction in the rate of glycolysis. Notably, glycolytically derived pyruvate is channeled to mitochondria and recycled to glucose via gluconeogenesis, contributing to rising circulatory glucose levels. In the liver of diabetic individuals, only about 20% pyruvate is oxidized to carbon dioxide. Taken together, reduced rate of glycolysis, recycling of glucose-6-phosphate and pyruvate to glucose contribute to hyperglycemia. More than three-fourths of glucose taken up by the liver gets recycled to glucose in the diabetic milieu. The metabolic adaptations in the settings of decreased insulin/glucagon ratio contribute to hyperglycemia predominantly by inhibition of hepatic glycolysis and an increase in hepatic gluconeogenesis, in addition to abnormalities in glucose absorption and metabolism in extra-hepatic organs (Fig. 3A).

Excess circulatory glucose (hyperglycemia) is implicated in developing diabetes-specific pathology in tissues such as the eye lens, retina, renal glomerulus, and peripheral neurons. The pathological events arising from

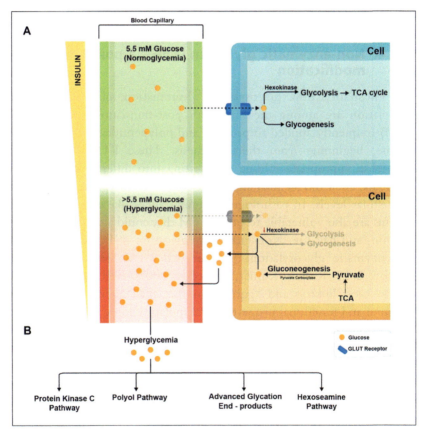

Fig. 3 (A) Adequate quantities of insulin help maintain strict glycemia. During inadequate insulin levels or insulin resistance conditions, glucose homeostasis will not be sufficiently regulated, leading to hyperglycemia. (B) Hyperglycemia triggers diabetic complications, majorly through increased polyol pathway flux; activation of protein kinase C (PKC) isoforms; elevated hexosamine flux; and formation of advanced glycation end-products (AGEs).

hyperglycemia manifest in microvascular complications that lead to cataract, retinopathy, CKD, and neuropathy. Macrovascular complications that result from deregulated glucose homeostasis include myocardial infarction and stroke. Aberrant activation of the complement system and consequent-immune complications were also observed in diabetic patients. Four major events put forward to explain hyperglycemia triggered diabetic complications are: increased polyol pathway flux; activation of protein kinase C (PKC) isoforms; elevated hexosamine flux; and accumulation of advanced glycation end-products (AGEs) formation due to aberrant non-enzymatic glycation (Fig. 3B).

4. Non-enzymatic glycation (NEG)- a post-translational modification

Phosphorylation, glycosylation, deamination, ubiquitination, and sumoylation are among the numerous post-translational modifications (PTM) frequently applied to proteins and polypeptides throughout their lifetime, beginning from their synthesis. These PTM help proteins accomplish appropriate folding and sorting into different cellular compartments, elicit intracellular signaling, confer stability, and regulate their intracellular levels. In pathological circumstances like diabetes or aging, proteins are also vulnerable to other PTM that could alter their structure, function, and half-life. Non-enzymatic glycation (NEG), also known as non-enzymatic glycosylation, is one such PTM (Jennings, Fritz, & Galligan, 2022; Kumar, Kumar, & Reddy, 2007). NEG of proteins may lead to cross-linking and aggregation that may adversely affect their structural and functional properties (Pasupulati, Chitra, & Reddy, 2016). NEG causes abnormal cellular signaling, activates transcription factors, changes gene expression profiles and alters the physicochemical properties of the affected proteins. AGEs refer to the cumulative byproducts of NEG. Though NEG happens at a slow rate during normal aging, it occurs more frequently and intensively in clinical conditions like diabetes. NEG is linked to the pathophysiology of several diabetes complications and other age-related diseases like cataracts, renal failure, cardiovascular complications, and neurodegenerative disorders, including Alzheimer's disease. Tissue AGEs correlate significantly with early kidney, eye, and nerve complications in patients with diabetes (Horton & Barrett, 2021). Besides proteins, fats and nucleic acids are also prone to NEG. However, the consequences of protein glycation are extensively investigated and implicated in the pathology of diabetic complications.

4.1 The chemistry of NEG

The non-enzymatic reaction between reducing sugars and amino acid glycine and the resultant formation of brown-colored substances was first reported in 1912 by Louis-Camille Maillard, a French Chemist. After him, the interrelated complex interactions that are triggered initially by the condensation of the free amino ($-NH_2$) group of amino acid and carbonyl ($-CHO-$ or $-C=O$) group of reducing sugars are known as Maillard reactions (Maillard, 1912) (Fig. 4). In these non-enzymatic reactions, the free $-NH_2$ group of N-terminal amino acids or the $-NH_2$ group of lysine/arginine react covalently with the carbonyl (aldehyde/keto) group of reducing sugars (predominantly glucose or fructose) and forms unstable Schiff's base (diamine). Mario Amadori, an Italian scientist, reported spontaneous rearrangement of this early unstable Schiff's base to a relatively stable intermediate glycated product, popularly known as Amadori adduct. The Schiffs base and Amadori products subsequently produce irreversible end-products, AGEs (Kumar Pasupulati, Chitra, & Reddy, 2016). Glycated hemoglobin (HbA1C) is the first glycated protein discovered by Kunkel and Wallenius in 1955 (Kunkel & Wallenius, 1955). Later, in 1981 Monnier and Cerami demonstrated the formation of AGEs by incubating eye lens proteins with reducing sugars, where they observed the formation of fluorescent yellow pigments (Monnier & Cerami, 1981). Cerami coined the term AGE to describe fluorescent colors produced by the hydrolysis of fructosamine, which binds to proteins (Monnier, Stevens, & Cerami, 1981).

4.2 AGEs

In brief, AGE refers to any protein-bound adducts following the formation of the initial Schiff's base/Amadori product, which could be considered the final product of the Maillard reaction. Even though AGEs formation occurs more intensively in hyperglycemic conditions, they build up in various tissues over time, including the skin and eye lens. The formation of reversible Schiff's bases takes hours to days, while the transformation into partially stable Amardori products and more stable AGEs takes weeks and months, respectively. Table 1 lists the most common AGEs identified in biological systems and the corresponding carbohydrate source, and the target amino acid involved in the formation of these AGEs. These AGEs can be derived from an Amadori product in a single step, as in the case of carboxymethyl-lysine (CML), or they can develop via a series of intricate

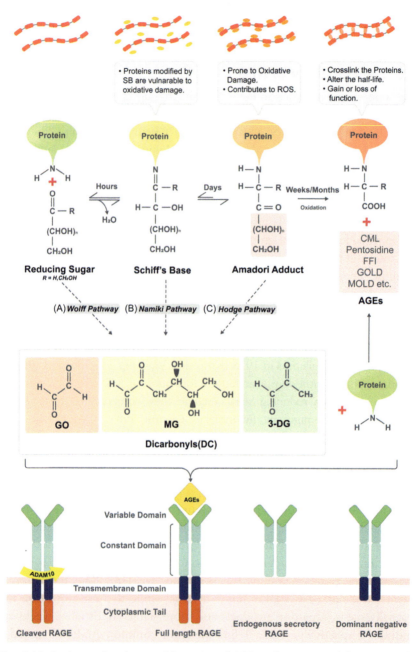

Fig. 4 Mechanism and pathways of formation of AGEs and activation of the receptor for AGE (RAGE). The initial non-enzymatic interaction between the highly reactive carbonyl group of glucose with a free amino group on proteins creates a reversible Schiff's base, which spontaneously undergoes rearrangement into a partially reversible Amadori product.

reactions that result in a cross-link, like pentosidine and glyoxal-lysine dimer (GOLD). Amadori products undergo oxidative degradation and produce reactive carbonyls such as glyoxal, methyl-glyoxal, and 3-deoxy-glucose (3DG). These carbonyl molecules can subsequently interact with free amino groups of proteins to create intermediate glycation products. Intermediate glycation and Amadori products eventually lead to the formation of irreversible inter or intra-protein cross-linking AGEs that may last for the duration of the modified substrate. John Hodge detailed the intricate web of NEG reactions that occur during the preparation and storage of food (Hodge, 1953). Hodge and Namiki pathways describe the production of AGEs from the autoxidation of Amadori products and the cleavage of dicarbonyl compounds from Schiff's base, respectively (Fig. 4). The 'Wolff pathway' describes the formation of dicarbonyls from the autoxidation of glucose, ribose, fructose, and glyceraldehyde (Wolff & Dean, 1987).

While carbohydrates like fructose and glucose-6-phosphate create AGEs at a considerably quicker rate and produce heterogeneous cross-linking products, glucose has the slowest rate of glycation among widely accessible sugars. The second generation of AGEs can be formed by binding the reactive intermediates created by the degradation of AGEs. Glyoxals, glycolaldehydes, and hydroxyl aldehydes are intermediate AGEs that accumulate during the oxidation of both carbohydrates and fatty acids. Diabetes causes an increase in glycoxidation and lipoxidation, which accelerate the accumulation of carbonyl chemicals in addition to NEG. This phenomenon is known as carbonyl stress, and the carbonyl stress intermediates also interact with the free amino groups of proteins to produce AGEs. Although AGEs are also formed at lesser rates during normal metabolic processes, they are produced endogenously at an accelerated pace during diabetes. However, in the former scenario, the host's anti-AGE cellular defense system can sequester them. Diet, smoking, and alcohol use are environmental factors that affect how quickly AGEs accumulate. AGEs can also be consumed exogenously through food, especially baked and foods processed at a higher

---

Amadori products rearrange to form heterogeneous AGEs (carboxymethyl lysine, pentosidine, Methylglyoxal lysine dimer, and glyoxal lysine dimer). Reactive carbonyl intermediates (glyoxal, methylglyoxal and 3-deoxyglucosone) can be formed from (A) the auto-oxidation of monosaccharides (Wolff pathway) or (B) Schiff's base (Namiki pathway) or (C) Amadori products (Hodge pathway). The highly reactive intermediates formed by these three pathways can react with free amino groups to form diverse AGEs. Glycation damages proteins to varying degrees and affects their function. AGEs serve as ligands for RAGE, whereas full-length (fl) RAGE transduces adverse cellular effects of AGEs.

temperature (Fallavena, Rodrigues, Marczak, & Mercali, 2022). After oral ingestion, AGEs in cooked food remain active in the bloodstream (Liang, Chen, Li, Li, & Yang, 2020).

4.3 Diversity in AGEs formation

As discussed above and presented in Fig. 4, AGEs represent several heterogeneous chemical structures; AGEs are mostly yellow-brown, relatively insoluble, and some of them are fluorescent in nature. Formation of AGEs may result in one step from Amadori product in the case of carboxymethyl lysine (CML) or a series of complex reactions leading to cross-links such as glyoxal lysine dimer (GOLD) and pentosidine (Pasupulati et al., 2016). Elevated tissue and serum AGEs are associated with adverse renal outcomes (Kumar Pasupulati, Chitra, & Reddy, 2016). CML represents the most prevalent AGE in vivo, used as an AGEs marker. Numerous AGEs have been identified in vivo and in vitro, classified into different groups based on their chemical structures and ability to emit fluorescence. Some of the commonly found AGEs in biological systems are shown in the Table 1.

5. Fate of AGEs

Accumulation of AGEs causes cumulative metabolic burden, oxidative stress, inflammatory response, and endothelial dysfunction upon binding to

Table 1 Commonly found AGEs in biological systems and the corresponding carbohydrate source and target amino acids.

AGEs	Carbohydrate source	Target amino acid
CML	Glucose, Threose	Lysine
Pentosidine	Ribose	Lysine + Arginine
Argypyrimidine	Methylglyoxal	Arginine
Imidazoline	3-Deoxyglucosone	Arginine
Carboxymethyl lysine	Methylglyoxal	Lysine
GOLD	Glyoxal	Lysine
MOLD	Methylglyoxal	Lysine
DOLD	3-Deoxyglucosone	Lysine

receptors for AGE (RAGEs). Several cell-surface RAGEs engage with AGEs, leading to their endocytosis and degradation. AGEs, on the other hand, cause cellular activation and elicit pro-oxidant and pro-inflammatory events. RAGE, an immunoglobulin superfamily member, expresses in a wide range of cells, including macrophages, endothelial cells, and brain cells. RAGE is expressed in various cells in the kidney, including endothelial cells, podocytes, mesangial cells, tubular epithelial cells, and infiltrating immune cells (Kierdorf & Fritz, 2013). RAGE expression and AGE levels are low in normal kidneys, whereas in glomerular, tubulointerstitial, and vascular cells of diabetic kidneys, increased AGE deposition is corroborated with increased RAGE expression (Li et al., 2004). RAGE is encoded by *AGER,* located on chromosome 6 within the MHC class III locus. RAGE has 394 amino acids and a molecular mass of approximately 45 kDa (Kumar Pasupulati et al., 2016). Although AGEs were the first reported ligands, RAGEs also interact with several other ligands, including lysophosphatidic acid, complement C1q, high-mobility group-1 (HMGB1-a transcription factor), S100A8/A9 (cytoskeletal proteins), and amyloid-β (Curran & Kopp, 2022). Expression of RAGE is minimal in physiological conditions, but it can be increased in response to stress and inflammation. RAGE expresses constitutively during the embryonic stage. In adults, RAGE ligands and inflammatory mediators control RAGE expression, except in the lungs, where RAGE is constitutively expressed at high levels (Oczypok, Perkins, & Oury, 2017). There is an increased RAGE expression in DKD (Liu et al., 2014). The symptoms observed in RAGE-transgenic mice, including kidney enlargement, glomerular hypertrophy, mesangial expansion, glomerulosclerosis, and proteinuria, are comparable with symptoms of advanced DKD (Yamamoto et al., 2001).

Four RAGE variants are categorized based on protein structure (Fig. 4). An extracellular domain, a transmembrane domain, and a cytosolic tail comprise the full-length RAGE (fl-RAGE). Dominant negative RAGE (dn-RAGE) is devoid of the cytoplasmic tail. Cleaved RAGE (c-RAGE) and endogenous secretory RAGE (es-RAGE) are produced by the proteolytic cleavage of fl-RAGE by ADAM10 and alternative splicing of RAGE mRNA, respectively. Together, these two RAGEs (c-RAGE and es-RAGE) make up soluble RAGE (s-RAGE) that possesses only an extracellular domain. This sRAGE comprises most of the serum species and serves as a sham receptor that blocks interactions between ligands and full-length RAGE and prevents interaction of AGEs with fl-RAGE vis-a-vis unfavorable signaling triggered by activation of the AGE-RAGE axis. Notably, rapid AGE buildup in the setting of diabetes depletes sRAGE levels and weakens innate defenses against AGEs. It should be noted that the activated AGE-RAGE axis stimulates RAGE production and further promote adverse events triggered by this axis.

6. Pathology of AGEs in DKD
6.1 AGEs contribute to ROS, inflammation and cell death

Several intracellular signaling cascades are triggered by the engagement of the AGEs with RAGE. The cytoplasmic tail of ligand-bound RAGE interacts with the formin homology (FH1) domain of Dia1 and activates Rho GTPases (Rac-1 and Cdc42). These Rho GTPases evoke JNK and MAPK signaling, activating the pro-inflammatory master transcription factor NF-kB (Taguchi & Fukami, 2023). NF-kB is accumulated by several means following RAGE activation. Ligand binding to RAGE activates AKT via adapter proteins TIRAP/MyD88 which, in turn, stimulate the activation of NF-κB. Alternatively, RAGE-induced ROS promotes the production of p21 RAS, which activates NF-kB (Tóbon-Velasco, Cuevas, & Torres-Ramos, 2014). Activation of NF-κB is well demonstrated in-vitro and in-vivo in glomerular cells such as podocytes and mesangial, tubular, and endothelial cells in renal injury (Gómez-Garre et al., 2001). Expression of IL-1, IL-6, and TNF-α, vascular cell adhesion molecule-1 (VCAM-1), intracellular cell adhesion molecule-1 (ICAM-1), and E-selectin is influenced by NF-kB, and all these molecules contribute to inflammation and fibrosis. NF-κB regulates the levels of inflammatory cytokines, including TNF-α and IL-6, by regulating fibrogenic mediators in the inflammatory responses in kidney injury (Zhang & Sun, 2015). Ample evidence suggests that NF-kB pathway is involved in the pathophysiology of DKD. It is active in several cell types, including kidney cells, through RAGE-mediated activation and subsequent generation of pro-inflammatory cytokines in DKD (White, Lin, & Hu, 2020). Glycation products promote ERK phosphorylation, stimulate TGF-β production, increase oxidative stress, and activate NF-κB signaling (Khalid, Petroianu, & Adem, 2022). AGEs induce NF-κB activation leading to increased IL-1β and TNF-α levels (Rea et al., 2018). AGE-induced expression of NF-kB stimulates transcription factor ZEB2 that evokes epithelial-mesenchymal transition (EMT) (Kumar et al., 2016).

Multiple studies revealed the AGE-RAGE axis mediated accumulation of pro-inflammatory cytokines by activating the MAPK, NF-kB, and PI3K-AKT signaling in mesangial and renal tubulointerstitial cells (Wu et al., 2021).

RAGE-induced cytosolic ROS elevates the mitochondrial oxygen levels that ensure the formation of the mitochondrial permeability transition pore (mPTP) and cell death (Coughlan et al., 2009). In diabetic settings, mitochondrial DNA malfunction is more common than nuclear DNA damage caused by ROS (Al-Ghamdi, Al-Shamrani, El-Shehawi, Al-Johani, & Al-Otaibi, 2022). Additionally, AGEs interact with metal ions like Cu^{2+} and Fe^{2+} and contribute to ROS. ROS stimulates the protein kinase C pathway in mesangial cells, thus increasing TGF-β1 levels and regulating kidney fibrosis (Kwan et al., 2005). RAGE-induced stimulation of activator protein (AP)-1 stimulates the synthesis of profibrotic TGF-β1.

Advanced glycated albumin promotes IL-1β and IL-18 secretion by upregulating NLRP3 inflammasome expression (Kelley, Jeltema, Duan, & He, 2019). Mature IL-1β and IL-18 cleave gasdermin D (GSDMD) into the N-terminal and C-terminal. The N-terminal GSDM forms pore in the cell membrane and mediate perforation, and the pro-inflammatory factor NLRP3 mediates pyroptosis (Feng, Fox, & Man, 2018). Recent reports suggested that NLRP3 inflammasome is involved in renal tubular injury by eliciting pyroptosis via cell expansion and rapid lysis with the extensive secretion of IL-1β and IL-18 (Kelley et al., 2019). Certain dicarbonyl AGEs (3,4 DGE and 3-DGal) stimulate NF-kB activation and pyroptosis-related NLRP3 inflammasome cascade (Cepas et al., 2021). It was also shown that RAGE induces cell death by upregulating p53-Bax expression and calcium-dependent caspase cascade activation.

The contribution of the AGE-RAGE axis towards progressive glomerulosclerosis and decline of renal function is evidenced from studies in the OVE26 type 1 mouse model (Reiniger et al., 2010). The homozygous RAGE knockout OVE26 mice were presented with reduced nephromegaly, GBM thickening, mesangial sclerosis, cast formation, podocyte effacement, and albuminuria. Deleting RAGE significantly prevented the reduction in GFR observed in OVE26 mice. Expression of profibrotic TGF-β1 was ameliorated in the kidneys of RAGE-depleted mice (Reiniger et al., 2010). Another study showed that RAGE-neutralizing antibodies reduced early hallmarks of DKD, such as GBM thickening and expansion of mesangium in db/db mice. In summary, AGEs via RAGE elicit an array of adverse renal events and implicate the pathology of DKD (Fig. 5).

6.2 AGEs and renin-angiotensin system (RAS)

The kidneys help maintain electrolyte balance, blood volume and pressure by synthesizing various RAS components. The renin-angiotensin system comprises angiotensinogen, renin, angiotensin-converting enzyme (ACE),

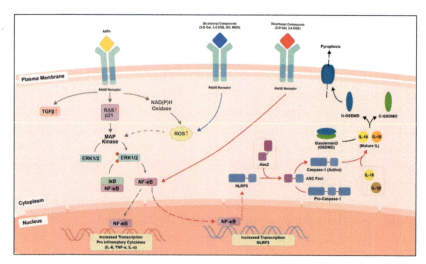

Fig. 5 Cellular manifestations of activated AGE-RAGE axis. Interaction of various ligands (AGEs) with RAGE results in the activation of NF-kB by several cascades. NF-kB forms the major elicitor of adverse effects of AGEs by inducing the secretion of inflammatory cytokines, inflammasome (NLRP3) synthesis, and evoking the cell death pathway.

angiotensin II, and two receptors of angiotensin II (AT1R and AT2R). Renin secreted by the kidney converts hepatic angiotensinogen to angiotensin I. Angiotensin-converting enzyme (ACE) catalyzes the conversion of angiotensin I to angiotensin II. Angiotensin II raises blood pressure by evoking vasoconstriction, sympathetic nervous stimulation, increased aldosterone biosynthesis and renal actions. Renal angiotensin II production is augmented during the early stages of DKD, which exerts feedback inhibition of systemic renin release and suppression of RAS activity.

On the contrary, activation of local RAS within glomeruli and proximal tubules was observed in diabetes. Angiotensin II exerts its effects by interacting with AT1R expressed in various cells, including endothelial cells, podocytes, mesangial cells, tubular cells, and interstitium. Aberrant activation of AT1R increases vascular resistance, reduces renal blood flow, and stimulates extracellular matrix accumulation in the mesangium and tubulointerstitum. Since elevated local angiotensin II production results in activation of pro-inflammatory response and subsequent renal injury, inhibitors of ACE and AT1R antagonists retard the progression of DKD (Malek et al., 2021). The enhanced hyperfiltration in the early stages of DKD could be explained by elevated angiotensin II levels and its activity. Angiotensin II is also implicated in another early manifestation in diabetic kidneys, including hypertrophy of mesangial cells and tubular epithelial cells (Wiecek, Chudek, & Kokot, 2003; Wolf & Ziyadeh, 1999).

The significant determinants of aberrant RAS activity are elevated renal angiotensin II production and its receptor AT1R. AGEs contributed to the pathology of DKD by contributing to the elevated expression of AT1R, through which angiotensin II ignites pro-inflammatory actions. Notably, angiotensin II induces RAGE expression in endothelial cells (Hudson & Lippman, 2018). RAS activation influences the expression of pro-inflammatory and profibrotic cytokines such as MCP-1 and TGF-β (Cheng et al., 2005; Li et al., 2015). A recent study reported that AT1R forms a heteromeric complex with RAGE, and the transactivation of RAGE mediates angiotensin II-induced inflammation (Pickering et al., 2019). Due to the physical interaction between RAGE and AT1R on the cell membrane and bidirectional cross-talk, binding of AGEs with RAGE results in the activation of AT1R and binding of angiotensin II to AT1R activates RAGE (Yokoyama et al., 2021). The intriguing cross-talk between the AGE-RAGE axis and angiotensin II-AT1R axis and the activation of both receptors by ligands of either, contribute to several adverse events related to the pathology of DKD.

6.3 AGEs and reactivation of developmental pathways

Several cells remain terminally differentiated and quiescent throughout to perform the specialized functions that include neurons and glomerular podocytes. These specialized cells' fate determination and differentiation are achieved by spatiotemporal activation of selective signaling events. For instance, Notch signaling is crucial in defining the podocyte fate during nephrogenesis (Pasupulati, 2022). Notch signaling is also warranted for establishing proximal convoluted tubules during kidney development (Pasupulati, 2022). Although there exists a distinct role for Notch signaling in maintaining progenitor niche during development, the Notch pathway was found to be less active in healthy adult kidneys. Notch signaling transduces short-range communication between cells, particularly juxtaposed cells. The Notch signaling components include four Notch receptors (Notch1–4) and five Notch ligands [Jagged (JAG) 1&2] and Delta-like ligands (DLL)-1, 3, & 4 in mammals. The interaction of the Notch receptor with the cognate ligand provokes γ-secretase to cleave and release the Notch intracellular domain (NICD) that serves as a transcription factor in association with other coactivators (RBP-Jk, MAML, and p300). Notch signaling often cross-talk with several pathways, including TGF-β/Smad, NF-κB, HIF-1, VEGF, and Wnt/β-catenin.

Owing to the importance of Notch signaling in kidney development, abnormalities of Notch signaling are associated with kidney diseases such as congenital kidney, urinary tract anomalies, impaired wound healing, and DKD (Mukherjee, Fogarty, Janga, & Surendran, 2019; Zhu et al., 2022). Unlike during nephrogenesis, the reactivation of Notch signaling in adult podocytes ensures their injury and depletion. Notch reactivation in adult kidneys manifests in glomerulosclerosis and albuminuria (Sanchez-Niño et al., 2015; Zhu et al., 2022). Glucose-derived AGEs induce NICD1 accumulation and the consequent expression of putative downstream targets Hes1 and Hey1 in human podocytes (Nishad, Meshram, Singh, Reddy, & Pasupulati, 2020). The involvement of AGE-RAGE in the reactivation of Notch signaling is evidenced by the suppression of Notch signaling by the RAGE inhibitor. Inhibition of RAGE and resultant suppression of Notch1 by RAGE inhibitor (FPS-ZM1) prevents AGE-induced glomerular fibrosis, thickening of the glomerular basement membrane, foot process effacement, and proteinuria (Nishad et al., 2020). Matrix metalloproteinase-9 (MMP-9) activation by cross-talk of AGEs-Notch signaling is involved in impaired wound healing in diabetic rats (Zhu et al., 2022). Conditional re-expression of NICD in vivo exclusively in podocytes caused proteinuria and glomerulosclerosis. Ectopic expression of Notch1 in the kidney presented with focal segmental glomerulosclerosis and correlated with proteinuria (Niranjan et al., 2008). Furthermore, kidney biopsy sections from people with DKD revealed the accumulation of AGEs in the glomerulus with elevated RAGE expression and activated Notch signaling (Niranjan et al., 2008; Nishad et al., 2020). These data suggest that AGEs re-activate developmental events, particularly Notch1 in the adult kidneys, culminating in kidney dysfunction (Fig. 6).

6.4 AGEs and phenotypic switch of renal cells

Epithelial cells of the glomerulus (podocytes and parietal epithelial cells) and tubular epithelial cells play a vital role in maintaining the nephron's normal function, including filtration and reabsorption. The metanephric mesenchyme undergoes differentiation during nephrogenesis and develops into morphologically diverse phases, including the peritubular aggregate, renal vesicle, comma-shaped body, S-shaped body, and capillary loop stage, to form a mature nephron. The metanephric mesenchymal cells undergo phenotypic conversion into glomerular and tubular epithelial cells via a highly programmed process known as a mesenchymal-epithelial transition (MET) (Pasupulati, 2022). Upon exposure to noxious stimuli, the nephron

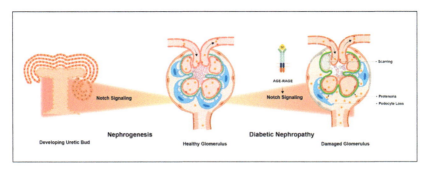

Fig. 6 Notch signaling though highly active during the early stages of nephrogenesis, is inactive in the healthy adult kidney. AGEs re-activate Notch signaling in the settings of diabetes and manifest in nephropathy.

epithelial cells tend to undergo epithelial-to-mesenchymal cell transition (EMT), a simple reversal of the embryonic process, MET. It should be noted that podocytes, though visceral epithelial cells express few mesenchymal markers (May, Saleem, & Welsh, 2014). After a phenotypic switch to mesenchymal cells, the epithelial cells gain enhanced migratory capacity and become more invasive. The EMT of cells is accompanied by the alteration of the underlying basement membrane that could facilitate the detachment of cells and also fibrosis. Besides EMT, epithelial cells also undergo a phenotypic transition to myofibroblasts (EMyT) which secrete extracellular matrix components and contribute to the profibrotic phenotype. Like epithelial cells, endothelial cells undergo a phenotypic switch known as endothelial-mesenchymal-transition (EndMT). A minor population of endothelial cells undergoes a phenotypic transition to fibroblasts in the setting of diabetes (Li, Qu, & Bertram, 2009).

AGEs are involved in the phenotypic switch of the kidney's epithelial and endothelial cells (Taguchi & Fukami, 2023). AGE–RAGE axis induces transforming growth factor (TGF)-β production and subsequent Smad3 activation in endothelial cells via down-regulation of SIRT1. Alternatively, a direct RAGE-dependent activation of Smad2 is implicated in the EndMT (He, Zhang, Gan, Xu, & Tang, 2015; Li et al., 2010). RAGE activation in endothelial cells elicits aberrant autophagy and, in turn, fibrosis (Zhang et al., 2021). Regarding epithelial cells, particularly podocytes, CML stimulates RAGE expression in podocytes, which by a series of events, induces zinc finger E-box binding homeobox 2 (ZEB2) (Kumar et al., 2016). ZEB2, a transcription factor, regulates EMT by ensuing a 'cadherin switch' characterized by E-cadherin depletion and N-cadherin augmentation. Activation of the AGE–RAGE axis in podocytes induces γ-secretase activity, which helps to cleave the Notch intracellular domain (NICD1) from its full-length receptor Notch1. In AGE-exposed podocytes and a mouse model administered with in vitro-prepared AGE, NICD1 accumulation elicited podocyte EMT (Fig. 7) (Nishad et al., 2020). The effect of RAGE-dependent reactivation of Notch signaling needs to be investigated in detail in endothelial and tubular epithelial cells. However, the reactivation of Notch in podocytes induces injury, suggesting that RAGE-induced γ-secretase activity may contribute to podocyte injury (Zhang et al., 2021). Activation of EMT of podocytes through any of the effector signaling is implicated in their depletion, as EMT drives podocyte detachment from the substratum. Studies in diabetic subjects and animal models of DKD revealed that the accumulation of AGEs is associated with the decreased density of podocytes (Nishad et al., 2021). RAGE-induced TGF-β is implicated in the EMT and EMyT of tubular epithelial cells contextually and thereby significantly contributes to pathological events such as interstitial fibrosis implicating in diabetic kidney disease (Oldfield et al., 2001; Wu et al., 2021).

6.5 AGEs and complement activation

The complement system is part of the innate immune system, and activation of the complement cascade enhances the ability of antibodies and phagocytic cells to combat microbes and clear damaged cells. The complement system is comprised of several soluble small proteins secreted predominantly by the liver that circulate in the blood as inactive precursors. These proteins interact with one another in three distinct enzymatic activation streams; classical, alternative, and lectin pathways. These three pathways converge to form C3 convertase, which cleaves and activates component C3, creating C3a and C3b; in turn, C3b binds to Factor B. Factor D releases Factor Ba from Factor B attached to C3b. The C3b(2)Bb complex is a protease that cleaves C5 into C5b and C5a. C5a is an inflammatory mediator, whereas C5b initiates the membrane attack complex (MAC) consisting of C5b, C6, C7, C8, and C9 (C5b-9) (Goldman & Prabhakar, 1996). MAC forms a transmembrane channel and ensures the osmotic lysis of the target cell (Morgan, Boyd, & Bubeck, 2017). The influx of salt and water through the MAC pore induces osmotic swelling and lysis of MAC-targeted cells such as gram-negative bacteria or heterologous erythrocytes (Qin et al., 2004).

The pores formed by the MAC on target cells are transient and serve as a bidirectional route for transporting autocrine and paracrine signals across

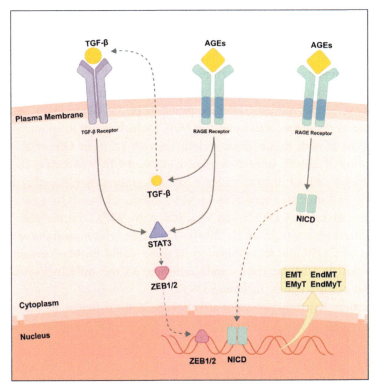

Fig. 7 AGE-RAGE axis induces phenotype switch of cells by activating transcription factors such as ZEB2 and NICD1. These transcription factors contextually regulate the epithelial-to-mesenchymal transition (EMT); epithelial-to-myofibroblast transition (EMyT); endothelial-to-mesenchymal transition (EndMT); and endothelial-to-myofibroblast transition (EndMyT). These phenotypic switches of glomerular epithelial cells, tubular epithelial cells, and glomerular endothelial cells may implicate the pathobiology of DKD.

human cell membranes (Acosta, Benzaquen, Goldstein, Tosteson, & Halperin, 1996). These transient MAC pores generated upon complement activation and nonlytic MAC formation mediate physiological and pathological responses. Nonlytic MAC deposition on endothelial cells releases several bioactive molecules, including bFGF, PDGF, IL-1, and MCP. The molecules released through the transient nonlytic MAC pore execute the proliferation of endothelial cells, fibroblasts, and smooth muscle cells; inflammation by attracting macrophages and inducing the expression of pro-inflammatory molecules such as VCAM-1, ICAM-1, and E- & P-selectin; thrombosis by tissue factor and clotting factor Va (Qin et al., 2004).

A regulatory system comprises several inhibitory proteins to prevent aberrant complement activation and protect self-injury from the catastrophic effect of complement amplification, including CD59. CD59 is a glycan phosphatidylinositol-linked membrane protein and a potent inhibitor of MAC formation (Meri et al., 1990). Notably, K41 and H44 residues of the active domain of human CD55 are prone to NEG. Glycated CD59 fails to execute its regulatory activity against MAC, which results in an uncontrolled release of growth factors from endothelial cells (Xuebin Qin et al., 2004). The urine of diabetic subjects consists of glycated CD59 (Acosta et al., 2000). Red blood cells of diabetic patients have a significantly reduced activity of hCD59 and increased sensitivity to MAC-mediated lysis (Qin et al., 2004). Inactive glycated CD59 is co-localized with MAC in diabetic tissues, including kidneys (Qin et al., 2004). Therefore, glycation-induced inactivation of CD59 and its elimination in urine could further increase MAC deposition in diabetic tissues and contribute to the pathology of diabetic vascular complications, including DKD.

Further, hyperglycemia also activates the lectin pathway, resulting in the glycation of pattern recognition molecules and complement regulatory proteins, causing uncontrolled activation of the complement system (Budge, Dellepiane, Yu, & Cravedi, 2020). Complement activation is implicated in tubulointerstitial injury in diabetic settings. Tubular deposition of C5a has also been shown to correlate with the severity of DKD and blocking the C5a–C5a1R axis ameliorated interstitial fibrosis in a murine diabetic (*db/db*) model (Yiu et al., 2017). Notably, uncontrolled complement activation is a crucial driving force in kidney diseases, including complement 3 glomerulopathy (C3G) and IgA nephropathy, caused by excessive activation of the alternative complement pathway (Poppelaars & Thurman, 2020). In summary, hyperactivation of complement components and their breakdown products (C3, C5a, and MAC) manifests in glomerular inflammation and injury that include endocapillary hypercellularity, mesangial hypercellularity, focal segmental sclerosis, interstitial fibrosis, tubular atrophy, crescent formation and all leading to proteinuria and CKD (Schena, Esposito, & Rossini, 2020; Xie et al., 2023).

7. Strategies to prevent AGEs formation and DKD

For diabetic people, maintaining strict normoglycemic levels is challenging. Intensive treatment methods for diabetic subjects with hyperglycemia raise mortality risk and are linked to severe hypoglycemia. Although stringent blood glucose management slows the evolution of albuminuria levels, clinical kidney disease outcomes may not significantly change. Therefore, more preventive and therapeutic choices are needed beyond those recommended for blood glucose normalization to prevent and manage diabetic renal problems. The balance between dietary intake, endogenous formation, and their clearance from the system reflect the steady-state abundance of AGEs, in addition to the status of glycemia. Many natural and synthetic substances have been tested for their ability to prevent AGE formation, breakdown of already formed AGEs, or slow down the manifestation of AGEs. Anti-glycation compounds are broadly categorized into five categories: (i) Hypoglycemic agents, (ii) RAGE-inhibitors, (iii) AGE-inhibitors; (iv) AGE-breakers; (v) neutraceuticals.

7.1 Hypoglycemic agents

Since hyperglycemia is a prime determinant of AGE formation, approaches to lowering blood glucose levels with insulin and metformin appear promising. In addition to traditional hypoglycemic approaches, lowering blood glucose levels with sodium-glucose cotransporter 2 inhibitors decreases glucose reabsorption in the renal tubule, thus increasing urinary glucose excretion. Eliciting glucose loss in the urine could lessen AGE formation and kidney toxicity (Fotheringham, Gallo, Borg, & Forbes, 2022; Sourris, Watson, & Jandeleit-Dahm, 2021). Other drugs that lower blood glucose levels, such as glucagon-like peptide-1 and dipeptidyl peptidase-4 inhibitors, are also promising approaches to combating hyperglycemia and, in turn, AGE formation (Gilbert & Pratley, 2020).

7.2 RAGE inhibitors

AGEs interact with RAGE and execute their cellular and molecular effects, and AGEs are also linked to RAGE overexpression. Hence, inhibition of RAGE is also a therapeutic option to combat DKD. By lessening AGE-mediated effects, anti-RAGE neutralizing antibodies show promising results (Ramya, Coral, & Bharathidevi, 2021). Another strategy is to use anti-sense RNA against RAGE mRNA to inhibit RAGE expression (Cai et al., 2014). Alternately, potential small-molecule antagonists can be used to target RAGE-mediated downstream signaling processes. Neutralizing the AGEs by sRAGE decoy receptors is another approach to reducing AGE toxicity. Statins (pravastatin and atorvastatin) also reduce the expression of RAGE and the effects of AGE on renal biology (Tsujinaka et al., 2017).

7.3 AGE-inhibitors

The first synthetic inhibitor that was employed to prevent the formation of AGEs was aminoguanidine (AG) (Yamagishi et al., 2008). AG improved AGE-mediated adverse events by preventing the formation of glomerular lesions in diabetic mice (Song, Liu, Dong, Wang, & Zhang, 2021). In addition to preventing the production of AGE, AG thwarts NO buildup in the diabetic environment by blocking inducible nitric oxide synthase (iNOS). Clinical studies revealed that AG improved renal function in diabetic individuals. Despite this, phase III clinical trials showed that AG had side effects that prevented it from being licensed for therapeutic use. Inhibitors that developed later, like tenilsetam and 2,3-diamino-phenazine work similar to that of AG to scavenge the reactive carbonyl intermediates (Hoffmann et al., 2006; Schrijvers, De Vriese, & Flyvbjerg, 2004).

OPB-9195 is another synthetic compound that traps reactive carbonyl intermediates. OPB-9195 prevented the proteinuria and several clinical symptoms of DKD in RAGE-transgenic mice (Schrijvers et al., 2004). The AGE-RAGE axis is a promising target for preventing the pathogenesis of DKD, and this data reiterate the role of RAGE in the pathogenesis of DKD. Glycated albumin levels were decreased, and urinary NO excretion was elevated in hypertensive rats after OPB-9195 administration (Nakamura et al., 1997). Pyridoxamine suppresses the post-Amadori stages of AGE production, scavenges ROS, and traps reactive carbonyl intermediates. Pyridoxamine given orally to diabetic rats decreased CML buildup. Phase II clinical trials showed that pyridoxamine was efficient in treating DKD. As exogenous arginine could interact with reactive carbonyls and spare the innate arginine residue of the proteins from NEG, arginine supplementation in diet is recommended as an anti-AGE agent.

7.4 AGE-breakers

Cells have developed internal detoxification mechanisms to prevent AGEs accumulation. The glutathione-dependent enzymes glyoxalase I, II, and III comprise the glyoxalase system (He et al., 2020). Glyoxal, methylglyoxal, and alpha-oxoaldehydes are carbonyl molecules that are converted to lactic acid by these enzymes. Increased glyoxalase expression prevented intracellular AGE production caused by hyperglycemia, which decreased intracellular oxidative stress and mesangial cell death (He et al., 2020). Fructosyl-amine oxidases and fructosyl-amine kinases are two different enzyme systems that phosphorylate and destabilize Amadori products

(Delanghe et al., 2023; Lin & Zheng, 2010). Together, the fructosyl-amine oxidase, fructosyl-amine kinase system and glyoxalase system offer defense against glycation by avoiding the buildup of highly reactive AGEs. To chemically disrupt Maillard reaction cross-links, small synthetic compounds like ALT-946, ALT-711, and N-phenacetylthiazolium bromide were synthesized. The therapeutic potential of these molecules in alleviating AGE-mediated diabetic complications was demonstrated by the administration of AGE breaker alagebrium, which decreased RAGE in diabetic rats (Kiland, Gabelt, Tezel, Lütjen-Drecoll, & Kaufman, 2009). ALT-711 prevented tubular AGEs accumulation and TGF-β1 expression in diabetic mice (Peppa et al., 2006).

7.5 Dietary/nutritional interventions

In addition to the endogenous AGEs, the diet we consume is also a significant source of AGEs. AGEs obtained through diet correlate with circulating AGEs (CML and methylglyoxal) in healthy subjects (Kumar Pasupulati et al., 2016; Uribarri et al., 2007). Dietary restriction of AGEs helps in combating CKD. A diet rich in antioxidants prevents the accumulation of AGEs. Green tea extract prevented the formation of AGE and collagen cross-linking in C57BL/6 mice (Rutter et al., 2003). Epigallocatechin-3-gallate served as a reactive carbonyl scavenger. The AGE content in cereals, legumes, vegetables, and fruits is very low, and on the other hand, foods with high protein and fat content are rich sources of AGEs. A diet heavy in AGEs results in proportional elevations in serum AGE levels in diabetic patients and evoked inflammatory response as evidenced by elevated levels of C-reactive protein, NF-κB, and VCAM-1 (Goldin, Beckman, Schmidt, & Creager, 2006; Kumar Pasupulati et al., 2016). By avoiding fatty meats, full-fat dairy products and solid dietary fats, the intake of AGEs can be restricted (Uribarri, Woodruff, et al., 2010; Kumar Pasupulati, Chitra, et al. 2016).

NEG is inhibited by natural antioxidants (vitamins C and E, alpha-tocopherol, niacinamide, pyridoxal, lipoic acid), chelating agents (sodium selenite, selenium yeast, riboflavin), and metals (zinc, and manganese) (Sadowska-Bartosz & Bartosz, 2015; Steiner, Marçon, & Sabban, 2018). Dietary treatment with alpha-lipoic acid in fructose-fed rats reduced collagen glycation (Thirunavukkarasu, Anitha Nandhini, & Anuradha, 2005). Reduced skin collagen glycation by vitamins C and E and a combination of N-acetylcysteine, taurine and oxerutin suggest the anti-glycation potential of these vitamins and bio-active compounds (Gkogkolou & Böhm, 2012). N-acetyl cysteine, an antioxidant, was reported to halt AGE-mediated abnormal cellular processes by inhibiting ROS-dependent NF-kB activation. Curcumin, ginger, cinnamon, cumin and cloves are some spices that were also proven to prevent NEG (Dearlove, Greenspan, Hartle, Swanson, & Hargrove, 2008; Kumar, Reddy, Srinivas, & Reddy, 2009; Starowicz & Zieliński, 2019). Their anti-glycation potential is linked to their polyphenolic content. It has been demonstrated that naturally occurring flavonoids like lutein, quercetin, and rutin suppress AGE production. It was discovered that administering curcumin, besides attenuating AGE production, reduced inflammation and oxidative stress, which improved DKD (Bhuiyan, Mitsuhashi, Sigetomi, & Ubukata, 2017; Tang & Chen, 2014). When given to diabetic rats, cinnamon and its active ingredient, procyanidin-B2, prevented CML from building up in the glomerulus and prevented the loss of essential podocyte proteins nephrin and podocin (Muthenna, Raghu, Kumar, Surekha, & Reddy, 2014). Intriguingly, supplementing diabetic rats with cinnamon and procyanidin-B2 showed systemic antiglycation capability, as seen by a decline in HbA1c levels and glycation-prone RBC-IgG cross-links (Muthenna et al., 2014). The combination of cinnamon and procyanidin-B2 reduced proteinuria in diabetic rats by reducing AGEs in the blood and kidneys. Diabetic rats were given ellagic acid, a polyphenol found in fruits and vegetables, to prevent RBC-IgG cross-links and the formation of HbA1c (Raghu, Jakhotia, Yadagiri Reddy, Anil Kumar, & Bhanuprakash Reddy, 2016). Ellagic acid reduced the accumulation of CML and improved kidney function in diabetic rats (Raghu et al., 2016).

8. Summary and future perspective

It is critical to have effective treatment plans to prevent and manage DKD, given the prevalence of diabetes, its toll on human health, and the significance of renal function. Additional factors burdening kidney patients are the lack of accessible and high-quality renal care for patients with DKD and CKD. A decline in kidney function is also a strong predictor of other diseases, including cardiovascular and cerebrovascular complications. Since kidneys are the critical organs for cleansing the blood and dealing with AGE metabolism, they are naturally vulnerable to AGE-induced manifestations. The components of the GFB, which serve as the blood-urine barrier, are more susceptible to AGE-induced cellular and molecular aberrations. By contributing to hypertrophy, phenotypic switch, fibrosis, dedifferentiation and resultant detachment and apoptosis of cells that represent kidney function, AGEs are implicated in the pathogenesis of

DKD (Fig. 8). Although the specific effects of AGEs on glomerular cells are extensively investigated, their impact on the tubular compartment must be studied extensively.

Since the levels of AGEs rise with normal aging in obese individuals, the list of adverse manifestations of AGEs on human health is expected to increase. Therefore, anti-AGE therapy is a preferred preventive and therapeutic approach for treating DKD and AGE-related adverse health outcomes. The studies on the efficacy and safety of anti-AGE drugs (both natural and synthetic), including AGE blockers and cross-link breakers,

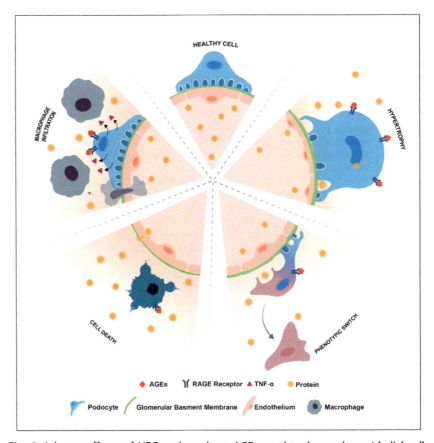

Fig. 8 Adverse effects of NEG and resultant AGEs on the glomerular epithelial cells (podocytes). AGEs could elicit hypertrophy of glomerular cells, including thickening of GBM and mesangial expansion; AGEs could evoke phenotypic switch of resident cells of the glomerulus; AGEs could induce cell death by pyroptosis or apoptosis; AGEs provoke accumulation of inflammatory cytokines that skew non-resident renal cells into the glomerulus. All these manifestations contribute to the pathogenesis fo DKD.

ACE inhibitors, sRAGE, and inhibitors of AT1R, are warranted. Successful anti-AGE therapy could be included in a treatment plan for patients with DKD. An efficient therapeutic intervention for DKD should invovlve identifying and developing a novel pharmacological target against AGE-mediated kidney disease. Besides internal AGEs, dietary and exogenous AGEs contribute to renal dysfunction and tissue damage. Antioxidants and anti-AGE compounds in the diet, especially people with diabetes, could protect them from AGE-mediated problems, improve kidney function, and manage blood glucose in the normal range.

Acknowledgments

GBR is supported by grants from the Indian Council of Medical Research (5/9/1447/2021-Nutr) and Department of Biotechnology (BT/PR36689/PFN/20/1524/2020). AKP is supported by grants from the Indian Council of Medical Research (5/4/7–12/Nephro/2022-NCD-II) and Science and Engineering Research Board (2019/5789).

References

Acosta, J., Hettinga, J., Flückiger, R., Krumrei, N., Goldfine, A., Angarita, L., & Halperin, J. (2000). Molecular basis for a link between complement and the vascular complications of diabetes. *Proceedings of the National Academy of Sciences of the United States of America, 97*(10), 5450–5455. https://doi.org/10.1073/pnas.97.10.5450.

Acosta, J. A., Benzaquen, L. R., Goldstein, D. J., Tosteson, M. T., & Halperin, J. A. (1996). The transient pore formed by homologous terminal complement complexes functions as a bidirectional route for the transport of autocrine and paracrine signals across human cell membranes. *Molecular Medicine, 2*(6), 755–765.

Al-Ghamdi, B. A., Al-Shamrani, J. M., El-Shehawi, A. M., Al-Johani, I., & Al-Otaibi, B. G. (2022). Role of mitochondrial DNA in diabetes Mellitus Type I and Type II. *Saudi Journal of Biological Sciences, 29*(12), 103434. https://doi.org/10.1016/j.sjbs.2022.103434.

Alallam, B., Choukaife, H., Seyam, S., Lim, V., & Alfatama, M. (2023). Advanced drug delivery systems for renal disorders. *Gels, 9*(2), 115.

Bakris, G. L. (2019). Major advancements in slowing diabetic kidney disease progression: focus on SGLT2 inhibitors. *American Journal of Kidney Diseases, 74*(5), 573–575. https://doi.org/10.1053/j.ajkd.2019.05.009.

Bello-Reuss, E., & Reuss, L. (1983). Homeostatic and excretory functions of the kidney. In S. Klahr (Ed.). *The kidney and body fluids in health and disease* (pp. 35–63)Boston, MA: Springer US.

Bhuiyan, M. N., Mitsuhashi, S., Sigetomi, K., & Ubukata, M. (2017). Quercetin inhibits advanced glycation end product formation via chelating metal ions, trapping methylglyoxal, and trapping reactive oxygen species. *Bioscience, Biotechnology, and Biochemistry, 81*(5), 882–890. https://doi.org/10.1080/09168451.2017.1282805.

Budge, K., Dellepiane, S., Yu, S. M., & Cravedi, P. (2020). Complement, a therapeutic target in diabetic kidney disease. *Frontiers in Medicine (Lausanne), 7*, 599236. https://doi.org/10.3389/fmed.2020.599236.

Cai, X. G., Xia, J. R., Li, W. D., Lu, F. L., Liu, J., Lu, Q., & Zhi, H. (2014). Anti-fibrotic effects of specific-siRNA targeting of the receptor for advanced glycation end products in a rat model of experimental hepatic fibrosis. *Molecular Medicine Reports, 10*(1), 306–314. https://doi.org/10.3892/mmr.2014.2207.

Cepas, V., Manig, F., Mayo, J. C., Hellwig, M., Collotta, D., Sanmartino, V., & Sainz, R. M. (2021). In vitro evaluation of the toxicological profile and oxidative stress of relevant diet-related advanced glycation end products and related 1,2-dicarbonyls. *Oxidative Medicine and Cellular Longevity, 2021,* 9912240. https://doi.org/10.1155/2021/9912240.

Chen, T. K., Knicely, D. H., & Grams, M. E. (2019). Chronic kidney disease diagnosis and management: A review. *JAMA, 322*(13), 1294–1304. https://doi.org/10.1001/jama.2019.14745.

Cheng, J., Diaz Encarnacion, M. M., Warner, G. M., Gray, C. E., Nath, K. A., & Grande, J. P. (2005). TGF-beta1 stimulates monocyte chemoattractant protein-1 expression in mesangial cells through a phosphodiesterase isoenzyme 4-dependent process. *American Journal of Physiology-Cell Physiology, 289*(4), C959–C970. https://doi.org/10.1152/ajpcell.00153.2005.

Coughlan, M. T., Thorburn, D. R., Penfold, S. A., Laskowski, A., Harcourt, B. E., Sourris, K. C., ... Forbes, J. M. (2009). RAGE-induced cytosolic ROS promote mitochondrial superoxide generation in diabetes. *Journal of the American Society of Nephrology, 20*(4), 742–752. https://doi.org/10.1681/asn.2008050514.

Curran, C. S., & Kopp, J. B. (2022). RAGE pathway activation and function in chronic kidney disease and COVID-19. *Frontiers in Medicine (Lausanne), 9,* 970423. https://doi.org/10.3389/fmed.2022.970423.

Curthoys, N. P., & Moe, O. W. (2014). Proximal tubule function and response to acidosis. *Clinical Journal of the American Society of Nephrology, 9*(9), 1627–1638. https://doi.org/10.2215/cjn.10391012.

Dearlove, R. P., Greenspan, P., Hartle, D. K., Swanson, R. B., & Hargrove, J. L. (2008). Inhibition of protein glycation by extracts of culinary herbs and spices. *Journal of Medicinal Food, 11*(2), 275–281. https://doi.org/10.1089/jmf.2007.536.

Delanghe, J. R., Beeckman, J., Beerens, K., Himpe, J., Bostan, N., Speeckaert, M. M., ... Van Aken, E. (2023). Topical application of deglycating enzymes as an alternative non-invasive treatment for presbyopia. *International Journal of Molecular Sciences, 24*(8), https://doi.org/10.3390/ijms24087343.

Fallavena, L. P., Rodrigues, N. P., Marczak, L. D. F., & Mercali, G. D. (2022). Formation of advanced glycation end products by novel food processing technologies: A review. *Food Chemistry,* 133338.

Feng, S., Fox, D., & Man, S. M. (2018). Mechanisms of gasdermin family members in inflammasome signaling and cell death. *Journal of Molecular Biology, 430*(18, Part B), 3068–3080. https://doi.org/10.1016/j.jmb.2018.07.002.

Fotheringham, A. K., Gallo, L. A., Borg, D. J., & Forbes, J. M. (2022). Advanced glycation end products (AGEs) and chronic kidney disease: Does the modern diet AGE the kidney? *Nutrients, 14*(13), https://doi.org/10.3390/nu14132675.

Gilbert, M. P., & Pratley, R. E. (2020). GLP-1 analogs and DPP-4 inhibitors in type 2 diabetes therapy: Review of head-to-head clinical trials. *Frontiers in Endocrinology (Lausanne), 11,* 178. https://doi.org/10.3389/fendo.2020.00178.

Gkogkolou, P., & Böhm, M. (2012). Advanced glycation end products. *Dermato-Endocrinology, 4*(3), 259–270. https://doi.org/10.4161/derm.22028.

Goldin, A., Beckman, J. A., Schmidt, A. M., & Creager, M. A. (2006). Advanced glycation end products. *Circulation, 114*(6), 597–605. https://doi.org/10.1161/Circulationaha.106.621854.

Goldman, A. S., & Prabhakar, B. S. (1996). Immunology overview. In S. Baron (Ed.). *Medical microbiology*Galveston, TX: University of Texas Medical Branch at Galveston Copyright © 1996, The University of Texas Medical Branch at Galveston.

Gómez-Garre, D., Largo, R., Tejera, N., Fortes, J., Manzarbeitia, F., & Egido, J. (2001). Activation of NF-κB in tubular epithelial cells of rats with intense proteinuria. *Hypertension, 37*(4), 1171–1178. https://doi.org/10.1161/01.HYP.37.4.1171.

Haider, M. Z., & Aslam, A. (2023). *Proteinuria StatPearls*. Treasure Island, FL: StatPearls Publishing Copyright © 2023, StatPearls Publishing LLC.

He, W., Zhang, J., Gan, T. Y., Xu, G. J., & Tang, B. P. (2015). Advanced glycation end products induce endothelial-to-mesenchymal transition via downregulating Sirt 1 and upregulating TGF-β in human endothelial cells. *BioMed Research International, 2015*, 684242. https://doi.org/10.1155/2015/684242.

He, Y., Zhou, C., Huang, M., Tang, C., Liu, X., Yue, Y., ... Liu, D. (2020). Glyoxalase system: A systematic review of its biological activity, related-diseases, screening methods and small molecule regulators. *Biomedicine & Pharmacotherapy, 131*, 110663. https://doi.org/10.1016/j.biopha.2020.110663.

Hodge, J. E. (1953). Dehydrated foods, chemistry of browning reactions in model systems. *Journal of Agricultural and Food Chemistry, 1*(15), 928–943.

Hoffmann, J., Alt, A., Lin, J., Lochnit, G., Schubert, U., Schleicher, E., ... Hammes, H. P. (2006). Tenilsetam prevents early diabetic retinopathy without correcting pericyte loss. *Thrombosis and Haemostasis, 95*(4), 689–695.

Horton, W. B., & Barrett, E. J. (2021). Microvascular dysfunction in diabetes mellitus and cardiometabolic disease. *Endocrine Reviews, 42*(1), 29–55. https://doi.org/10.1210/endrev/bnaa025.

Hudson, B. I., & Lippman, M. E. (2018). Targeting RAGE signaling in inflammatory disease. *Annual Review of Medicine, 69*, 349–364. https://doi.org/10.1146/annurev-med-041316-085215.

Jennings, E. Q., Fritz, K. S., & Galligan, J. J. (2022). Biochemical genesis of enzymatic and non-enzymatic post-translational modifications. *Molecular Aspects of Medicine, 86*, 101053. https://doi.org/10.1016/j.mam.2021.101053.

KDIGO. (2020). Clinical practice guideline for diabetes management in chronic kidney disease. *Kidney International, 98*(4s), S1–s115. https://doi.org/10.1016/j.kint.2020.06.019.

KDIGO. (2021). Clinical practice guideline for the management of glomerular diseases. *Kidney International, 100*(4s), S1–s276. https://doi.org/10.1016/j.kint.2021.05.021.

Kelley, N., Jeltema, D., Duan, Y., & He, Y. (2019). The NLRP3 inflammasome: An overview of mechanisms of activation and regulation. *International Journal of Molecular Sciences, 20*(13), https://doi.org/10.3390/ijms20133328.

Khalid, M., Petroianu, G., & Adem, A. (2022). Advanced glycation end products and diabetes mellitus: Mechanisms and perspectives. *Biomolecules, 12*(4), 542.

Kierdorf, K., & Fritz, G. (2013). RAGE regulation and signaling in inflammation and beyond. *Journal of Leukocyte Biology, 94*(1), 55–68. https://doi.org/10.1189/jlb.1012519.

Kiland, J. A., Gabelt, B. T., Tezel, G., Lütjen-Drecoll, E., & Kaufman, P. L. (2009). Effect of the age cross-link breaker alagebrium on anterior segment physiology, morphology, and ocular age and rage. *Transactions of the American Ophthalmological Society, 107*, 146–158.

Kumar, P. A., Kumar, M. S., & Reddy, G. B. (2007). Effect of glycation on α-crystallin structure and chaperone-like function. *Biochemical Journal, 408*(2), 251–258. https://doi.org/10.1042/bj20070989.

Kumar, P. A., Reddy, P. Y., Srinivas, P. N., & Reddy, G. B. (2009). Delay of diabetic cataract in rats by the antiglycating potential of cumin through modulation of alpha-crystallin chaperone activity. *The Journal of Nutritional Biochemistry, 20*(7), 553–562. https://doi.org/10.1016/j.jnutbio.2008.05.015.

Kumar, P. A., Welsh, G. I., Raghu, G., Menon, R. K., Saleem, M. A., & Reddy, G. B. (2016). Carboxymethyl lysine induces EMT in podocytes through transcription factor ZEB2: Implications for podocyte depletion and proteinuria in diabetes mellitus. *Archives of Biochemistry and Biophysics, 590*, 10–19. https://doi.org/10.1016/j.abb.2015.11.003.

Kumar Pasupulati, A., Chitra, P. S., & Reddy, G. B. (2016). Advanced glycation end products mediated cellular and molecular events in the pathology of diabetic nephropathy. *Biomolecular Concepts, 7*(5-6), 293–309. https://doi.org/10.1515/bmc-2016-0021.

Kumar Pasupulati, A., Chitra, P. S., & Reddy, G. B. (2016). Advanced glycation end products mediated cellular and molecular events in the pathology of diabetic nephropathy. *Biomolecular Concepts, 7*(5-6), 293–309.

Kunkel, H. G., & Wallenius, G. (1955). New hemoglobin in normal adult blood. *Science, 122*(3163), 288. https://doi.org/10.1126/science.122.3163.288.

Kwan, J., Wang, H., Munk, S., Xia, L., Goldberg, H. J., & Whiteside, C. I. (2005). In high glucose protein kinase C-zeta activation is required for mesangial cell generation of reactive oxygen species. *Kidney International, 68*(6), 2526–2541. https://doi.org/10.1111/j.1523-1755.2005.00660.x.

Li, A., Wang, J., Zhu, D., Zhang, X., Pan, R., & Wang, R. (2015). Arctigenin suppresses transforming growth factor-β1-induced expression of monocyte chemoattractant protein-1 and the subsequent epithelial-mesenchymal transition through reactive oxygen species-dependent ERK/NF-κB signaling pathway in renal tubular epithelial cells. *Free Radical Research, 49*(9), 1095–1113. https://doi.org/10.3109/10715762.2015.1038258.

Li, J., Qu, X., & Bertram, J. F. (2009). Endothelial-myofibroblast transition contributes to the early development of diabetic renal interstitial fibrosis in streptozotocin-induced diabetic mice. *The American Journal of Pathology, 175*(4), 1380–1388. https://doi.org/10.2353/ajpath.2009.090096.

Li, J., Qu, X., Yao, J., Caruana, G., Ricardo, S. D., Yamamoto, Y., ... Bertram, J. F. (2010). Blockade of endothelial-mesenchymal transition by a Smad3 inhibitor delays the early development of streptozotocin-induced diabetic nephropathy. *Diabetes, 59*(10), 2612–2624. https://doi.org/10.2337/db09-1631.

Li, J. H., Huang, X. R., Zhu, H. J., Oldfield, M., Cooper, M., Truong, L. D., ... Lan, H. Y. (2004). Advanced glycation end products activate Smad signaling via TGF-beta-dependent and independent mechanisms: Implications for diabetic renal and vascular disease. *The FASEB Journal, 18*(1), 176–178. https://doi.org/10.1096/fj.02-1117fje.

Liang, Z., Chen, X., Li, L., Li, B., & Yang, Z. (2020). The fate of dietary advanced glycation end products in the body: from oral intake to excretion. *Critical Reviews in Food Science and Nutrition, 60*(20), 3475–3491. https://doi.org/10.1080/10408398.2019.1693958.

Lin, Z., & Zheng, J. (2010). Occurrence, characteristics, and applications of fructosyl amine oxidases (amadoriases). *Applied Microbiology and Biotechnology, 86*(6), 1613–1619. https://doi.org/10.1007/s00253-010-2523-5.

Liu, J., Huang, K., Cai, G.-Y., Chen, X.-M., Yang, J.-R., Lin, L.-R., ... He, Y.-N. (2014). Receptor for advanced glycation end-products promotes premature senescence of proximal tubular epithelial cells via activation of endoplasmic reticulum stress-dependent p21 signaling. *Cellular Signalling, 26*(1), 110–121.

Maillard, L. (1912). Action of amino acids on sugars. Formation of melanoidins in a methodical way. *Compte-Rendu de l'Academie des Science, 154*, 66–68.

Malek, V., Suryavanshi, S. V., Sharma, N., Kulkarni, Y. A., Mulay, S. R., & Gaikwad, A. B. (2021). Potential of renin-angiotensin-aldosterone system modulations in diabetic kidney disease: Old players to new hope!. *Reviews of Physiology, Biochemistry and Pharmacology, 179*, 31–71. https://doi.org/10.1007/112_2020_50.

May, C. J., Saleem, M., & Welsh, G. I. (2014). Podocyte dedifferentiation: A specialized process for a specialized cell. *Frontiers in Endocrinology (Lausanne), 5*, 148. https://doi.org/10.3389/fendo.2014.00148.

Meri, S., Morgan, B. P., Davies, A., Daniels, R. H., Olavesen, M. G., Waldmann, H., & Lachmann, P. J. (1990). Human protectin (CD59), an 18,000-20,000 MW complement lysis restricting factor, inhibits C5b-8 catalysed insertion of C9 into lipid bilayers. *Immunology, 71*(1), 1–9.

Mise, K., Ueno, T., Hoshino, J., Hazue, R., Sumida, K., Yamanouchi, M., ... Ubara, Y. (2017). Nodular lesions in diabetic nephropathy: Collagen staining and renal prognosis. *Diabetes Research and Clinical Practice, 127*, 187–197. https://doi.org/10.1016/j.diabres.2017.03.006.

Monnier, V. M., & Cerami, A. (1981). Non-enzymatic browning in vivo: Possible process for aging of long-lived proteins. *Science, 211*(4481), 491–493. https://doi.org/10.1126/science.6779377.

Monnier, V. M., Stevens, V. J., & Cerami, A. (1981). Maillard reactions involving proteins and carbohydrates in vivo: Relevance to diabetes mellitus and aging. *Progress in Food & Nutrition Science, 5*(1-6), 315–327.

Morgan, B. P., Boyd, C., & Bubeck, D. (2017). Molecular cell biology of complement membrane attack. *Seminars in Cell & Developmental Biology, 72*, 124–132. https://doi.org/10.1016/j.semcdb.2017.06.009.

Mukherjee, M., Fogarty, E., Janga, M., & Surendran, K. (2019). Notch signaling in kidney development, maintenance, and disease. *Biomolecules, 9*(11), https://doi.org/10.3390/biom9110692.

Mukoyama, M., & Nakao, K. (2005). Hormones of the kidney. *Endocrinology: Basic and Clinical Principles*, 353–365.

Muthenna, P., Raghu, G., Kumar, P. A., Surekha, M. V., & Reddy, G. B. (2014). Effect of cinnamon and its procyanidin-B2 enriched fraction on diabetic nephropathy in rats. *Chemico-Biological Interactions, 222*, 68–76. https://doi.org/10.1016/j.cbi.2014.08.013.

Nakamura, S., Makita, Z., Ishikawa, S., Yasumura, K., Fujii, W., Yanagisawa, K., ... Koike, T. (1997). Progression of nephropathy in spontaneous diabetic rats is prevented by OPB-9195, a novel inhibitor of advanced glycation. *Diabetes, 46*(5), 895–899.

Niranjan, T., Bielesz, B., Gruenwald, A., Ponda, M. P., Kopp, J. B., Thomas, D. B., & Susztak, K. (2008). The Notch pathway in podocytes plays a role in the development of glomerular disease. *Nature Medicine, 14*(3), 290–298. https://doi.org/10.1038/nm1731.

Nishad, R., Meshram, P., Singh, A. K., Reddy, G. B., & Pasupulati, A. K. (2020). Activation of Notch1 signaling in podocytes by glucose-derived AGEs contributes to proteinuria. *BMJ Open Diabetes Research & Care, 8*(1), https://doi.org/10.1136/bmjdrc-2020-001203.

Nishad, R., Tahaseen, V., Kavvuri, R., Motrapu, M., Singh, A. K., Peddi, K., & Pasupulati, A. K. (2021). Advanced-glycation end-products induce podocyte injury and contribute to proteinuria. *Frontiers in Medicine (Lausanne), 8*, 685447. https://doi.org/10.3389/fmed.2021.685447.

Oczypok, E. A., Perkins, T. N., & Oury, T. D. (2017). All the "RAGE" in lung disease: The receptor for advanced glycation end-products (RAGE) is a major mediator of pulmonary inflammatory responses. *Paediatric Respiratory Reviews, 23*, 40–49.

Oldfield, M. D., Bach, L. A., Forbes, J. M., Nikolic-Paterson, D., McRobert, A., Thallas, V., ... Cooper, M. E. (2001). Advanced glycation end products cause epithelial-myofibroblast transdifferentiation via the receptor for advanced glycation end products (RAGE). *Journal of Clinical Investigation, 108*(12), 1853–1863. https://doi.org/10.1172/jci11951.

Oshima, M., Shimizu, M., Yamanouchi, M., Toyama, T., Hara, A., Furuichi, K., & Wada, T. (2021). Trajectories of kidney function in diabetes: A clinicopathological update. *Nature Reviews Nephrology, 17*(11), 740–750. https://doi.org/10.1038/s41581-021-00462-y.

Pasupulati, A. K. (2022). Podocyte developmental pathways in diabetic nephropathy: A spotlight on Notch signaling. *Diabetic Nephropathy, 2*(1), 1–6. https://doi.org/10.2478/dine-2022-0015.

Pasupulati, A. K., Chitra, P. S., & Reddy, G. B. (2016). Advanced glycation end products mediated cellular and molecular events in the pathology of diabetic nephropathy. *Biomolecular Concepts, 7*(5-6), 293–309. https://doi.org/10.1515/bmc-2016-0021.

Pavkov, M. E., Collins, A. J., Coresh, J., & Nelson, R. G. (2018). Kidney disease in diabetes. In C. C. Cowie, S. S. Casagrande, A. Menke, M. A. Cissell, M. S. Eberhardt, J. B. Meigs, E. W. Gregg, W. C. Knowler, E. Barrett-Connor, D. J. Becker, F. L. Brancati, E. J. Boyko, W. H. Herman, B. V. Howard, K. M. V. Narayan, M. Rewers, & J. E. Fradkin (Eds.), *Diabetes in America*. Bethesda (MD) conflicts of interest, with the following potential exceptions. Dr. Coresh received grant support from the National Kidney Foundation. Dr. Coresh possesses rights to the following intellectual property: PCT/US2015/044567 Provisional patent (Coresh, Inker, and Levey) filed August 15, 2014—Precise estimation of glomerular filtration rate from multiple biomarkers. The technology is not licensed in whole or in part to any company. Tufts Medical Center, Johns Hopkins University, and Metabolon, Inc. have a collaboration agreement to develop a product to estimate glomerular filtration rate from a panel of markers (June 25, 2016). Dr. Coresh is a member of the Global Hyperkalemia Council (sponsored by Relypsa, no personal compensation). National Institute of Diabetes and Digestive and Kidney Diseases (US).

Peppa, M., Brem, H., Cai, W., Zhang, J. G., Basgen, J., Li, Z., ... Uribarri, J. (2006). Prevention and reversal of diabetic nephropathy in db/db mice treated with alagebrium (ALT-711). *American Journal of Nephrology, 26*(5), 430–436. https://doi.org/10.1159/000095786.

Pickering, R. J., Tikellis, C., Rosado, C. J., Tsorotes, D., Dimitropoulos, A., Smith, M., ... Thomas, M. C. (2019). Transactivation of RAGE mediates angiotensin-induced inflammation and atherogenesis. *Journal of Clinical Investigation, 129*(1), 406–421. https://doi.org/10.1172/jci99987.

Poppelaars, F., & Thurman, J. M. (2020). Complement-mediated kidney diseases. *Molecular Immunology, 128*, 175–187. https://doi.org/10.1016/j.molimm.2020.10.015.

Qin, X., Goldfine, A., Krumrei, N., Grubissich, L., Acosta, J., Chorev, M., ... Halperin, J. A. (2004). Glycation inactivation of the complement regulatory protein CD59: A possible role in the pathogenesis of the vascular complications of human diabetes. *Diabetes, 53*(10), 2653–2661. https://doi.org/10.2337/diabetes.53.10.2653.

Raghu, G., Jakhotia, S., Yadagiri Reddy, P., Anil Kumar, P., & Bhanuprakash Reddy, G. (2016). Ellagic acid inhibits non-enzymatic glycation and prevents proteinuria in diabetic rats. *Food & Function, 7*(3), 1574–1583. https://doi.org/10.1039/C5FO01372K.

Ramya, R., Coral, K., & Bharathidevi, S. R. (2021). RAGE silencing deters CML-AGE induced inflammation and TLR4 expression in endothelial cells. *Experimental Eye Research, 206*, 108519. https://doi.org/10.1016/j.exer.2021.108519.

Rea, I. M., Gibson, D. S., McGilligan, V., McNerlan, S. E., Alexander, H. D., & Ross, O. A. (2018). Age and age-related diseases: Role of inflammation triggers and cytokines. *Frontiers in Immunology, 9*, 586. https://doi.org/10.3389/fimmu.2018.00586.

Reiniger, N., Lau, K., McCalla, D., Eby, B., Cheng, B., Lu, Y., ... Schmidt, A. M. (2010). Deletion of the receptor for advanced glycation end products reduces glomerulosclerosis and preserves renal function in the diabetic OVE26 mouse. *Diabetes, 59*(8), 2043–2054. https://doi.org/10.2337/db09-1766.

Rutter, K., Sell, D., Fraser, N., Obrenovich, M., Zito, M., Starke-Reed, P., & Monnier, V. (2003). Green tea extract suppresses the age-related increase in collagen cross-linking and fluorescent products in C57BL/6 mice. *International Journal for Vitamin and Nutrition Research. Internationale Zeitschrift für Vitamin- und Ernährungsforschung. Journal international de vitaminologie et de nutrition, 73*, 453–460. https://doi.org/10.1024/0300-9831.73.6.453.

Sadowska-Bartosz, I., & Bartosz, G. (2015). Prevention of protein glycation by natural compounds. *Molecules, 20*(2), 3309–3334.

Sanchez-Niño, M. D., Carpio, D., Sanz, A. B., Ruiz-Ortega, M., Mezzano, S., & Ortiz, A. (2015). Lyso-Gb3 activates Notch1 in human podocytes. *Human Molecular Genetics, 24*(20), 5720–5732. https://doi.org/10.1093/hmg/ddv291.

Santoro, D., Torreggiani, M., Pellicanò, V., Cernaro, V., Messina, R. M., Longhitano, E., ... Piccoli, G. B. (2021). Kidney biopsy in type 2 diabetic patients: Critical reflections on present indications and diagnostic alternatives. *International Journal of Molecular Sciences, 22*(11), https://doi.org/10.3390/ijms22115425.

Schena, F. P., Esposito, P., & Rossini, M. (2020). A narrative review on C3 glomerulopathy: A rare renal disease. *International Journal of Molecular Sciences, 21*(2), https://doi.org/10.3390/ijms21020525.

Schrijvers, B. F., De Vriese, A. S., & Flyvbjerg, A. (2004). From hyperglycemia to diabetic kidney disease: The role of metabolic, hemodynamic, intracellular factors and growth factors/cytokines. *Endocrine Reviews, 25*(6), 971–1010. https://doi.org/10.1210/er.2003-0018.

Song, Q., Liu, J., Dong, L., Wang, X., & Zhang, X. (2021). Novel advances in inhibiting advanced glycation end product formation using natural compounds. *Biomedicine & Pharmacotherapy, 140*, 111750. https://doi.org/10.1016/j.biopha.2021.111750.

Sourris, K. C., Watson, A., & Jandeleit-Dahm, K. (2021). Inhibitors of advanced glycation end product (AGE) formation and accumulation. *Handbook of Experimental Pharmacology, 264*, 395–423. https://doi.org/10.1007/164_2020_391.

Starowicz, M., & Zieliński, H. (2019). Inhibition of advanced glycation end-product formation by high antioxidant-leveled spices commonly used in European cuisine. *Antioxidants (Basel), 8*(4), https://doi.org/10.3390/antiox8040100.

Steiner, D., Marçon, C. R., & Sabban, E. N. C. (2018). Diabetes, non-enzymatic glycation, and aging. In E. N. Cohen Sabban, F. M. Puchulu, & K. Cusi (Eds.). *Dermatology and diabetes* (pp. 243–279) Cham: Springer International Publishing.

Taguchi, K., & Fukami, K. (2023). RAGE signaling regulates the progression of diabetic complications. *Frontiers in Pharmacology, 14*, 1128872. https://doi.org/10.3389/fphar.2023.1128872.

Tan, J., Xu, Y., Jiang, Z., Pei, G., Tang, Y., Tan, L., ... Qin, W. (2020). Global glomerulosclerosis and segmental glomerulosclerosis could serve as effective markers for prognosis and treatment of IgA vasculitis with nephritis. *Frontiers in Medicine (Lausanne), 7*, 588031. https://doi.org/10.3389/fmed.2020.588031.

Tang, Y., & Chen, A. (2014). Curcumin eliminates the effect of advanced glycation end-products (AGEs) on the divergent regulation of gene expression of receptors of AGEs by interrupting leptin signaling. *Laboratory Investigation, 94*(5), 503–516. https://doi.org/10.1038/labinvest.2014.42.

Thirunavukkarasu, V., Anitha Nandhini, A. T., & Anuradha, C. V. (2005). Lipoic acid improves glucose utilisation and prevents protein glycation and AGE formation. *Pharmazie, 60*(10), 772–775.

Tóbon-Velasco, J. C., Cuevas, E., & Torres-Ramos, M. A. (2014). Receptor for AGEs (RAGE) as mediator of NF-kB pathway activation in neuroinflammation and oxidative stress. *CNS & Neurological Disorders – Drug Targets, 13*(9), 1615–1626. https://doi.org/10.2174/1871527313666140806144831.

Tsujinaka, H., Itaya-Hironaka, A., Yamauchi, A., Sakuramoto-Tsuchida, S., Shobatake, R., Makino, M., ... Ogata, N. (2017). Statins decrease vascular epithelial growth factor expression via down-regulation of receptor for advanced glycation end-products. e00401 *Heliyon, 3*(9), https://doi.org/10.1016/j.heliyon.2017.e00401.

Uribarri, J., Cai, W., Peppa, M., Goodman, S., Ferrucci, L., Striker, G., & Vlassara, H. (2007). Circulating glycotoxins and dietary advanced glycation end-products: Two links to inflammatory response, oxidative stress, and aging. *Journals of Gerontology Series A: Biological Sciences and Medical Sciences, 62*(4), 427–433. https://doi.org/10.1093/gerona/62.4.427.

White, S., Lin, L., & Hu, K. (2020). NF-κB and tPA signaling in kidney and other diseases. *Cells, 9*(6), 1348.

Wiecek, A., Chudek, J., & Kokot, F. (2003). Role of angiotensin II in the progression of diabetic nephropathy-therapeutic implications. v16-20 *Nephrology Dialysis Transplantation, 18(Suppl. 5)*, https://doi.org/10.1093/ndt/gfg1036.

Wolf, G., & Ziyadeh, F. N. (1999). Molecular mechanisms of diabetic renal hypertrophy. *Kidney International, 56*(2), 393–405.

Wolff, S. P., & Dean, R. (1987). Glucose autoxidation and protein modification. The potential role of 'autoxidative glycosylation' in diabetes. *Biochemical Journal, 245*(1), 243–250.

Wu, H. Y., Peng, Y. S., Chiang, C. K., Huang, J. W., Hung, K. Y., Wu, K. D., ... Chien, K. L. (2014). Diagnostic performance of random urine samples using albumin concentration vs ratio of albumin to creatinine for microalbuminuria screening in patients with diabetes mellitus: A systematic review and meta-analysis. *JAMA Internal Medicine, 174*(7), 1108–1115. https://doi.org/10.1001/jamainternmed.2014.1363.

Wu, X.-Q., Zhang, D.-D., Wang, Y.-N., Tan, Y.-Q., Yu, X.-Y., & Zhao, Y.-Y. (2021). AGE/RAGE in diabetic kidney disease and ageing kidney. *Free Radical Biology and Medicine, 171*, 260–271. https://doi.org/10.1016/j.freeradbiomed.2021.05.025.

Xie, M., Zhu, Y., Wang, X., Ren, J., Guo, H., Huang, B., ... Zhang, J. (2023). Predictive prognostic value of glomerular C3 deposition in IgA nephropathy. *Journal of Nephrology, 36*(2), 495–505. https://doi.org/10.1007/s40620-022-01363-4.

Yamagishi, S.-I., Nakamura, K., Matsui, T., Ueda, S., Noda, Y., & Imaizumi, T. (2008). Inhibitors of advanced glycation end products (AGEs): Potential utility for the treatment of cardiovascular disease. *Cardiovascular Drug Reviews, 26*(1), 50–58. https://doi.org/10.1111/j.1527-3466.2007.00038.x.

Yamamoto, Y., Kato, I., Doi, T., Yonekura, H., Ohashi, S., Takeuchi, M., ... Yamamoto, H. (2001). Development and prevention of advanced diabetic nephropathy in RAGE-overexpressing mice. *Journal of Clinical Investigation, 108*(2), 261–268. https://doi.org/10.1172/jci11771.

Yiu, W. H., Li, R. X., Wong, D. W. L., Wu, H. J., Chan, K. W., Chan, L. Y. Y., ... Tang, S. C. W. (2017). Complement C5a inhibition moderates lipid metabolism and reduces tubulointerstitial fibrosis in diabetic nephropathy. *Nephrology Dialysis Transplantation, 33*(8), 1323–1332. https://doi.org/10.1093/ndt/gfx336.

Yokoyama, S., Kawai, T., Yamamoto, K., Yibin, H., Yamamoto, H., Kakino, A., ... Rakugi, H. (2021). RAGE ligands stimulate angiotensin II type I receptor (AT1) via RAGE/AT1 complex on the cell membrane. *Scientific Reports, 11*(1), 5759. https://doi.org/10.1038/s41598-021-85312-4.

Zhang, H., & Sun, S. C. (2015). NF-κB in inflammation and renal diseases. *Cell & Bioscience, 5*, 63. https://doi.org/10.1186/s13578-015-0056-4.

Zhang, L., He, J., Wang, J., Liu, J., Chen, Z., Deng, B., ... Wang, L. (2021). Knockout RAGE alleviates cardiac fibrosis through repressing endothelial-to-mesenchymal transition (EndMT) mediated by autophagy. *Cell Death & Disease, 12*(5), 470. https://doi.org/10.1038/s41419-021-03750-4.

Zhu, P., Chen, C., Wu, D., Chen, G., Tan, R., & Ran, J. (2022). AGEs-induced MMP-9 activation mediated by Notch1 signaling is involved in impaired wound healing in diabetic rats. *Diabetes Research and Clinical Practice, 186*, 109831. https://doi.org/10.1016/j.diabres.2022.109831.

CHAPTER TEN

Nanoparticles in prevention of protein glycation

Aruna Sivaram and Nayana Patil*
School of Bioengineering Sciences and Research, MIT ADT University, Pune, India
*Corresponding author. e-mail address: nayana.patil@mituniversity.edu.in

Contents

1. Introduction	288
2. Chemistry of the glycation reaction	288
3. Glycation of biological macromolecules	290
3.1 Protein glycation	290
4. DNA glycation	292
5. RNA glycation	293
6. Lipid glycation	293
7. Clinical relevance of glycation and advanced glycation end products	294
8. Osteoporosis	295
9. Cancer	296
10. Neurodegenerative diseases	296
11. Cardiovascular diseases	296
12. Prevention of glycation	297
13. Nanotechnology approach for combating glycation	298
14. Nanoparticles	299
15. Gold nanoparticles	299
16. Silver nanoparticles	301
17. Zinc nanoparticles	302
18. Challenges and future perspectives	304
19. Challenges	304
20. Future perspectives	305
21. Conclusion	305
References	306

Abstract

Advanced glycation end products (AGEs) are formed by the non-enzymatic attachment of carbohydrates to a biological macromolecule. These AGEs bind to their cognate receptor called receptor for AGEs (RAGEs), which becomes one of the important causal factors for the initiation and progression of several diseases. A deep understanding into the pathways of RAGEs will help in identifying novel intervention modalities as a part of new therapeutic strategies. Although several approaches exist to target this pathway using small molecules, compounds of plant origin etc, nanoparticles have proven to be a critical method, given its several advantages. A high bioavailability, biocompatibility, ability to cross blood brain barrier and modifiable surface properties give nanoparticles an upper edge over other strategies. In this chapter, we will discuss AGEs, their involvement in diseases and the nanoparticles used for targeting this pathway.

1. Introduction

Glycation refers to a chemical reaction where sugar molecules (glucose or fructose) bind to proteins or fats or nucleic acids without the involvement of enzymes. This process can happen spontaneously in the body, particularly when blood sugar levels are elevated over a prolonged period (Lima & Baynes, 2013).

As our body lacks the enzymes to hydrolyse glycated compounds, advanced glycation end products (AGEs) can accumulate in various tissues throughout the body, including the skin, blood vessels, kidneys, and nerves (Zawada et al., 2022). At these sites, functional and structural changes in glycated proteins lead to impairing their normal activity. Additionally, AGEs can generate oxidative stress and inflammation, contributing to the acceleration of the glycation reaction and the development of several chronic diseases (Nakamura & Kawaharada, 2021).

2. Chemistry of the glycation reaction

The glycated residues known as Amadori products subsequently proceed through a series of intricate rearrangements to produce abnormal structures known as advanced glycation end products AGEs. This process known as Maillard reaction was first introduced by French chemist Louis-Camille Maillard initially in the cooking process (Maillard, 1912). The formation of AGEs is a complex process (Fig. 1) involving multiple pathways, including the Hodge pathway, the Namiki pathway, and the Wolff pathway (Ott et al., 2014).
- Hodge pathway: The Hodge pathway, also known as the classic pathway initiates with a chemical reaction between a lone pair of donatable electrons in an amino group of a protein and reactive carbonyl group of a sugar or an aldehyde, forming a glycosylamine intermediate and unstable Schiff base (imine). The glycosylamine rapidly undergoes rearrangement to form a reversible and more stable Amadori product. Future Amadori

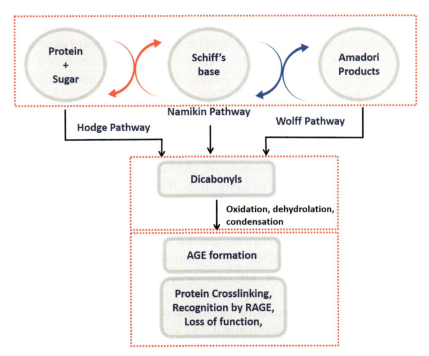

Fig. 1 Formation of advanced glycation end products (AGE) through the Maillard reaction by three mechanisms Hodge pathway, Namiki pathway and Wolff pathway. In the initiation stages, Schiff base formation occurs when the amino acid carbonyl amine and the carbonyl group of the reducing sugar combine. During the propagation stage, the dicarbonyl compounds stabilizes to form Amadori products. Later, in advanced stage, AGEs development occurs which further cross-link with macromolecules.

products through a series of polycondensation and complex rearrangements (dehydration, fragmentation, and condensation) produce a diverse range of compounds. These compounds include furans, pyrazines, aldehydes, and ketones, and eventually progress into the formation of structures collectively known as advance glycation end products (AGEs).
- Namiki pathway: The Namiki pathway is a non-enzymatic pathway also known as the 3-deoxyglucosone pathway, involves the reaction of 3-deoxyglucosone (3-DG), a highly reactive alpha-dicarbonyl compound, with proteins or amino acids. It starts with the initial Maillard reaction, where a reducing sugar reacts with the amino group of a protein, forming a reversible Schiff base or imine adduct. In this pathway, the Schiff base can undergo an intramolecular rearrangement known as

the Namiki rearrangement. The rearrangement involves the migration of the carbon atom next to the carbonyl group of the reducing sugar to the nitrogen atom of the Schiff base, resulting in the formation of a more stable product called a Namiki product. The Namiki product can then undergo further reactions, such as dehydration and cyclization, leading to the formation of 3-DG-derived AGEs, such as carboxyethyllysine (CEL) and methylglyoxal (MGO)-derived hydroimidazolones.

- Wolff pathway: The Wolff pathway, also known as the glyoxal/oxidative pathway, involves the reaction of glyoxal, another alpha-dicarbonyl compound, with proteins or amino acids. It starts with the initial Maillard reaction, where a reducing sugar reacts with the amino group of a protein, forming a Schiff base. In this pathway, the Schiff base undergoes oxidative cleavage by reactive oxygen species (ROS), such as free radicals or reactive carbonyl species. The oxidative cleavage of the Schiff base generates highly reactive carbonyl compounds, such as glyoxal and MGO. These carbonyl compounds can then react with nearby amino groups in proteins, leading to the formation of glyoxal-derived AGEs, such as N-(1-carboxyethyl)lysine (CEL) and N-(1-carboxymethyl)lysine (CML).

This same reaction of AGEs formation is an integral part of metabolism in the body, but occurs at a slower rate leading to continuous glycation of macromolecules, and gradual accumulation results in aging. Their formation is accelerated in conditions such as high blood sugar levels, such as in diabetes, intake of a diet high in sugar or processed foods, smoking and oxidative stress (Nakamura & Kawaharada, 2021; Singh et al., 2014). Glycation's effects and the buildup of AGEs in the body have been associated with numerous diseases, including diabetes, cardiovascular disease, Alzheimer's disease, and kidney dysfunction (Tripathi et al., 2023; Twarda-Clapa et al., 2022).

3. Glycation of biological macromolecules
3.1 Protein glycation

The amino acids prone to get glycated are lysine residues. The α amino group of the N- terminal amino acid or the ε amino groups of lysines reacts with the aldehyde or keto group of the reducing sugar leading to glycation of protein (Fig. 2A). AGEs formation has been also discovered on the side

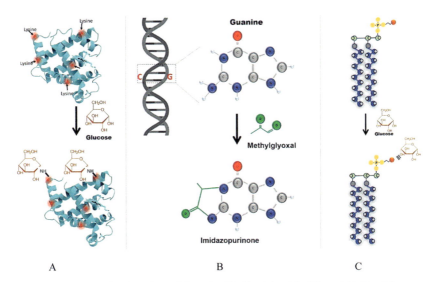

Fig. 2 Non-enzymatic glycation of Protein (A), Nuclei acids (B) and Lipids (C) at corresponding target residues.

chains of arginines and histidines in addition to lysine residues (Rabbani, Ashour, & Thornalley, 2016).

The process of protein glycation is influenced by several factors, including the concentration of glucose or other reducing sugars, the duration of exposure, and the presence of oxidative stress. Structural proteins with a long biological half-life and low turnover number (collagen and crystallins) exposed to elevated sugar levels for longer duration tend to accumulate AGEs over time (Shin et al., 2023; Twarda-Clapa et al., 2022). Given its abundance, accumulation of AGEs in these proteins can lead to structural and functional alterations in proteins. Cellular and extracellular proteins that have undergone glycation can lose their function or take on new pathogenic functions (Rabbani & Paul, 2021). Some of the key effects of protein glycation are as follows:

- Cross-linking of proteins: AGEs can cause proteins to become cross-linked, leading to changes in their structure and function. Cross-linking can affect the elasticity and integrity of tissues, impairing their normal functioning.
- Oxidative stress: AGEs can generate ROS and induce oxidative stress, which can damage cells and contribute to inflammation, tissue injury, and disease progression.

- Alteration of protein function: Protein glycation can disrupt the normal functions of proteins. It can interfere with enzymatic activity, receptor-ligand interactions, and protein–protein interactions, leading to functional impairments.
- Inflammation: AGEs can trigger an inflammatory response by activating various signaling pathways and promoting the production of pro-inflammatory molecules. Chronic inflammation is implicated in the development of many diseases, including diabetes, cardiovascular disease, and neurodegenerative disorders.

4. DNA glycation

DNA glycation refers to the process by which advanced glycation end products (AGEs) form on DNA molecules. In the case of DNA glycation, this reaction occurs between reducing sugars, such as glucose or fructose, and the nucleotide bases or the sugar-phosphate backbone of DNA (Shuck et al., 2018). The amino groups of guanine in DNA are more prone to react with MGO and glyoxal (Fig. 2B). At earlier stages, DNA glycation leads to induction of DNA repair but at excessive glycation level, it leads to induction of apoptosis (Jeong & Lee, 2021; Thornalley, 2008).

The glycation of DNA can lead to several consequences (Tripath et al., 2023):
- DNA Damage: Glycation can cause structural changes in DNA, leading to DNA damage. It can result in the formation of DNA adducts, where sugar molecules are covalently attached to DNA bases or backbone. These adducts can interfere with DNA replication, transcription, and repair processes.
- Crosslinking: Glycation can lead to the formation of crosslinks between DNA molecules or between DNA and other proteins. These crosslinks can affect DNA stability and function and may impair the binding of DNA-binding proteins, such as transcription factors.
- Oxidative Stress: Glycation can induce oxidative stress, promoting the production of ROS within cells. Increased ROS levels can cause further DNA damage and contribute to cellular dysfunction.
- Altered Gene Expression: DNA glycation can influence gene expression by modifying the DNA structure and interfering with the binding of transcription factors or other regulatory proteins. This alteration in gene expression patterns can have profound effects on cellular function.

5. RNA glycation

While the process of RNA glycation is not as extensively studied as DNA glycation, similar glycation reactions can occur on RNA molecules. Glycation of RNA involves the non-enzymatic reaction between reducing sugars and the amino groups present in the nucleotides of RNA. The consequences of RNA glycation are not yet fully understood, but it is believed that glycation can affect the structure, stability, and function of RNA (Weng et al., 2020).

Like DNA glycation, RNA glycation can impair important cellular processes such as transcription and translation, leading to disruptions in gene expression and protein synthesis, potentially impacting their functionality and contributing to cellular dysfunction (Zheng et al., 2020).

6. Lipid glycation

Also known as lipid peroxidation or lipid glycoxidation, it is a process that involves the reaction of lipids, particularly polyunsaturated fatty acids, with reducing sugars, leading to the formation of advanced glycation end products (AGEs) within lipid molecules (Moldogazieva, Mokhosoev, Mel'nikova, Porozov, & Terentiev, 2019; Zheng et al., 2020). The glucose residues are attached to the aminophospholipids such as phosphatidylethanolamine (PE) and form an intermediate Schiff base which progresses to rearrange into PE-linked Amadori products (Fig. 2C). The autoxidation of the Amadori product generates ROS that carry out peroxidation of unsaturated fatty acid residues, accelerating the free radical reactions and the formation of phosphatidylcholine hydroperoxide (Moldogazieva et al., 2019). This reaction produces imine adducts, which can undergo rearrangements and modifications to form stable AGEs. This process can occur in various lipid-rich tissues, including cell membranes and lipoproteins.

The consequences of lipid glycation can be detrimental to cellular and physiological functions. Glycation of lipids can result in the production of ROS and oxidative stress, which can further promote lipid peroxidation and inflammation. AGE- modified lipids can also induce the formation of foam cells and contribute to the development and progression of atherosclerosis (Bao et al., 2020).

7. Clinical relevance of glycation and advanced glycation end products

As depicted in Fig. 3, the elevated level of sugar in diabetes generates a conducive environment for glycation of macromolecules and AGEs accumulation which in turn is associated with impairment in the kidney, eyes, nerves, and blood vessels (Savateev, Spasov, & Rusinov, 2022).

The interaction between advanced glycation end products (AGEs) and the receptor for AGEs (RAGE) triggers several pathways through the generation of oxidative stress and activation of transcription factors. These pathways contribute to upregulation of cytokines, inflammation, immune cell recruitment, and the production of ROS. The detrimental consequences of these cascades result in the disruption of the blood-retinal

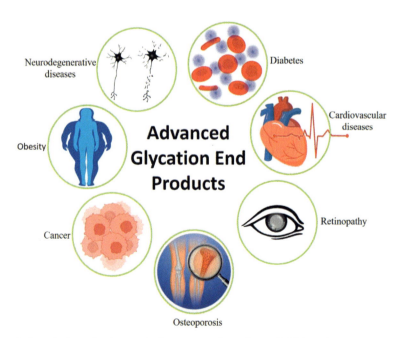

Fig. 3 Association of advanced glycation end products (AGEs) with progression of human diseases and its accompanying symptoms.

barrier (BRB) and increased permeability, contributing to the development of retinopathy and nephropathy (Oshitari, 2023). Diseases like diabetes affect both the BRB and the kidneys, highlighting the interconnectedness of various physiological systems. The BRB and nephropathy (kidney disease) are two distinct physiological barriers in the body, and disruption of the BRB itself does not directly contribute to nephropathy. However, both the BRB and the blood-brain barrier share similarities. They share some common features in terms of tight junctions and selective permeability. Diseases or conditions that lead to disruption of these tight junctions and increased permeability in the BRB serve as analogies for understanding similar processes in the kidneys. Along with this, AGEs accumulation aid in degradation of α-crystallin by altering the surface charge and hydrophobicity of α-crystallin. As a consequence of the loss of chaperone activity in α-crystallin, β- and γ-crystallins become unstable, forming disulphide bonds and aggregating, rendering them insoluble. The accumulation of these insoluble fibers induces opacity in the eye lens, impairs its ability to scatter light effectively, and eventually hampers vision, leading to the formation of cataract (Fan & Monnier, 2021). Another common complication in diabetes is nerve damage wherein glycation of proteins in the nerves can result in peripheral neuropathy. It can cause symptoms such as numbness, tingling, and pain in the extremities, leading to impaired sensation and mobility (Parwani & Mandal, 2023).

8. Osteoporosis

The role of glycation in osteoporosis, a condition characterized by low bone density and increased risk of fractures, is an area of ongoing research. The most abundant structural protein collagen acquires glycation promoting cross-linking of collagen fibers in bone, making them stiffer and more brittle. Glycation is also known to interfere with the normal process of bone remodelling and bone healing, which involves the continuous turnover of bone tissue through the actions of bone-forming cells (osteoblasts) and bone-resorbing cells (osteoclasts). AGEs is involved in inhibiting the differentiation and function of osteoblasts and promoting osteoclast activity, leading to a net loss of bone mass (Ge et al., 2022). AGEs can inhibit the formation of hydroxyapatite, the mineral component of bone, leading to reduced bone density and increased fracture risk.

9. Cancer

The role of glycation in cancer is a complex and multifaceted area of research. Glycation can cause DNA damage by generating ROS and promoting oxidative stress. This oxidative damage to DNA can lead to mutations and genomic instability, which are key factors in the development of cancer (Purushothaman, Mohajeri, & Lele, 2023). AGEs activate inflammatory pathways and signalling cascades that are known to play a critical role in cancer development by creating a microenvironment that supports tumor growth, angiogenesis (formation of new blood vessels), and metastasis (spread of cancer to distant sites). Recently, it has been reported that glycation can influence the drug uptake, metabolism, and efflux of drugs in cancer cells, affecting the efficacy of chemotherapy agents (Dawood, Younus, Alnori, & Mahmood, 2022).

10. Neurodegenerative diseases

Advanced glycation end products (AGEs) tend to accumulate in the brain and disrupt normal cellular functions. AGEs can trigger oxidative stress, inflammation, and neuronal damage, leading to the impairment and death of nerve cells (D'cunha et al., 2022). Additionally, AGEs can contribute to the aggregation of misfolded proteins, such as amyloid beta and tau, which are characteristic of diseases like Alzheimer's and Parkinson's (Fournet, Bonté, & Desmoulière, 2018; Sharma et al., 2020). In addition, AGEs can impair the function of glial cells, which are responsible for clearing amyloid-beta plaques, playing a critical role in maintaining brain homeostasis and neuronal health. AGEs can interfere and or disrupt important cellular signalling pathways involved in neuronal survival, synaptic plasticity etc. leading to altered signalling and impaired cellular responses (Ko et al., 2015).

11. Cardiovascular diseases

Lipid glycation enhances the uptake of glycated lipids by macrophages in the arterial wall transforming them into lipid-laden foam cells. Foam cells are a hallmark of early atherosclerotic plaques and play a critical role in plaque formation which can obstruct blood flow, potentially leading to ischemic events such as heart attacks or strokes. The accumulation of

AGEs is also involved in disrupting the production and availability of nitric oxide, a molecule that regulates blood vessel dilation thereby regulating blood pressure and maintaining vascular health. The protein in the arterial walls which acquire glycation undergo deformation which compromises the elasticity of proteins, leading to increased stiffness and reduced flexibility placing an additional stress on the heart, contributing to hypertension and cardiac hypertrophy. Over a period of time, excess of AGEs triggers platelet aggregation and clot formation, increasing the risk of blood clot formation within the vessels (Wang et al., 2023). This can lead to blockages in the arteries, causing heart attacks or strokes.

12. Prevention of glycation

As the mitigating effects of the glycation process are now better understood, effective management of this process through prevention of AGE formation and controlling glycation is extremely desirable. The most efficient method to manage and prevent the above mentioned problems is the utilization of prospective antiglycation medicines of natural and synthetic origins (Fig. 4) with enhanced inhibitory action and lower toxicity (Sharma & Roy, 2022).

As the process of AGE formation is a multistep reaction, a range of different compounds have been identified which target different stages of AGE formation and accumulation. The antiglycating agents can be grouped into different types (Abbas et al., 2016; Lehman & Beryl, 2001). For example, Type A inhibitors called "sugar competitors", mask the

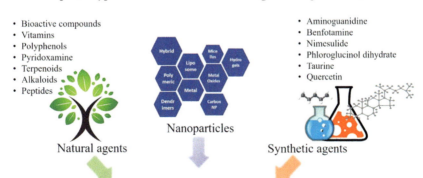

Fig. 4 Compounds that affect the formation of AGEs and show antiglycation activity.

amino groups thus preventing sugar attachment. Type B targets multiple steps by interacting with sugars and deflecting them from proteins, Type C work as carbonyl trapping agents, Type D inhibitors act by trapping methyl glyoxal, while inhibitors of Type E inhibit the cross-linking of AGE to proteins.

Several antiglycation agents play a crucial role in preventing and managing conditions associated with glycation, including diabetes, cardiovascular diseases, neurodegenerative disorders, and aging-related complications. Here are some examples of antiglycation agents:

- Natural antiglycation agents: As represented in Fig. 4, various medicinal plants and herbs, including *Gymnema sylvestre*, *Pueraria lobata*, *Cinnamomum* spp, *Zingiber officinale*, garlic (*Allium sativum*) and *Curcuma longa* (turmeric), contain bioactive compounds that can inhibit glycation reactions and reduce AGE formation (Tariq & Ali, 2023). Similarly, polyphenolic compounds found in various plants, such as green tea, cocoa, and berries, trap reactive carbonyl species and prevent the cross-linking of glycated proteins (Favre et al., 2022). Antioxidants such as vitamin C, vitamin E, pyridoxamine and alpha-lipoic acid can neutralize ROS, reduce oxidative stress leading to inhibition of the formation of AGEs (Odjakova, Popova, Al Sharif, & Mironova, 2012). Not only plants but carnosine, abundantly present in meat and fish has been found to possess antiglycation activity. It can scavenge reactive carbonyl species and inhibit the formation of AGEs (Caruso, Di Pietro, Cardaci, Maugeri, & Caraci, 2023).
- Synthetic antiglycation agents: Aminoguanidine (AG) ALT-711, benfotamine, are synthetic compounds that inhibit the formation of AGEs by blocking the reaction between glucose and proteins. Similarly, nimesulide and phloroglucinol dihydrate, aspirin, taurine and quercetin (Fig. 4) have been studied for their antiglycation effects and has shown potential in reducing AGE accumulation (Savateev et al., 2022).

13. Nanotechnology approach for combating glycation

Nanoparticles (NPs) with antiglycation activity have emerged as potential therapeutic agents for mitigating the harmful effects of glycation (Patil, Anushka, & Sivaram, 2022; Rashid, Naqash, Bader, & Sheikh, 2020). These NPs can target and inhibit the formation of advanced glycation end products (AGEs) or aid in their removal. As shown in Fig. 4, there are different types of NPs which are emerging as an antiglycating agent.

- Polyphenol-based NPs: Polyphenols, natural compounds found in various fruits, vegetables, and plant extracts, possess antiglycation properties. By encapsulating polyphenols within NPs, their stability and bioavailability can be improved. These NPs can scavenge reactive carbonyl species (precursors of AGEs) and inhibit glycation reactions, thereby reducing the formation of AGEs (Islam et al., 2022).
- Metal-based NPs: Certain metal NPs, such as gold, titanium, zinc and silver nanoparticles (AgNPs), have demonstrated antiglycation activity. These NPs can inhibit the glycation process by binding to reactive carbonyl species or AGEs, preventing their interaction with proteins or other biomolecules (Patil et al., 2022). Additionally, metal NPs can possess antioxidant properties, helping to counteract oxidative stress associated with glycation.
- Nanocarriers for antiglycation agents: NPs can serve as carriers for antiglycation agents, enhancing their stability, solubility, and targeted delivery. For example, NPs encapsulating antiglycation drugs or enzymes (such as AG or glyoxalase-I) can be designed to specifically target glycation sites in blood vessels or tissues, facilitating the inhibition or degradation of AGEs (Alenazi et al., 2022).
- NPs for AGE clearance: A more recent approach is to develop NPs capable of selectively binding and removing existing AGEs. These NPs, often functionalized with specific ligands or antibodies, can recognize and bind AGEs, facilitating their clearance from the body (Woźniak et al., 2021). This approach aims to reduce the burden of accumulated AGEs in tissues and organs.

14. Nanoparticles

NPs have been widely studied for their anti-glycation properties. In this section we will discuss about NPs as a potential treatment opportunity for disease mitigation through modulation of the protein glycation pathway.

15. Gold nanoparticles

Gold nanoparticles (AuNPs) have become increasingly important in the field of therapeutics due to their remarkable biocompatibility, surface reactivity, and plasmon resonance. The outstanding anti-inflammatory and antioxidant properties of AuNPs have made them highly desirable for therapeutic purposes. Additionally, gold compounds exhibit distinctive optical properties and offer advantageous surface-to-volume ratios. Their easily modifiable surface properties enable effortless conjugation with other compounds. Consequently, extensive research has been conducted on the application of AuNPs in therapies targeting a wide range of diseases such as cancer and diabetes.

The extent of glycation was found to be influenced by the total surface area of the colloidal suspension of NPs. Increasing the surface area of the suspension resulted in a more pronounced antiglycation effect, whereas decreasing the surface area led to a reduction in this effect. Suspensions with comparable surface areas exhibited similar levels of glycation, highlighting the relationship between NP surface area and their activity. Several literature reports suggest that in addition to concentration, the surface area of AuNPs plays a crucial role in determining the degree of inhibition (Alomari, Salehhuddin, & Bahaa, 2021).

The AuNPs synthesised using *Couroupita guianensis* extract, known for its high polyphenol content, exhibited an IC_{50} value of 48.99 and 61.09 µg/ml for inhibiting the activities of alpha amylase and alpha glucosidase, respectively. To assess the antiglycation ability of the nano supplements, carbonyl protein content and reacted lysine content were measured. Upon incubation of the nano supplements with the HSA-glucose system, the carbonyl protein content significantly decreased to 18.82 ± 1.10 nmol/mg, and only 49.32% of the lysine residues became glycated. This study successfully demonstrated the remarkable antioxidant property of gold nano supplements, as they effectively reduced blood glucose levels and inhibited the interaction between MGO and lysine (Sengani & Rajeswari, 2017). Furthermore, AuNPs synthesized with beta galactosidase were also found to inhibit HSA glycation using the HSA-glucose model. Nephrotoxic compounds like AG have been encapsulated into AuNPs. In contrast to AG, these AuNPs demonstrated a notable therapeutic advantage by not exhibiting cytotoxicity. This characteristic enhances their potential for safe and effective therapeutic applications (Wang et al., 2021). A comparative analysis of curcumin-AuNP and curcumin NPs confirmed that the curcumin-AuNP conjugates exhibited both antiglycation and antioxidant activity. Furthermore, both the gold-curcumin conjugates and curcumin NPs demonstrated anti-cataract activity by inhibiting the reduction of chaperone activity in crystalline protein. This was achieved through the inhibition of advanced glycation end product (AGE) formation in the lens (Azharuddin, Dasgupta, & Datta, 2015).

16. Silver nanoparticles

The remarkable properties of AgNPs, such as optical, thermal, and electrical conductivity, combined with their biological characteristics, make them exceptionally versatile for a range of applications. Numerous studies in the literature have highlighted the diverse uses of AgNPs in pharmaceutical and related sectors. AgNPs have been extensively investigated for their potential as antibacterial, antifungal, antiviral, anti- inflammatory, anti-angiogenic, and anticancer agents.

AgNPs were synthesized using Aloe vera leaf extract, and their ability to inhibit advanced glycation end product (AGE) formation in the HSA-MG system was investigated. The incubation of AgNPs with HSA-MG resulted in a significant decrease in the absorbance of the system, reducing it to 25%. The fluorescence emission intensities of the three systems, namely native HSA, HSA-MG, and HSA-MG with varying concentrations of AgNPs, were found to align with the results obtained from high-performance liquid chromatography (HPLC) analysis. The treatment of HSA-MG with suitable concentrations of AgNPs effectively mitigated structural distortions in the protein (Ashraf, Ansari, Khan, Alzohairy, & Choi, 2016). The antiglycation properties of AgNPs capped with gum arabic were investigated using multiple methods, including ultraviolet-visible spectroscopy, fluorescence spectrometry, and HPLC. As the concentration of AgNPs increased, the chromophoric properties (at 280 nm) of MG-glycated BSA were significantly reduced by fourfold. Additionally, incubation of AgNPs with native BSA resulted in a remarkable 78.4% reduction in the fluorophoric properties (at 440 nm) compared to the control. These findings clearly indicate the inhibition of advanced glycation end (AGE) formation by AgNPs (Ashraf, Ansari, Choi, Khan, & Alzohairy, 2014).

Another study aimed to investigate the impact of AgNP on the glycation process of BSA in the presence of glucose using osteogenic cells as model systems. The findings revealed that AgNP exerted a significant antiglycating effect, leading to reductions of 45.5% and 34.02% in the fructosamine and protein carbonyl content of BSA, respectively, when exposed to glucose. Notably, as the concentration of AgNP increased, the extent of glycation decreased, indicating a dose-dependent relationship. Moreover, the presence of AgNP demonstrated remarkable benefits by enhancing cell viability by nearly 50% (Nayak et al., 2016). A remarkable reduction in glycation and fructosamine formation in the BSA-glucose system was observed upon treatment with bark extracts of *Eysenhardtia polystachya* encapsulated into AgNPs. This reduction was observed in a concentration-dependent manner over a duration of 4 weeks. The inhibitory effect on advanced glycation end product (AGE) formation ranged from 38.65% to 52.95%, and the inhibition of fructosamine formation ranged from 45.3% to 58.6% at AgNP concentrations of 0.5–2.0 mg/ml (Garcia Campoy, Perez Gutierrez, Manriquez-Alvirde, & Muñiz, 2018).

Furthermore, similar activities were observed using in vivo studies in the zebrafish model. These findings highlight the potential application of these AgNPs in the treatment of diseases associated with AGE formation. AgNPs were also used for evaluation in in vivo models. In the study, AgNPs were synthesized using an extract derived from *Nigella sativa*. These AgNPs were then assessed in mouse models with diabetic neuropathy. The group receiving the treatment exhibited a notable reduction in glucose levels, advanced glycation end products (AGEs), and aldose reductase activity, while insulin levels increased. Moreover, the treated group displayed a significant decrease in the expression levels of biomarkers associated with inflammation, oxidative stress, and nerve growth factor. The histological examination of brain tissue also demonstrated improvements in the disease condition due to the treatment (Alkhalaf, Hussein, & Hamza, 2020). In yet another study, the NPs were subjected to incubation with an AGE-BSA system, and the results revealed their capability to regulate the excessive growth of retinal endothelial cells, which would otherwise be stimulated by AGE production. Treatment of the system with the NPs effectively reduced the permeability of the endothelial barrier induced by AGE-BSA, while simultaneously preserving its structural integrity (Sheikhpranbabu et al., 2010).

17. Zinc nanoparticles

Zinc, the primary micronutrient, plays a crucial role in vital biological processes. It aids in the efficient transportation of glucose by enhancing insulin signaling and secretion (Bjørklund et al., 2020). Numerous zinc oxide nanoparticles (ZnO NPs) have demonstrated promising therapeutic advantages in preclinical investigations. These NPs have undergone comprehensive characterization and have exhibited notable properties, including anticancer, antibacterial, antidiabetic, and anti-inflammatory effects. Additionally, researchers have explored the potential of these NPs in drug delivery and theranostics.

In the presence of ZnO NPs, the major markers of glycation, including browning, fructosamine, and carbonyl content in glycated samples, experienced a substantial reduction of approximately 50%. These NPs not only exhibited antiglycation properties but also preserved the protein's native conformation and intrinsic activities. Furthermore, the addition of the NPs resulted in a remarkable 58% inhibition of amyloid beta aggregation, while simultaneously preventing alterations in its secondary structure (Kumar, Bhatkalkar, Sachar, & Ali, 2020). ZnO NPs prepared using leaf extracts of *Morus indica* showed beneficial effects in cell line and animal models. The NPs exhibited a dose-dependent reduction in the glycation of BSA mediated by MGO, leading to a decrease in HbA1c formation. Moreover, the treatment with these NPs effectively prevented the MGO-induced morphological distortion of red blood cells. Such structural distortions often result in functional loss and contribute to oxidative stress. Additionally, the NPs derived from *M. indica* demonstrated the ability to protect against oxidative stress associated with glycation (Anandan et al., 2019). In an in vivo study involving streptozotocin-induced diabetic rats, the administration of these NPs at a dose of 100 mg/kg for a duration of 6 weeks resulted in significant reductions in glucose levels. Furthermore, biochemical and histopathological examinations indicated improved hepatic and renal functions. Molecular docking studies revealed that the NPs exhibited a strong affinity through hydrogen bonding, binding to and shielding lysine and arginine residues, which are commonly targeted sites of glycation (Afify et al., 2019). In another research study, ZnNPs were synthesized using *Aloe vera* leaf extract and their effects were examined using biophysical, biochemical, and immunological methods in the MG-IgG glycation reaction system. The findings revealed that the presence of ZnO NPs led to a significant inhibition of advanced glycation end (AGE) formation, with a notable reduction of up to 68.01% observed at a concentration of 100 μg/ml. Additionally, the ZnO NPs exhibited protective effects on specific amino acid residues, including tryptophan (74.13% protection), lysine (65.3% protection), and arginine (63.36% protection) against damage induced by MGO. This protective action made the protein target sites less susceptible to glycation and oxidation, while also safeguarding protein structures that may otherwise be compromised by MGO-induced damage (Ashraf et al., 2018).

Though lesser number of reports are available, Platinum NPs, palladium NPs, Cadmium NPs, iron oxide NPs, Cerium NPs, Selenium NPs have shown potential antiglycation activities. Extensive research has focused on the remarkable capability of NPs to traverse challenging physiological barriers, including the blood-brain barrier and the placental barrier. These barriers pose significant obstacles to drug delivery, making NPs particularly valuable in this regard. Exploring the pharmacodynamics and pharmacokinetics of NPs, with a specific emphasis on their potential as antiglycation agents, holds great promise for improving patient outcomes. Conducting comprehensive laboratory investigations into the properties of NPs will shed further light on their therapeutic potential in managing diseases by effectively regulating protein glycation.

18. Challenges and future perspectives

Antiglycation NPs have emerged as a promising approach in the field of nanomedicine for combating glycation-related diseases. These NPs offer potential therapeutic benefits by targeting and inhibiting the glycation process, which plays a significant role in the pathogenesis of several chronic diseases. However, like any emerging technology, antiglycation NPs face specific challenges that need to be addressed to realize their full potential.

19. Challenges

Design and Synthesis: One of the primary challenges is the design and synthesis of antiglycation NPs with optimal properties. Achieving precise control over their size, shape, surface properties, and stability is crucial for effective drug delivery and target specificity. Additionally, developing efficient and scalable synthesis methods for these NPs remains a challenge, limiting their widespread application.

- Targeting and Specificity: To maximize therapeutic efficacy, antiglycation NPs must selectively target and accumulate at glycation sites in the body. Overcoming biological barriers and achieving site-specific delivery pose significant challenges. Improving targeting strategies, such as active targeting through ligand- receptor interactions or stimuli-responsive mechanisms, is essential to enhance the specificity and effectiveness of antiglycation NPs.
- Biocompatibility and Safety: Biocompatibility is a critical aspect of any nanoparticle- based therapy. The potential toxicity, immunogenicity, and long-term effects of antiglycation NPs on the human body need to be

thoroughly investigated. Ensuring the safety of these NPs is essential for their successful translation into clinical applications.
- Lack of studies in cell line and animal models: The majority of research studies have primarily focused on utilizing protein and sugar models to investigate the glycation pathway. Only a limited number of studies have explored the targeted intervention of the glycation pathway in cell lines and animal models. Furthermore, there is a scarcity of research examining the role of antiglycation compounds/agents in disease models specifically. Moreover, the long-term effects of administering these agents have not been thoroughly characterized.

20. Future perspectives

- Improved Drug Delivery: Antiglycation NPs hold great promise in enhancing drug delivery to glycation sites, thereby improving therapeutic outcomes. The future development of multifunctional NPs capable of carrying multiple drugs or therapeutic agents will allow for synergistic effects and personalized treatments for glycation-related diseases.
- Enhanced Targeting Strategies: Advancements in targeting strategies, such as surface modifications with specific ligands or the use of nanoscale devices for active targeting, will facilitate improved site-specific accumulation of antiglycation NPs. This will enhance treatment efficacy while minimizing off-target effects.
- Combination Therapies: Combining antiglycation NPs with other therapeutic approaches, such as antioxidants or enzyme inhibitors, can potentially create synergistic effects and overcome drug resistance. The future of antiglycation therapy lies in exploring and optimizing such combination therapies for better disease management.
- Biocompatible Formulations: Further research is needed to develop biocompatible and biodegradable formulations for antiglycation NPs. This will ensure their safe administration, minimize potential adverse effects, and facilitate efficient clearance from the body after treatment.

21. Conclusion

Antiglycation NPs offer a promising avenue for the treatment of glycation-related diseases. While several challenges exist in their design, synthesis, targeting, and safety, ongoing research and technological advancements will likely overcome these hurdles. Future perspectives include improved drug delivery, enhanced targeting strategies, combination therapies, and the development of biocompatible formulations. By addressing these challenges and capitalizing on future opportunities, antiglycation NPs hold tremendous potential in revolutionizing the field of nanomedicine and improving the outcomes for patients affected by glycation-related diseases. Conducting comprehensive laboratory investigations that focus on the pharmacodynamics and pharmacokinetic properties of these NPs, particularly in relation to their antiglycation potential, would greatly benefit patients. Such studies would provide valuable insights into the efficacy of NPs as a treatment modality for glycation-associated diseases by effectively controlling protein glycation.

References

Abbas, G., Al-Harrasi, A. S., Hussain, H., Hussain, J., Rashid, R., & Choudhary, M. I. (2016). Antiglycation therapy: Discovery of promising antiglycation agents for the management of diabetic complications. *Pharmaceutical Biology, 54*(2), 198–206.

Afify, M., Samy, N., Hafez, N., Alazzouni, A., Mahdy, E., Sayed, E. L., ... Kelany, M. (2019). Evaluation of zinc-oxide nanoparticles effect on treatment of diabetes in streptozotocin-induced diabetic rats. *Egyptian Journal of Chemistry. 62*(10), 1771–1783.

Alenazi, F., Saleem, M., Khaja, A. S., Zafar, M., Alharbi, M. S., Hagbani, T. A., ... Ahmad, S. (2022). Metformin encapsulated gold nanoparticles (MTF-GNPs): A promising antiglycation agent. *Cell Biochemistry and Function, 40*(7), 729–741.

Alkhalaf, M. I., Hussein, R. H., & Hamza, A. (2020). Green synthesis of silver nanoparticles by *Nigella sativa* extract alleviates diabetic neuropathy through anti- inflammatory and antioxidant effects. *Saudi Journal of Biological Sciences. 27*(9), 2410–2419.

Alomari, G., Salehhuddin, H., & Bahaa, T. (2021). Gold nanoparticles as a promising treatment for diabetes and its complications: Current and future potentials. *Brazilian Journal of Pharmaceutical Sciences. 57*, e19040.

Anandan, S., Mahadevamurthy, M., Ansari, M. A., Alzohairy, M., Alomary, M. N., FarhaSiraj, S., ... Urooj, A. (2019). Biosynthesized ZnO-NPs from *Morus indica* attenuates methylglyoxal-induced protein glycation and RBC damage: In-vitro, in-vivo and molecular docking study. *Biomolecules, 9*(12), 882.

Ashraf, J. M., Ansari, M. A., Choi, I., Khan, H. M., & Alzohairy, M. A. (2014). Antiglycating potential of gum arabic capped-silver nanoparticles. *Applied Biochemistry and Biotechnology, 174*(1), 398–410.

Ashraf, J. M., Ansari, M. A., Fatma, S., Abdullah, S. M., Iqbal, J., Madkhali, A., ... Ashraf, G. M. (2018). Inhibiting effect of zinc oxide nanoparticles on advanced glycation products and oxidative modifications: A potential tool to counteract oxidative stress in neurodegenerative diseases. *Molecular Neurobiology, 55*(9), 7438–7452.

Ashraf, J. M., Ansari, M. A., Khan, H. M., Alzohairy, M. A., & Choi, I. (2016). Green synthesis of silver nanoparticles and characterization of their inhibitory effects on AGEs formation using biophysical techniques. *Scientific Reports, 6*, 20414.

Azharuddin, M., Dasgupta, A., & Datta, H. (2015). Gold nanoparticle conjugated with curcumin and curcumin nanoparticles as a possible nano-therapeutic drug in cataract. *Current Eye Research, 2*(2), 71–73.

Bao, Z., Li, L., Geng, Y., Yan, J., Dai, Z., Shao, C., & Wang, Z. (2020). Advanced glycation end products induce vascular smooth muscle cell-derived foam cell formation and transdifferentiate to a macrophage-like state. *Mediators of Inflammation. 2020*, 6850187.

Bjørklund, G., Dadar, M., Pivina, L., Doşa, M. D., Semenova, Y., & Aaseth, J. (2020). The role of zinc and copper in insulin resistance and diabetes mellitus. *Current Medicinal Chemistry, 27*(39), 6643–6657.

Caruso, G., Di Pietro, L., Cardaci, V., Maugeri, S., & Caraci, F. (2023). The therapeutic potential of carnosine: Focus on cellular and molecular mechanisms. *Current Research in Pharmacology and Drug Discovery, 4*, 100153.

Dawood, M., Younus, Z. M., Alnori, M., & Mahmood, S. (2022). The biological role of advanced glycation end products in the development and progression of colorectal cancer. *Open Access Macedonian Journal of Medical Sciences, 10*(F), 487–494.

D'cunha, N. M., Sergi, D., Lane, M. M., Naumovski, N., Gamage, E., Rajendran, A., & Travica, N. (2022). The effects of dietary advanced glycation end-products on neurocognitive and mental disorders. *The Journal of Nutrition, 14*(12), 2421.

Fan, X., & Monnier, V. M. (2021). Protein posttranslational modification (PTM) by glycation Role in lens aging and age-related cataractogenesis. *Experimental Eye Research, 210*, 108705.

Favre, L. C., López-Fernández, M. P., dos Santos Ferreira, C., Mazzobre, M. F., Mshicileli, N., van Wyk, J., & del Pilar Buera, M. (2022). The antioxidant and antiglycation activities of selected spices and other edible plant materials and their decay in sugar-protein systems under thermal stress. *Food Chemistry, 371*, 131199.

Fournet, M., Bonté, F., & Desmoulière, A. (2018). Glycation damage: A possible hub for major pathophysiological disorders and aging. *Aging and Disease. 9*(5), 880.

Garcia Campoy, A. H., Perez Gutierrez, R. M., Manriquez-Alvirde, G., & Muñiz, R. A. (2018). Protection of silver nanoparticles using *Eysenhardtia polystachya* in peroxide-induced pancreatic β-Cell damage and their antidiabetic properties in zebrafish. *International Journal of Nanomedicine, 13*, 2601–2612.

Ge, W., Jie, J., Yao, J., Li, W., Cheng, Y., & Lu, W. (2022). Advanced glycation end products promote osteoporosis by inducing ferroptosis in osteoblasts. *Molecular Medicine Reports. 25*(4), 1–9.

Islam, F., Khadija, J. F., Islam, M. R., Shohag, S., Mitra, S., Alghamdi, S., & Akter, A. (2022). Investigating polyphenol nanoformulations for therapeutic targets against diabetes mellitus. *Evidence-Based Complementary and Alternative Medicine.*

Jeong, S. R., & Lee, K. W. (2021). Methylglyoxal-derived advanced glycation end product (AGE4)-induced apoptosis leads to mitochondrial dysfunction and endoplasmic reticulum stress through the RAGE/JNK pathway in kidney cells. *International Journal of Molecular Sciences, 22*(12), 6530.

Ko, S. Y., Ko, H. A., Chu, K. H., Shieh, T. M., Chi, T. C., Chen, H. I., & Chang, S. S. (2015). The possible mechanism of advanced glycation end products (AGEs) for Alzheimer's disease. *PLoS One, 10*(11), e0143345.

Kumar, D., Bhatkalkar, S. G., Sachar, S., & Ali, A. (2020). Studies on the antiglycating potential of zinc oxide nanoparticle and its interaction with BSA. *Journal of Biomolecular Structure and Dynamics. 39*(18), 6918–6925.

Lehman, T. D., & Beryl, J. O. (2001). Inhibitors of advanced glycation end product-associated protein cross-linking. *Biochimica et Biophysica Acta. Molecular Basis of Disease, 1535*(2), 110–119.

Lima, M., & Baynes, J.W. (2013). Encycl. Biol. Chem (Second Ed., pp. 405–411). Academic Press. https://doi.org/10.1016/B978-0-12-378630-2.00120-1.

Maillard, L. C. (1912). Action des acides amines sur les sucres: Formation des melanoidines par voie methodique. *CR Seances Soc Biol Fil, 154*, 66–68.

Moldogazieva, N. T., Mokhosoev, I. M., Mel'nikova, T. I., Porozov, Y. B., & Terentiev, A. A. (2019). Oxidative stress and advanced lipoxidation and glycation end products (ALEs and AGEs) in aging and age-related diseases. *Oxidative Medicine and Cellular Longevity* p. 3085756.

Nakamura, A., & Kawaharada, R. (2021). Advanced glycation end products and oxidative stress in a hyperglycaemic environment. In *Fundamentals of Glycosylation*. London, UK: IntechOpen.

Nayak, A., Kumari, D., Nayak, B., & Nayak, B. (2016). Ameliorating effects of green synthesized silver nanoparticles on glycated end product induced reactive oxygen species production and cellular toxicity in osteogenic Saos-2 cells. *ACS Appl Mater Interfaces*, 8(44), 30005–30016.

Odjakova, M., Popova, E., Al Sharif, M., & Mironova, R. (2012). Plant-derived agents with anti-glycation activity. *Glycosylation*, 10, 48186.

Oshitari, T. (2023). Advanced glycation end-products and diabetic neuropathy of the retina. *International Journal of Molecular Sciences*. 24(3), 2927.

Ott, C., Jacobs, K., Haucke, E., Santos, A. N., Grune, T., & Simm, A. (2014). Role of advanced glycation end products in cellular signaling. *Redox Biology*. 2, 411–429.

Parwani, K., & Mandal, P. (2023). Role of advanced glycation end products and insulin resistance in diabetic nephropathy. *Archives of Physiology and Biochemistry*, 129(1), 95–107.

Patil, N., Anushka, K., & Sivaram, A. (2022). Prevention of protein glycation by nanoparticles: Potential applications in T2DM and associated neurodegenerative diseases. *Journal of BionanoScience,,* 12(2), 607–619.

Purushothaman, A., Mohajeri, M., & Lele, T. P. (2023). The role of glycans in the mechanobiology of cancer. *The Journal of Biological Chemistry*, 299, 3.

Rabbani, N., Ashour, A., & Thornalley, P. J. (2016). Mass spectrometric determination of early and advanced glycation in biology. *Glycoconjugate Journal*, 33, 553–568.

Rabbani, N., & Paul, J. T. (2021). Protein glycation–biomarkers of metabolic dysfunction and early-stage decline in health in the era of precision medicine. *Redox Biology*, 42, 101920.

Rashid, R., Naqash, A., Bader, G. N., & Sheikh, F. A. (2020). *Nanotechnology and diabetes management: Recent advances and future perspectives. Application of nanotechnology in Biomedical Sciences*, 99–117.

Savateev, K. V., Spasov, A., & Rusinov, V. L. (2022). Small synthetic molecules with antiglycation activity. Structure–activity relationship. *Russian Chemical Reviews*. 91(6), RCR5041.

Sengani, M., & Rajeswari, D. (2017). Gold nanosupplement in selective inhibition of methylglyoxal and key enzymes linked to diabetes. *IET Nanobiotechnology*. 11(7), 861–865.

Sharma, A., Weber, D., Raupbach, J., Dakal, T. C., Fließbach, K., Ramirez, A., & Wüllner, U. (2020). Advanced glycation end products and protein carbonyl levels in plasma reveal sex-specific differences in Parkinson's and Alzheimer's disease. *Redox Biology*. 34, 101546.

Sharma, V., Gautam, D. N. S., Radu, A. F., Behl, T., Bungau, S. G., & Vesa, C. M. (2022). Reviewing the Traditional/Modern Uses, Phytochemistry, Essential Oils/Extracts and Pharmacology of Embelia ribes Burm. *Antioxidants,* 11(7), 1359.

Sheikpranbabu, S., Kalishwaralal, K., Lee, K. J., Vaidyanathan, R., Eom, S. H., & Gurunathan, S. (2010). The inhibition of advanced glycation end-products-induced retinal vascular permeability by silver nanoparticles. *Biomaterials*. 31(8), 2260–2271.

Shuck, S. C., Wuenschell, G. E., & Termini, J. S. (2018). Product studies and mechanistic analysis of the reaction of methylglyoxal with deoxyguanosine. *Chemical Research in Toxicology*, 31, 105–115.

Singh, V. P., Bali, A., Singh, N., & Jaggi, A. S. (2014). Advanced glycation end products and diabetic complications. *The Korean Journal of Physiology & Pharmacology. 18*(1), 1–14.

Tariq, M., & Ali, A. (2023). *Plant-derived natural products as antiglycating agents. Biologically active small molecules.* Apple Academic Press, 17–44.

Thornalley, P. J. (2008). Protein and nucleotide damage by glyoxal and methylglyoxal in physiological systems-role in ageing and disease. *Drug Metabolism and Drug Interactions, 23*(1–2), 125–150.

Tian, Z., Chen, S., Shi, Y., Wang, P., Wu, Y., & Li, G. (2023). Dietary advanced glycation end products (dAGEs): An insight between modern diet and health. *Food Chemistry, 415*, 135735.

Tripathi, D., Oldenburg, D. J., & Bendich, A. J. (2023). Oxidative and glycation damage to mitochondrial DNA and plastid DNA during plant development. *Antioxidants, 12*(4), 891.

Wang, H. F., Wang, Y. X., Zhou, Y. P., Wei, Y. P., Yan, Y., Zhang, Z. J., & Jing, Z. C. (2023). Protein O-GlcNAcylation in cardiovascular diseases. *Acta Pharmacologica Sinica, 44*(1), 8–18.

Wang, Y., Wang, H., Khan, M. S., Husain, F. M., Ahmad, S., & Bian, L. (2021). Bioconjugation of gold nanoparticles with aminoguanidine as a potential inhibitor of non-enzymatic glycation reaction. *Biomolecular Structure and Dynamics, 39*(6), 2014–2020.

Weng, X., Gong, J., Chen, Y., Wu, T., Wang, F., Yang, S., & He, C. (2020). Keth-seq for transcriptome-wide RNA structure mapping. *Nature Chemical Biology, 16*(5), 489–492.

Woźniak, M., Konopka, C. J., Płoska, A., Hedhli, J., Siekierzycka, A., Banach, M., & Dobrucki, I. T. (2021). Molecularly targeted nanoparticles: An emerging tool for evaluation of expression of the receptor for advanced glycation end products in a murine model of peripheral artery disease. *Cellular & Molecular Biology Letters, 26*, 1–17.

Zawada, A., Machowiak, A., Rychter, A. M., Ratajczak, A. E., Szymczak-Tomczak, A., Dobrowolska, A., & Krela-Kaźmierczak, I. (2022). Accumulation of advanced glycation end-products in the body and dietary habits. *Nutrients, 14*(19), 3982.

Zheng, Q., Maksimovic, I., Upad, A., & David, Y. (2020). Non-enzymatic covalent modifications: A new link between metabolism and epigenetics. *Protein Cell, 11*(6), 401–416.

CHAPTER ELEVEN

Breath of fresh air: Investigating the link between AGEs, sRAGE, and lung diseases

Charlotte Delrue[a], Reinhart Speeckaert[b], Joris R. Delanghe[c], and Marijn M. Speeckaert[a,d,*]

[a]Department of Nephrology, Ghent University Hospital, Ghent, Belgium
[b]Department of Dermatology, Ghent University Hospital, Ghent, Belgium
[c]Department of Diagnostic Sciences, Ghent University, Ghent, Belgium
[d]Research Foundation-Flanders (FWO), Brussels, Belgium
*Corresponding author. e-mail address: Marijn.Speeckaert@ugent.be

Contents

1. Introduction	314
2. Chronic obstructive pulmonary disease	318
2.1 Advanced glycation end products and other RAGE ligands	319
2.2 (Soluble) receptor for advanced glycation end products	322
3. Asthma	327
3.1 Advanced glycation end products and other RAGE ligands	327
3.2 (Soluble) receptor for advanced glycation end products	331
4. Lung fibrosis	334
4.1 Advanced glycation end products and other RAGE ligands	335
4.2 (Soluble) receptor for advanced glycation end products	336
5. Cystic fibrosis	338
5.1 Advanced glycation end products and other RAGE ligands	338
5.2 (Soluble) receptor for advanced glycation end products	338
6. Acute lung injury/acute respiratory distress syndrome	339
6.1 Advanced glycation end products and other RAGE ligands	339
6.2 (Soluble) receptor for advanced glycation end products	341
7. Lung cancer	342
7.1 Advanced glycation end products and other RAGE ligands	343
7.2 (Soluble) receptor for advanced glycation end products	343
8. Unmasking RAGE's complex role in lung cancer progression	343
9. Unlocking the impact of AGER polymorphisms on lung cancer	345
10. sRAGE: Illuminating lung cancer diagnosis and beyond	346
11. Conclusions and future directions	347
References	348

Vitamins and Hormones, Volume 125
ISSN 0083-6729, https://doi.org/10.1016/bs.vh.2024.01.003
Copyright © 2024 Elsevier Inc. All rights are reserved, including those for text and data mining, AI training, and similar technologies.

Abstract

Advanced glycation end products (AGEs) are compounds formed via non-enzymatic reactions between reducing sugars and amino acids or proteins. AGEs can accumulate in various tissues and organs and have been implicated in the development and progression of various diseases, including lung diseases. The receptor of advanced glycation end products (RAGE) is a receptor that can bind to advanced AGEs and induce several cellular processes such as inflammation and oxidative stress. Several studies have shown that both AGEs and RAGE play a role in the pathogenesis of lung diseases, such as chronic obstructive pulmonary disease, asthma, idiopathic pulmonary fibrosis, cystic fibrosis, and acute lung injury. Moreover, the soluble form of the receptor for advanced glycation end products (sRAGE) has demonstrated its ability to function as a decoy receptor, possessing beneficial characteristics such as anti-inflammatory, antioxidant, and anti-fibrotic properties. These qualities make it an encouraging focus for therapeutic intervention in managing pulmonary disorders. This review highlights the current understanding of the roles of AGEs and (s)RAGE in pulmonary diseases and their potential as biomarkers and therapeutic targets for preventing and treating these pathologies.

Abbreviations

ADAM10	ADAM metalloproteinase domain-containing protein 10.
AGEs	advanced glycation end products.
AHR	airway hyperresponsiveness.
ALI	acute lung injury.
ALK	anaplastic lymphoma kinase.
AM	alveolar macrophage.
ARDS	acute respiratory distress syndrome.
AT2	alveolar type 2.
BAL	bronchoalveolar lavage.
CC16	club cell secretory protein.
CEL	carboxyethyl-lysine.
CF	cystic fibrosis.
CFRD	cystic fibrosis-related diabetes.
CML	carboxymethyl-lysine.
COPD	chronic obstructive pulmonary disease.
CR	cockroach.
CRP	C-reactive protein.
cRAGE	cleaved form of the receptor for advanced glycation end products.
CSE	cigarette smoke extract.
CTGF	connective tissue growth factor.
DLCO	carbon monoxide diffusing capacity.
ECLIPSE	evaluation of COPD to longitudinally identify predictive surrogate endpoints.
ECM	extracellular matrix.
EFF	excessive intake of free fructose.
EMT	epithelial-to-mesenchymal transition.
ERK	extracellular signal-regulated kinases.

esRAGE	endogenous secretory form of the receptor for advanced glycation end products.
FEV1	forced expiratory volume in one second.
FVC	forced vital capacity.
GSH	glutathione.
HDM	house dust mite.
HIF1α	hypoxia inducible factor 1 subunit alpha.
HMGB1	high-motility group protein B1.
HSP47	heat shock protein 47.
ICAM-1	intercellular adhesion molecule 1.
IFN-γ	interferon-gamma.
IL	interleukin.
IL-1RA	interleukin-1 receptor antagonist.
ILC2s	group 2 innate lymphoid cells.
ILDs	interstitial lung diseases.
IPF	idiopathic pulmonary fibrosis.
JNK	c-Jun-N-terminal kinase.
KO	knockout.
LOX-1	lectin-like oxidized low-density lipoprotein receptor-1.
LTC4	leukotriene C4.
MAP	mitogen-activated protein.
MAPKs	mitogen-activated protein kinases.
MC	mast cell.
MCP-1	monocyte chemoattractant protein-1.
MIP-2	macrophage inflammatory protein-2.
NO	nitric oxide.
PI3K	phosphatidyl-inositol 3 kinase.
MMP	matrix metalloproteinases.
mRANGE	membrane-bound form of the receptor for advanced glycation end products.
NADPH	nicotinamide adenine dinucleotide phosphate.
NSCLC	non-small-cell lung cancer.
NF-κB	nuclear factor kappa-light-chain-enhancer of activated B cells.
PKC	protein kinase C.
RAGE	receptor of advanced glycation end products.
ROC	Receiver Operating Characteristic.
ROS	reactive oxygen species.
RT-PCR	reverse transcription-polymerase chain reaction.
SAPK	stress-activated protein kinase.
siRAGE	siRNA targeting the receptor for advanced glycation end products.
SNP	single nucleotide polymorphism.
SOD	superoxide dismutase.
SP-D	surfactant protein D.
sRAGE	soluble form of the receptor for advanced glycation end products.
STAT3	signal transducer and activator of transcription 3.
TESRA	treatment of emphysema with a selective retinoid agonist.
TGF-β1	transforming growth factor-beta 1.
Th	T-helper.

TLR4	Toll-like receptor 4.
TNF-α	tumor necrosis factor-alpha.
TNFR2	tumor necrosis factor receptor 2.
VCAM-1	vascular cell adhesion protein 1.
WHO	World Health Organization.
WTC-OAD	World Trade Center-occupational airways disease (WTC-OAD).
WTC-PM	World Trade Center-particulate matter.

1. Introduction

Advanced glycation end products (AGEs), first described by Louis-Camille Maillard in 1912, are the end products of irreversible molecular adducts formed through non-enzymatic reactions between reducing sugars such as fructose and glucose with proteins or lipids (Gkogkolou & Böhm, 2012; Singh, Barden, Mori, & Beilin, 2001). A comprehensive evaluation of the diverse precursors and intricate mechanisms involved in the heterogeneous pathways leading to the chemical synthesis of AGEs has already been published (Ahmed, Thorpe, & Baynes, 1986; Gkogkolou & Böhm, 2012; Henning & Glomb, 2016; Vistoli et al., 2013). Over 20 different types of AGEs have been detected in human tissues and blood, including carboxymethyl-lysine (CML), carboxyethyl-lysine, pyrraline, pentosidine and methylglyoxal-lysine dimer. Despite having varying chemical structures, these AGEs share the common feature of containing a lysine residue within their molecular makeup (Perrone, Giovino, Benny, & Martinelli, 2020). AGEs damage human tissues through two distinct mechanisms: first, they alter the structures of proteins through chemical cross-linking, and second, they activate various receptors, thereby stimulating several intracellular pathways that lead to an increase in the production of pro-inflammatory cytokines and reactive oxygen species (ROS) (Ott et al., 2014). These include scavenger receptors, such as CD36 (Horiuchi et al., 1996), as well as receptors that are predominantly found on macrophages, such as galectin-3 (Pricci et al., 2000), OST-48 (Y. M. Li et al., 1996), lectin-like oxidized low-density lipoprotein receptor-1 (LOX-1) (X. Chen, Zhang, & Du, 2008), and the receptor of advanced glycation end products (RAGE), which is the most extensively researched and best-known receptor for AGEs (Schmidt et al., 1992). RAGE, a 35 kDa protein belonging to the immunoglobulin superfamily, is encoded by the *AGER* gene located near major histocompatibility complex III on chromosome 6 (Bierhaus et al., 2005; Fleming, Humpert, Nawroth, & Bierhaus, 2011; Gkogkolou & Böhm, 2012). Genetic variations within the *AGER* gene, such as single nucleotide polymorphisms (SNPs) and mutations, can significantly influence its expression and function, ultimately affecting the progression of lung disease. One key aspect of RAGE's role of RAGE in lung diseases is its involvement in the regulation of inflammation. When RAGE binds to ligands such as AGEs or high-mobility group box 1 (HMGB1), it activates pro-inflammatory signaling pathways, including the nuclear factor kappa-light-chain-enhancer of activated B cells (NF-κB). This activation leads to the production of proinflammatory cytokines such as tumor necrosis factor-alpha (TNF-α), interleukin (IL)-1β, IL-8, interferon-γ, and monocyte chemoattractant protein-1 (MCP-1) (Wang et al., 2019). Excessive inflammation is a common feature in conditions such as acute respiratory distress syndrome (ARDS) and asthma, and genetic variants that enhance RAGE-mediated inflammation can worsen disease severity (Jones et al., 2020; Niu et al., 2019). RAGE signaling is also associated with lung tissue damage. When activated, RAGE can induce oxidative stress and stimulate the release of proinflammatory molecules, leading to the breakdown of lung tissue and impairment of lung function (Sanders et al., 2019). Genetic factors that increase RAGE expression or ligand-binding can amplify these damaging effects. In diseases characterized by fibrosis, such as idiopathic pulmonary fibrosis (IPF), RAGE activation plays a critical role in promoting fibroblast differentiation and collagen production, thereby contributing to the fibrotic process (Perkins & Oury, 2021; Yamaguchi et al., 2022). Genetic variations that enhance RAGE-mediated fibrotic pathways may accelerate IPF progression (Kinjo et al., 2020). Furthermore, RAGE signaling can lead to the generation of ROS within lung tissue, contributing to oxidative stress. Oxidative stress can damage lung cells and exacerbate conditions, such as chronic obstructive pulmonary disease (COPD) (Taniguchi, Tsuge, Miyahara, & Tsukahara, 2021) or pulmonary hypertension (Prasad, 2019). Genetic factors that promote RAGE-mediated oxidative stress may aggravate oxidative damage in lungs (Malik, Hoidal, & Mukherjee, 2021). Understanding the intricate relationship between *AGER* genetics and lung disease is crucial for tailoring treatments and improving patient outcomes. Genetic profiling of individuals with lung diseases can potentially identify those who are more likely to respond to therapies targeting RAGE or its downstream pathways, such as RAGE-blocking drugs, sRAGE, or small-molecule inhibitors. Ongoing research is essential to identify specific genetic markers and their functional consequences in various lung diseases, enabling the development of more precise and effective therapeutic interventions.

RAGE can be found in the body in two forms: membrane-bound RAGE (mRAGE) and soluble RAGE (sRAGE). Three domains comprise membrane-bound RAGE: an extracellular domain that identifies and attaches RAGE ligands, a hydrophobic transmembrane domain, and a charged cytoplasmic domain that participates in intracellular signaling (Oczypok, Perkins, & Oury, 2017). RAGE is strongly expressed in many tissues of the developing embryo, but as the organism matures, it loses its expression in all tissues but the lung (Bierhaus et al., 2005). mRAGE functions as a pattern recognition receptor that can bind to various molecules, including AGEs, S-100/calgranulins, HMGB1, β-amyloid peptides, and β-sheet fibrils (Bierhaus et al., 2005; Fleming et al., 2011; Gkogkolou & Böhm, 2012). Increasing evidence suggests that RAGE activation triggers inflammatory reactions and oxidative stress (Lubitz et al., 2016; Sharma, Kaur, Sarkar, Sarin, & Changotra, 2021). When ligands bind to RAGE, they trigger several signaling pathways, such as mitogen-activated protein kinases (MAPKs), extracellular signal-regulated kinases (ERK) 1 and 2, phosphatidyl-inositol 3 kinase (PI3K), p21Ras, stress-activated protein kinase/c-Jun-N-terminal kinase (SAPK/JNK), and Janus kinase (JAK). This stimulation of RAGE leads to the activation of NF-κB and subsequent transcription of multiple pro-inflammatory genes (Bierhaus et al., 2005; Fleming et al., 2011; Gkogkolou & Böhm, 2012). In addition to its pro-inflammatory character, RAGE activation can also induce oxidative stress through both direct and indirect mechanisms. It can directly activate nicotinamide adenine dinucleotide phosphate (NADPH) oxidase and lower the activity of superoxide dismutase (SOD), catalase, and other pathways. RAGE activation can directly diminish cellular antioxidant defenses such as glutathione (GSH) and ascorbic acid (Bierhaus et al., 2005; Gkogkolou & Böhm, 2012; Loughlin & Artlett, 2010; Ramasamy et al., 2005). sRAGE is incapable of triggering cellular signaling owing to the absence of intracellular domains. sRAGE exists in two distinct isoforms, namely cleaved RAGE (cRAGE) and endogenous secretory RAGE (esRAGE), although the mechanisms regulating these isoforms remain unclear. cRAGE production is mediated by proteases, including matrix metalloproteinases (MMP), disintegrin, and ADAM metalloproteinase domain-containing protein 10 (ADAM10), which act on the cell surface. The *RAGE* gene uses alternative splicing to generate the second isoform, esRAGE (Libby, Ridker, & Hansson, 2011; Park et al., 2004; Pinto, Minanni, de Araújo Lira, & Passarelli, 2022; Yonekura et al., 2003).

Chronic pulmonary diseases, such as COPD, asthma, and pulmonary fibrosis, are the third leading cause of death worldwide and are increasing in prevalence over time (Somayaji & Chalmers, 2022). According to the World Health Organization (WHO), chronic respiratory disease causes 4.6 million premature deaths annually, accounting for over 5% of all global fatalities. Notably, almost 90% of these fatalities occur in low-income and middle-income nations (Byrne, Marais, Mitnick, Lecca, & Marks, 2015; Racanelli, Kikkers, Choi, & Cloonan, 2018). The pathogenesis of chronic pulmonary illnesses is characterized by persistent respiratory tract inflammation. Genetic predisposition and environmental factors, such as exposure to bacteria, atmospheric particles, irritants, pollutants, allergens, and toxic molecules, may contribute to persistent lung inflammation (Racanelli et al., 2018). RAGE is a pro-inflammatory mediator, and numerous studies have identified RAGE as a key component of numerous pulmonary illnesses, including lung cancer, COPD, asthma, pulmonary fibrosis, cystic fibrosis (CF), and acute lung injury (Oczypok et al., 2017). Lung diseases are characterized by matrix remodeling, such as elastosis and fibrosis (Ito et al., 2019), in which RAGE plays a significant role (Oczypok et al., 2017). Elastosis, characterized by the abnormal accumulation of elastic fibers in the lung tissue, is one of the processes influenced by RAGE. RAGE activation triggers signaling pathways that lead to the production and activation of MMPs (Hergrueter, Nguyen, & Owen, 2011). These MMPs, notably elastase-type MMPs such as MMP-2 and MMP-9, are enzymes responsible for degrading elastin, a crucial component of elastic fibers within the lung (Corbel, Belleguic, Boichot, & Lagente, 2002). Consequently, MMP-mediated elastin degradation contributes to the disruption of the elastic recoil properties of the lungs, resulting in reduced lung compliance and impaired respiratory function. On the other hand, fibrosis, a hallmark of several lung diseases, involves the excessive deposition of collagen and other extracellular matrix components in lung tissue (Todd, Luzina, & Atamas, 2012). RAGE signaling has also been implicated in this process (Perkins & Oury, 2021). RAGE activation stimulates fibroblast activation, prompting these cells to transition into myofibroblasts, which possess enhanced contractile properties and increased capacity for collagen production. RAGE-mediated signaling activates intracellular pathways, including the MAPK and NF-κB pathways, which when triggered by RAGE, promote the transcriptional upregulation of genes associated with collagen synthesis (Thakur et al., 2022; Tian et al., 2019). This results in the accumulation of collagen, which contributes to the stiffening of lung tissue and disruption of normal lung architecture. These fibrotic changes lead to a decreased oxygen exchange and impaired lung function. This review highlights the current understanding of the role of AGEs and (s)RAGE in pulmonary diseases and its potential as a biomarker and therapeutic target for preventing and treating these diseases.

2. Chronic obstructive pulmonary disease

COPD is a worldwide health burden that affects 10% of the global population, leading to 3 million deaths and $44 billion in healthcare expenses annually (Zemans et al., 2017). It is an avoidable and treatable illness characterized by persistent respiratory symptoms and airflow restriction caused by abnormalities in the airways and/or alveoli (Barnes et al., 2015; Sharma et al., 2021; Vogelmeier et al., 2017). Chronic inflammation, which affects both the central and peripheral airways, lung parenchyma, and alveoli, as well as the pulmonary vasculature, is the defining characteristic of COPD. The cycle of repeated injury and subsequent repair results in alterations in the structure and function of affected areas (Barnes et al., 2015; Sharma et al., 2021; Vogelmeier et al., 2017). The most important cells contributing to this inflammatory state are neutrophils, which are also thought to be markers of severity in patients with COPD (Sharma et al., 2021; Wang Wang et al., 2018). Furthermore, destruction of the lung parenchyma (emphysema) may be a factor in the hallmark of mucociliary malfunction and airflow restriction (Barnes et al., 2015; Sharma et al., 2021; Vogelmeier et al., 2017). The most frequent causes of COPD are smoking, exposure to cigarette smoke from the surroundings, and a patient's history of tuberculosis (Barnes et al., 2015; Sharma et al., 2021). Other characteristics of COPD include comorbid conditions, such as cardiovascular diseases, osteoporosis, skeletal muscle atrophy, anemia, and mental symptoms. Comorbidities must be diagnosed and treated to effectively manage the illness because they lower the quality of life, which increases the likelihood of hospitalization and death (Huber, Wacker, Vogelmeier, & Leidl, 2015; Sharma et al., 2021). Various subtypes and inconsistent disease progression define the heterogeneous characteristics of COPD. This heterogeneity extends to a molecular level. Therefore, biomarkers for measuring disease heterogeneity and forecasting development are becoming increasingly popular (Zemans et al., 2017). Currently, a vast amount of literature is available on the different biomarkers used for diagnosing COPD, including blood and sputum biomarkers [e.g. fibrinogen, C-reactive protein (CRP), surfactant protein D (SP-D), and club cell secretory protein (CC16)]. However, each of these biomarkers has certain limitations (Sharma et al., 2021; Zemans et al., 2017). Extensive research has established the pivotal role of the AGE-RAGE axis in various signaling networks implicated in inflammatory diseases like COPD (Fig. 1) (Sharma et al., 2021). Building on this understanding, the subsequent section delves into the significance of Advanced glycation end products and other RAGE ligands in COPD, as well as the role of (s)RAGE in this complex pulmonary condition.

2.1 Advanced glycation end products and other RAGE ligands

Multiple studies have indicated that in individuals with COPD, AGEs tend to accumulate in various compartments of the human body (Caram et al., 2017; Hoonhorst et al., 2016; Wu, Ma, Nicholson, & Black, 2011). Besides the small airways and lung tissue (L. Wu et al., 2011), COPD patients have higher levels of AGEs in their epidermis than healthy smokers and non-smokers (Gopal, Reynaert, Scheijen, Engelen, et al., 2014; Hoonhorst et al., 2016). Also the serum levels of AGEs are significantly elevated in patients with COPD compared with controls (Caram et al., 2017). AGEs attached to RAGE may promote inflammation by activating NF-κB. Therefore, it can be assumed that AGEs may contribute to the development of COPD through inflammation (Sharma et al., 2021; Wu et al., 2011).

The role of diet and obesity in the emergence of obstructive airway disease is significant (Hanson, Rutten, Wouters, & Rennard, 2014; Napier et al., 2017; Wood, 2017). Diets that prioritize the reduction of inflammation and incorporate a higher intake of vegetables and fish have been found to lower the risk of COPD. Conversely, diets that contain elevated levels of pro-inflammatory AGEs have been associated with an increased likelihood of developing the disease (DeChristopher, Uribarri, & Tucker, 2015; Guo et al., 2012; Scoditti, Massaro, Garbarino, & Toraldo, 2019; Zheng et al., 2016). Experiments using in vitro and in vivo laboratory models have indicated that RAGE is linked to lung impairment following exposure to World Trade Center-particulate matter (WTC-PM). Notably, mice lacking RAGE and exposed to WTC-PM demonstrated a safeguarding effect against WTC-occupational airways disease (WTC-OAD) (Caraher et al., 2017; Haider et al., 2020; Veerappan et al., 2020). Both dietary and naturally occurring AGEs can influence the signaling pathways involved in inflammatory conditions (Aragno & Mastrocola, 2017). Another human study that involved observing and tracking dietary patterns found that individuals who consumed sugary drinks and processed meats more frequently while reducing their intake of vegetables and whole grains were also at a higher risk of developing WTC-OAD. Additionally, subjects who followed diets rich in AGEs were significantly more prone to developing WTC-OAD (Lam et al., 2021). These findings align with those of other

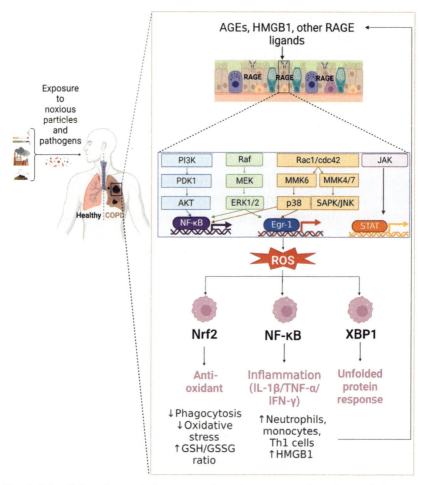

Fig. 1 Role of the advanced glycation end product-receptor for advanced glycation end products (AGE-RAGE) axis in chronic obstructive pulmonary disease (COPD) pathophysiology. Cigarette smoke contains RAGE ligands attached to RAGE found in the alveolar epithelium. This interaction triggers oxidative stress via distinct signaling pathways, such as the Raf- extracellular signal-regulated kinases (ERK)1/2, stress-activated protein kinase/c-Jun-N-terminal kinase (SAPK/JNK) p38 mitogen-activated protein kinases (MAPK), phosphatidyl-inositol 3 kinase (PI3K)/Akt, and Janus kinase (JAK) pathways. Consequently, several transcription factors such as nuclear factor kappa-light-chain-enhancer of activated B cells (NF-κB), signal transducer and activator of transcription 3 (STAT3), and early growth response protein 1 (Egr-1) are activated, stimulating oxidative stress. Macrophages play a crucial role in COPD inflammation as they release chemokines that attract neutrophils, monocytes, and T cells. This subsequently induces compensatory reactions such as the activation of Nrf2 and XBP1.

studies that have demonstrated the detrimental impact of processed meat on the elevated risk of developing COPD. This association is attributed to the high concentration of pro-inflammatory AGEs found in processed meats (Salari-Moghaddam, Milajerdi, Larijani, & Esmaillzadeh, 2019; Varraso & Camargo, 2015; Varraso et al., 2019; Vlassara & Uribarri, 2014). Despite evidence highlighting the risks associated with diet and obesity, some studies have suggested that obesity might have a protective effect on survival and lung function in individuals with COPD (Guo Jiang et al., 2016; Guo Zhang et al., 2016; Spelta, Fratta Pasini, Cazzoletti, & Ferrari, 2018).

According to several studies (Ferhani et al., 2010; Hou et al., 2011; Oczypok et al., 2017; Reynolds et al., 2008; Sharma et al., 2021; Waseda et al., 2015), other RAGE ligand levels (HMGB1, S100A4, and amyloid A) are increased in the lungs, mediating the inflammatory response. HMGB1 appears to be related to the development of pulmonary emphysema (Waseda et al., 2015). Active secretion of pro-inflammatory cytokines induces the expression of HMGB1 in monocytes/macrophages as well as in epithelial and mesenchymal cells (Barnes, 2004; Liu et al., 2006; Ogawa et al., 2006). Compared to never smokers or smokers without COPD, smokers with COPD have higher bronchoalveolar lavage (BAL) levels of tumor necrosis factor receptor 2 (TNFR2) and IL-1β and lower levels of IL-1 receptor antagonist protein (IL-1RA). The idea that these mediators control the formation of HMGB1 in the airways of COPD patients is supported by the BAL levels of HMGB1, which were positively correlated with those of IL-1β and TNFR2, and negatively correlated with IL-1RA. Notably, HMGB1 induces the synthesis of IL-1β and TNF-α in immunological and inflammatory cells, indicating a reciprocal regulation of the synthesis of these mediator (Ferhani et al., 2010). Furthermore, it has been demonstrated that HMGB1 promotes stromal fibroblast proliferation, chemotaxis, and activation of metalloproteinases (Palumbo et al., 2004; Rossini et al., 2008). Overall, these results may contribute to the significant histopathological aspects of tissue remodeling that are linked to airflow obstruction and emphysema, in addition to inflammation. These findings imply an autocrine/paracrine loop in which binding of HMGB1 to IL-1β, and probably to RAGE and/or other HMGB1 receptors, enhances inflammatory and remodeling signals in the pathogenesis of COPD. Targeting the interaction between HMGB1 and its partners, which is prevalent in COPD airways, may therefore be considered a unique therapeutic approach for enhancing clinical outcomes and slowing the course of the illness (Song et al., 2011).

Amyloid A, another RAGE ligand, was increased in patients with stable COPD, but plasma enRAGE (S100A12) levels in patients with COPD were similar to those in healthy controls (Barberà, Peinado, & Santos, 2003; Weitzenblum et al., 1981). However, the S100 protein family appears to be associated with pulmonary hypertension, the most common complication of COPD. The causes of pulmonary hypertension in COPD are unknown. Remodeling of the pulmonary arteries is one of the primary pathological alterations. In COPD patients with modified intrapulmonary arteries, even in the non-hypoxic COPD-stage there was increased expression of S100A4. Therefore, the inhibition of vascular remodeling in COPD patients may be achieved by targeting S100A4 (Hou et al., 2011). These findings suggest that distinct ligands may be responsible for RAGE activation in various disease conditions (Miniati et al., 2011; Reynaert, Gopal, Rutten, Wouters, & Schalkwijk, 2016; Smith et al., 2011).

2.2 (Soluble) receptor for advanced glycation end products

COPD patients seem to have elevated RAGE levels in various parts of the body (sputum, bronchial biopsies, plasma, and skin) (Hoonhorst et al., 2016). RAGE overexpression in the airways of COPD smokers results in higher levels of nitric oxide (NO), higher NO synthase activity, and lower levels of total GSH, all of which enhance NF-κB activation (L. Chen et al., 2014). The presence of tobacco smoke, a well-known risk factor, has been shown to increase RAGE expression in the pulmonary epithelium (Reynolds et al., 2008; Reynolds, Kasteler, Schmitt, & Hoidal, 2011). Tobacco smoke exposure stimulates RAGE expression, leading to the activation of various cellular pathways. These pathways include PI3K/Akt, Raf-ERK1/2, SAPK/JNK, and p38 pathways. Consequently, several transcription factors such as NF-κB, Egr-1, and signal transducer and activator of transcription 3 (STAT3) are activated, resulting in a state of oxidative stress (Hoonhorst et al., 2016; Reynolds et al., 2008; Sukkar et al., 2012). Macrophages play a crucial role in inflammation in COPD. They release the chemokines IL-1β, TNF-α, and IFN-γ, which attract neutrophils, monocytes, and T cells. This subsequently induces compensatory reactions, such as the activation of *Nrf2* and *XBP1* genes (Ferhani et al., 2010; Lee, 2017; Reynaert et al., 2016; Reynolds et al., 2008; Sanders et al., 2019).

The extremely polymorphic *RAGE* gene, located at 6p.21.3, encodes RAGE and has been associated with the pathogenesis of COPD (Li et al., 2014; Miller et al., 2016; Sharma et al., 2021). The main inducers of

RAGE expression are oxidative and inflammatory stress. Compared to other tissues, lung tissue has a high level of *RAGE* expression, which is increased in cases of COPD owing to the buildup of RAGE proteins (Fehrenbach et al., 1998; Gaens et al., 2009). Multiple animal studies (Reynolds et al., 2011; Sambamurthy, Leme, Oury, & Shapiro, 2015; Sanders et al., 2019; Wolf, Herr, Niederstraßer, Beisswenger, & Bals, 2017) have reported that mice lacking RAGE exhibit decreased levels of lung inflammatory mediators and are protected against cigarette smoke-induced emphysema. *RAGE* gene deletion affects the expression of numerous essential genes implicated in the pathogenesis of COPD-associated inflammation and oxidative stress (Sanders et al., 2019). According to one study, the RAGE-Ras-NF-κB axis probably plays a role in inflammation associated with a number of smoking-related inflammatory lung illnesses. Ras activation is reduced in cells treated with RAGE siRNA (siRAGE). Ras was also dramatically decreased in the lungs of RAGE-deficient mice exposed to long-term tobacco smoke when compared to wild-type mice. Additionally, there was an increased NF-κB activation upon cigarette smoke extract (CSE) stimulation and decreased NF-κB activation in cells transfected with siRAGE prior to CSE exposure using a luciferase reporter incorporating NF-κB binding sites (Reynolds et al., 2011). In another animal study (Robinson, Johnson, Bennion, & Reynolds, 2012), it was observed that when wild-type alveolar macrophages (AMs) were exposed to CSE, the levels of active Ras were elevated. Conversely, AMs lacking RAGE showed reduced Ras activation compared to wild-type AMs following exposure to CSE. Examination of p38 MAPK and NF-κB, which are essential intracellular signaling molecules involved in inflammation, indicated that CSE-induced inflammation in RAGE-deficient AMs may occur, at least in part, through RAGE signaling. Additionally, quantitative reverse transcription-polymerase chain reaction (RT-PCR) analysis revealed noticeable decreases in the expression of pro-inflammatory cytokines, such as TNF-α and IL-1β, in RAGE-deficient AMs exposed to CSE compared to CSE-exposed wild-type AMs.

According to various studies (Cheng et al., 2013; Coxson et al., 2013; Gopal, Reynaert, Scheijen, Schalkwijk, et al., 2014; Miniati et al., 2011; Pouwels et al., 2018; Reynaert et al., 2016; Smith et al., 2011; Sukkar et al., 2012), patients with COPD appear to have lower serum sRAGE levels than control participants, particularly those with severe illness. In contrast, other studies (Boschetto et al., 2013; Iwamoto et al., 2014) found that COPD patients with mild to severe airflow limitation and healthy volunteers did not have significantly different plasma sRAGE levels. These contradictory results may still point to the possibility that sRAGE deficiency may be linked to more severe COPD (Iwamoto et al., 2014). sRAGE may also serve as a marker for emphysema (Cheng et al., 2013; Miniati et al., 2011). Lower sRAGE levels in the treatment of emphysema with a selective retinoid agonist (TESRA) cohort were a biomarker of lung density and carbon monoxide diffusing capacity (DLCO), whereas lower sRAGE levels in the evaluation of COPD to longitudinally identify predictive surrogate endpoints (ECLIPSE) cohort were linked to CT-defined emphysema (global initiative for chronic obstructive lung disease) GOLD stage and COPD state. These correlations held true even after accounting for spirometric and demographic covariates (Cheng et al., 2013). As mentioned above, compared to other types of tissue, the normal adult human lung has a high level of expression of membrane-bound RAGE, particularly in alveolar epithelial cells (Brett et al., 1993; Fehrenbach et al., 1998; Morbini et al., 2006). It is believed that RAGE plays a role in maintaining lung homeostasis by facilitating the binding of type I epithelial cells to the extracellular matrix (Fehrenbach et al., 1998) and contributing to the transformation of type II epithelial cells into type I cells (Demling et al., 2006). This transformation is a crucial step in alveolar repair. The lower levels of sRAGE detected in patients with moderate-to-severe emphysema could be attributed to extensive damage to the alveolar walls and alveoli, which are characteristic of emphysema. Another possibility is that the reduced sRAGE levels in patients with emphysema could be due to exposure to a large amount of RAGE ligands, which might be the result of the release of pro-inflammatory cytokines and inflammatory mediators commonly observed in COPD (Cosio, Saetta, & Agusti, 2009; Miniati et al., 2011). In light of the correlation between circulating sRAGE in patients with COPD and the severity of emphysema, decreased diffusion capacity, and airway neutrophilic inflammation, a supporting function for RAGE in maintaining alveolar integrity and anti-inflammatory qualities of sRAGE have been proposed (Iwamoto et al., 2014).

sRAGE might be utilized to predict COPD exacerbations and determine the phenotype of frequent exacerbations (Miłkowska-Dymanowska et al., 2018; Smith et al., 2011). A cutoff point of 850.407 pg/mL for serum sRAGE levels was found to differentiate frequent and non-frequent exacerbators with a sensitivity of 0.80 [95% confidence interval (CI): 0.28–1.0] and a specificity of 0.93 (95% CI: 0.66–1.0) (Miłkowska-Dymanowska et al., 2018). The most common causes of COPD exacerbation are bacterial or viral infections, which are also linked to pulmonary inflammation, hypoxia, and the release of cytokines and pro-inflammatory mediators. The release of pro-inflammatory mediators may be the primary cause of the low plasma sRAGE concentrations observed in the acute exacerbation of COPD (Pinto-Plata et al., 2007). Another way that acute exacerbation of COPD can cause alterations in plasma sRAGE is by stimulating the production of RAGE ligands (Chang et al., 2008; Smith et al., 2011). COPD is caused by a combination of long-term exposure to toxic chemicals, such as cigarette smoke inhalation, and genetic predisposition (Pouwels et al., 2018). First, results regarding how smoking affects sRAGE levels in the blood remain inconsistent (Caram et al., 2017; Iwamoto et al., 2014; Pouwels et al., 2018; Prasad, 2013; Smith et al., 2011). On one hand, minimal discernible distinctions in serum sRAGE levels have been found between smokers and non-smokers (Caram et al., 2017). Other studies have shown that the levels of sRAGE in smokers with and without COPD were significantly lower than those in non-smokers (Gopal, Reynaert, Scheijen, Schalkwijk, et al., 2014; Iwamoto et al., 2014). Further research is needed before designating sRAGE as a potential biomarker for COPD to establish consistency in serum sRAGE levels as a result of smoking among various groups and to clarify the acute and chronic effects (Pouwels, Klont, Bischoff, & Ten Hacken, 2019). Second, SNPs in *RAGE* are associated with a decreased forced expiratory volume in one second (FEV1)/forced vital capacity (FVC) ratio, which is suggestive of airflow obstruction (Hancock et al., 2010; Repapi et al., 2010). According to a case-control analysis (Li et al., 2014) of 216 patients in the Chinese population to determine the association of three *RAGE* variants, 374T/A, 429T/C, and G82S, only G82S of the RAGE protein product is related to COPD. The GS genotype among smokers serves as a greater risk factor and G82S contributes to the development of COPD. In several research populations (Cheng et al., 2013; Hancock et al., 2010; Hobbs et al., 2017; Miller et al., 2016; Repapi et al., 2010), the *RAGE* variant rs2070600 (Ser82) has a substantial association with COPD. This variant involves a change in the genetic code that leads to a different amino acid (glycine to serine) in the protein region that binds to its ligand. This change is considered functional and affects protein activity. Despite being a functional genetic variant, rs2070600 was not significantly correlated with emphysema or other clinical characteristics related to COPD, except for a minor link with the distance walked in 6 min in the ECLIPSE study. As rs2070600 is a relatively rare genetic variant, its lack of correlation with specific clinical characteristics may be due to limitations in the sample size of the studies (minor allele frequency: 3.9% in TESRA and 3.4% in

ECLIPSE) (Cheng et al., 2013). Soler et al. (Artigas et al., 2011) found a new association between the FEV1/FVC ratio and a different genetic variant, rs2857595, located in the same genomic region as rs2070600. This finding suggests that the genetic influence of this locus on COPD and lung function is complex. Interestingly, the T allele of rs2070600 is linked to a higher FEV1/FVC ratio and a lower risk of developing COPD (Castaldi et al., 2011). Similarly, the same T allele was associated with protection against COPD in smokers defined by prebronchodilator spirometry, including those who were considered normal (Young, Hay, & Hopkins, 2011; Young, Hopkins, et al., 2011). This allele has consistently a negative correlation with sRAGE levels, which were closely associated with emphysema. Therefore, it is highly likely that the *RAGE* locus affects both lung function and emphysema and that different genetic variants may contribute to these associations. In the TESRA study, another SNP (rs2071288) of *RAGE* was correlated with sRAGE levels ($P=0.01$) and carbon monoxide diffusing ability ($P=0.01$). However, it remains unclear how *RAGE* mutations, sRAGE, and the onset of emphysema interact mechanistically (Cheng et al., 2013). These polymorphisms may affect the extent of conversion of membrane-bound RAGE to sRAGE. Alternatively, decreased levels of circulating sRAGE may indicate that a person has been exposed to high concentrations of RAGE ligands and other inflammatory mediators, in which case sRAGE has been "mopped-up" in the lungs or circulation. Another hypothesis is that low sRAGE levels are a direct result of alveolar cell loss caused by emphysema, given that alveolar cells are a significant source of RAGE in the lungs (Smith et al., 2011).

Finally, no significant correlations were observed between esRAGE levels and lung function parameters (FEV1, FEV1/VC, and DLCO) (Gopal, Reynaert, Scheijen, Schalkwijk, et al., 2014). Most studies that describe plasma sRAGE levels have used an enzyme-linked immunosorbent assay that is marketed for detecting total sRAGE, which also includes esRAGE. Sukkar et al. (Sukkar et al., 2012) individually assessed esRAGE levels in patients with COPD. They found that esRAGE significantly contributed to overall sRAGE and correlated positively with sRAGE in both asthmatic and COPD patients with asthma and COPD. Further investigation is required to determine whether changes in gene expression, alternative splicing, and/or sheddase activity are the cause of decreasing sRAGE levels in circulation (Gopal, Reynaert, Scheijen, Schalkwijk, et al., 2014).

3. Asthma

Asthma is a long-term inflammatory condition characterized by narrowing of the airways and airway hyperresponsiveness (AHR), affecting approximately 300 million individuals worldwide (An et al., 2007; James & Wenzel, 2007; Perkins, Oczypok, Milutinovic, Dutz, & Oury, 2019). Asthma can be induced by aeroallergens, including pollen, molds, dust mite antigens, and cockroach antigens. It can also be triggered by irritants such as cigarette smoke, spray cleaners, and colognes, as well as by nonspecific stimuli such as fluctuations in weather conditions (A. W. James, 2001). Inflammation of the airways seems to be the driver of disease exacerbation and progression, especially in the development of AHR and airway remodeling. These structural changes can result in fixed airway obstruction and severe disease, possibly owing to alterations in the structural and functional properties of the airways (An et al., 2007; James & Wenzel, 2007). Asthma can manifest in both allergic and non-allergic forms, as determined by the presence or absence of IgE antibodies against common environmental allergens. In both variants, there is infiltration of T-helper (Th) cells into the airways, which primarily produce Th2 cytokines such as IL-4, IL-5, and IL-13. These cytokines stimulate mast cells (MC), resulting in eosinophilia, leukocytosis, and enhanced B-cell production of IgE antibodies (Cohn, Elias, & Chupp, 2004; Maslan & Mims, 2014).

3.1 Advanced glycation end products and other RAGE ligands

There is anecdotal evidence suggesting a connection between the consumption of foods and beverages that are high in fructose sweeteners and asthma. Excessive intake of free fructose (EFF) has been proposed to lead to in situ intestinal formation of AGEs (enFruAGE) within the intestines, which may serve as a potential mechanism for the development of asthma in association with EFF consumption (Dechristopher, 2012). Multiple studies have established a link between regular soda consumption and asthma in high school children (DeChristopher, Uribarri, & Tucker, 2016; Park, Blanck, Sherry, Jones, & Pan, 2013) and adults (Shi et al., 2012). According to the national health and nutrition examination survey 2003–2006, the consumption of high-EFF beverages, such as apple juice and drinks sweetened with high-fructose corn syrup, is linked to asthma in children. After adjusting for non-diet fruit drinks and soft drinks, the analysis revealed that children between the ages of 2–9 year who reported consuming apple juice one to four times a week had almost triple the odds of developing asthma compared to those who consumed low or no apple juice. For those who consumed apple juice five or more times per week, the odds remained more than twice as high. Furthermore, the odds of asthma were found to be more than five times higher in children who reported consuming all EFF beverages (apple juice, non-diet fruit drinks, and non-diet soft drinks) five times or more per week than in those who consumed low or no excess free fructose beverages. The more pronounced dose-response observed with the cumulative intake of all EFF beverages suggests that the association is with the total amount of exposure to EFF rather than any specific effect of apple juice alone. These findings support the mechanistic hypothesis that enFruAGE stimulated by elevated pH levels in the jejunum may play a role as an underestimated factor contributing to asthma in children (DeChristopher et al., 2016). Under physiological pH conditions, a greater proportion of fructose than glucose exists in the open-chain form (Wrolstad, 2012). This explains why fructose is considerably more reactive than glucose. It is possible that exposure to non-diet soft drinks alone is not sufficient to contribute to the formation of enFruAGE, which is necessary to reach the lowest observed adverse effect level and to trigger an observable immune response (DeChristopher et al., 2016). The "intestinal enFruAGE fructositis" hypothesis suggests that the consumption of high levels of EFF results in the formation of enFruAGE which may enter the systemic circulation and eventually reach the lungs, which have a high concentration of RAGE. This can lead to the activation of pro-inflammatory signaling pathways associated with asthma. These findings align with similar patterns observed between the consumption of high-fructose corn syrup in the US and the unexplained increase in childhood asthma since the 1980s (DeChristopher, 2013). The Healthy Start study (Venter et al., 2021), which followed a group of 1410 mothers in Colorado from before birth, found no association between maternal AGEs intake and asthma or allergy outcomes in their offspring. Exposure to AGEs during pregnancy may not have the same effects on child development as exposure after birth.

Other RAGE ligands (such as HMGB1, S100A8/A9, and enRAGE) have also been linked with asthma. Asthma-related airway remodeling and inflammation have been associated with the heterodimer complex S100A8/A9, which binds RAGE (Halayko & Ghavami, 2009). Additionally, eosinophil enRAGE communicates with RAGE to encourage MC degranulation and IgE-mediated reactions in the lung. Notably, compared to non-asthmatic controls, asthmatic patient sputum has higher enRAGE concentrations and asthmatic patient lungs have more enRAGE-positive eosinophils. The morphological changes observed in murine bone marrow-derived mast cells after exposure to enRAGE suggest that the cells were activated, and different types of MCs also responded to enRAGE in vitro. It is important to note that various types of MCs, including those found in mucosal and connective tissues, exhibit responsiveness that reflects their functional diversity. enRAGE did not induce a significant amount of leukotriene C4 production, which is typically associated with immediate-early allergic reactions. However, it enhanced the responses triggered by the cross-linking of FcεRI, a high-affinity IgE receptor. The colocalization of enRAGE -positive leukocytes with MCs in the airways and eosinophils in biopsy specimens from patients with asthma suggests that enRAGE is involved in MC activation mediated by IgE and antigens. In patients with allergic asthma, S100A12 was found inside eosinophils in lung biopsy samples and in patients with eosinophilic asthma. The levels of enRAGE in their sputum were significantly higher than those in patients with airway neutrophilia. This suggests that enRAGE could worsen asthma outcomes by enhancing the activation of MCs by allergens (Yang et al., 2007). MC products play a crucial role in initiating host defense against bacterial and parasitic infections (Echtenacher, Männel, & Hültner, 1996; Malaviya, Ikeda, Ross, & Abraham, 1996) and in cell-mediated immune responses in chronic diseases (Askenase, Bursztajn, Gershon, & Gershon, 1980), despite their harmful effects on allergies and asthma. When MCs are activated by microbial products, neutrophils are rapidly recruited from circulation, leading to the release of constitutive enRAGE, which could amplify responses (Echtenacher et al., 1996; Malaviya et al., 1996). Along with histamine, TNF-α from activated MCs can also promote neutrophil influx, and since TNF-α upregulates enRAGE in monocytes/macrophages, this could be a critical feedback loop for MC activation during chronic inflammation (Yang et al., 2001). However, results from studies with mice lacking RAGE suggest that RAGE has only a limited role in adaptive immune responses (Liliensiek et al., 2004). Although sRAGE was shown to reduce the immune response triggered by enRAGE, it had a similar effect in RAGE-deficient mice (Hofmann et al., 1999; Liliensiek et al., 2004), suggesting that it functions through mechanisms other than blocking cell-surface RAGE function. So, there may be another receptor on MCs responsible for the effects of enRAGE (Yang et al., 2007). Finally, HMGB1 levels seem to be increased in asthmatic sputum, and favorably correlate with both the severity of the condition and the number of inflammatory cells (eosinophils and neutrophils) in the lungs (Hou et al., 2011; Shim et al., 2012; Watanabe et al., 2011). According to one study (Shim et al., 2012), HMGB1 levels are favorably correlated with the expression of TNF-α, IL-5, and IL-13 in human sputum samples and promote the recruitment of eosinophils to the lungs in asthma. In patients with severe asthma, there is a noticeably higher percentage of neutrophils present in sputum than in healthy controls or in patients with milder forms of asthma. A significant positive correlation exists between the HMGB1 level and the percentage of neutrophils in the sputum samples in these subgroups (Watanabe et al., 2011). Some studies (Holgate & Polosa, 2006; MacDowell & Peters, 2007; Shaw et al., 2007) have suggested that persistent airflow limitation in severe asthma may be linked to neutrophilic airway inflammation. HMGB1 may function as a neutrophilic chemoattractant and contribute to a new inflammatory pathway in severe asthma. The origin of HMGB1 in sputum is not fully understood, as asthma involves various inflammatory and structural cells (e.g. epithelial and smooth muscle cells), any of which could be a source of HMGB1 (Watanabe et al., 2011). Finally, the interaction between HMGB1 and RAGE seems to contribute to the progression of allergic sensitization as well.

Toll-like receptor 4 (TLR4), which is a crucial mediator of house dust mite (HDM, Dermatophagoides pteronyssinus) sensitization, has been suggested to play a crucial role because RAGE and TLR4 have shared ligands (e.g. HMGB1) and signaling pathways in common. Sensitization to HDM caused the airway epithelium to release HMGB1 in a two-phase pattern, with the activation of TLR4 followed by RAGE activation. Similarly, in response to cockroach (CR) sensitization, the release of HMGB1 was dependent on RAGE. Importantly, the release of HMGB1 occurred after the induction of IL-1α by TLR4 and prior to the production of IL-25 and IL-33. When comparing the effects of lacking either TLR4 or RAGE receptors with the absence of both receptors, it was observed that additional protection against allergic inflammation induced by HDM or CR was not achieved. This suggests that RAGE and TLR4 do not functionally interact to enhance the effector phase of the allergic inflammatory response (Ullah et al., 2014). Another animal study (Akirav et al., 2014) found that RAGE is involved in the stimulation of T-cells in ovalbumin-induced allergic airway sensitization. Mice lacking RAGE have lower levels of T-cell infiltration in their lungs and reduced accumulation of granulocytes and antigen-specific T-cells in their BAL fluid.

3.2 (Soluble) receptor for advanced glycation end products

According to two genome-wide association analyses (Hancock et al., 2010; Repapi et al., 2010), RAGE may play a significant role in asthma pathogenesis. An association with rs207060, a SNP in the RAGE ligand-binding domain, was discovered in individuals with decreased FEV1. Compared to wild-type RAGE, this sequence variant causes a glycine-to-serine substitution at amino acid 82 (G82S), increasing RAGE's affinity of RAGE for ligands and amplifying inflammatory reactions. This suggests that the AGE-RAGE axis may be crucial in the pathogenesis of asthma (Hofmann et al., 2002; Osawa et al., 2007).

esRAGE and total sRAGE levels were elevated in both adult and pediatric asthma patients' sputum samples, and in some instances, these levels and disease severity were linked (Bediwy, Hassan, & El-Najjar, 2016; El-Seify, Fouda, & Nabih, 2014; Ullah et al., 2014; Watanabe et al., 2011). sRAGE levels of 1733.5 pg/mL or higher were indicative of a poor response to initial therapy in the emergency room, suggesting that hospitalization may be necessary. The sensitivity and specificity of this level as a predictor of a poor response were 90.5% and 83.35%, respectively (Bediwy et al., 2016). However, there were no noticeable variations in esRAGE levels among patients with mild-to-severe persistent asthma. The factors that control and influence the expression of esRAGE in asthmatic airways are not completely clear and may have a significant impact on its regulation. Nonetheless, it is suggested that the esRAGE feedback mechanism may not function properly in patients with more severe airflow limitation, such as those with moderate or severe persistent stages of the disease (Watanabe et al., 2011). These results are in contrast with the findings of Sukkar et al. (Sukkar et al., 2012), who reported decreased esRAGE levels in asthmatic sputum. Although these results may be inconsistent, the aforementioned studies did not offer a detailed understanding of the mechanisms underlying the functions of mRAGE versus sRAGE in relation to asthma. Additionally, these studies did not provide clear insights into how cytokines and chemokines crucial to allergic disease are differentially modulated in the presence or absence of RAGE. To better understand the molecular mechanisms by which RAGE promotes asthma pathogenesis (Fig. 2), mouse models of asthma have been used in recent years (Akirav et al., 2014; Milutinovic, Alcorn, Englert, Crum, & Oury, 2012; Oczypok et al., 2015; Ullah et al., 2014). The airway was normal from a physiological and histological standpoint in both wild-type and

RAGE knockout (KO) mouse models, protecting them from an HDM-induced asthma-like condition. Both RAGE KO and wild-type rodents were administered HDM and experienced a normal rise in IL-4, a factor crucial for T-cell activation. The production of the type 2 cytokines IL-5 and IL-13, which control eosinophil recruitment and mucus secretion, respectively, was raised in wild-type mice but not in RAGE KO mice. This suggests that RAGE is essential for the production of these type 2 cytokines. Eotaxin, an eosinophilic chemokine, was also decreased in RAGE KO mice than in wild-type mice. Using an ovalbumin model, all of these experiments were replicated with comparable outcomes (Milutinovic et al., 2012). These findings raised awareness of group 2 innate lymphoid cells (ILC2s). Due to their capacity to secrete significant amounts of IL-5 and IL-13, ILC2s have become crucial new actors in the pathogenesis of allergic asthma (Cayrol & Girard, 2014; Klein Wolterink et al., 2012; Lloyd & Saglani, 2015; Perkins et al., 2019). Recent research has demonstrated that RAGE is required for the IL-33-induced buildup of ILC2s in mouse lungs in reaction to allergens. RAGE may not directly attract ILC2s to the lungs, but it could encourage the production of a downstream mediator. Activation of RAGE could induce the translocation of NF-κB into the nucleus and facilitate the amplification of downstream signaling pathways, some of which promote the expression of adhesion molecules such as vascular cell adhesion protein 1 and intercellular adhesion molecule 1. However, additional research is required to establish whether RAGE plays a role in attracting ILC2s to the lungs or whether it is significant in promoting the proliferation and development of resident ILC2 populations during an allergic airway reaction. Additional research using RAGE-KO animals has also revealed that RAGE supports IL-33 expression in the lungs and plays a crucial role in coordinating the downstream inflammatory signaling effects of IL-33. Therefore, RAGE may play a significant role in the early development of allergic airway inflammation (Oczypok et al., 2015). Further investigation of the role of RAGE in the pathogenesis of asthma has revealed that it can function as early as the sensitization stage of allergic airway reactions. RAGE has ligands and signaling pathways (e.g. HMGB1) similar to TLR4, which is a crucial mediator of HDM sensitization. In addition to eosinophils, neutrophils are also important inflammatory cells in the pathophysiology of asthma. The presence of neutrophilic asthma can lead to irreversible airflow limitation. Evidence suggests that sRAGE is associated with inflammation in the airway caused by neutrophils. Hypersecretion of airway mucus was observed in a mouse

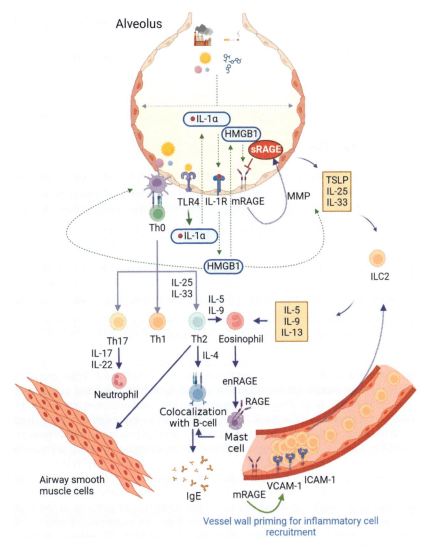

Fig. 2 Role of (soluble)receptor for advanced glycation end products [(s)RAGE] in the pathophysiology of asthma. During the initiation of asthma response, the airway lining produces a large number of signaling molecules called cytokines and chemokines. These substances stimulate immature dendritic cells present in the airways, which then migrate to the lymph nodes and activate naive CD4[+] T helper cells (Th0). Additionally, a type of immune cell called ILC2s (innate lymphoid cell type 2), when activated by certain alarm signals [thymic stromal lymphopoietin (TSLP), interleukin (IL)-25, and IL-33] caused by damaged epithelium, release type 2 cytokines (IL-5, IL-9, and IL-13), which further contribute to the development of allergic responses. In the presence of these type 2 cytokines, Th2-cells are generated in response to the allergens. These Th2 cells move toward the site of inflammation, and upon exposure to the allergen, they primarily

(Continued)

model of neutrophilic asthma. However, injection of sRAGE suppressed collagen deposition, reduced pediatric asthma score, and limited mucus expression. This suggests that sRAGE may be a promising therapeutic target for the treatment of mucus hypersecretion in neutrophilic asthma (Zhang, Xie, Sun, Wei, & Nong, 2022).

4. Lung fibrosis

When the lung tissue is injured, improper remodeling, repair, and regenerative mechanisms can lead to pulmonary fibrosis (Bargagli et al., 2009; Buckley, Medina, Kasper, & Ehrhardt, 2011; Englert et al., 2008; Queisser et al., 2008). Idiopathic pulmonary fibrosis (IPF) is the most common type of fibrotic interstitial lung diseases (ILDs). IPF is characterized by a series of events that start with alveolar epithelial micro-injuries, leading to the development of fibroblastic foci, and end with an excessive deposition of the extracellular matrix (ECM) that causes the loss of lung parenchymal architecture (Selman & Pardo, 2003). It presents with varying outcomes and is typically fatal within 2–4 years of diagnosis (Buendía-Roldán, Mejía, Navarro, & Selman, 2017).

Fig. 2—Cont'd produce IL-4, IL-5, and IL-9. Among these, IL-5 and IL-9 play crucial roles in promoting eosinophil accumulation in tissues. IL-4, on the other hand, stimulates the activation of B-cells. Several hypotheses have been proposed regarding the role of RAGE in the pathophysiology of asthma. First, mRAGE is highly expressed in type 1 alveolar cells in the lungs. RAGE has ligands and signaling pathways [e.g. high mobility group box 1 (HMGB1)] similar to Toll-like receptor 4 (TLR4), which is a crucial mediator of house dust mite (HDM) sensitization. Interaction of HDM with TLR4 causes IL-1α release, and the interaction of IL-1α with its receptor induces the generation of HMGB1, an RAGE ligand. Interaction of IL-1α with his receptor induces the generation of HMGB1, a RAGE ligand. The binding of HMGB1 to RAGE can affect the dendritic cell pathway in the pathophysiology of asthma. Furthermore, HMGB1 stimulates the expression of matrix metalloproteinases (MMP), which in turn induce the cleavage of RAGE, resulting in the formation of sRAGE. It is hypothesized that sRAGE may function as a decoy, mitigating the detrimental effects triggered by RAGE-stimulated signals. Second, mRAGE is expressed in the vascular endothelial cells. RAGE may not directly attract ILC2s to the lungs; however, it may encourage the production of a downstream mediator. Activation of RAGE could induce the translocation of NF-κB into the nucleus and facilitate the amplification of downstream signaling pathways, some of which promote the expression of adhesion molecules, such as vascular cell adhesion protein 1 (VCAM-1) and intercellular adhesion molecule 1 (ICAM-1). Additionally, eosinophil enRAGE communicates with RAGE to encourage mast cell (MC) degranulation and IgE-mediated reactions in the lung. On the other hand, sRAGE can act as a decoy receptor and serve as an anti-inflammatory mediator, protecting the lungs from inflammation involved in asthma pathophysiology.

4.1 Advanced glycation end products and other RAGE ligands

The potential involvement of AGEs in pulmonary fibrosis has been suggested, but its precise role in this process remains incompletely comprehended (Abuelezz, Hendawy, & Osman, 2016; Chen et al., 2009). In a study (Machahua et al., 2016) with 16 IPF and 9 control patients, lung samples were obtained through surgical lung biopsy. An increase in AGEs with a decrease in RAGEs was observed in the lungs of patients with IPF compared to controls. Specifically, two AGEs commonly associated with aging, pentosidine and CML, were found to be significantly elevated in IPF samples. Immunohistochemistry showed that AGEs were more prominently stained in the ECM proteins and the apical surface of alveolar epithelial cells surrounding fibroblast foci in fibrotic lungs. *In vitro* studies have demonstrated that the effect of AGEs on cell viability differs between alveolar epithelial cells and fibrotic fibroblasts. In contrast, RAGE was observed to be localized at the cell membrane of alveolar epithelial cells in healthy lungs, whereas it was largely absent in pulmonary fibrotic tissue. AGEs decreased cell viability in alveolar epithelial cells, even at low concentrations, but had less of an effect on fibroblast viability. Additionally, ECM glycation enhances fibroblast-to-myofibroblast transformation. These findings imply that the increased ratio of AGEs to RAGEs in IPF may play a significant role in accelerating the aging of lung tissue and contribute to the abnormal wound-healing process that leads to fibrosis. Nevertheless, the role of the AGE-RAGE interaction in the onset of pulmonary fibrosis is not yet fully understood (Song et al., 2011). In rats with bleomycin-induced fibrosis, there was a significant increase in AGEs in the lungs, as well as in lung hydroxyproline content and Ashcroft score. However, these effects were prevented by treatment with aminoguanidine, a well-known AGE formation inhibitor. Additionally, the administration of aminoguanidine resulted in a reduction in bleomycin-induced expression of heat shock protein 47 (HSP47) as well as downregulation of transforming growth factor-beta 1 (TGF-β1), phosphorylated Smad2, and phosphorylated Smad3 expression. Based on these findings, it can be concluded that AGEs may play a crucial role in bleomycin-induced pulmonary fibrosis and potentially exert regulatory effects on HSP47 expression and the TGF-β/Smad signaling pathway (Chen et al., 2009). It was proposed that AGE boosted Smad7 expression and blocked TGF-β-induced ERK1/2 and Smad2 phosphorylation by interacting with RAGE (Song et al., 2011). On the other hand, it has been reported that AGEs have dose- and time-dependent effects on enhancing collagen production and increasing the expression of connective tissue growth factor (CTGF) mRNA and protein in NRK-49F (normal rat kidney fibroblast) cells (Lee et al., 2004). Incubation of human foreskin fibroblasts with AGEs increased the mRNA levels of CTGF, TGF-β1, and procollagen-alpha 1 (Lohwasser, Neureiter, Weigle, Kirchner, & Schuppan, 2006). Furthermore, treating rats with AGEs resulted in a significant increase in fibronectin and type IV collagen accumulation in the renal glomeruli. Additionally, there was a notable induction of renal expression of TGF-β1 and CTGF in response to AGE treatment (Zhou, Li, & Cai, 2004).

Specimens from patients with pulmonary fibrosis exhibited high expression of the RAGE ligand S100A8/A9. S100A8/A9 has been shown to be active not only in fibroblast proliferation but also in fibroblast differentiation into myofibroblasts, which are the active form of fibroblasts. S100A8/A9 also induced the upregulation of collagen production in fibroblasts by acting through RAGE in these cells. Activation of RAGE in fibroblasts induced by extracellular S100A8/A9 is associated with the robust activation of NF-κB. This is essential for inducing a range of inflammatory cytokines, promoting growth, accelerating differentiation into myofibroblasts, and increasing the expression of collagen and fibronectin, which are major components of the stroma (Araki et al., 2021).

4.2 (Soluble) receptor for advanced glycation end products

The involvement of RAGE in IPF etiology remains unclear (Bargagli et al., 2009). RAGE protein levels in human lung homogenates and BAL fluid seem to be reduced in pulmonary fibrosis, in contrast to other pulmonary inflammatory diseases (Bargagli et al., 2009; Buckley et al., 2011; Englert et al., 2008; Queisser et al., 2008). In addition, *RAGE* gene expression is altered in the airways of patients with IPF and during acute exacerbation of the condition (Englert et al., 2008; Konishi et al., 2009). RAGE may play a role in the prevention of pulmonary fibrosis because in old mice lacking RAGE, where the lungs change spontaneously and exhibit characteristics of fibrosis (Englert et al., 2008). In a number of experimental models of pulmonary injury/fibrosis (Englert et al., 2008; Hanford et al., 2003; Ramsgaard et al., 2010), including bleomycin, asbestos, and silica, *RAGE* expression was markedly decreased in the lungs of mice. RAGE appears to play a different role in the onset of fibrosis in each model. A lack of RAGE had no impact on fibrosis caused by silica (Ramsgaard et al., 2010), a negative, aggravating effect on fibrosis caused by asbestos (Englert et al., 2008), and a protective effect on fibrosis caused by bleomycin (Englert et al., 2011; He et al., 2007). Although the reason for the variation in RAGE function between these animal models is unknown, it may be related to the role of RAGE in cellular adhesion and regeneration. After bleomycin-induced injury, RAGE KO alveolar cells re-epithelialized more effectively than wild-type cells in vitro. This is probably because RAGE KO cells exhibit superior migration and reparative processes. Despite the observed effects in animals with genetic modifications, the administration of external sRAGE had no impact on fibrosis induced by bleomycin or asbestos (Englert et al., 2011). The alveolar epithelium of the lung comprises two different types of cells: flat alveolar type 1 (AT1) cells and cuboidal alveolar type 2 (AT2) cells (Shirasawa et al., 2004). RAGE attaches to collagen at AT1 to promote cell spreading. Alveolar adhesion to the basement membrane is compromised by RAGE deficiency, which may increase alveolar cell migration. RAGE is also implicated in the conversion of AT2 cells into AT1 cells and is less abundant in AT2 cells obtained from fibrotic human and murine lungs (Queisser et al., 2008; Shirasawa et al., 2004). *In vitro*, the absence of RAGE in alveolar epithelial cells improved cell growth and migration while impairing cellular adhesion. Depending on the type of injury to which the cell is exposed, these modifications may result in cellular dysfunction or better repair mechanisms. Whether a lack of RAGE is detrimental to the cell may depend on that injury (Oczypok et al., 2017; Queisser et al., 2008). The ability of RAGE to facilitate contact between cells and the ECM seems to be similar to that of integrin-mediated adhesion in lung epithelial cells, and disruption of *RAGE* expression is linked to impaired cellular connections (Queisser et al., 2008). Epithelial-to-mesenchymal transition (EMT), which has been identified as a critical mechanism in interstitial lung disease for promoting a profibrotic environment, has also been linked to RAGE (Willis & Borok, 2007). RAGE KO animals are resistant to the harmful effects of bleomycin and their AT2 cells do not experience HMGB1-induced EMT (He et al., 2007). In contrast, TGF-β and pro-inflammatory cytokines stimulate AT2-like cells, causing cytoskeletal changes indicative of EMT and lower RAGE expression (Buckley et al., 2011). The role of RAGE in EMT is still unclear because it is not clear whether RAGE is necessary for EMT or if it is not encouraged. These data imply that this may rely on HMGB1 or TGF-β driving factors linked to the experimental models used (Oczypok et al., 2017). However, RAGE overexpression appears to be present in fibroblastic foci and other sites of active fibrosis (Morbini et al., 2006).

5. Cystic fibrosis

CF is a prevalent autosomal recessive condition among Caucasians that causes bronchiectasis, airway inflammation, and an elevated risk of diabetes (Rosenstein & Cutting, 1998). The prevalence of cystic fibrosis-related diabetes (CFRD) increases with age, with over 50% of patients with CF developing diabetes mellitus by the age of 40 years. Patients with CFRD have a higher mortality rate than those with CF alone, and there is a significant correlation between CFRD, deterioration in lung function, and overall clinical status (Brennan, Geddes, Gyi, & Baker, 2004).

5.1 Advanced glycation end products and other RAGE ligands

Research on AGEs and CF is scant. In a cohort of CF patients with relatively well-controlled hyperglycemia, patients with CFRD exhibited a significant increase of approximately 10% in the plasma levels of AGEs compared to their normoglycemic counterparts. Moreover, there was a notable negative correlation between plasma AGE levels and pulmonary function, which included both FEV1 and FVC percent predicted (Hunt, Helfman, McCarty, & Hansen, 2016).

5.2 (Soluble) receptor for advanced glycation end products

Although the association between RAGE and CF has only been reported in a limited number of studies, cystic fibrosis airway neutrophils have shown higher RAGE levels than peripheral blood neutrophils. Additionally, cystic fibrosis airway fluid has shown elevated levels of enRAGE and a lack of sRAGE (Foell et al., 2003; Makam et al., 2009). This indicates that RAGE may play a significant role in DF, particularly in CFRD, given the connection between RAGE, inflammation, the lungs, and diabetes mellitus. A pilot study (Mulrennan et al., 2015) evaluated the expression of flRAGE, sRAGE, esRAGE, enRAGE, and AGEs in the serum, white blood cells, and sputum of patients with CF, diabetes mellitus, CFRD, and healthy individuals. Sputum enRAGE/sRAGE ratios were significantly elevated in CF patients, particularly in those with CFRD, and were negatively correlated with the % predicted FEV1. Serum AGE and AGE/sRAGE ratios were high in diabetics but not in CF patients. A comprehensive and multifaceted approach was used to examine the role of RAGE and its ligands, which are essential for determining their impact on airway inflammation. This study found a clear association between RAGE activity in the airways of CF and CFRD patients, which is not evident in the vascular compartment and correlates with lung function, unlike diabetes mellitus. This strongly suggests that RAGE contributes to the inflammatory response observed in CF and to a greater extent in CFRD. The progression of CF can differ in individuals with the same *CFTR* mutations, indicating the potential influence of modifier genes. Among the 967 CF patients, the presence of the minor allele (C) at position-429 of the *RAGE* gene was associated with poorer lung function ($P = 0.03$). In the cell experiments, the RAGE-429C allele demonstrated higher promoter activity than the RAGE-429 T allele ($P = 0.016$ in BEAS-2B cells). This suggests that RAGE acts as a modifier gene that influences the severity of lung disease in CF and could serve as a potentially valuable biomarker for CF-related airway inflammation. The functional RAGE-429C variant in the promoter region is associated with increased *RAGE* expression, which in turn can contribute to elevated lung inflammation and more severe progression of lung disease (Beucher et al., 2012).

6. Acute lung injury/acute respiratory distress syndrome

Acute lung injury (ALI) and ARDS lead to acute respiratory failure due to severe inflammation and alveolar damage to the lungs. They are characterized by the rapid onset of respiratory failure and can be caused by various factors, including direct lung injury from sources such as bacterial or viral pneumonia, aspiration of gastric contents, lung contusion, inhalation of toxic substances, or near-drowning (Guo et al., 2016). Despite advancements in intensive care, the mortality rate of ALI/ARDS remains high at approximately 29%–42% over the last decade (Johnson & Matthay, 2010). The development of this condition involves an increase in the production of inflammatory mediators such as TNF-α and IL-1β. Histologically, it is characterized by the accumulation of numerous neutrophils and the appearance of interstitial edema (Abraham et al., 1995; Kollef & Schuster, 1995; Repine, 1992).

6.1 Advanced glycation end products and other RAGE ligands

The reason why some patients develop ALI or ARDS after aspiration, while others do not, is not yet understood (Guo et al., 2012). Both animal and human studies have shown that a high-AGE diet worsens ALI following gastric aspiration (Guo et al., 2012; Ottosen et al., 2010). This evidence was apparent from increased neutrophil infiltration, airway albumin leakage, decreased pulmonary compliance, and myeloperoxidase activity in the lung parenchyma. Neutrophils have been identified as the cellular agents involved in the development of ALI in patients with ALI/ARDS. Both clinical studies and animal models of ARDS have shown the sequestration of a significant quantity of neutrophils in the lung microvasculature and the buildup of neutrophil proteases in BAL fluid and serum. Myeloperoxidase is a glycoprotein present in all cells originating from the myeloid lineage, but is primarily found in the azurophilic granules of neutrophils. Activated neutrophils release myeloperoxidase. Therefore, assessing myeloperoxidase levels in whole lung homogenates indicates the accumulation of neutrophils in the lungs (Guo et al., 2012). In conclusion, it may be possible to reduce lung injury after gastric aspiration by targeting AGEs in the circulatory system as a therapeutic strategy.

Patients diagnosed with ARDS displayed considerably increased levels of pulmonary enRAGE, a well-known RAGE ligand, as well as higher BAL levels of enRAGE protein and sRAGE in comparison to the control groups. Even among healthy individuals, inhalation of lipopolysaccharide led to an increase in enRAGE levels in their BAL (Wittkowski et al., 2007). Patients who developed ALI following emergency surgery for lower gastrointestinal tract perforation had significantly higher levels of enRAGE in their blood during the immediate postoperative period than those who did not develop ALI (Kikkawa et al., 2010). Moreover, intratracheal administration of HMGB1 in mice led to the development of ALI, characterized by elevated production of pro-inflammatory cytokines, such as IL-1β and TNF-α, increased accumulation of neutrophils, and the formation of lung edema. The migration of neutrophils to the lungs and the development of lung edema were reduced by the administration of anti-HMGB1 antibodies. The contribution of HMGB1 to the development of ALI seems to differ from its impact on pro-inflammatory cytokines that act earlier in the process. Although anti-HMGB1 antibodies can reduce pulmonary damage and prevent the accumulation of neutrophils, the levels of IL-1β, TNF-α, and macrophage inflammatory protein-2 (MIP-2) in the lungs were not affected by HMGB1 inhibition. Additionally, while treatment with anti-HMGB1 antibodies decreased lung edema and neutrophil accumulation, it did not fully prevent the development of lung injury compared to the control group. This suggests that besides HMGB1, other mediators are also involved in the pathogenesis (Abraham, Arcaroli, Carmody, Wang, & Tracey, 2000). Following severe trauma, the plasma levels of HMGB1 in human subjects were found to increase within 30 min. These elevated levels correlated with the severity of injury, tissue hypoperfusion, early post-traumatic coagulopathy, and hyperfibrinolysis. Non-survivors had higher plasma levels of HMGB1 than survivors (Cohen et al., 2009). Mechanically ventilated patients requiring long-term ventilation due to respiratory failure had higher levels of HMGB1 in BAL than those requiring short-term ventilation for less than 5 h for an elective surgical procedure (van Zoelen et al., 2009). These findings suggest that HMGB1 acts as a prototype mediator of cellular damage-associated molecular patterns after stress and ALI (Guo et al., 2016).

6.2 (Soluble) receptor for advanced glycation end products

RAGE appears to be a significant contributor in the development of ALI/ARDS, as it is found abundantly in the lungs and is primarily located on the basal membranes of alveolar type 1 epithelial cells (Newman, Gonzalez, Matthay, & Dobbs, 2000). Multiple animal studies (Izushi et al., 2016; Su, Looney, Gupta, & Matthay, 2009; Uchida et al., 2006; Zhang et al., 2008) have provided evidence supporting the use of sRAGE as a valuable biomarker for type 1 alveolar epithelial cell injury and have shown that sRAGE levels are positively associated with the severity of ALI, making it a significant factor in the pathogenesis of ALI. As mentioned above, sRAGE is a marker of inflammation. This is significant because both epithelial injury and inflammation are key components in the development of ALI and sRAGE is involved in both processes (Guo et al., 2016). In a RAGE-KO mouse model, the absence of RAGE significantly enhanced the clearance of fluid in the alveoli and reduced the leakage of albumin into pulmonary vessels after exposure to lipopolysaccharide. RAGE signaling may cause a decrease in the expression of ion channels (epithelial sodium channel, sodium-potassium ATPase) and tight junction proteins (zonula occluden-1), which could be the mechanism underlying bacterial endotoxin-induced lung injury (Downs, Kreiner, Johnson, Brown, & Helms, 2015; Hirose, Tanikawa, Mori, Okada, & Tanaka, 2010; Wang, Xu, Meng, Adcock, & Yao, 2018). Molecular and biochemical methods have revealed that suppressing the activity of RAGE and protein kinase C (PKC) reduces the activation of epithelial sodium channel caused by AGEs. AGEs triggered the phosphorylation of p47phox and increased the generation of ROS dependent on gp91phox. This reaction is nullified when RAGE and PKC are inhibited. Introducing AGEs through tracheal instillation facilitated the removal of lung fluid. However, this effect was completely eliminated when RAGE, PKC, and gp91phox were simultaneously inhibited (Downs et al., 2015). Others have proposed that the increased permeability of blood vessels could be attributed to the activation of RAGE/Rho, which leads to the formation of gaps and rearrangement of actin structures (Hirose et al., 2010). Studies in humans (Christie et al., 2009; Jabaudon et al., 2011; Uchida et al., 2006) have also indicated that sRAGE can serve as an indicator of the extent of ALI/ARDS. sRAGE levels in the fluid of patients with pulmonary edema were found to be higher in patients with ALI than in those with hydrostatic pulmonary edema, and the plasma level of sRAGE was significantly higher in patients with ALI than in healthy volunteers or those with hydrostatic pulmonary edema. This indicates that sRAGE in these samples mainly originates from the lung, suggesting that sRAGE can serve as a biomarker for acute pulmonary inflammatory responses (Uchida et al., 2006). The levels of sRAGE were also found to be useful in predicting the outcome or future course of the disease in patients with ALI (Calfee et al., 2008; Cohen et al., 2010). Although RAGE has been identified as a potential biomarker of ALI severity, its ability to predict ALI onset in at-risk patients still needs to be evaluated. Further animal and clinical studies are necessary to determine the accuracy and usefulness of RAGE as a diagnostic and prognostic tool for ALI. Identifying patient subgroups that would benefit from targeted therapies based on relevant pathogenetic pathways could be facilitated by the sensitivity and specificity of RAGE measurement (Guo et al., 2016). Finally, dysregulation of MMP-9 is also been implicated in the development of ALI (Abuelezz et al., 2016; Wang Cui Zeng et al., 2012). MMPs facilitate RAGE shedding and release of sRAGE. In a mouse model (Zhang et al., 2021), MMP-9-mediated RAGE shedding plays a role in reducing the severity of sepsis-induced pulmonary edema, inflammation, oxidative stress, and lung injury. This occurs by suppressing the RAGE/NF-κB signaling pathway through decoy receptor functions of sRAGE. Therefore, MMP-9-mediated production of sRAGE may act as a mechanism for self-limitation, controlling and resolving excessive inflammation and oxidative stress in the lungs during sepsis.

7. Lung cancer

Lung cancer ranks as the world's second most common malignancy and stands out as the most deadly cancer globally (Siegel, Miller, Fuchs, & Jemal, 2022). Among various tissues in the body, the adult lung is distinguished by its elevated RAGE expression levels in the absence of any pathological conditions, in contrast to the lower baseline expression levels observed in other tissues (Lizotte et al., 2007).

7.1 Advanced glycation end products and other RAGE ligands

The role of AGEs and other RAGE ligands, such as amphoterin (Huttunen & Rauvala, 2004) and S100 proteins (Roth, Vogl, Sorg, & Sunderkötter, 2003) in cancer invasion and metastasis holds significant importance. When AGEs or other ligands bind to RAGE, it triggers the activation of critical signaling mediators, including p21ras, mitogen-activated protein (MAP) kinases, NF-κB, and cdc42/rac. These mediators, in turn, initiate downstream processes that promote increased cell proliferation, invasion, and metastatic potential. Studies in mice have demonstrated that inhibiting the amphoterin/RAGE system, achieved through the suppression of p44/p42, p38, and SAP/JNK MAP protein kinases, effectively reduces the migration and invasiveness of tumor cells and may also influence cell proliferation and the production of tissue metalloproteinases (Taguchi et al., 2000). Additionally, the interaction between AGE/RAGE and Ras has been shown to activate hypoxia-inducible factor 1 subunit alpha, further enhancing tumor aggressiveness (Kang et al., 2014). Furthermore, the AGE/RAGE axis plays a crucial role in governing the reversible transition to quiescent states in anaplastic lymphoma kinase-rearranged non-small-cell lung cancer (NSCLC) cells within monolayer cultures (Kadonosono et al., 2022).

7.2 (Soluble) receptor for advanced glycation end products

The involvement of RAGE in lung cancer progression has been the subject of intense research. Multiple studies have explored variations in RAGE expression in NSCLC tissues (Hofmann et al., 2004; Jing, Cui, Wang, & Wang, 2010; Schraml, Shipman, Colombi, & Ludwig, 1994; Schraml, Bendik, & Ludwig, 1997; Stav, Bar, & Sandbank, 2007). However, the precise relationship between RAGE and NSCLC development remains unclear.

8. Unmasking RAGE's complex role in lung cancer progression

RAGE may exhibit a dual role, potentially leading to growth inhibition during the early stages of tumor formation while simultaneously promoting epithelial-to-mesenchymal transition and creating a favorable tumor microenvironment for tumorigenesis in lung adenocarcinoma (Chen et al., 2020). Some studies have suggested that RAGE might function protectively in lung tissues, potentially reducing vulnerability to tumor development. This perspective raises the possibility that suppressing RAGE could pave the way for the onset of NSCLC (Bartling, Hofmann, Weigle, Silber, & Simm, 2005; Bartling et al., 2011). Moreover, reduced RAGE expression correlates with an unfavorable prognosis as well as impaired immune infiltration and cellular senescence in cases of lung adenocarcinoma (Lin et al., 2022). Conversely, other studies have indicated that RAGE may play a contributory role not only in lung cancer (Yu, Pan, & Cheng, 2017). In an in vitro model, RAGE significantly enhanced the proliferation, migration, and invasion of lung cancer cells through a protein kinase B-dependent pathway primarily mediated by lysophosphatidic acid (Ray, Jangde, Singh, Sinha, & Rai, 2020). Heat shock protein 70 has been recognized as a RAGE agonist with the capacity to affect communication between cancer cells and immune system cells (Somensi et al., 2017). DNA methylation of RAGE is linked to immune infiltration and prognosis in both lung adenocarcinoma and lung squamous cell carcinoma (Yang et al., 2023). The presence of dominant-negative RAGE expression could play a role in promoting metastasis and progression of advanced lung adenocarcinoma (Downs, 2021). Apoptotic cancer cells promote the metastatic growth of surviving cells through a process mediated by peptidyl arginine deiminase 4, resulting in the expulsion of the cell nucleus. This nuclear expulsion, which is associated with poor prognosis in lung cancer patients, leads to the formation of an extracellular DNA-protein complex that is rich in RAGE ligands. Among these chromatin-bound RAGE ligands, S100a4 activates RAGE receptors in adjacent surviving tumor cells, ultimately triggering ERK activation (Park et al., 2023). Blocking the interaction between RAGE and its ligands offers a promising approach to impeding cancer development, potentially serving as an effective strategy for cancer treatment (Faruqui et al., 2022).

Considering RAGE's role of RAGE in the development of lung cancer, it is crucial to highlight the central role of redox regulation in cellular signaling processes. Disruption of this regulation can lead to oxidative stress, which, in turn, may affect various lung pathologies, including NSCLC. It is worth noting that ROS not only influences the behavior of cancer cells but also their responsiveness to therapeutic interventions (Goldkorn, Filosto, & Chung, 2014). Previous studies (Mukherjee, Mukhopadhyay, & Hoidal, 2005; Sanders et al., 2019) indicated that RAGE plays a role in redox regulation. Recent research has further reinforced this connection, demonstrating that

RAGE knockdown significantly reduces key redox signaling regulators, including thioredoxin-1, superoxide dismutase 1, thioredoxin-like protein 1, and thioredoxin domain-containing protein 17 (Downs, Johnson, Tsaprailis, & Helms, 2018). This suggests that RAGE has a profound effect on intrinsic redox regulation. Whether oxidative stress is a critical factor for RAGE-associated complications in NSCLC requires further investigation.

A significant challenge in understanding the relevance of reduced RAGE expression in NSCLC pathogenesis lies in the uncertainty surrounding the cellular origin of NSCLC (Sutherland & Berns, 2010). Identifying the specific cell or cells from which various types of NSCLC originate could shed light on whether the diminished activity of RAGE and sRAGE in NSCLC cells represents a departure from the non-neoplastic state. The multifaceted roles of RAGE in cancer require further investigation for a more comprehensive understanding.

9. Unlocking the impact of AGER polymorphisms on lung cancer

Over the past decade, several studies have linked *AGER* polymorphisms with the development and complications of NSCLC. These genetic variations have been associated with the diagnosis, prognosis, and even the response to chemotherapy in NSCLC patients (Pan et al., 2013; Pikor, Ramnarine, Lam, & Lam, 2013; Schenk, Schraml, Bendik, & Ludwig, 2001; Stav et al., 2007; Wang et al., 2015; Wang, Li, Qian, Wang, & Jing, 2012; Wu et al., 2018; Yamaguchi et al., 2017). For example, the rs2070600 variant A allele potentially diminishes the expression of the tumor suppressor gene *AGER*, consequently increasing the risk of developing lung cancer (Wu et al., 2018). The *AGER* rs2070600 polymorphism was also independently associated with systemic inflammation and unfavorable prognosis among patients diagnosed with metastatic lung adenocarcinoma (Yamaguchi et al., 2017).

Exploring anti-RAGE therapy in specific NSCLC patients based on their unique *AGER* genetic polymorphic variants could provide valuable insights into the relationship between RAGE expression and NSCLC development. Furthermore, given that certain *AGER* polymorphic variants may reduce the incidence of NSCLC, it is essential to investigate whether experimentally augmenting RAGE expression or introducing these polymorphic *AGER* variants into NSCLC cells can mitigate the development or progression of NSCLC (Mukherjee, Malik, & Hoidal, 2021).

10. sRAGE: Illuminating lung cancer diagnosis and beyond

In a small study ($n = 45$), serum sRAGE levels were significantly reduced in lung cancer patients compared to controls (in comparison to healthy donors, $P = 0.034$; and pulmonary tuberculosis patients, $P = 0.010$). The lower serum concentration of sRAGE was negatively correlated with lymph node involvement (N0 vs N1–2, $P = 0.028$). Downregulation of membranous and cytoplasmic RAGE expression was also observed in lung cancer tissues compared to that in nearby normal lung tissues. Collectively, these results suggest that serum sRAGE levels decrease during the progression of lung cancer and may serve as a reliable and convenient diagnostic biomarker for this condition. Elevated serum sRAGE levels may interact with various RAGE ligands, such as AGEs, amphoterin, and S100 proteins, potentially hindering RAGE-mediated stimulation of tumor growth and invasion (Jing et al., 2010).

In a recent study involving 81 patients, comprising 20 individuals with tuberculosis, 30 with pneumonia, and 31 with lung cancer, no statistically significant differences were observed in serum sRAGE levels among these groups. However, a noteworthy contrast emerged when examining bronchial sRAGE levels, especially in patients with lung cancer, where the levels were notably lower than those in individuals with other respiratory conditions. The analysis of Receiver Operating Characteristic (ROC) curves for bronchial sRAGE values in lung cancer patients identified an optimal cutoff point at 118.9 pg/mL, resulting in a sensitivity of 76%, specificity of 58%, and an area under the ROC curve of 0.695 ($P = 0.005$) (Kim et al., 2023). The underlying reason for this reduction in sRAGE levels in patients with lung cancer remains unclear. Current research suggests that the binding of RAGE to AGEs stimulates the production of cytokines and growth factors, potentially leading to an inflammatory response that could contribute to cancer development. Conversely, an increase in sRAGE may inhibit intracellular signal transduction by AGEs, thereby mitigating inflammation and oxidative stress linked to cancer progression (Ahmad et al., 2018; Erusalimsky, 2021; Perrone et al., 2020). The observed difference in sRAGE levels between serum and bronchial samples can be attributed to the high expression of RAGE in type 1 alveolar epithelial cells, given that lung cancer typically originates in the lungs. Consequently, alterations in bronchial sRAGE levels may precede changes in the serum sRAGE levels. In cases of respiratory infections such as tuberculosis and pneumonia, serum and bronchial sRAGE levels tend to correlate closely due to the rapid progression of these diseases. Conversely, lung cancer progresses at a slower rate than respiratory infections, resulting in delayed or diminished changes in serum sRAGE levels compared to bronchial sRAGE levels. To assess the clinical utility of bronchial sRAGE as a supplementary diagnostic marker for lung cancer, external validation and longitudinal analyses in cohorts of lung cancer patients will be indispensable in future research (Kim et al., 2023).

11. Conclusions and future directions

In conclusion, accumulating evidence suggests that AGEs and sRAGE play important roles in the development and progression of lung diseases. The implications of these findings open promising avenues for future research and therapeutic interventions. sRAGE has shown potential as a biomarker for lung diseases, with altered levels observed in different conditions such as COPD, asthma, and IPF. Further research is needed to establish the diagnostic value of sRAGE in various pulmonary diseases and its utility for monitoring disease progression and treatment response. Identifying the specific cutoff levels of sRAGE associated with different stages or severity of lung diseases can aid in risk stratification and inform personalized treatment approaches.

While existing evidence provides valuable insights, many of the studies included in this review employed a cross-sectional methodology. Conducting longitudinal studies is crucial to better understand the temporal relationship between AGEs, sRAGE, and pulmonary diseases. Long-term observational studies and intervention trials can help to elucidate cause-and-effect relationships and determine the prognostic significance of AGEs and sRAGE in different lung diseases. These studies could also aid in identifying specific time points or stages of disease progression when targeting AGEs, with sRAGE being the most effective.

Future research should focus on unraveling the underlying molecular and cellular mechanisms through which AGEs and sRAGE influence lung disease development and progression. Elucidating the signaling pathways and downstream effects of AGEs and sRAGE in various lung diseases can provide insights into potential therapeutic targets. This deeper understanding of the molecular basis will facilitate the development of more precise interventions that specifically modulate AGEs and sRAGE-related pathways while minimizing off-target effects.

Targeting AGEs and their receptors may offer innovative therapeutic strategies for the prevention and treatment of various pulmonary diseases, including COPD, asthma, pulmonary fibrosis, CF, and ALI/ARDS. As previously mentioned, sRAGE exhibits anti-inflammatory, antioxidant, and anti-fibrotic properties, making it a promising therapeutic target. Future investigations should focus on developing sRAGE-based therapies, including the administration of recombinant sRAGE, agonists that enhance endogenous sRAGE production, or strategies to stabilize sRAGE in the extracellular environment. Preclinical and clinical studies are necessary to evaluate the efficacy, safety, and long-term outcomes of such interventions.

References

Abraham, E., Arcaroli, J., Carmody, A., Wang, H., & Tracey, K. J. (2000). HMG-1 as a mediator of acute lung inflammation. *Journal of Immunology (Baltimore, Md.: 1950), 165*(6), 2950–2954. https://doi.org/10.4049/jimmunol.165.6.2950.

Abraham, E., Bursten, S., Shenkar, R., Allbee, J., Tuder, R., Woodson, P., ... Repine, J. E. (1995). Phosphatidic acid signaling mediates lung cytokine expression and lung inflammatory injury after hemorrhage in mice. *The Journal of Experimental Medicine, 181*(2), 569–575. https://doi.org/10.1084/jem.181.2.569.

Abuelezz, S. A., Hendawy, N., & Osman, W. M. (2016). Aliskiren attenuates bleomycin-induced pulmonary fibrosis in rats: Focus on oxidative stress, advanced glycation end products, and matrix metalloproteinase-9. *Naunyn-Schmiedeberg's Archives of Pharmacology, 389*(8), 897–909. https://doi.org/10.1007/s00210-016-1253-3.

Ahmad, S., Khan, H., Siddiqui, Z., Khan, M. Y., Rehman, S., Shahab, U., ... Moinuddin (2018). AGEs, RAGEs and s-RAGE; friend or foe for cancer. *Semin Cancer Biology, 49*, 44–55. https://doi.org/10.1016/j.semcancer.2017.07.001.

Ahmed, M. U., Thorpe, S. R., & Baynes, J. W. (1986). Identification of N epsilon-carboxymethyllysine as a degradation product of fructoselysine in glycated protein. *The Journal of Biological Chemistry, 261*(11), 4889–4894.

Akirav, E. M., Henegariu, O., Preston-Hurlburt, P., Schmidt, A. M., & Herold, K. C. (2014). The receptor for advanced glycation end products (RAGE) affects T cell differentiation in OVA induced asthma. *PLoS One, 9*(4), e95678. https://doi.org/10.1371/journal.pone.0095678.

An, S. S., Bai, T. R., Bates, J. H. T., Black, J. L., Brown, R. H., Brusasco, V., ... Wang, L. (2007). Airway smooth muscle dynamics: A common pathway of airway obstruction in asthma. *The European Respiratory Journal, 29*(5), 834–860. https://doi.org/10.1183/09031936.00112606.

Aragno, M., & Mastrocola, R. (2017). Dietary sugars and endogenous formation of advanced glycation endproducts: Emerging mechanisms of disease. *Nutrients, 9*(4), 385. https://doi.org/10.3390/nu9040385.

Araki, K., Kinoshita, R., Tomonobu, N., Gohara, Y., Tomida, S., Takahashi, Y., ... Sakaguchi, M. (2021). The heterodimer S100A8/A9 is a potent therapeutic target for idiopathic pulmonary fibrosis. *Journal of Molecular Medicine (Berlin, Germany), 99*(1), 131–145. https://doi.org/10.1007/s00109-020-02001-x.

Artigas, M. S., Loth, D. W., Wain, L. V., Gharib, S. A., Obeidat, M., Tang, W., ... Tobin, M. D. (2011). Genome-wide association and large-scale follow up identifies 16 new loci influencing lung function. *Nature Genetics, 43*(11), https://doi.org/10.1038/ng.941 Article 11.

Askenase, P. W., Bursztajn, S., Gershon, M. D., & Gershon, R. K. (1980). T cell-dependent mast cell degranulation and release of serotonin in murine delayed-type hypersensitivity. *Journal of Experimental Medicine, 152*(5), 1358–1374. https://doi.org/10.1084/jem.152.5.1358.

Barberà, J. A., Peinado, V. I., & Santos, S. (2003). Pulmonary hypertension in chronic obstructive pulmonary disease. *The European Respiratory Journal, 21*(5), 892–905. https://doi.org/10.1183/09031936.03.00115402.

Bargagli, E., Penza, F., Bianchi, N., Olivieri, C., Bennett, D., Prasse, A., & Rottoli, P. (2009). Controversial role of RAGE in the pathogenesis of idiopathic pulmonary fibrosis. *Respiratory Physiology & Neurobiology, 165*(2), 119–120. https://doi.org/10.1016/j.resp.2008.10.017.

Barnes, P. J. (2004). Mediators of chronic obstructive pulmonary disease. *Pharmacological Reviews, 56*(4), 515–548. https://doi.org/10.1124/pr.56.4.2.

Barnes, P. J., Burney, P. G. J., Silverman, E. K., Celli, B. R., Vestbo, J., Wedzicha, J. A., & Wouters, E. F. M. (2015). Chronic obstructive pulmonary disease. *Nature Reviews. Disease Primers, 1*, 15076. https://doi.org/10.1038/nrdp.2015.76.

Bartling, B., Hofmann, H.-S., Weigle, B., Silber, R.-E., & Simm, A. (2005). Down-regulation of the receptor for advanced glycation end-products (RAGE) supports non-small cell lung carcinoma. *Carcinogenesis, 26*(2), 293–301. https://doi.org/10.1093/carcin/bgh333.

Bartling, B., Hofmann, H.-S., Sohst, A., Hatzky, Y., Somoza, V., Silber, R.-E., & Simm, A. (2011). Prognostic potential and tumor growth-inhibiting effect of plasma advanced glycation end products in non-small cell lung carcinoma. *Molecular Medicine (Cambridge, Mass.), 17*(9-10), 980–989. https://doi.org/10.2119/molmed.2011.00085.

Bediwy, A. S., Hassan, S. M., & El-Najjar, M. R. (2016). Receptor of advanced glycation end products in childhood asthma exacerbation. *Egyptian Journal of Chest Diseases and Tuberculosis, 65*(1), 15–18. https://doi.org/10.1016/j.ejcdt.2015.10.008.

Beucher, J., Boëlle, P.-Y., Busson, P.-F., Muselet-Charlier, C., Clement, A., & Corvol, H. (2012). AGER-429T/C is associated with an increased lung disease severity in cystic fibrosis. *PLoS One, 7*(7), e41913. https://doi.org/10.1371/journal.pone.0041913.

Bierhaus, A., Humpert, P. M., Morcos, M., Wendt, T., Chavakis, T., Arnold, B., ... Nawroth, P. P. (2005). Understanding RAGE, the receptor for advanced glycation end products. *Journal of Molecular Medicine (Berlin, Germany), 83*(11), 876–886. https://doi.org/10.1007/s00109-005-0688-7.

Boschetto, P., Campo, I., Stendardo, M., Casimirri, E., Tinelli, C., Gorrini, M., ... Luisetti, M. (2013). Plasma sRAGE and N-(carboxymethyl) lysine in patients with CHF and/or COPD. *European Journal of Clinical Investigation, 43*(6), 562–569. https://doi.org/10.1111/eci.12079.

Brennan, A. L., Geddes, D. M., Gyi, K. M., & Baker, E. H. (2004). Clinical importance of cystic fibrosis-related diabetes. *Journal of Cystic Fibrosis: Official Journal of the European Cystic Fibrosis Society, 3*(4), 209–222. https://doi.org/10.1016/j.jcf.2004.08.001.

Brett, J., Schmidt, A. M., Yan, S. D., Zou, Y. S., Weidman, E., Pinsky, D., ... Stern, D. (1993). Survey of the distribution of a newly characterized receptor for advanced glycation end products in tissues. *The American Journal of Pathology, 143*(6), 1699–1712.

Buckley, S. T., Medina, C., Kasper, M., & Ehrhardt, C. (2011). Interplay between RAGE, CD44, and focal adhesion molecules in epithelial-mesenchymal transition of alveolar epithelial cells. *American Journal of, 300*(4), L548–L559. https://doi.org/10.1152/ajplung.00230.2010.

Buendía-Roldán, I., Mejía, M., Navarro, C., & Selman, M. (2017). Idiopathic pulmonary fibrosis: Clinical behavior and aging associated comorbidities. *Respiratory Medicine, 129*, 46–52. https://doi.org/10.1016/j.rmed.2017.06.001.

Byrne, A. L., Marais, B. J., Mitnick, C. D., Lecca, L., & Marks, G. B. (2015). Tuberculosis and chronic respiratory disease: A systematic review. *International Journal of Infectious Diseases, 32*, 138–146. https://doi.org/10.1016/j.ijid.2014.12.016.

Calfee, C. S., Ware, L. B., Eisner, M. D., Parsons, P. E., Thompson, B. T., Wickersham, N., & Matthay, M. A. (2008). Plasma receptor for advanced glycation end products and clinical outcomes in acute lung injury. *Thorax, 63*(12), 1083–1089. https://doi.org/10.1136/thx.2008.095588.

Caraher, E. J., Kwon, S., Haider, S. H., Crowley, G., Lee, A., Ebrahim, M., ... Nolan, A. (2017). Receptor for advanced glycation end-products and World Trade Center particulate induced lung function loss: A case-cohort study and murine model of acute particulate exposure. *PLoS One, 12*(9), e0184331. https://doi.org/10.1371/journal.pone.0184331.

Caram, L. M., de, O., Ferrari, R., Nogueira, D. L., Oliveira, M. R. M., Francisqueti, F. V., ... Godoy, I. (2017). Tumor necrosis factor receptor 2 as a possible marker of COPD in smokers and ex-smokers. *International Journal of Chronic Obstructive Pulmonary Disease, 12*, 2015–2021. https://doi.org/10.2147/COPD.S138558.

Castaldi, P. J., Cho, M. H., Litonjua, A. A., Bakke, P., Gulsvik, A., Lomas, D. A., ... Silverman, E. K. (2011). The Association of Genome-wide significant spirometric loci with chronic obstructive pulmonary disease susceptibility. *American Journal of Respiratory Cell and Molecular Biology, 45*(6), 1147–1153. https://doi.org/10.1165/rcmb.2011-0055OC.

Cayrol, C., & Girard, J.-P. (2014). IL-33: An alarmin cytokine with crucial roles in innate immunity, inflammation and allergy. *Current Opinion in Immunology, 31*, 31–37. https://doi.org/10.1016/j.coi.2014.09.004.

Chang, J. S., Wendt, T., Qu, W., Kong, L., Zou, Y. S., Schmidt, A. M., & Yan, S.-F. (2008). Oxygen deprivation triggers upregulation of early growth response-1 by the receptor for advanced glycation end products. *Circulation Research, 102*(8), 905–913. https://doi.org/10.1161/CIRCRESAHA.107.165308.

Chen, L., Wang, T., Wang, X., Sun, B.-B., Li, J.-Q., Liu, D.-S., ... Wen, F.-Q. (2009). Blockade of advanced glycation end product formation attenuates bleomycin-induced pulmonary fibrosis in rats. *Respiratory Research, 10*(1), 55. https://doi.org/10.1186/1465-9921-10-55.

Chen, L., Wang, T., Guo, L., Shen, Y., Yang, T., Wan, C., ... Wen, F. (2014). Overexpression of RAGE contributes to cigarette smoke-induced nitric oxide generation in COPD. *Lung, 192*(2), 267–275. https://doi.org/10.1007/s00408-014-9561-1.

Chen, M.-C., Chen, K.-C., Chang, G.-C., Lin, H., Wu, C.-C., Kao, W.-H., ... Yang, T.-Y. (2020). RAGE acts as an oncogenic role and promotes the metastasis of human lung cancer. *Cell Death & Disease, 11*(4), 265. https://doi.org/10.1038/s41419-020-2432-1.

Chen, X., Zhang, T., & Du, G. (2008). Advanced glycation end products serve as ligands for lectin-like oxidized low-density lipoprotein receptor-1(LOX-1): Biochemical and binding characterizations assay. *Cell Biochemistry and Function, 26*(7), 760–770. https://doi.org/10.1002/cbf.1502.

Cheng, D. T., Kim, D. K., Cockayne, D. A., Belousov, A., Bitter, H., Cho, M. H., ... TESRA and ECLIPSE Investigators (2013). Systemic soluble receptor for advanced glycation endproducts is a biomarker of emphysema and associated with AGER genetic variants in patients with chronic obstructive pulmonary disease. *American Journal of Respiratory and Critical Care Medicine, 188*(8), 948–957. https://doi.org/10.1164/rccm.201302-0247OC.

Christie, J. D., Shah, C. V., Kawut, S. M., Mangalmurti, N., Lederer, D. J., Sonett, J. R., ... Ware, L. B. (2009). Plasma levels of receptor for advanced glycation end products, blood transfusion, and risk of primary graft dysfunction. *American Journal of Respiratory and Critical Care Medicine, 180*(10), 1010–1015. https://doi.org/10.1164/rccm.200901-0118OC.

Cohen, M. J., Brohi, K., Calfee, C. S., Rahn, P., Chesebro, B. B., Christiaans, S. C., ... Pittet, J.-F. (2009). Early release of high mobility group box nuclear protein 1 after severe trauma in humans: Role of injury severity and tissue hypoperfusion. *Critical Care, 13*(6), R174. https://doi.org/10.1186/cc8152.

Cohen, M. J., Carles, M., Brohi, K., Calfee, C. S., Rahn, P., Call, M. S., ... Pittet, J.-F. (2010). Early release of soluble receptor for advanced glycation endproducts after severe trauma in humans. *Journal of Trauma and Acute Care Surgery, 68*(6), 1273. https://doi.org/10.1097/TA.0b013e3181db323e.

Cohn, L., Elias, J. A., & Chupp, G. L. (2004). Asthma: Mechanisms of disease persistence and progression. *Annual Review of Immunology, 22*(1), 789–815. https://doi.org/10.1146/annurev.immunol.22.012703.104716.

Corbel, M., Belleguic, C., Boichot, E., & Lagente, V. (2002). Involvement of gelatinases (MMP-2 and MMP-9) in the development of airway inflammation and pulmonary fibrosis. *Cell Biology and Toxicology, 18*(1), 51–61. https://doi.org/10.1023/a:1014471213371.

Cosio, M. G., Saetta, M., & Agusti, A. (2009). Immunologic aspects of chronic obstructive pulmonary disease. *The New England Journal of Medicine, 360*(23), 2445–2454. https://doi.org/10.1056/NEJMra0804752.

Coxson, H. O., Dirksen, A., Edwards, L. D., Yates, J. C., Agusti, A., Bakke, P., ... Vestbo, J. (2013). The presence and progression of emphysema in COPD as determined by CT scanning and biomarker expression: A prospective analysis from the ECLIPSE study. *The Lancet Respiratory Medicine, 1*(2), 129–136. https://doi.org/10.1016/S2213-2600(13)70006-7.

Dechristopher, L. (2012). *Consumption of fructose and high fructose corn syrup: Is "Fructositis" triggered bronchitis, asthma, & auto-immune reactivity merely a side bar in the etiology of metabolic syndrome II (to be defined)?—Evidence and a hypothesis.*

DeChristopher, L. R. (2013). *Consumption of fructose and high fructose corn syrup: Is fructositis triggered bronchitis, asthma, auto-immune reactivity merely a side bar in the etiology of metabolic syndrome II Kindle edition.*

DeChristopher, L. R., Uribarri, J., & Tucker, K. L. (2015). Intake of high fructose corn syrup sweetened soft drinks is associated with prevalent chronic bronchitis in U.S. Adults, ages 20–55 y. *Nutrition Journal, 14*, 107. https://doi.org/10.1186/s12937-015-0097-x.

DeChristopher, L. R., Uribarri, J., & Tucker, K. L. (2016). Intakes of apple juice, fruit drinks and soda are associated with prevalent asthma in US children aged 2-9 years. *Public Health Nutrition, 19*(1), 123–130. https://doi.org/10.1017/S1368980015000865.

Demling, N., Ehrhardt, C., Kasper, M., Laue, M., Knels, L., & Rieber, E. P. (2006). Promotion of cell adherence and spreading: A novel function of RAGE, the highly selective differentiation marker of human alveolar epithelial type I cells. *Cell and Tissue Research, 323*(3), 475–488. https://doi.org/10.1007/s00441-005-0069-0.

Downs, C. A. (2021). Analysis of RAGE proteome and interactome in lung adenocarcinoma using PANTHER and STRING databases. *Biological Research for Nursing, 23*(4), 698–707. https://doi.org/10.1177/10998004211021496.

Downs, C. A., Johnson, N. M., Tsaprailis, G., & Helms, M. N. (2018). RAGE-induced changes in the proteome of alveolar epithelial cells. *Journal of Proteomics, 177*, 11–20. https://doi.org/10.1016/j.jprot.2018.02.010.

Downs, C. A., Kreiner, L. H., Johnson, N. M., Brown, L. A., & Helms, M. N. (2015). Receptor for advanced glycation end-products regulates lung fluid balance via protein kinase C-gp91(phox) signaling to epithelial sodium channels. *American Journal of Respiratory Cell and Molecular Biology, 52*(1), 75–87. https://doi.org/10.1165/rcmb.2014-0002OC.

Echtenacher, B., Männel, D. N., & Hültner, L. (1996). Critical protective role of mast cells in a model of acute septic peritonitis. *Nature, 381*(6577), https://doi.org/10.1038/381075a0 Article 6577.

El-Seify, M. Y. H., Fouda, E. M., & Nabih, E. S. (2014). Serum level of soluble receptor for advanced glycation end products in asthmatic children and its correlation to severity and pulmonary functions. *Clinical Laboratory, 60*(6), 957–962. https://doi.org/10.7754/clin.lab.2013.130418.

Englert, J. M., Hanford, L. E., Kaminski, N., Tobolewski, J. M., Tan, R. J., Fattman, C. L., ... Oury, T. D. (2008). A role for the receptor for advanced glycation end products in idiopathic pulmonary fibrosis. *The American Journal of Pathology, 172*(3), 583–591. https://doi.org/10.2353/ajpath.2008.070569.

Englert, J. M., Kliment, C. R., Ramsgaard, L., Milutinovic, P. S., Crum, L., Tobolewski, J. M., & Oury, T. D. (2011). Paradoxical function for the receptor for advanced glycation end products in mouse models of pulmonary fibrosis. *International Journal of Clinical and Experimental Pathology, 4*(3), 241–254.

Erusalimsky, J. D. (2021). The use of the soluble receptor for advanced glycation-end products (sRAGE) as a potential biomarker of disease risk and adverse outcomes. *Redox Biology, 42*, 101958. https://doi.org/10.1016/j.redox.2021.101958.

Faruqui, T., Khan, M. S., Akhter, Y., Khan, S., Rafi, Z., Saeed, M., ... Yadav, D. K. (2022). RAGE inhibitors for targeted therapy of cancer: A comprehensive review. *International Journal of Molecular Sciences, 24*(1), 266. https://doi.org/10.3390/ijms24010266.

Fehrenbach, H., Kasper, M., Tschernig, T., Shearman, M. S., Schuh, D., & Müller, M. (1998). Receptor for advanced glycation endproducts (RAGE) exhibits highly differential cellular and subcellular localisation in rat and human lung. *Cellular and Molecular Biology (Noisy-le-Grand, France), 44*, 1147–1157.

Ferhani, N., Letuve, S., Kozhich, A., Thibaudeau, O., Grandsaigne, M., Maret, M., ... Pretolani, M. (2010). Expression of high-mobility group box 1 and of receptor for advanced glycation end products in chronic obstructive pulmonary disease. *American Journal of Respiratory and Critical Care Medicine, 181*(9), 917–927. https://doi.org/10.1164/rccm.200903-0340OC.

Fleming, T. H., Humpert, P. M., Nawroth, P. P., & Bierhaus, A. (2011). Reactive metabolites and AGE/RAGE-mediated cellular dysfunction affect the aging process: A mini-review. *Gerontology, 57*(5), 435–443. https://doi.org/10.1159/000322087.

Foell, D., Seeliger, S., Vogl, T., Koch, H.-G., Maschek, H., Harms, E., ... Roth, J. (2003). Expression of S100A12 (EN-RAGE) in cystic fibrosis. *Thorax, 58*(7), 613–617. https://doi.org/10.1136/thorax.58.7.613.

Gaens, K. H. J., Ferreira, I., Van Der Kallen, C. J. H., Van Greevenbroek, M. M. J., Blaak, E. E., Feskens, E. J. M., ... Schalkwijk, C. G. (2009). Association of polymorphism in the receptor for advanced glycation end products (RAGE) gene with circulating RAGE levels. *The Journal of Clinical Endocrinology & Metabolism, 94*(12), 5174–5180. https://doi.org/10.1210/jc.2009-1067.

Gkogkolou, P., & Böhm, M. (2012). Advanced glycation end products. *Dermato-endocrinology, 4*(3), 259–270. https://doi.org/10.4161/derm.22028.

Goldkorn, T., Filosto, S., & Chung, S. (2014). Lung injury and lung cancer caused by cigarette smoke-induced oxidative stress: Molecular mechanisms and therapeutic opportunities involving the ceramide-generating machinery and epidermal growth factor receptor. *Antioxidants & Redox Signaling, 21*(15), 2149–2174. https://doi.org/10.1089/ars.2013.5469.

Gopal, P., Reynaert, N. L., Scheijen, J. L. J. M., Engelen, L., Schalkwijk, C. G., Franssen, F. M. E., ... Rutten, E. P. A. (2014). Plasma advanced glycation end-products and skin autofluorescence are increased in COPD. *The European Respiratory Journal, 43*(2), 430–438. https://doi.org/10.1183/09031936.00135312.

Gopal, P., Reynaert, N. L., Scheijen, J. L. J. M., Schalkwijk, C. G., Franssen, F. M. E., Wouters, E. F. M., & Rutten, E. P. A. (2014). Association of plasma sRAGE, but not esRAGE with lung function impairment in COPD. *Respiratory Research, 15*(1), 24. https://doi.org/10.1186/1465-9921-15-24.

Guo, C., Jiang, X., Zeng, X., Wang, H., Li, H., Du, F., & Chen, B. (2016). Soluble receptor for advanced glycation end-products protects against ischemia/reperfusion-induced myocardial apoptosis via regulating the ubiquitin proteasome system. *Free Radical Biology and Medicine, 94*, 17–26. https://doi.org/10.1016/j.freeradbiomed.2016.02.011.

Guo, W. A., Davidson, B. A., Ottosen, J., Ohtake, P. J., Raghavendran, K., Mullan, B. A., ... Knight, P. R. (2012). Effect of high advanced glycation end product diet on pulmonary inflammatory response and pulmonary function following gastric aspiration. *Shock (Augusta, Ga.), 38*(6), 677–684. https://doi.org/10.1097/SHK.0b013e318273982e.

Guo, Y., Zhang, T., Wang, Z., Yu, F., Xu, Q., Guo, W., ... He, J. (2016). Body mass index and mortality in chronic obstructive pulmonary disease. *Medicine, 95*(28), e4225. https://doi.org/10.1097/MD.0000000000004225.

Haider, S. H., Veerappan, A., Crowley, G., Caraher, E. J., Ostrofsky, D., Mikhail, M., ... Nolan, A. (2020). Multiomics of World Trade Center particulate matter–induced persistent airway hyperreactivity. Role of receptor for advanced glycation end products. *American Journal of Respiratory Cell and Molecular Biology, 63*(2), 219–233. https://doi.org/10.1165/rcmb.2019-0064OC.

Halayko, A. J., & Ghavami, S. (2009). S100A8/A9: A mediator of severe asthma pathogenesis and morbidity? *Canadian Journal of Physiology and Pharmacology, 87*(10), 743–755. https://doi.org/10.1139/Y09-054.

Hancock, D. B., Eijgelsheim, M., Wilk, J. B., Gharib, S. A., Loehr, L. R., Marciante, K. D., ... London, S. J. (2010). Meta-analyses of genome-wide association studies identify multiple novel loci associated with pulmonary function. *Nature Genetics, 42*(1), 45–52. https://doi.org/10.1038/ng.500.

Hanford, L. E., Fattman, C. L., Shaefer, L. M., Enghild, J. J., Valnickova, Z., & Oury, T. D. (2003). Regulation of receptor for advanced glycation end products during bleomycin-induced lung injury. *American Journal of Respiratory Cell and Molecular Biology, 29*(3 Suppl.), S77–S81.

Hanson, C., Rutten, E. P., Wouters, E. F., & Rennard, S. (2014). Influence of diet and obesity on COPD development and outcomes. *International Journal of Chronic Obstructive Pulmonary Disease, 9*, 723–733. https://doi.org/10.2147/COPD.S50111.

He, M., Kubo, H., Ishizawa, K., Hegab, A. E., Yamamoto, Y., Yamamoto, H., & Yamaya, M. (2007). The role of the receptor for advanced glycation end-products in lung fibrosis. *American Journal of Physiology. Lung Cellular and Molecular Physiology, 293*(6), L1427–L1436. https://doi.org/10.1152/ajplung.00075.2007.

Henning, C., & Glomb, M. A. (2016). Pathways of the Maillard reaction under physiological conditions. *Glycoconjugate Journal, 33*(4), 499–512. https://doi.org/10.1007/s10719-016-9694-y.

Hergrueter, A. H., Nguyen, K., & Owen, C. A. (2011). Matrix metalloproteinases: All the RAGE in the acute respiratory distress syndrome. *American Journal of Physiology. Lung Cellular and Molecular Physiology, 300*(4), L512–L515. https://doi.org/10.1152/ajplung.00023.2011.

Hirose, A., Tanikawa, T., Mori, H., Okada, Y., & Tanaka, Y. (2010). Advanced glycation end products increase endothelial permeability through the RAGE/Rho signaling pathway. *FEBS Letters, 584*(1), 61–66. https://doi.org/10.1016/j.febslet.2009.11.082.

Hobbs, B. D., De Jong, K., Lamontagne, M., Bossé, Y., Shrine, N., Artigas, M. S., ... Cho, M. H. (2017). Genetic loci associated with chronic obstructive pulmonary disease overlap with loci for lung function and pulmonary fibrosis. *Nature Genetics, 49*(3), 426–432. https://doi.org/10.1038/ng.3752.

Hofmann, H.-S., Hansen, G., Burdach, S., Bartling, B., Silber, R.-E., & Simm, A. (2004). Discrimination of human lung neoplasm from normal lung by two target genes. *American Journal of Respiratory and Critical Care Medicine, 170*(5), 516–519. https://doi.org/10.1164/rccm.200401-127OC.

Hofmann, M. A., Drury, S., Fu, C., Qu, W., Taguchi, A., Lu, Y., ... Schmidt, A. M. (1999). RAGE mediates a novel proinflammatory axis: A central cell surface receptor for S100/calgranulin polypeptides. *Cell, 97*(7), 889–901. https://doi.org/10.1016/S0092-8674(00)80801-6.

Hofmann, M. A., Drury, S., Hudson, B. I., Gleason, M. R., Qu, W., Lu, Y., ... Schmidt, A. M. (2002). RAGE and arthritis: The G82S polymorphism amplifies the inflammatory response. *Genes and Immunity, 3*(3), 123–135. https://doi.org/10.1038/sj.gene.6363861.

Holgate, S. T., & Polosa, R. (2006). The mechanisms, diagnosis, and management of severe asthma in adults. *The Lancet, 368*(9537), 780–793. https://doi.org/10.1016/S0140-6736(06)69288-X.

Hoonhorst, S. J. M., Lo Tam Loi, A. T., Pouwels, S. D., Faiz, A., Telenga, E. D., Van den Berge, M., ... Ten Hacken, N. H. T. (2016). Advanced glycation endproducts and their receptor in different body compartments in COPD. *Respiratory Research, 17*(1), 46. https://doi.org/10.1186/s12931-016-0363-2.

Horiuchi, S., Higashi, T., Ikeda, K., Saishoji, T., Jinnouchi, Y., Sano, H., ... Araki, N. (1996). Advanced glycation end products and their recognition by macrophage and macrophage-derived cells. *Diabetes, 45*(Suppl. 3), S73–S76. https://doi.org/10.2337/diab.45.3.s73.

Hou, C., Zhao, H., Liu, L., Li, W., Zhou, X., Lv, Y., ... Zou, F. (2011). High mobility group protein B1 (HMGB1) in asthma: Comparison of patients with chronic obstructive pulmonary disease and healthy controls. *Molecular Medicine, 17*(7-8), 807–815. https://doi.org/10.2119/molmed.2010.00173.

Huber, M. B., Wacker, M. E., Vogelmeier, C. F., & Leidl, R. (2015). Comorbid influences on generic health-related quality of life in COPD: A systematic review. *PLoS One, 10*(7), e0132670. https://doi.org/10.1371/journal.pone.0132670.

Hunt, W. R., Helfman, B. R., McCarty, N. A., & Hansen, J. M. (2016). Advanced glycation end products are elevated in cystic fibrosis-related diabetes and correlate with worse lung function. *Journal of Cystic Fibrosis: Official Journal of the European Cystic Fibrosis Society, 15*(5), 681–688. https://doi.org/10.1016/j.jcf.2015.12.011.

Huttunen, H. J., & Rauvala, H. (2004). Amphoterin as an extracellular regulator of cell motility: From discovery to disease. *Journal of Internal Medicine, 255*(3), 351–366. https://doi.org/10.1111/j.1365-2796.2003.01301.x.

Ito, J. T., Lourenço, J. D., Righetti, R. F., Tibério, I. F. L. C., Prado, C. M., & Lopes, F. D. T. Q. S. (2019). Extracellular matrix component remodeling in respiratory diseases: What has been found in clinical and experimental studies? *Cells, 8*(4), 342. https://doi.org/10.3390/cells8040342.

Iwamoto, H., Gao, J., Pulkkinen, V., Toljamo, T., Nieminen, P., & Mazur, W. (2014). Soluble receptor for advanced glycation end-products and progression of airway disease. *BMC Pulmonary Medicine, 14*(1), 68. https://doi.org/10.1186/1471-2466-14-68.

Izushi, Y., Teshigawara, K., Liu, K., Wang, D., Wake, H., Takata, K., ... Nishibori, M. (2016). Soluble form of the receptor for advanced glycation end-products attenuates inflammatory pathogenesis in a rat model of lipopolysaccharide-induced lung injury. *Journal of Pharmacological Sciences, 130*(4), 226–234. https://doi.org/10.1016/j.jphs.2016.02.005.

Jabaudon, M., Futier, E., Roszyk, L., Chalus, E., Guerin, R., Petit, A., ... Constantin, J.-M. (2011). Soluble form of the receptor for advanced glycation end products is a marker of acute lung injury but not of severe sepsis in critically ill patients. *Critical Care Medicine, 39*(3), 480. https://doi.org/10.1097/CCM.0b013e318206b3ca.

James, A. L., & Wenzel, S. (2007). Clinical relevance of airway remodelling in airway diseases. *The European Respiratory Journal, 30*(1), 134–155. https://doi.org/10.1183/09031936.00146905.

James, A. W. (2001). ASTHMA. *Obstetrics and Gynecology Clinics of North America, 28*(2), 305–320. https://doi.org/10.1016/S0889-8545(05)70202-3.

Jing, R., Cui, M., Wang, J., & Wang, H. (2010). Receptor for advanced glycation end products (RAGE) soluble form (sRAGE): A new biomarker for lung cancer. *Neoplasma, 57*(1), 55–61. https://doi.org/10.4149/neo_2010_01_055.

Johnson, E. R., & Matthay, M. A. (2010). Acute lung injury: Epidemiology, pathogenesis, and treatment. *Journal of Aerosol Medicine and Pulmonary Drug Delivery, 23*(4), 243–252. https://doi.org/10.1089/jamp.2009.0775.

Jones, T. K., Feng, R., Kerchberger, V. E., Reilly, J. P., Anderson, B. J., Shashaty, M. G. S., ... Meyer, N. J. (2020). Plasma sRAGE acts as a genetically regulated causal intermediate in sepsis-associated acute respiratory distress syndrome. *American Journal of Respiratory and Critical Care Medicine, 201*(1), 47–56. https://doi.org/10.1164/rccm.201810-2033OC.

Kadonosono, T., Miyamoto, K., Sakai, S., Matsuo, Y., Kitajima, S., Wang, Q., ... Kizaka-Kondoh, S. (2022). AGE/RAGE axis regulates reversible transition to quiescent states of ALK-rearranged NSCLC and pancreatic cancer cells in monolayer cultures. *Scientific Reports, 12*(1), 9886. https://doi.org/10.1038/s41598-022-14272-0.

Kang, R., Hou, W., Zhang, Q., Chen, R., Lee, Y. J., Bartlett, D. L., ... Zeh, H. J. (2014). RAGE is essential for oncogenic KRAS-mediated hypoxic signaling in pancreatic cancer. *Cell Death & Disease, 5*(10), e1480. https://doi.org/10.1038/cddis.2014.445.

Kikkawa, T., Sato, N., Kojika, M., Takahashi, G., Aoki, K., Hoshikawa, K., ... Endo, S. (2010). Significance of measuring S100A12 and sRAGE in the serum of sepsis patients with postoperative acute lung injury. *Digestive Surgery, 27*(4), 307–312. https://doi.org/10.1159/000313687.

Kim, T., Kim, S. J., Choi, H., Shin, T. R., & Sim, Y. S. (2023). Diagnostic utility and tendency of bronchial and serum soluble receptor for advanced glycation end products (sRAGE) in lung cancer. *Cancers (Basel), 15*, 2819. https://doi.org/10.3390/cancers15102819.

Kinjo, T., Kitaguchi, Y., Droma, Y., Yasuo, M., Wada, Y., Ueno, F., ... Hanaoka, M. (2020). The Gly82Ser mutation in AGER contributes to pathogenesis of pulmonary fibrosis in combined pulmonary fibrosis and emphysema (CPFE) in Japanese patients. *Scientific Reports, 10*(1), 12811. https://doi.org/10.1038/s41598-020-69184-8.

Klein Wolterink, R. G. J., Kleinjan, A., Van Nimwegen, M., Bergen, I., De Bruijn, M., Levani, Y., & Hendriks, R. W. (2012). Pulmonary innate lymphoid cells are major producers of IL-5 and IL-13 in murine models of allergic asthma. *European Journal of Immunology, 42*(5), 1106–1116. https://doi.org/10.1002/eji.201142018.

Kollef, M. H., & Schuster, D. P. (1995). The acute respiratory distress syndrome. *The New England Journal of Medicine, 332*(1), 27–37. https://doi.org/10.1056/NEJM199501053320106.

Konishi, K., Gibson, K. F., Lindell, K. O., Richards, T. J., Zhang, Y., Dhir, R., ... Kaminski, N. (2009). Gene expression profiles of acute exacerbations of idiopathic pulmonary fibrosis. *American Journal of Respiratory and Critical Care Medicine, 180*(2), 167–175. https://doi.org/10.1164/rccm.200810-1596OC.

Lam, R., Kwon, S., Riggs, J., Sunseri, M., Crowley, G., Schwartz, T., ... Nolan, A. (2021). Dietary phenotype and advanced glycation end-products predict WTC-obstructive airways disease: A longitudinal observational study. *Respiratory Research, 22*, 19. https://doi.org/10.1186/s12931-020-01596-6.

Lee, C.-I., Guh, J.-Y., Chen, H.-C., Lin, K.-H., Yang, Y.-L., Hung, W.-C., ... Chuang, L.-Y. (2004). Leptin and connective tissue growth factor in advanced glycation end-product-induced effects in NRK-49F cells. *Journal of Cellular Biochemistry, 93*(5), 940–950. https://doi.org/10.1002/jcb.20222.

Lee, J.-H. (2017). Pathogenesis of COPD. In S.-D. Lee (Ed.). *COPD: Heterogeneity and personalized treatment* (pp. 35–54). Springer https://doi.org/10.1007/978-3-662-47178-4_4.

Li, Y., Yang, C., Ma, G., Gu, X., Chen, M., Chen, Y., ... Li, K. (2014). Association of polymorphisms of the receptor for advanced glycation end products gene with COPD in the Chinese population. *DNA and Cell Biology, 33*(4), 251–258. https://doi.org/10.1089/dna.2013.2303.

Li, Y. M., Mitsuhashi, T., Wojciechowicz, D., Shimizu, N., Li, J., Stitt, A., ... Vlassara, H. (1996). Molecular identity and cellular distribution of advanced glycation endproduct receptors: Relationship of p60 to OST-48 and p90 to 80K-H membrane proteins. *Proceedings of the National Academy of Sciences of the United States of America, 93*(20), 11047–11052. https://doi.org/10.1073/pnas.93.20.11047.

Libby, P., Ridker, P. M., & Hansson, G. K. (2011). Progress and challenges in translating the biology of atherosclerosis. *Nature, 473*(7347), https://doi.org/10.1038/nature10146 Article 7347.

Liliensiek, B., Weigand, M. A., Bierhaus, A., Nicklas, W., Kasper, M., Hofer, S., ... Arnold, B. (2004). Receptor for advanced glycation end products (RAGE) regulates sepsis but not the adaptive immune response. *The Journal of Clinical Investigation, 113*(11), 1641–1650. https://doi.org/10.1172/JCI18704.

Lin, Z., Yu, B., Yuan, L., Tu, J., Shao, C., & Tang, Y. (2022). RAGE is a potential biomarker implicated in immune infiltrates and cellular senescence in lung adenocarcinoma. *Journal of Clinical Laboratory Analysis, 36*(5), e24382. https://doi.org/10.1002/jcla.24382.

Liu, S., Stolz, D. B., Sappington, P. L., Macias, C. A., Killeen, M. E., Tenhunen, J. J., ... Fink, M. P. (2006). HMGB1 is secreted by immunostimulated enterocytes and contributes to cytomix-induced hyperpermeability of Caco-2 monolayers. *American Journal of Physiology-Cell Physiology, 290*(4), C990–C999. https://doi.org/10.1152/ajpcell.00308.2005.

Lizotte, P.-P., Hanford, L. E., Enghild, J. J., Nozik-Grayck, E., Giles, B.-L., & Oury, T. D. (2007). Developmental expression of the receptor for advanced glycation end-products (RAGE) and its response to hyperoxia in the neonatal rat lung. *BMC Developmental Biology, 7*, 15. https://doi.org/10.1186/1471-213X-7-15.

Lloyd, C. M., & Saglani, S. (2015). Epithelial cytokines and pulmonary allergic inflammation. *Current Opinion in Immunology, 34*, 52–58. https://doi.org/10.1016/j.coi.2015.02.001.

Lohwasser, C., Neureiter, D., Weigle, B., Kirchner, T., & Schuppan, D. (2006). The receptor for advanced glycation end products is highly expressed in the skin and upregulated by advanced glycation end products and tumor necrosis factor-alpha. *The Journal of Investigative Dermatology, 126*(2), 291–299. https://doi.org/10.1038/sj.jid.5700070.

Loughlin, D. T., & Artlett, C. M. (2010). Precursor of advanced glycation end products mediates ER-stress-induced caspase-3 activation of human dermal fibroblasts through NAD(P)H oxidase 4. *PLoS One, 5*(6), e11093. https://doi.org/10.1371/journal.pone.0011093.

Lubitz, I., Ricny, J., Atrakchi-Baranes, D., Shemesh, C., Kravitz, E., Liraz-Zaltsman, S., ... Schnaider-Beeri, M. (2016). High dietary advanced glycation end products are associated with poorer spatial learning and accelerated Aβ deposition in an Alzheimer mouse model. *Aging Cell, 15*(2), 309–316. https://doi.org/10.1111/acel.12436.

MacDowell, A. L., & Peters, S. P. (2007). Neutrophils in asthma. *Current Allergy and Asthma Reports, 7*(6), 464–468. https://doi.org/10.1007/s11882-007-0071-6.

Machahua, C., Montes-Worboys, A., Llatjos, R., Escobar, I., Dorca, J., Molina-Molina, M., & Vicens-Zygmunt, V. (2016). Increased AGE-RAGE ratio in idiopathic pulmonary fibrosis. *Respiratory Research, 17*(1), 144. https://doi.org/10.1186/s12931-016-0460-2.

Makam, M., Diaz, D., Laval, J., Gernez, Y., Conrad, C. K., Dunn, C. E., ... Tirouvanziam, R. (2009). Activation of critical, host-induced, metabolic and stress pathways marks neutrophil entry into cystic fibrosis lungs. *Proceedings of the National Academy of Sciences of the United States of America, 106*(14), 5779–5783. https://doi.org/10.1073/pnas.0813410106.

Malaviya, R., Ikeda, T., Ross, E., & Abraham, S. N. (1996). Mast cell modulation of neutrophil influx and bacterial clearance at sites of infection through TNF-α. *Nature, 381*(6577), https://doi.org/10.1038/381077a0.

Malik, P., Hoidal, J. R., & Mukherjee, T. K. (2021). Implication of RAGE polymorphic variants in COPD complication and Anti-COPD therapeutic potential of sRAGE. *COPD, 18*(6), 737–748. https://doi.org/10.1080/15412555.2021.1984417.

Maslan, J., & Mims, J. W. (2014). What is asthma? Pathophysiology, demographics, and health care costs. *Otolaryngologic Clinics of North America, 47*(1), 13–22. https://doi.org/10.1016/j.otc.2013.09.010.

Miłkowska-Dymanowska, J., Białas, A. J., Szewczyk, K., Kurmanowska, Z., Górski, P., & Piotrowski, W. J. (2018). The usefulness of soluble receptor for advanced glycation end-products in the identification of COPD frequent exacerbator phenotype. *International Journal of Chronic Obstructive Pulmonary Disease, 13*, 3879–3884. https://doi.org/10.2147/COPD.S186170.

Miller, S., Henry, A. P., Hodge, E., Kheirallah, A. K., Billington, C. K., Rimington, T. L., ... Sayers, I. (2016). The Ser82 RAGE variant affects lung function and serum RAGE in smokers and sRAGE production in vitro. *PLoS One, 11*(10), e0164041. https://doi.org/10.1371/journal.pone.0164041.

Milutinovic, P. S., Alcorn, J. F., Englert, J. M., Crum, L. T., & Oury, T. D. (2012). The receptor for advanced glycation end products is a central mediator of asthma pathogenesis. *The American Journal of Pathology, 181*(4), 1215–1225. https://doi.org/10.1016/j.ajpath.2012.06.031.

Miniati, M., Monti, S., Basta, G., Cocci, F., Fornai, E., & Bottai, M. (2011). Soluble receptor for advanced glycation end products in COPD: Relationship with emphysema and chronic cor pulmonale: A case-control study. *Respiratory Research, 12*(1), 37. https://doi.org/10.1186/1465-9921-12-37.

Morbini, P., Villa, C., Campo, I., Zorzetto, M., Inghilleri, S., & Luisetti, M. (2006). The receptor for advanced glycation end products and its ligands: A new inflammatory pathway in lung disease? *Modern Pathology, 19*(11), 1437–1445. https://doi.org/10.1038/modpathol.3800661.

Mukherjee, T. K., Mukhopadhyay, S., & Hoidal, J. R. (2005). The role of reactive oxygen species in TNFalpha-dependent expression of the receptor for advanced glycation end products in human umbilical vein endothelial cells. *Biochimica et Biophysica Acta, 1744*(2), 213–223. https://doi.org/10.1016/j.bbamcr.2005.03.007.

Mukherjee, T. K., Malik, P., & Hoidal, J. R. (2021). Receptor for Advanced Glycation End Products (RAGE) and its polymorphic variants as predictive diagnostic and prognostic markers of NSCLCs: A perspective. *Current Oncology Reports, 23*(1), 12. https://doi.org/10.1007/s11912-020-00992-x.

Mulrennan, S., Baltic, S., Aggarwal, S., Wood, J., Miranda, A., Frost, F., ... Thompson, P. J. (2015). The role of receptor for advanced glycation end products in airway inflammation in CF and CF related diabetes. *Scientific Reports, 5*, 8931. https://doi.org/10.1038/srep08931.

Napier, C. O., Mbadugha, O., Bienenfeld, L. A., Doucette, J. T., Lucchini, R., Luna-Sánchez, S., & De La Hoz, R. E. (2017). Obesity and weight gain among former World Trade Center workers and volunteers. *Archives of Environmental & Occupational Health, 72*(2), 106–110. https://doi.org/10.1080/19338244.2016.1197174.

Newman, V., Gonzalez, R. F., Matthay, M. A., & Dobbs, L. G. (2000). A novel alveolar type I cell-specific biochemical marker of human acute lung injury. *American Journal of Respiratory and Critical Care Medicine, 161*(3 Pt 1), 990–995. https://doi.org/10.1164/ajrccm.161.3.9901042.

Niu, H., Niu, W., Yu, T., Dong, F., Huang, K., Duan, R., ... Wang, C. (2019). Association of RAGE gene multiple variants with the risk for COPD and asthma in northern Han Chinese. *Aging, 11*(10), 3220–3237. https://doi.org/10.18632/aging.101975.

Oczypok, E. A., Perkins, T. N., & Oury, T. D. (2017). All the "RAGE" in lung disease: The receptor for advanced glycation endproducts (RAGE) is a major mediator of pulmonary inflammatory responses. *Paediatric Respiratory Reviews, 23*, 40–49. https://doi.org/10.1016/j.prrv.2017.03.012.

Oczypok, E. A., Milutinovic, P. S., Alcorn, J. F., Khare, A., Crum, L. T., Manni, M. L., ... Oury, T. D. (2015). Pulmonary receptor for advanced glycation endproducts promotes asthma pathogenesis via IL-33 and accumulation of group 2 innate lymphoid cells. *The Journal of Allergy and Clinical Immunology, 136*(3), 747–756.e4. https://doi.org/10.1016/j.jaci.2015.03.011.

Ogawa, E. N., Ishizaka, A., Tasaka, S., Koh, H., Ueno, H., Amaya, F., ... Takeda, J. (2006). Contribution of High-mobility Group Box-1 to the development of ventilator-induced lung injury. *American Journal of Respiratory and Critical Care Medicine, 174*(4), 400–407. https://doi.org/10.1164/rccm.200605-699OC.

Osawa, M., Yamamoto, Y., Munesue, S., Murakami, N., Sakurai, S., Watanabe, T., ... Yamamoto, H. (2007). De-N-glycosylation or G82S mutation of RAGE sensitizes its interaction with advanced glycation endproducts. *Biochimica et Biophysica Acta, 1770*(10), 1468–1474. https://doi.org/10.1016/j.bbagen.2007.07.003.

Ott, C., Jacobs, K., Haucke, E., Navarrete Santos, A., Grune, T., & Simm, A. (2014). Role of advanced glycation end products in cellular signaling. *Redox Biology, 2*, 411–429. https://doi.org/10.1016/j.redox.2013.12.016.

Ottosen, J. M., Mullan, B., Ohtake, P. J., Davidson, B. A., Guo, W., & Knight, P. R. (2010). Diet high in advanced glycation end products exacerbates pulmonary inflammatory response and impairs lung compliance in mice following gastric aspiration. *Journal of Surgical Research, 158*(2), 214. https://doi.org/10.1016/j.jss.2009.11.125.

Palumbo, R., Sampaolesi, M., De Marchis, F., Tonlorenzi, R., Colombetti, S., Mondino, A., ... Bianchi, M. E. (2004). Extracellular HMGB1, a signal of tissue damage, induces mesoangioblast migration and proliferation. *The Journal of Cell Biology, 164*(3), 441–449. https://doi.org/10.1083/jcb.200304135.

Pan, H., Niu, W., He, L., Wang, B., Cao, J., Zhao, F., ... Wu, H. (2013). Contributory role of five common polymorphisms of RAGE and APE1 genes in lung cancer among Han Chinese. *PLoS One, 8*(7), e69018. https://doi.org/10.1371/journal.pone.0069018.

Park, I. H., Yeon, S. I., Youn, J. H., Choi, J. E., Sasaki, N., Choi, I.-H., & Shin, J.-S. (2004). Expression of a novel secreted splice variant of the receptor for advanced glycation end products (RAGE) in human brain astrocytes and peripheral blood mononuclear cells. *Molecular Immunology, 40*(16), 1203–1211. https://doi.org/10.1016/j.molimm.2003.11.027.

Park, S., Blanck, H. M., Sherry, B., Jones, S. E., & Pan, L. (2013). Regular-soda intake independent of weight status is associated with asthma among US high school students. *Journal of the Academy of Nutrition and Dietetics, 113*(1), 106–111. https://doi.org/10.1016/j.jand.2012.09.020.

Park, W.-Y., Gray, J. M., Holewinski, R. J., Andresson, T., So, J. Y., Carmona-Rivera, C., ... Yang, L. (2023). Apoptosis-induced nuclear expulsion in tumor cells drives S100a4-mediated metastatic outgrowth through the RAGE pathway. *Nature Cancer, 4*(3), 419–435. https://doi.org/10.1038/s43018-023-00524-z.

Perkins, T. N., & Oury, T. D. (2021). The perplexing role of RAGE in pulmonary fibrosis: Causality or casualty? *Therapeutic Advances in Respiratory Disease, 15*, 17534666211016071. https://doi.org/10.1177/17534666211016071.

Perkins, T. N., Oczypok, E. A., Milutinovic, P. S., Dutz, R. E., & Oury, T. D. (2019). RAGE-dependent VCAM-1 expression in the lung endothelium mediates IL-33-induced allergic airway inflammation. *Allergy, 74*(1), 89–99. https://doi.org/10.1111/all.13500.

Perrone, A., Giovino, A., Benny, J., & Martinelli, F. (2020). Advanced glycation end products (AGEs): Biochemistry, signaling, analytical methods, and epigenetic effects. *Oxidative Medicine and Cellular Longevity, 2020*, 3818196. https://doi.org/10.1155/2020/3818196.

Pikor, L. A., Ramnarine, V. R., Lam, S., & Lam, W. L. (2013). Genetic alterations defining NSCLC subtypes and their therapeutic implications. *Lung Cancer (Amsterdam, Netherlands), 82*(2), 179–189. https://doi.org/10.1016/j.lungcan.2013.07.025.

Pinto, R. S., Minanni, C. A., de Araújo Lira, A. L., & Passarelli, M. (2022). Advanced glycation end products: A sweet flavor that embitters cardiovascular disease. *International Journal of Molecular Sciences, 23*(5), 2404. https://doi.org/10.3390/ijms23052404.

Pinto-Plata, V., Toso, J., Lee, K., Park, D., Bilello, J., Mullerova, H., ... Celli, B. (2007). Profiling serum biomarkers in patients with COPD: Associations with clinical parameters. *Thorax, 62*(7), 595–601. https://doi.org/10.1136/thx.2006.064428.

Pouwels, S. D., Klont, F., Bischoff, R., & Ten Hacken, N. H. T. (2019). Confounding factors affecting sRAGE as a biomarker for chronic obstructive pulmonary disease. *American Journal of Respiratory and Critical Care Medicine, 200*(1), 114. https://doi.org/10.1164/rccm.201902-0356LE.

Pouwels, S. D., Klont, F., Kwiatkowski, M., Wiersma, V. R., Faiz, A., Van Den Berge, M., ... Ten Hacken, N. H. T. (2018). Cigarette smoking acutely decreases serum levels of the chronic obstructive pulmonary disease biomarker sRAGE. *American Journal of Respiratory and Critical Care Medicine, 198*(11), 1456–1458. https://doi.org/10.1164/rccm.201807-1249LE.

Prasad, K. (2013). Do statins have a role in reduction/prevention of post-PCI restenosis? *Cardiovascular Therapeutics, 31*(1), 12–26. https://doi.org/10.1111/j.1755-5922.2011.00302.x.

Prasad, K. (2019). AGE-RAGE stress in the pathophysiology of pulmonary hypertension and its treatment. *The International Journal of Angiology: Official Publication of the International College of Angiology, Inc, 28*(2), 71–79. https://doi.org/10.1055/s-0039-1687818.

Pricci, F., Leto, G., Amadio, L., Iacobini, C., Romeo, G., Cordone, S., ... Pugliese, G. (2000). Role of galectin-3 as a receptor for advanced glycosylation end products. *Kidney International. Supplement, 77*, S31–S39. https://doi.org/10.1046/j.1523-1755.2000.07706.x.

Queisser, M. A., Kouri, F. M., Königshoff, M., Wygrecka, M., Schubert, U., Eickelberg, O., & Preissner, K. T. (2008). Loss of RAGE in pulmonary fibrosis: Molecular relations to functional changes in pulmonary cell types. *American Journal of Respiratory Cell and Molecular Biology, 39*(3), 337–345. https://doi.org/10.1165/rcmb.2007-0244OC.

Racanelli, A. C., Kikkers, S. A., Choi, A. M. K., & Cloonan, S. M. (2018). Autophagy and inflammation in chronic respiratory disease. *Autophagy, 14*(2), 221–232. https://doi.org/10.1080/15548627.2017.1389823.

Ramasamy, R., Vannucci, S. J., Yan, S. S. D., Herold, K., Yan, S. F., & Schmidt, A. M. (2005). Advanced glycation end products and RAGE: A common thread in aging, diabetes, neurodegeneration, and inflammation. *Glycobiology, 15*(7), 16R–28R. https://doi.org/10.1093/glycob/cwi053.

Ramsgaard, L., Englert, J. M., Tobolewski, J., Tomai, L., Fattman, C. L., Leme, A. S., ... Oury, T. D. (2010). The role of the receptor for advanced glycation end-products in a murine model of silicosis. *PLoS One, 5*(3), e9604. https://doi.org/10.1371/journal.pone.0009604.

Ray, R., Jangde, N., Singh, S. K., Sinha, S., & Rai, V. (2020). Lysophosphatidic acid-RAGE axis promotes lung and mammary oncogenesis via protein kinase B and regulating tumor microenvironment. *Cell Communication and Signaling: CCS, 18*(1), 170. https://doi.org/10.1186/s12964-020-00666-y.

Repapi, E., Sayers, I., Wain, L. V., Burton, P. R., Johnson, T., Obeidat, M., ... Tobin, M. D. (2010). Genome-wide association study identifies five loci associated with lung function. *Nature Genetics, 42*(1), 36–44. https://doi.org/10.1038/ng.501.

Repine, J. E. (1992). Scientific perspectives on adult respiratory distress syndrome. *Lancet (London, England), 339*(8791), 466–469. https://doi.org/10.1016/0140-6736(92)91067-i.

Reynaert, N. L., Gopal, P., Rutten, E. P. A., Wouters, E. F. M., & Schalkwijk, C. G. (2016). Advanced glycation end products and their receptor in age-related, non-communicable chronic inflammatory diseases; Overview of clinical evidence and potential contributions to disease. *The International Journal of Biochemistry & Cell Biology, 81*, 403–418. https://doi.org/10.1016/j.biocel.2016.06.016.

Reynolds, P. R., Kasteler, S. D., Schmitt, R. E., & Hoidal, J. R. (2011). Receptor for advanced glycation end-products signals through Ras during tobacco smoke-induced pulmonary inflammation. *American Journal of Respiratory Cell and Molecular Biology, 45*(2), 411–418. https://doi.org/10.1165/rcmb.2010-0231OC.

Reynolds, P. R., Kasteler, S. D., Cosio, M. G., Sturrock, A., Huecksteadt, T., & Hoidal, J. R. (2008). RAGE: Developmental expression and positive feedback regulation by Egr-1 during cigarette smoke exposure in pulmonary epithelial cells. *American Journal of Physiology-Lung Cellular and Molecular Physiology, 294*(6), L1094–L1101. https://doi.org/10.1152/ajplung.00318.2007.

Robinson, A. B., Johnson, K. D., Bennion, B. G., & Reynolds, P. R. (2012). RAGE signaling by alveolar macrophages influences tobacco smoke-induced inflammation. *American Journal of Physiology-Lung Cellular and Molecular Physiology, 302*(11), L1192–L1199. https://doi.org/10.1152/ajplung.00099.2012.

Rosenstein, B. J., & Cutting, G. R. (1998). The diagnosis of cystic fibrosis: A consensus statement. Cystic Fibrosis Foundation Consensus Panel. *The Journal of Pediatrics, 132*(4), 589–595. https://doi.org/10.1016/s0022-3476(98)70344-0.

Rossini, A., Zacheo, A., Mocini, D., Totta, P., Facchiano, A., Castoldi, R., ... Germani, A. (2008). HMGB1-stimulated human primary cardiac fibroblasts exert a paracrine action on human and murine cardiac stem cells. *Journal of Molecular and Cellular Cardiology, 44*(4), 683–693. https://doi.org/10.1016/j.yjmcc.2008.01.009.

Roth, J., Vogl, T., Sorg, C., & Sunderkötter, C. (2003). Phagocyte-specific S100 proteins: A novel group of proinflammatory molecules. *Trends in Immunology, 24*(4), 155–158. https://doi.org/10.1016/s1471-4906(03)00062-0.

Salari-Moghaddam, A., Milajerdi, A., Larijani, B., & Esmaillzadeh, A. (2019). Processed red meat intake and risk of COPD: A systematic review and dose-response meta-analysis of prospective cohort studies. *Clinical Nutrition (Edinburgh, Scotland), 38*(3), 1109–1116. https://doi.org/10.1016/j.clnu.2018.05.020.

Sambamurthy, N., Leme, S., Oury, T. D., & Shapiro, S. D. (2015). The receptor for advanced glycation end products (RAGE) contributes to the progression of emphysema in mice. *PLoS One, 10*(3), e0118979. https://doi.org/10.1371/journal.pone.0118979.

Sanders, K. A., Delker, D. A., Huecksteadt, T., Beck, E., Wuren, T., Chen, Y., ... Hoidal, J. R. (2019). RAGE is a critical mediator of pulmonary oxidative stress, alveolar macrophage activation and emphysema in response to cigarette smoke. *Scientific Reports, 9*(1), 231. https://doi.org/10.1038/s41598-018-36163-z.

Schenk, S., Schraml, P., Bendik, I., & Ludwig, C. U. (2001). A novel polymorphism in the promoter of the RAGE gene is associated with non-small cell lung cancer. *Lung Cancer (Amsterdam, Netherlands), 32*(1), 7–12. https://doi.org/10.1016/s0169-5002(00)00209-9.

Schmidt, A. M., Vianna, M., Gerlach, M., Brett, J., Ryan, J., Kao, J., ... Clauss, M. (1992). Isolation and characterization of two binding proteins for advanced glycosylation end products from bovine lung which are present on the endothelial cell surface. *The Journal of Biological Chemistry, 267*(21), 14987–14997.

Schraml, P., Bendik, I., & Ludwig, C. U. (1997). Differential messenger RNA and protein expression of the receptor for advanced glycosylated end products in normal lung and non-small cell lung carcinoma. *Cancer Research, 57*(17), 3669–3671.

Schraml, P., Shipman, R., Colombi, M., & Ludwig, C. U. (1994). Identification of genes differentially expressed in normal lung and non-small cell lung carcinoma tissue. *Cancer Research, 54*(19), 5236–5240.

Scoditti, E., Massaro, M., Garbarino, S., & Toraldo, D. M. (2019). Role of diet in chronic obstructive pulmonary disease prevention and treatment. *Nutrients, 11*(6), 1357. https://doi.org/10.3390/nu11061357.

Selman, M., & Pardo, A. (2003). The epithelial/fibroblastic pathway in the pathogenesis of idiopathic pulmonary fibrosis. *American Journal of Respiratory Cell and Molecular Biology, 29*(3 Suppl), S93–S97.

Sharma, A., Kaur, S., Sarkar, M., Sarin, B. C., & Changotra, H. (2021). The AGE-RAGE axis and rage genetics in chronic obstructive pulmonary disease. *Clinical Reviews in Allergy & Immunology, 60*(2), 244–258. https://doi.org/10.1007/s12016-020-08815-4.

Shaw, D. E., Berry, M. A., Hargadon, B., McKenna, S., Shelley, M. J., Green, R. H., ... Pavord, I. D. (2007). Association between neutrophilic airway inflammation and airflow limitation in adults with asthma. *Chest, 132*(6), 1871–1875. https://doi.org/10.1378/chest.07-1047.

Shi, Z., Dal Grande, E., Taylor, A. W., Gill, T. K., Adams, R., & Wittert, G. A. (2012). Association between soft drink consumption and asthma and chronic obstructive pulmonary disease among adults in Australia. *Respirology (Carlton, Vic.), 17*(2), 363–369. https://doi.org/10.1111/j.1440-1843.2011.02115.x.

Shim, E.-J., Chun, E., Lee, H.-S., Bang, B.-R., Kim, T.-W., Cho, S.-H., ... Park, H.-W. (2012). The role of high-mobility group box-1 (HMGB1) in the pathogenesis of asthma. *Clinical and Experimental Allergy: Journal of the British Society for Allergy and Clinical Immunology, 42*(6), 958–965. https://doi.org/10.1111/j.1365-2222.2012.03998.x.

Shirasawa, M., Fujiwara, N., Hirabayashi, S., Ohno, H., Iida, J., Makita, K., & Hata, Y. (2004). Receptor for advanced glycation end-products is a marker of type I lung alveolar cells. *Genes to Cells, 9*(2), 165–174. https://doi.org/10.1111/j.1356-9597.2004.00712.x.

Siegel, R. L., Miller, K. D., Fuchs, H. E., & Jemal, A. (2022). Cancer statistics, 2022. *CA: A Cancer Journal for Clinicians, 72*(1), 7–33. https://doi.org/10.3322/caac.21708.

Singh, R., Barden, A., Mori, T., & Beilin, L. (2001). Advanced glycation end-products: A review. *Diabetologia, 44*(2), 129–146. https://doi.org/10.1007/s001250051591.

Smith, D. J., Yerkovich, S. T., Towers, M. A., Carroll, M. L., Thomas, R., & Upham, J. W. (2011). Reduced soluble receptor for advanced glycation end-products in COPD. *European Respiratory Journal, 37*(3), 516–522. https://doi.org/10.1183/09031936.00029310.

Somayaji, R., & Chalmers, J. D. (2022). Just breathe: A review of sex and gender in chronic lung disease. *European Respiratory Review, 31*(163), 210111. https://doi.org/10.1183/16000617.0111-2021.

Somensi, N., Brum, P. O., De Miranda Ramos, V., Gasparotto, J., Zanotto-Filho, A., Rostirolla, D. C., ... Pens Gelain, D. (2017). Extracellular HSP70 activates ERK1/2, NF-kB and pro-inflammatory gene transcription through binding with RAGE in A549 human lung cancer cells. *Cellular Physiology and Biochemistry: International Journal of Experimental Cellular Physiology, Biochemistry, and Pharmacology, 42*(6), 2507–2522. https://doi.org/10.1159/000480213.

Song, J. S., Kang, C. M., Park, C. K., Yoon, H. K., Lee, S. Y., Ahn, J. H., & Moon, H.-S. (2011). Inhibitory effect of receptor for advanced glycation end products (RAGE) on the TGF-β-induced alveolar epithelial to mesenchymal transition. Article 9 *Experimental & Molecular Medicine, 43*(9), https://doi.org/10.3858/emm.2011.43.9.059.

Spelta, F., Fratta Pasini, A. M., Cazzoletti, L., & Ferrari, M. (2018). Body weight and mortality in COPD: Focus on the obesity paradox. *Eating and Weight Disorders: EWD, 23*(1), 15–22. https://doi.org/10.1007/s40519-017-0456-z.

Stav, D., Bar, I., & Sandbank, J. (2007). Usefulness of CDK5RAP3, CCNB2, and RAGE genes for the diagnosis of lung adenocarcinoma. *The International Journal of Biological Markers, 22*(2), 108–113. https://doi.org/10.1177/172460080702200204.

Su, X., Looney, M. R., Gupta, N., & Matthay, M. A. (2009). Receptor for advanced glycation end-products (RAGE) is an indicator of direct lung injury in models of experimental lung injury. *American Journal of Physiology-Lung Cellular and Molecular Physiology, 297*(1), L1–L5. https://doi.org/10.1152/ajplung.90546.2008.

Sukkar, M. B., Wood, L. G., Tooze, M., Simpson, J. L., McDonald, V. M., Gibson, P. G., & Wark, P. a B. (2012). Soluble RAGE is deficient in neutrophilic asthma and COPD. *European Respiratory Journal, 39*(3), 721–729. https://doi.org/10.1183/09031936.00022011.

Sutherland, K. D., & Berns, A. (2010). Cell of origin of lung cancer. *Molecular Oncology, 4*(5), 397–403. https://doi.org/10.1016/j.molonc.2010.05.002.

Taguchi, A., Blood, D. C., del Toro, G., Canet, A., Lee, D. C., Qu, W., ... Schmidt, A. M. (2000). Blockade of RAGE-amphoterin signalling suppresses tumour growth and metastases. *Nature, 405*(6784), 354–360. https://doi.org/10.1038/35012626.

Taniguchi, A., Tsuge, M., Miyahara, N., & Tsukahara, H. (2021). Reactive oxygen species and antioxidative defense in chronic obstructive pulmonary disease. *Antioxidants (Basel, Switzerland), 10*(10), 1537. https://doi.org/10.3390/antiox10101537.

Thakur, D., Taliaferro, O., Atkinson, M., Stoffel, R., Guleria, R. S., & Gupta, S. (2022). Inhibition of nuclear factor κB in the lungs protect bleomycin-induced lung fibrosis in mice. *Molecular Biology Reports, 49*(5), 3481–3490. https://doi.org/10.1007/s11033-022-07185-8.

Tian, M., Chang, X., Zhang, Q., Li, C., Li, S., & Sun, Y. (2019). TGF-β1 mediated MAPK signaling pathway promotes collagen formation induced by Nano NiO in A549 cells. *Environmental Toxicology, 34*(6), 719–727. https://doi.org/10.1002/tox.22738.

Todd, N. W., Luzina, I. G., & Atamas, S. P. (2012). Molecular and cellular mechanisms of pulmonary fibrosis. *Fibrogenesis & Tissue Repair, 5*(1), 11. https://doi.org/10.1186/1755-1536-5-11.

Uchida, T., Shirasawa, M., Ware, L. B., Kojima, K., Hata, Y., Makita, K., ... Matthay, M. A. (2006). Receptor for advanced glycation end-products is a marker of type I cell injury in acute lung injury. *American Journal of Respiratory and Critical Care Medicine, 173*(9), 1008–1015. https://doi.org/10.1164/rccm.200509-1477OC.

Ullah, M. A., Loh, Z., Gan, W. J., Zhang, V., Yang, H., Li, J. H., ... Sukkar, M. B. (2014). Receptor for advanced glycation end products and its ligand high-mobility group box-1 mediate allergic airway sensitization and airway inflammation. *Journal of Allergy and Clinical Immunology, 134*(2), 440–450.e3. https://doi.org/10.1016/j.jaci.2013.12.1035.

Van Zoelen, M. A. D., Yang, H., Florquin, S., Meijers, J. C. M., Akira, S., Arnold, B., ... Van, D. (2009). Role of toll-like receptors 2 and 4, and the receptor for advanced glycation end products in high-mobility group box 1-induced inflammation in vivo. *Shock (Augusta, Ga.), 31*(3), 280. https://doi.org/10.1097/SHK.0b013e318186262d.

Varraso, R., & Camargo, C. A. (2015). The influence of processed meat consumption on chronic obstructive pulmonary disease. *Expert Review of Respiratory Medicine, 9*(6), 703–710. https://doi.org/10.1586/17476348.2015.1105743.

Varraso, R., Dumas, O., Boggs, K. M., Willett, W. C., Speizer, F. E., & Camargo, C. A. (2019). Processed meat intake and risk of chronic obstructive pulmonary disease among middle-aged women. *EClinicalMedicine, 14*, 88–95. https://doi.org/10.1016/j.eclinm.2019.07.014.

Veerappan, A., Oskuei, A., Crowley, G., Mikhail, M., Ostrofsky, D., Gironda, Z., ... Nolan, A. (2020). World Trade Center-cardiorespiratory and vascular dysfunction: Assessing the phenotype and metabolome of a murine particulate matter exposure model. *Scientific Reports, 10*, 3130. https://doi.org/10.1038/s41598-020-58717-w.

Venter, C., Pickett, K., Starling, A., Maslin, K., Smith, P. K., Palumbo, M. P., ... Dabelea, D. (2021). Advanced glycation end product intake during pregnancy and offspring allergy outcomes: A prospective cohort study. *Clinical & Experimental Allergy, 51*(11), 1459–1470. https://doi.org/10.1111/cea.14027.

Vistoli, G., De Maddis, D., Cipak, A., Zarkovic, N., Carini, M., & Aldini, G. (2013). Advanced glycoxidation and lipoxidation end products (AGEs and ALEs): An overview of their mechanisms of formation. *Free Radical Research, 47*(Suppl. 1), 3–27. https://doi.org/10.3109/10715762.2013.815348.

Vlassara, H., & Uribarri, J. (2014). Advanced glycation end products (AGE) and diabetes: Cause, effect, or both? *Current Diabetes Reports, 14*(1), 453. https://doi.org/10.1007/s11892-013-0453-1.

Vogelmeier, C. F., Criner, G. J., Martinez, F. J., Anzueto, A., Barnes, P. J., Bourbeau, J., ... Agustí, A. (2017). Global strategy for the diagnosis, management, and prevention of chronic obstructive lung disease 2017 report. GOLD executive summary. *American Journal of Respiratory and Critical Care Medicine, 195*(5), 557–582. https://doi.org/10.1164/rccm.201701-0218PP.

Wang, C., Li, D., Qian, Y., Wang, J., & Jing, H. (2012). Increased matrix metalloproteinase-9 activity and mRNA expression in lung injury following cardiopulmonary bypass. *Laboratory Investigation; A Journal of Technical Methods and Pathology, 92*(6), 910–916. https://doi.org/10.1038/labinvest.2012.50.

Wang, H., Li, Y., Yu, W., Ma, L., Ji, X., & Xiao, W. (2015). Expression of the receptor for advanced glycation end-products and frequency of polymorphism in lung cancer. *Oncology Letters, 10*(1), 51–60. https://doi.org/10.3892/ol.2015.3200.

Wang, H., Wang, T., Yuan, Z., Cao, Y., Zhou, Y., He, J., ... Chen, L. (2018). Role of receptor for advanced glycation end products in regulating lung fluid balance in lipopolysaccharide-induced acute lung injury and infection-related acute respiratory distress syndrome. *Shock (Augusta, Ga.), 50*(4), 472. https://doi.org/10.1097/SHK.0000000000001032.

Wang, M., Gauthier, A., Daley, L., Dial, K., Wu, J., Woo, J., ... Mantell, L. L. (2019). The role of HMGB1, a nuclear damage-associated molecular pattern molecule, in the pathogenesis of lung diseases. *Antioxidants & Redox Signaling, 31*(13), 954–993. https://doi.org/10.1089/ars.2019.7818.

Wang, X., Cui, E., Zeng, H., Hua, F., Wang, B., Mao, W., & Feng, X. (2012). RAGE genetic polymorphisms are associated with risk, chemotherapy response and prognosis in patients with advanced NSCLC. *PLoS One, 7*(10), e43734. https://doi.org/10.1371/journal.pone.0043734.

Wang, Y., Xu, J., Meng, Y., Adcock, I. M., & Yao, X. (2018). Role of inflammatory cells in airway remodeling in COPD. *International Journal of Chronic Obstructive Pulmonary Disease, 13*, 3341–3348. https://doi.org/10.2147/COPD.S176122.

Waseda, K., Miyahara, N., Taniguchi, A., Kurimoto, E., Ikeda, G., Koga, H., ... Kanehiro, A. (2015). Emphysema requires the receptor for advanced glycation end-products triggering on structural cells. *American Journal of Respiratory Cell and Molecular Biology, 52*(4), 482–491. https://doi.org/10.1165/rcmb.2014-0027OC.

Watanabe, T., Asai, K., Fujimoto, H., Tanaka, H., Kanazawa, H., & Hirata, K. (2011). Increased levels of HMGB-1 and endogenous secretory RAGE in induced sputum from asthmatic patients. *Respiratory Medicine, 105*(4), 519–525. https://doi.org/10.1016/j.rmed.2010.10.016.

Weitzenblum, E., Hirth, C., Ducolone, A., Mirhom, R., Rasaholinjanahary, J., & Ehrhart, M. (1981). Prognostic value of pulmonary artery pressure in chronic obstructive pulmonary disease. *Thorax, 36*(10), 752–758. https://doi.org/10.1136/thx.36.10.752.

Willis, B. C., & Borok, Z. (2007). TGF-beta-induced EMT: Mechanisms and implications for fibrotic lung disease. *American Journal of, 293*(3), L525–L534. https://doi.org/10.1152/ajplung.00163.2007.

Wittkowski, H., Sturrock, A., van Zoelen, M. A. D., Viemann, D., van der Poll, T., Hoidal, J. R., ... Foell, D. (2007). Neutrophil-derived S100A12 in acute lung injury and respiratory distress syndrome. *Critical Care Medicine, 35*(5), 1369. https://doi.org/10.1097/01.CCM.0000262386.32287.29.

Wolf, L., Herr, C., Niederstraßer, J., Beisswenger, C., & Bals, R. (2017). Receptor for advanced glycation endproducts (RAGE) maintains pulmonary structure and regulates the response to cigarette smoke. *PLoS One, 12*(7), e0180092. https://doi.org/10.1371/journal.pone.0180092.

Wood, L. G. (2017). Diet, obesity, and asthma. *Annals of the American Thoracic Society, 14*(Suppl. 5), S332–S338. https://doi.org/10.1513/AnnalsATS.201702-124AW.

Wrolstad, R. (2012). *Food carbohydrate chemistry*. John Wiley & Sons, Inc.

Wu, L., Ma, L., Nicholson, L. F. B., & Black, P. N. (2011). Advanced glycation end products and its receptor (RAGE) are increased in patients with COPD. *Respiratory Medicine, 105*(3), 329–336. https://doi.org/10.1016/j.rmed.2010.11.001.

Wu, S., Mao, L., Li, Y., Yin, Y., Yuan, W., Chen, Y., ... Wu, J. (2018). RAGE may act as a tumour suppressor to regulate lung cancer development. *Gene, 651*, 86–93. https://doi.org/10.1016/j.gene.2018.02.009.

Yamaguchi, K., Iwamoto, H., Sakamoto, S., Horimasu, Y., Masuda, T., Miyamoto, S., ... Hattori, N. (2022). Association of the RAGE/RAGE-ligand axis with interstitial lung disease and its acute exacerbation. *Respiratory Investigation, 60*(4), 531–542. https://doi.org/10.1016/j.resinv.2022.04.004.

Yamaguchi, K., Iwamoto, H., Sakamoto, S., Horimasu, Y., Masuda, T., Miyamoto, S., ... Hattori, N. (2017). AGER rs2070600 polymorphism elevates neutrophil-lymphocyte ratio and mortality in metastatic lung adenocarcinoma. *Oncotarget, 8*(55), 94382–94392. https://doi.org/10.18632/oncotarget.21764.

Yang, J., Lin, M., Zhang, M., Wang, Z., Lin, H., Yu, Y., ... Li, J. (2023). Advanced glycation end products' receptor DNA methylation associated with immune infiltration and prognosis of lung adenocarcinoma and lung squamous cell carcinoma. *Genetics Research, 2023*, 7129325. https://doi.org/10.1155/2023/7129325.

Yang, Z., Tao, T., Raftery, M. J., Youssef, P., Di Girolamo, N., & Geczy, C. L. (2001). Proinflammatory properties of the human S100 protein S100A12. *Journal of Leukocyte Biology, 69*(6), 986–994. https://doi.org/10.1189/jlb.69.6.986.

Yang, Z., Yan, W. X., Cai, H., Tedla, N., Armishaw, C., Di Girolamo, N., ... Geczy, C. L. (2007). S100A12 provokes mast cell activation: A potential amplification pathway in asthma and innate immunity. *The Journal of Allergy and Clinical Immunology, 119*(1), 106–114. https://doi.org/10.1016/j.jaci.2006.08.021.

Yonekura, H., Yamamoto, Y., Sakurai, S., Petrova, R. G., Abedin, M. J., Li, H., ... Yamamoto, H. (2003). Novel splice variants of the receptor for advanced glycation endproducts expressed in human vascular endothelial cells and pericytes, and their putative roles in diabetes-induced vascular injury. *Biochemical Journal, 370*(Pt 3), 1097–1109. https://doi.org/10.1042/BJ20021371.

Young, R. P., Hay, B. A., & Hopkins, R. J. (2011). Does RAGE protect smokers from COPD? *European Respiratory Journal, 38*(3), 743–744. https://doi.org/10.1183/09031936.00041711.

Young, R. P., Hopkins, R. J., Whittington, C. F., Hay, B. A., Epton, M. J., & Gamble, G. D. (2011). Individual and cumulative effects of GWAS susceptibility loci in lung cancer: Associations after sub-phenotyping for COPD. *PLoS One, 6*(2), e16476. https://doi.org/10.1371/journal.pone.0016476.

Yu, Y. X., Pan, W. C., & Cheng, Y. F. (2017). Silencing of advanced glycosylation and glycosylation and product-specific receptor (RAGE) inhibits the metastasis and growth of non-small cell lung cancer. *American Journal of Translational Research, 9*(6), 2760–2774.

Zemans, R. L., Jacobson, S., Keene, J., Kechris, K., Miller, B. E., Tal-Singer, R., & Bowler, R. P. (2017). Multiple biomarkers predict disease severity, progression and mortality in COPD. *Respiratory Research, 18*, 117. https://doi.org/10.1186/s12931-017-0597-7.

Zhang, H., Mao, Y.-F., Zhao, Y., Xu, D.-F., Wang, Y., Xu, C.-F., ... Liu, Y.-J. (2021). Upregulation of matrix metalloproteinase-9 protects against sepsis-induced acute lung injury via promoting the release of soluble receptor for advanced glycation end products. *Oxidative Medicine and Cellular Longevity, 2021*, 8889313. https://doi.org/10.1155/2021/8889313.

Zhang, H., Tasaka, S., Shiraishi, Y., Fukunaga, K., Yamada, W., Seki, H., ... Ishizaka, A. (2008). Role of soluble receptor for advanced glycation end products on endotoxin-induced lung injury. *American Journal of Respiratory and Critical Care Medicine, 178*(4), 356–362. https://doi.org/10.1164/rccm.200707-1069OC.

Zhang, X., Xie, J., Sun, H., Wei, Q., & Nong, G. (2022). sRAGE inhibits the mucus hypersecretion in a mouse model with neutrophilic asthma. *Immunological Investigations, 51*(5), 1243–1256. https://doi.org/10.1080/08820139.2021.1928183.

Zheng, P.-F., Shu, L., Si, C.-J., Zhang, X.-Y., Yu, X.-L., & Gao, W. (2016). Dietary patterns and chronic obstructive pulmonary disease: A meta-analysis. *COPD: Journal of Chronic Obstructive Pulmonary Disease, 13*(4), 515–522. https://doi.org/10.3109/15412555.2015.1098606.

Zhou, G., Li, C., & Cai, L. (2004). Advanced glycation end-products induce connective tissue growth factor-mediated renal fibrosis predominantly through transforming growth factor β-independent pathway. *The American Journal of Pathology, 165*(6), 2033–2043.

CHAPTER TWELVE

Antioxidant and antibrowning properties of Maillard reaction products in food and biological systems

Majid Nooshkam and Mehdi Varidi*

Department of Food Science and Technology, Faculty of Agriculture, Ferdowsi University of Mashhad, Mashhad, Iran
*Corresponding author. e-mail address: m.varidi@um.ac.ir

Contents

1. Introduction	368
2. Chemistry of the Maillard reaction	369
3. MRPs as antioxidants	371
3.1 Antioxidant mechanisms	371
3.2 Antioxidant methods	372
3.3 Applications	384
4. MRPs as antibrowning agents	387
4.1 Antibrowning mechanism	387
4.2 Antibrowning effect	387
5. Deleterious effect of MRPs as glycation end products in biological systems	391
6. Conclusions	392
Acknowledgments	392
References	392

Abstract

Oxidative damage refers to the harm caused to biological systems by reactive oxygen species such as free radicals. This damage can contribute to a range of diseases and aging processes in organisms. Moreover, oxidative deterioration of lipids is a serious problem because it reduces the shelf life of food products, degrades their nutritional value, and produces reaction products that could be toxic. Antioxidants are effective compounds for preventing lipid oxidation, and synthetic antioxidants are frequently added to foods due to their high effectiveness and low cost. However, the safety of these antioxidants is a subject that is being discussed in the public more and more. Synthetic antioxidants have been found to have potential negative effects on health due to their ability to accumulate in tissues and disrupt natural antioxidant systems. During thermal processing and storage, foods containing reducing sugars and amino compounds frequently produce Maillard reaction products (MRPs). Through the chelation of metal ions, scavenging of reactive oxygen species, destruction of hydrogen peroxide, and suppression of radical chain reaction, MRPs exhibit excellent antioxidant properties in a variety of food products and biological systems. Also, the capacity of MRPs to chelate metals makes them as a potential inhibitor of the enzymatic browning in fruits and vegetables. In this book chapter, the methods used for the evaluation of antioxidant activity of MRPs are provided. Moreover, the antioxidant and antibrowning activities of MRPs in food and biological systems is discussed. MRPs can generally be isolated and used as commercial preparations of natural antioxidants.

1. Introduction

Oxidative damage refers to the harmful effects caused by reactive oxygen species (ROS) in biological systems, which can damage cellular components such as lipids, proteins, and DNA, leading to various diseases and aging (Shahidi & Ambigaipalan, 2018). To counteract the damaging effects of ROS, cells have developed various antioxidant defence mechanisms, including enzymes such as superoxide dismutase (SOD), catalase (CAT), and glutathione peroxidase (GPX), as well as non-enzymatic molecules such as vitamins C and E and glutathione. However, when ROS production exceeds the capacity of antioxidant defences, oxidative damage can occur (Anik et al., 2022; Halliwell, 2006).

Moreover, oxidative deterioration of lipids is a significant issue because it shortens the shelf life of food products and causes unfavorable changes to the flavor, taste, and appearance of foods. Additionally, it generates substances that may have questionable metabolic effects (e.g., hydroperoxides and reactive aldehydes), which could damage food's nutritional value and make it difficult to incorporate good polyunsaturated fatty acids (PUFA) into foods (Feng, Berton-Carabin, Fogliano, & Schroën, 2022). Lipid oxidation can produce by-products that have toxic properties and may contribute to the development of diseases (Zhong et al., 2019).

Synthetic antioxidants are widely used in the food and cosmetic industries to prevent oxidation and extend the shelf life of products. Due to their high effectiveness and low cost, synthetic antioxidants (like butylated hydroxytoluene (BHT) and tertiary butylhydroquinone (TBHQ)) are frequently added to foods (Augustyniak et al., 2010; Pokorný, 2007). However, there is growing concern about the potential negative effects of these compounds on human health. Some studies have suggested that synthetic antioxidants may have toxic and carcinogenic effects, particularly when consumed in high doses or over long periods of time (Shahidi & Ambigaipalan, 2018). As a result, European Food Safety Authority (EFSA) has established low acceptable daily intake (ADI) values (e.g., 0.25 mg/kg bw/day for BHT and 0.7 mg/kg bw/day for TBHQ) (Feng et al., 2022). Therefore, it is important to carefully evaluate the safety of synthetic antioxidants and consider alternative natural sources of antioxidants. Alternatives that are label-friendly and natural are clearly in demand. One way to delay or prevent lipid oxidation is by using Maillard reaction products (MRPs) as natural, endogenous antioxidants. These MRPs are commonly formed during food thermal processing and storage (Nooshkam & Varidi, 2021; Nooshkam, Varidi, & Verma, 2020; Nooshkam, Varidi, Zareie, & Alkobeisi, 2023; Nooshkam, Varidi, & Bashash, 2019). The mechanisms of MRPs' antioxidant and antibrowning activities are discussed in this book chapter. This chapter also emphasizes the role of MRPs as antioxidants and antibrowning compounds in food and biological systems.

2. Chemistry of the Maillard reaction

Louis Maillard, a French chemist, is the name given to the reaction after he first described it (Maillard, 1912), but it wasn't until 1953 that Hodge proposed the first coherent scheme (Fig. 1) (Hodge, 1953). The Maillard reaction is supported by a very intricate chemical structure. It includes a vast network of various reactions rather than just one specific reaction pathway. In this reaction, a reducing sugar like glucose condenses with a compound that has a free amino group (of an amino acid or in proteins, primarily the lysine ϵ-amino group, but also the α-amino groups of N-terminal) in an early stage to produce N-substituted glycosilamine, which then rearranges to produce the Amadori rearrangement product. The pH of the system affects the subsequent degradation of the Amadori product. It mainly goes through 1,2-enolisation at pH 7 or lower, resulting in the formation of hydroxymethylfurfural (HMF) (when hexoses are involved) and furfural (when pentoses are involved). At pH levels above 7, it is believed that the Amadori compound degrades primarily through 2,3 enolization, which produces reductones like 4-hydroxy-5-methyl-2,3-dihydrofuran-3-one (HMF^{one}) and a number of fission products like acetol, pyruvaldehyde, and diacetyl. These substances all participate in subsequent reactions and are highly reactive. It is possible for carbonyl groups to condense with free amino groups, incorporating nitrogen into the reaction

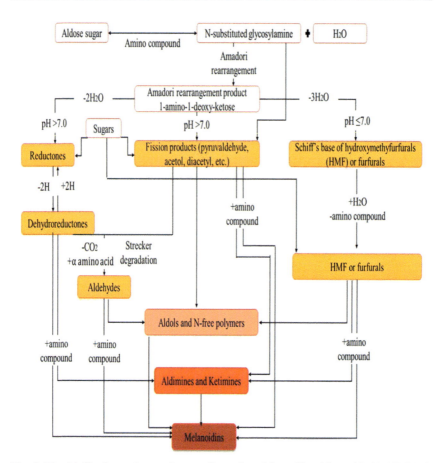

Fig. 1 The Maillard reaction pathways. *Reproduced from Nooshkam, M., Varidi, M., & Bashash, M. (2019). The Maillard reaction products as food-born antioxidant and antibrowning agents in model and real food systems. Food Chemistry, 275, 644–660. https://doi.org/10.1016/j.foodchem.2018.09.083, with permission from Elsevier Science Ltd. Copyright 2019.*

products as a result. Aldehydes and α-aminoketones are produced when amino acids and dicarbonyl compounds interact. The Strecker degradation is the name of this reaction. Following this, a number of reactions, including cyclizations, dehydrations, retroaldolisations, rearrangements, isomerizations, and additional condensations occur in an advanced stage. These reactions ultimately result in the formation of brown nitrogenous polymers and copolymers, also known as melanoidins, in a final stage (Martins, Jongen, & van Boekel, 2000; Nooshkam & Varidi, 2020a). Anionic chromophoric compounds called food melanoidins have antioxidative, antimicrobial, prebiotic, and antihypertensive properties. The foods we consume every day—cooked, fried, and roasted—contain a lot of melanoidins (Nooshkam, Babazadeh, & Jooyandeh, 2018; Nooshkam, Varidi, et al., 2019).

3. MRPs as antioxidants
3.1 Antioxidant mechanisms

MRPs, especially melanoidins, could have antioxidant properties comparable to phenolic compounds and secondary antioxidants. MRPs, particularly melanoidins (M-H) have the ability to donate hydrogen atoms and neutralize lipid oxidation-derived free radicals like alkyl (R$^\bullet$), alkoxyl (RO$^\bullet$), and peroxyl (ROO$^\bullet$), according to the following mechanisms (Morales & Jiménez-Pérez, 2004; Nooshkam, Varidi, et al., 2019):

(1) R$^\bullet$ + M-H → RH + M$^\bullet$
(2) RO$^\bullet$ + M-H → ROH + M$^\bullet$
(3) ROO$^\bullet$ + M-H → ROOH + M$^\bullet$
(4) RO$^\bullet$ + M$^\bullet$ → ROM
(5) ROO$^\bullet$ + M$^\bullet$ → ROOM

Melanoidins, due to their conjugated systems and the presence of electron-donating and -withdrawing groups, are believed to act as chromophores (Moss, 2002). The conjugated structure of melanoidins allows for the delocalization of unpaired electrons, which stabilizes the resulting melanoidin radicals (as shown in mechanisms 4 and 5). These stabilized radicals are less reactive and less likely to form new radicals, which can help to break free radical chains in the propagation step. Moreover, melanoidin radicals have the ability to neutralize free radicals, which can help to stop the oxidation process. In this context, by simultaneously donating hydrogen and accepting electrons, melanoidins are able to entrap two free radicals. Melanoidins have the ability to bind metal ions such as iron (Fe), which can catalyze oxidation reactions and produce free radicals (Morales, Fernandez-Fraguas, & Jiménez-Pérez, 2005). By binding these metal ions, melanoidins can help to prevent the formation of free radicals and reduce the rate of oxidation. Melanoidins may have the same ability to quench singlet oxygen as carotenoids because of their conjugated structures. Melanoidins can work in synergy with other antioxidants present in the system by donating hydrogen atoms, which can help to regenerate these antioxidants. This can enhance the overall antioxidant capacity of the system and further reduce the rate of oxidation (Nooshkam, Varidi, et al., 2019).

3.2 Antioxidant methods
3.2.1 Chemical-based antioxidant methods
3.2.1.1 DPPH radical scavenging

DPPH, also known as 2,2-diphenyl-1-picrylhydrazyl, is a chemical compound that has a chromogen-radical structure. When antioxidants scavenge the DPPH radical and convert it into a stable DPPH-H molecule, the color of the solution changes from purple to yellow. This change in color can be measured using a spectrophotometer at a wavelength of 517 nm (Nooshkam & Madadlou, 2016a, 2016b; Nooshkam, Falah, et al., 2019). This method has frequently been used to evaluate the antioxidant activity of MRPs (Table 1). Nasrollahzadeh and colleagues (2017) demonstrated that the interaction between hydrophobic patches in bovine serum albumin-maltodextrin conjugate molecules and hydrophobic radicals like DPPH is enhanced by protein ornamentation during heating. This interaction increases the DPPH-radical scavenging capability of the conjugate molecules. Additionally, they noted that melanoidins have strong antioxidant properties and can effectively inhibit free radicals (Nasrollahzadeh, Varidi, Koocheki, & Hadizadeh, 2017). Chen et al. (2021) conducted a study to investigate the impact of hydroxyl groups on the antioxidant properties of 2,3-dihydro-3,5-dihydroxy-6-methyl-4H-pyran-4-one (DDMP), a compound typically formed in the Maillard reaction that contributes to the antioxidant properties of MRPs. The study found that introducing protecting groups to the free hydroxyl groups of DDMP decreased their reducing capacities. Specifically, the hydroxyl group at the olefin position had a significant impact on DDMP's antioxidant activity, indicating that the unstable enol structure in the DDMP moiety is a key factor for its antioxidant activity (Chen et al., 2021). The ability of the Maillard reaction in improving the antioxidant effect of glucose-lysine model has also been reported by Cao, Yan, and Liu (2022). Another study by Ullah et al. (2019) investigated the effects of the Maillard reaction on the technological functionality of walnut protein isolate. The study found that MRPs created using glucose and heated for varying lengths of time exhibited promising antioxidant effects. In contrast, after the conjugation reaction, the antioxidant activity of soy protein isolate-low acyl gellan gum physical mixture was found to be decreased significantly. This was likely due to the deterioration of and/or the formation of complex MRPs (Lavaei, Varidi, & Nooshkam, 2022). Similarly, it was found that as the temperature and duration of the Maillard reaction increased, there was a corresponding decrease in the ability

Table 1 Antioxidant activity of MRPs based on chemical-based antioxidant methods.

Antioxidant method	Carbohydrate source	Amino group source	Heating mode	References
• DPPH radical scavenging	Maltodextrin	*Spirulina* protein concentrate	Wet heating	Zhang, Holden, Wang, and Adhikari (2023)
• DPPH radical scavenging • ABTS radical scavenging • Reducing power	Xylose	*Harpadon nehereus* protein hydrolysate	Wet heating	Ren et al. (2023)
• DPPH radical scavenging • Reducing power	Carboxymethyl cellulose	Whey protein	Wet heating	Jiang et al. (2023)
• ABTS radical scavenging • Metal chelation	Pectin/arabinogalactan	Casein	Wet heating	Siqi Yang et al. (2023)
• DPPH radical scavenging • ABTS radical scavenging • Reducing power	Inulin/fructooligosaccharides	Cricket protein	Wet heating	Chailangka et al. (2022)
• DPPH radical scavenging • ABTS radical scavenging • Reducing power	Glucose	Lysine	Wet heating	Cao et al. (2022)
• DPPH radical scavenging • Oxygen radical absorbing	Glucose, xylose, fructose, sucrose	Lysine, glycine	Wet heating	Kitts (2021)
• Oxygen radical absorbing	Glucose	Histidine	Wet heating	Yilmaz and Toledo (2005)

(*continued*)

Table 1 Antioxidant activity of MRPs based on chemical-based antioxidant methods. (cont'd)

Antioxidant method	Carbohydrate source	Amino group source	Heating mode	References
• DPPH radical scavenging	Polydextrose	Whey protein isolate	Dry heating	Liu et al. (2021)
• DPPH radical scavenging • ABTS radical scavenging • Metal chelation	Glucose	*Hermetia illucens* protein	Wet heating	Mshayisa and Van Wyk (2021)
• DPPH radical scavenging • Metal chelation	Dextrin	Whey protein hydrolysate	Dry heating	Pan et al. (2020)
• ABTS radical scavenging • Peroxyl radical scavenging • Hydroxyl radical scavenging • Metal chelation	Galactomannan/xyloglucan	Casein phosphopeptide	Dry heating	Zhang, Nakamura, and Kitts (2020)
• ABTS radical scavenging • Hydroxy radical scavenging	Galactose	Whey protein hydrolysate	Wet heating	Zhang et al. (2020)
• DPPH radical scavenging • Reducing power • Metal chelation	Xylose	Whey protein isolate	Wet heating	Shang et al. (2020)
• DPPH radical scavenging	Flaxseed gum	Whey protein isolate	Dry heating	Dong et al. (2020)

• Reducing power • Metal chelation	Low acyl gellan gum	Whey protein isolate	Wet heating	Nooshkam and Varidi (2020)
• Hydroxyl radical scavenging	Glucose	Chicken myofibrillar proteins	Dry heating	Nishimura, Suzuki, and Saeki (2019)
• DPPH radical scavenging • Reducing power	Inulin	Chitosan	Wet heating	Nooshkam, Falah, et al. (2019)
• ABTS radical scavenging	Dextran	Casein	Dry heating	Steiner et al. (2019)

of the resulting MRPs to scavenge DPPH radicals. This may be due to the oxidation of proteins during heat processing; the formation of protein aggregates can bury electron−dense areas, leading to a decrease in their ability to donate electrons and hydrogen (Yan et al., 2023). In summary, MRPs have been shown to have antioxidant properties, including the ability to scavenge free radicals such as DPPH. Further research is needed to fully understand the mechanisms behind this effect and to identify individual active components of MRPs that display this capacity.

3.2.1.2 ABTS radical scavenging

The principle of the widely used ABTS method to assess antioxidant capacity is that when blue-green ABTS$^{•+}$ react with antioxidants, the solution turns colorless (Zhu et al., 2022). The ABTS radical scavenging capacity of different MRPs has been reported (Table 1). The ability to scavenge ABTS radicals was reported to be enhanced by an increase in the intermediates and end products of the Maillard reaction (Cao et al., 2022). Similar to this, Ren et al. (2023) found that the ABTS scavenging ability of *Harpadon nehereus* protein hydrolysate (HNPH) and HNPH-xylose conjugates increased gradually with concentration, and the antioxidant activity was significantly higher for the latter than for the former. It was suggested that the intermediate or brown polymer compounds from the final stage of the Maillard reaction can act as hydrogen donors to scavenge ABTS free radicals. It's interesting to note that in the determined samples, the ABTS assay's results for scavenging activity were higher than those from the DPPH assay, and this was probably due to the fact that, in aqueous solutions, hydrophilic groups (e.g., ABTS) interact with the MRPs more favorably than hydrophobic groups (e.g., DPPH) (Ren et al., 2023). Yang et al. (2023) carried out research on the ability of casein conjugates, prepared through the Maillard reaction with arabinogalactan or pectin, to scavenge ABTS radicals. They found that, within a concentration range of $2 \, \text{mg mL}^{-1}$ to $16 \, \text{mg mL}^{-1}$, the ABTS radical scavenging ability of the conjugate was significantly higher than that of casein alone, and increased with increasing concentration. Specifically, at a concentration of $16 \, \text{mg mL}^{-1}$, the ABTS radical scavenging activities of casein-pectin and casein-arabinogalactan conjugates were 56.96% and 51.63%, respectively (Yang et al., 2023). Moreover, it has been found that dextran-*Cinnamomum camphora* seed kernel protein MRPs have a significantly higher ABTS free radical scavenging activity compared to the native protein (Yan et al., 2023). The ABTS radical scavenging mechanisms of Maillard conjugates are complex and depend on various factors such as their chemical structure and composition.

3.2.1.3 Hydroxyl radical scavenging

Hydroxyl radical is the most active and dangerous form of reactive oxygen free radicals because it can easily attack biological macromolecules, including DNA, proteins, and lipids (Su & Li, 2020). When hydrogen peroxide exists in a system, the Fenton reaction results in the production of a hydroxyl radical ($Fe^{2+} + H_2O_2 \rightarrow Fe^{3+} + OH^- + {}^{\bullet}OH$). Deoxyribose is oxidized by a hydroxyl radical to malondialdehyde (MDA), which is then converted into a pink chromogen that has a maximum absorbance at 532 nm when MDA is combined with thiobarbituric acid (TBA). By competing with deoxyribose to scavenge the hydroxyl radical, antioxidants like MRPs prevent the formation of MDA. Conjugates have the ability to scavenge hydroxyl radicals, which is likely a result of the compounds' chelating activity toward the metal ions that joined H_2O_2 to form the radical, which can then slow down the Fenton reaction (Nooshkam, Varidi, et al., 2019). In 1999, Wijewickreme and colleagues conducted an experiment in which they reacted lysine with glucose, fructose, and ribose to create nondialyzable MRPs. The reaction was carried out under various conditions, including different reaction times, pH levels, temperatures, and water activities. In the deoxyribose assay, it was found that each of the MRPs exhibited varying levels of non-site-specific activity in scavenging OH radicals, ranging from 30% to 90%. It was also found that ribose-lysine MRPs were highly effective at neutralizing hydroxyl radicals, ranging from 80% to 90%, in both deoxyribose and DNA nicking assays. All MRPs in linoleic acid emulsions, when used at higher concentrations of 0.2 mg mL^{-1}, were effective at reducing lipid peroxidation (Wijewickreme, Krejpcio, & Kitts, 1999). The relative efficiency of various MRPs to scavenge free radicals can change depending on the substrates and reaction conditions. Similarly, glucose–, fructose– and ribose–casein conjugates were found to have antioxidant activity against hydroxyl free radicals produced by the Fenton reaction (Jing & Kitts, 2002). The improved hydroxyl free radical scavenging activity of Jiuzao glutelin after conjugation with pullulan was also demonstrated by Jiang et al. (2022).

3.2.1.4 Superoxide radical anion scavenging

Superoxide radical, a highly toxic species, is produced by many biological reactions. Superoxide radical anions ($\bullet O_2^-$) are potential precursors of highly reactive species, such as the hydroxyl radical, even though they cannot directly cause lipid oxidation (Li, Jiang, Zhang, Mu, & Liu, 2008). Superoxide radical anions could be scavenged by MRPs. The effectiveness of rutin-lysine conjugates in scavenging superoxide radical anions has been observed to be significantly dose-dependent (i.e., antioxidant activity increased with an increased dose) due to the formation of melanoidins (Liu et al., 2020). Zhao and colleagues (2021) have also reported that rice protein-exopolysaccharide MRPs have a potent capacity to scavenge superoxide radicals (Zhao, Ye, Wan, Zhang, & Sun, 2021). Moreover, Suzuki, Matsumiya, Saeki, Matsumura, and Nishimura (2022) developed glucose-conjugated chicken myofibrillar protein with the strongest superoxide anion radical scavenging activity.

3.2.1.5 Peroxyl radical scavenging

The peroxyl radical (ROO$^{\bullet}$) is a biologically relevant active species that can damage cellular components. It is produced when a carbon-centered radical reacts with oxygen. Additionally, in relation to the chain-propagation mechanism of lipid peroxidation, the pathological effects of the peroxyl radical have drawn a lot of attention (Grzesik, Naparło, Bartosz, & Sadowska-Bartosz, 2018). It has been reported that MRPs have the ability to scavenge peroxyl radicals. In 2016, Vhangani and Van Wyk (2016) conducted a study in which they found that the high molecular weight (HMW) fraction of MRPs had a greater ability to scavenge peroxyl radicals than the low molecular weight (LMW) fraction. The HMW-MRPs' greater complexity, which enabled them to more successfully neutralize peroxyl radicals, was attributed to this effect. In the final stage of the Maillard reaction, LMW compounds are transformed into more complex polymers. These polymers are known as melanoidins or the HMW fraction. In a different study, Kitts (2021) found that the relative peroxyl free-radical scavenging capacity in glycine-sugar MRPs models followed a pattern of glucose > fructose ≈ xylose >> sucrose. When lysine was used, the pattern was changed to glucose ≈ xylose > fructose >> sucrose.

3.2.1.6 Reducing capacity

The reducing capacity is a common antioxidative method based on the reduction of Fe^{3+} to Fe^{2+} to assess individual antioxidants and overall antioxidative activity of MRPs (Table 1) (Cao et al., 2022). In this context, whey protein isolate-gellan gum conjugates were found to have a higher reducing power than whey protein isolate, by 1.53–28.86 folds. This shows that the Maillard reaction was effective in enhancing the antioxidant activity of whey proteins (Nooshkam & Varidi, 2020b). The remarkable reducing activity of β-lactoglobulin-polydextrose MRPs has also been shown in the literature (Luo, Tu, Ren, & Zhang, 2022). MRPs derived from bovine bone marrow extract hydrolysate exhibited enhanced antioxidant activities against DPPH, ABTS and hydroxyl free radicals, and presented improved reducing power (Begum et al., 2020). Another study found that chitosan-inulin conjugates with high browning intensity had a significantly greater reducing power compared to other pairs and chitosan solutions, indicating a positive correlation between browning intensity and antioxidant potential (Nooshkam, Falah, et al., 2019). The mechanisms behind the reducing power/capacity of MRPs are complex and not yet fully understood. However, it is thought that the reducing power of MRPs is related to their ability to donate electrons or hydrogen atoms to free radicals, thus neutralizing them and preventing oxidative damage. According to research by Ren et al. (2023), the reduction activity of constituents in MRPs is largely due to the presence of hydroxyl and pyrrole groups, which facilitate electron transfer redox potential. Additionally, the intermediate reducing ketone compounds of MRPs can terminate free radical chains by providing hydrogen atoms, which also function as reducing agents (Ren et al., 2023). Further research is needed to fully understand the mechanisms behind the reducing power/capacity of MRPs.

3.2.1.7 Ion chelation capacity

There are reports that the production of free radicals via Fenton reactions requires transition metals, particularly iron and copper. The lipid peroxidation can also be initiated and propagated by the transition metals. Since the Fe^{2+} ion has the strongest pro-oxidant effect of all known species of metal ions, compounds with metal chelation activity like MRPs are crucial in reducing the effective concentration of the transition metals that are involved in lipid peroxidation (Nooshkam, Varidi, et al., 2019). The metal chelating effect of MRPs is summarized in Table 1. In 2010, Gu and colleagues conducted a study in which they found that HMW MRPs, with a molecular weight greater than 50 kDa, produced during the final stage of the Maillard reaction, had a greater ability to chelate metal ions than LMW MRPs (Gu et al., 2010). The MRPs obtained from glycine/diglycine/triglycine-glucose were found to be more effective at chelating iron than copper (Kim & Lee, 2009). In another study, it was found that MRPs created using glucose-lysine and fructose-lysine were able to bind to copper ions. The amount of copper bound to the MRPs varied, with glucose-lysine MRPs binding between 0.031 and 1.574 mol of copper per mg of MRP, and fructose-lysine MRPs binding between 0.016 and 2.267 mol of copper per mg of MRP (Wijewickreme & Kitts, 1998). We recently reported that whey protein's iron chelating activity improved upon conjugation with gellan gum. This effect was attributed to (i) the anionic nature of melanoidins and (ii) the ability of the hydroxyl and/or ketone groups in pyranone, pyridone, and Amadori products to act as chelating donors (Fig. 2) (Nooshkam & Varidi, 2020b).

3.2.2 In-vitro antioxidant cell-based methods
3.2.2.1 Cell viability
The antioxidant effect of MRPs on human intestinal cells has also been evaluated in the literature (Table 2). In this context, it was discovered that HepG2 cells exposed to H_2O_2 could still maintain over 90% of their cell viability when treated with $0.4\,mg\,mL^{-1}$ of the 30–50 kDa rutin-lysine conjugate fraction (Liu et al., 2020). Furthermore, the viability of human intestinal cells (CCD 841 CON and Caco-2 cells) was examined to determine the applicability of Jiuzao glutelin-pullulan MRPs in the food systems (Jiang et al., 2022). The survival rate of CCD 841 cells is

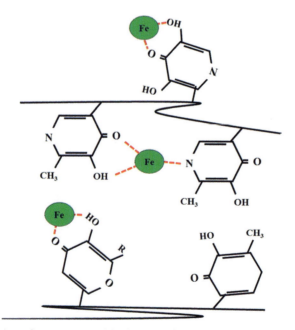

Fig. 2 The ability of pyranone and hydroxypyridone groups of melanoidins in chelating metal ions. *Reproduced from Nooshkam, M., Varidi, M., & Bashash, M. (2019). The Maillard reaction products as food-born antioxidant and antibrowning agents in model and real food systems. Food Chemistry, 275, 644–660. https://doi.org/10.1016/j.foodchem.2018.09.083, with permission from Elsevier Science Ltd. Copyright 2019.*

Table 2 Antioxidant activity of MRPs based on in-vitro antioxidant cell-based methods.

Amino group source	Carbonyl group source	Results	References
Bread crust's melanoidins	Bread crust's melanoidins	Evaluations were conducted on the bioaccessibility, bioactivity, and genoprotective effect of melanoidins extracted from bread crust. The bioaccessibility of the colonic fermentation, intestinal digestion, and gastric fractions was evaluated. The outcomes showed a correlation between the bioaccessible melanoidins and their type (soft or common bread). The bioaccessible fractions were found to be cytotoxic-free and to have no effect on the distribution of E-cadherin in Caco-2 cells, indicating that they are able to preserve membrane integrity. The investigation additionally revealed antioxidant activity on the basolateral side of the cell monolayer and showed that the gastrointestinal and colonic fermentation fractions successfully crossed the intestinal barrier without compromising cell permeability. It's noteworthy that in Caco-2 cells, both fractions significantly exhibited a genoprotective effect.	Temiño et al. (2023)
Scallop female gonad hydrolysates (SFGHs)	Ribose	Following digestion, the viability of HepG2 cells in the presence of MRPs-methanol extracts increased to 89.58%. This was 1.49, 1.26, and 1.14 times higher than the groups treated with H_2O_2, SFGHs, and MRPs, respectively. As such, it was proposed that the antioxidant potential of SFGHs-ribose MRPs can be enhanced through methanol extraction during digestion. The resulting extracts could potentially be used as antioxidants in food.	Han et al. (2021)
Lysine, glycine	Glucose, xylose, fructose, sucrose	In an inflamed model of Caco-2 cells, glucose–amino acid MRPs demonstrated an ability to inhibit nitric oxide. This was in contrast to the nitric oxide boosting activity observed with fructose–amino acid MRPs, particularly when glycine was the chosen amino acid. Moreover, when Caco-2 cells were pre-treated with fructose-glycine MRPs, it protected them from a decrease in trans-epithelial resistance.	Kitts (2021)

(continued)

Table 2 Antioxidant activity of MRPs based on in-vitro antioxidant cell-based methods. (cont'd)

Amino group source	Carbonyl group source	Results	References
Lysine, glycine	Glucose, fructose, ribose	The components of MRPs exhibited a potential for bioactivity, particularly in terms of mitigating oxidative stress and inflammation in Caco-2 cells that were induced by interferon γ and phorbol ester.	Kitts, Chen, and Jing (2012)
Glycine	Rice starch	The most intensely colored MRPs were found to curb oxidative stress within Caco-2 cells. The decline in cell viability and the activity of antioxidant enzymes, which was brought about by reactive oxygen species and apoptosis, was reversed by the MRPs. In fact, the incorporation of MRPs led to a reduction in apoptosis by lowering the percentage of cells in the sub-G1 phase.	Chung, Lee, Han, Lee, and Rhee (2012)
Lysine	Glucose, fructose, ribose	All MRPs composed of sugar-lysine demonstrated a comparable protective effect on Caco-2 cells cultured in the presence of cytotoxic agents such as H_2O_2, 2,2′-azobis-(2-amidinopropane) dihydrochloride (AAPH), ferrous (Fe^{2+}), and cupric (Cu^{2+}). This was assessed based on both redox reactions (MTT response) and the integrity of the cell membrane (lactate dehydrogenase secretion). It was observed that HMW-MRPs provided a superior protective effect against cytotoxicity induced by both Fe^{2+} and Cu^{2+} in Caco-2 cells compared to their LMW-MRP counterparts.	Jing and Kitts (2004)

significantly influenced by environmental factors. The survival rate of these cells was consequently affected by the addition of Jiuzao glutelin-pullulan MRPs (0.15, 0.625, and 5 mg mL^{-1}). The cell's survival rate was markedly elevated because Jiuzao glutelin, a protein, plays a role in the growth of cells at these three concentrations. While Jiuzao glutelin content in Jiuzao glutelin-pullulan MRPs was lower than native protein at the same concentrations. In contrast to native Jiuzao glutelin, the survival rate of the cells was markedly higher in the other three concentrations of Jiuzao glutelin-pullulan MRPs (1.25, 2.5, and 10 mg mL^{-1}) and the cell viability was raised by 10.35%. This result might have been brought about by the high concentration of Jiuzao glutelin, which can cause water loss and an imbalance in the osmotic pressure outside and inside the cell. After receiving treatment with Jiuzao glutelin-pullulan MRPs, this imbalance was lessened. Caco-2 cells have higher viability because they are cancer cells. Jiuzao glutelin-pullulan MRPs significantly outperformed Jiuzao glutelin in tests that measured the viability of Caco-2 cells at all concentrations. This outcome is the result of Jiuzao glutelin's inability to significantly promote the growth of Caco-2 cells at the appropriate concentrations. Therefore, cell viability is unaffected when Jiuzao glutelin concentrations in Jiuzao glutelin-pullulan MRPs decrease. Although both types of cells are affected by Jiuzao glutelin-pullulan MRPs, the cells are typically able to survive at concentrations lower than 5 mg mL^{-1}. The aforementioned finding supports Jiuzao glutelin-pullulan MRPs' low toxicity (Jiang et al., 2022). MRPs have also been shown to have an effect on the viability of intestinal cells. In a study that examined the effects of MRPs created using glucose-amino acids and fructose-amino acids on an inflamed Caco-2 cell model, it was found that the glucose-amino acid MRPs had the ability to inhibit the production of nitric oxide, while the fructose-amino acid MRPs had the opposite effect, stimulating the production of nitric oxide. This effect was particularly pronounced when glycine was applied as a source of amino group in the MRPs. Pre-treating Caco-2 cells with MRPs created using fructose and glycine helped to protect against the loss of trans-epithelial resistance and reduced the disruption of tight-junction protein cells in the Caco-2 intestinal epithelium upon exposure to 7.5% ethanol. This means that the fructose-glycine MRPs were able to help maintain the integrity of the intestinal epithelium, even when exposed to potentially damaging substances (Kitts, 2021). Although it may be challenging, further research is needed to isolate and characterize individual active components of MRPs that have this capacity.

3.2.2.2 Antioxidant enzymes

In addition to direct antioxidative effects, MRPs could increase the antioxidative capacity of cells through the induction of cytoprotective enzymes. CAT, SOD, and GPX are the three key enzymes in the antioxidant system. Superoxide anion disproportionation to hydrogen peroxide is caused by SOD, and hydrogen peroxide can be reduced by CAT and GPX to stop the production of dangerous hydroxyl radicals. They have the capacity to shield the cell membrane's composition and operation from peroxide's interference and damage. In addition, GPX can reduce the formation of hydroperoxides from polyunsaturated fatty acids, which helps to mitigate the potentially harmful effects of lipid peroxidation (Jiang et al., 2020). On this point, Jiuzao glutelin-pullulan MRPs effects on three enzyme activities were compared to untreated, Trolox-, and AAPH-treated cells in two cell lines (CON 841 and Caco-2 cells) by Jiang et al. (2022). In this study where AAPH was used to initiate oxidative stress, it was found that the levels of these three enzymes were lower than in other groups. This indicates that the enzymes were being used excessively. Trolox is a substance with antioxidant properties. When added to a system, it can help to stimulate the production of antioxidant enzymes. As a result, the levels of these enzymes may be higher in the presence of Trolox than in the AAPH group, which did not receive Trolox treatment. Three enzymes can be increased in content by both glutenin and Jiuzao glutelin-pullulan MRPs, but Jiuzao glutelin-pullulan MRPs increased the content of the three enzymes more than glutenin did (Jiang et al., 2022). The dose-dependent protective effects of rutin-lysine MRPs against H_2O_2-induced oxidative stress have also been reported to involve the inhibition of reactive oxygen species generation, enhancement of SOD and CAT activities, activation of the nuclear factor E2-related factor 2 (Nrf2)-dependent pathway, and upregulation of phase II antioxidant genes (such as NAD(P)H quinone dehydrogenase 1, heme oxygenase 1, glutamate—cysteine ligase catalytic subunit, and glutamate-cysteine ligase modifier subunit) (Liu et al., 2020). This outcome demonstrates the improvement in antioxidant capacity following the Maillard reaction. The resulting MRPs can be isolated and utilized as industrial preparations of natural antioxidants.

3.3 Applications

3.3.1 In food systems

There has been extensive research into the antioxidant activity of MRPs in food systems, such as meat, cereal, dairy, edible oils, and potato-based products. In a study, it was found that the addition of 0.1 or 0.2 mg/g of MRPs created by heating glucose with valine, arginine, or histidine was effective in reducing the oxidation of ground chicken breast (Miranda, Rakovski, & Were, 2012). Fernández and co-workers (2016) examined the antioxidant effects of bovine plasma proteins-based MRPs on lipid oxidation of n-3 fatty acids fortified beef patties. Both stages of lipid oxidation were significantly inhibited by MRPs at a concentration of 3.0%; the percentage of peroxidation that was inhibited was > 70% for the formation of hydroperoxides and > 45% for the formation of TBARS (Fernández, Fogar, Doval, Romero, & Judis, 2016). In 2008, Michalska and colleagues were interested in understanding how the baking process affects the formation of MRPs in rye bread, and how these MRPs contribute to the bread's antioxidant properties. According to their findings, the baking process caused MRPs (primarily melanoidins) to form in the product, which had the ability to scavenge ABTS and peroxyl radicals and reduce Folin-Ciocalteu reagent. According to the information provided, during the baking process, the water content on the surface of the dough rapidly decreases. This creates ideal conditions for the formation of more MRPs and the development of color in the crust (Michalska, Amigo-Benavent, Zielinski, & del Castillo, 2008). The ability of glucose-histidine MRPs in improving the oxidate stability of lard- and vegetable oil-based cookies was also reported in the literature (Lingnert, 1980). In 2013, Zhu and colleagues conducted a study to examine the effect of MRPs created using chitosan and xylose on the quality and preservation of semi-dried noodles. The researchers found that incorporating MRPs at a level of 0.35% into the noodles extended their shelf life by more than 7 days in comparison to control pair stored at room temperature. The presence of MRPs had an inhibitory effect on browning enzymes, resulting in a noticeably less intense dark color in the noodles (Zhu et al., 2013). The antioxidant activity of MRPs in ice creams was also demonstrated by Puangmanee, Hayakawa, Sun, and Ogawa (2008). Furthermore, Giroux, Houde, and Britten (2010) found that MRPs made from milk protein concentrate and sugars in linseed oil-enriched dairy beverages reduced lipid oxidation during sterilization. The presence of MRPs resulted in a significant reduction in propanal and hexanal concentrations. Specifically, the concentration of hexanal was reduced by 100%, while the concentration of propanal was reduced by 78% (Giroux et al., 2010). There have been successful reports of adding MRPs directly into oils to slow down or prevent the oxidation reaction (Chiu, Tanaka, Nagashima, & Taguchi, 1991; Elizalde, Rosa, & Lerici, 1991; Tanaka, Kuei, Nagashima, & Taguchi, 1988). Reports indicate that coffee beverages contain approximately 25% melanoidins.

These melanoidins are in charge of the product's exceptional properties, such as *ex vivo* protective, antioxidant, metal-chelating, and antimicrobial activities (Borrelli, Visconti, Mennella, Anese, & Fogliano, 2002; Delgado-Andrade, Rufián-Henares, & Morales, 2005; Feng et al., 2023; Yang, Fan, & Xu, 2022). As a result, MRPs are a type of natural antioxidant that can be used to lessen or prevent the oxidation of lipids in a variety of food products.

3.3.2 In biological systems

The in-vivo antioxidant activity of MRPs has also been evaluated. A two-week randomized two-period crossover trial using two diets, high and low in MRPs, was conducted to determine whether MRPs intake affects the antioxidant defense system in male adolescents (11–14 years old, n = 18). To assess oxidative status following the dietary intervention periods, fasting blood samples were taken. The brown diet (breakfast cereal, chocolate, baked products, and fried/toasted/breaded food) was rich in MRPs and the white diet was poor in MRPs. The brown diet had greater capacity to inhibit lipid peroxidation and had stronger in vitro antioxidant activity to scavenge free radicals. In the in vivo assay, it was found that indicators of oxidative damage, such as serum thiobarbituric acid-reactive substances and erythrocyte hydroperoxides, as well as antioxidant defense parameters, such as serum antioxidants and enzymatic activities of CAT, SOD, and GPX, remained unaffected by the dietary interventions. However, diets high in MRPs demonstrated a protective effect against triggered oxidation male adolescents (Seiquer et al., 2008). It was also investigated how thermally processed foods high in MRPs affected the in vitro oxidation of human low-density lipoprotein (LDL) caused by copper. In addition, the oxidative resistance of LDL (OR) in the blood plasma of eight healthy subjects was observed after they consumed diets high and low in MRPs alternately every week for three weeks. Compared to pale beer, bread crumb, and raw coffee, dark beer, bread crust, and roasted coffee produced a statistically significant higher OR in vitro. Consuming a diet high in MRPS significantly raised plasma OR by 35.5% when compared to a diet low in MRPs. The thermally processed foods high in MRPs can inhibit LDL oxidation in vitro and can lessen the oxidative modification of LDL in vivo (Dittrich et al., 2009). Similarly, when mice were given half-fin anchovy hydrolysates-glucose MRPs, their GPX and SOD activities increased and their levels of lipid peroxidation were decreased, especially in the liver (Song, Shi, Yang, & Wei, 2018).

4. MRPs as antibrowning agents
4.1 Antibrowning mechanism

A polyphenol oxidase (PPO)-catalyzed oxidation reaction known as "enzymatic browning" of raw fruits and vegetables results in the formation of polymerized dark-colored pigments from the oxidation of o-quinones (Singh et al., 2018; Walker, 1995). Endogenous monophenols are converted to ortho-dihydroxyaryl compounds by PPOs, which are copper-containing enzymes. These compounds are then converted to ortho-quinones (Fig. 3). Phenols, amino acids, and proteins then undergo nucleophilic addition reactions with the electrophilic ortho-quinones to create secondary products that are brown, black, or red in color and are linked to the unintended discoloration of fruit and vegetables (Parveen, Threadgill, Moorby, & Winters, 2010). This reaction has a significant impact on the product's functional, nutritive, and organoleptic properties. Sulfites are the most cost-effective and efficient agents for preventing enzymatic browning; however, their use is limited due to the adverse effects they can have on individuals with asthma (Mogol, Yıldırım, & Gökmen, 2010). Thus, research on alternative enzymatic browning inhibitors is crucial. In real foods, PPO, peroxidase (POD), and tyrosinase activity have been shown to be decreased or inhibited by MRPs (Mogol et al., 2010; Xu, Zhang, & Karangwa, 2016). The chelating activity of MRPs on enzymes that contain copper ions at their active sites, such as oxidoreductases like PPO, POD, and tyrosinase, is the primary reason for the antibrowning effect of these products (Fig. 3) (Xu et al., 2016).

4.2 Antibrowning effect

MRPs have successfully been employed to decrease or prevent the enzymatic browning development in fruits and vegetables (Table 3). In this context, equimolar mixtures of hexose/cysteine were tested on purified enzyme activity by Billaud, Roux, Brun-Mérimee, Maraschin, and Nicolas (2003) to show whether MRPs could inhibit enzymatic browning and/or inactivate apple PPO. The effects of temperature (80–110 °C) and heating time (0–48 h) on inhibition were assessed. As the mixture was heated for a longer period of time and at a higher temperature, the obtained MRPs demonstrated a very strong inhibitory potency (Billaud et al., 2003). Similarly, Lee and Park (2005) investigated the inhibition of potato PPO by MRPs derived from amino acids (glycine, valine, asparagine, serine, cysteine, histidine, arginine, and lysine) and glucose. The inhibitory effect of the generated MRPs against potato PPO increased as the reaction time

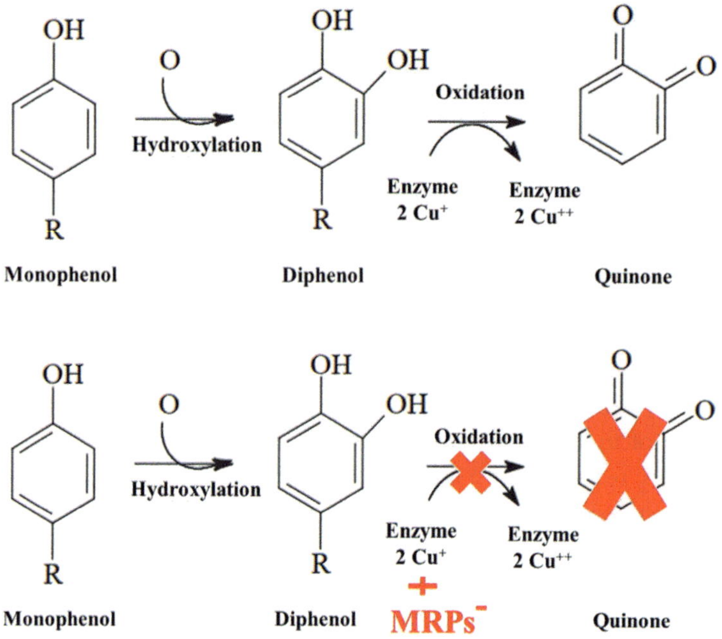

Fig. 3 The ability of MRPs in inhibiting enzymatic browning.

of glucose/glycine was prolonged at 90 °C. Potato PPO was significantly inhibited by the MRPs made from arginine, cysteine, histidine, and lysine. Potato PPO was non-competitively inhibited by MRPs made from glucose and glycine. In comparison to MRPs made from disaccharides (e.g., lactose, sucrose, and maltose), those made from monosaccharides, such as fructose and glucose, and glycine were more inhibitory against potato PPO. The MRPs obtained by raising the concentrations of glycine and glucose had a stronger inhibitory effect on the PPO enzyme (Lee & Park, 2005). Additionally, it was investigated how glucose-cysteine MRPs inhibited the activity of tyrosinase. When mushroom tyrosinase was pre-incubated with MRPs, enzyme activity decreased as reaction time increased, and MRPs permanently blocked the enzyme's active site by chelating the copper ions (Xu et al., 2016).

MRPs have also been used to prevent enzymatic browning in actual foods. For instance, Billaud et al. (2005) investigated the impact of MRPs on the activity of PPO enzyme eggplant, mushroom, and apple. They reacted various mono- and di-saccharides with glutathione and cysteine, which are natural antibrowning agents. The enzyme's activity was suppressed by each conjugate. The researchers used MRPs, including cysteine-xylose, cysteine-glucose, and

Table 3 Antibrowning activity of MRPs.

Amino group source	Carbonyl group source	Results	References
Glycine	Glucose	The presence of MRPs led to a significant reduction, almost to the point of complete inhibition, in the activity of PPO and POD.	Nicoli, Elizalde, Pitotti, and Lerici (1991)
Cysteine, glutathione	Glucose, fructose	The Maillard reaction resulted in the formation of MRPs that exhibited strong inhibitory effects on the apple PPO.	Billaud et al. (2003); Billaud, Brun-Merimee, Louarme, and Nicolas (2004)
Cysteine, glutathione	Glucose, galactose, fructose, ribose, xylose, arabinose, mannose	The MRPs demonstrated a suppressive impact on PPO enzyme activity. In many instances, MRPs that exhibited greater inhibitory efficiency, such as cysteine-glucose/xylose and glutathione-glucose, had an antibrowning activity comparable to that of metabisulfite.	Billaud et al. (2005)
Glutathione	Glucose, galactose, ribose, xylose, dextran, xanthan gum, tragacanth gum, potato starch	MRPs derived from monosaccharide-glutathione were found to be more effective than glutathione itself in inhibiting mushroom tyrosinase, while those derived from polysaccharide-glutathione were not. Interestingly, in the case of fresh-cut apple slices, glutathione outperformed MRPs derived from sucrose-glutathione when the slices were kept at room temperature for 24 h. A subsequent time-course study revealed a decline in the inhibitory activity of sugar-glutathione derived	Wu, Cheng, Li, Wang, and Ye (2008)

(continued)

Table 3 Antibrowning activity of MRPs. (cont'd)

Amino group source	Carbonyl group source	Results	References
		MRPs against mushroom tyrosinase over time, indicating that the tyrosinase inhibitors formed in MRPs are not stable. In addition to the instability of the main inhibitors, an unpleasant odor was observed from apple slices treated with MRPs, raising concerns about the potential adverse effects of these inhibitors on the sensory quality of food products.	
Histidine, arginine	Glucose	The enzymatic browning in apple purée and potato cubes was notably suppressed by MRPs, as evidenced by colorimetric analysis.	Mogol et al. (2010)
Chitosan	Maltose	MRPs derived from maltose and high molecular weight chitosan demonstrated superior performance in reducing PPO activity and discoloration. They also mitigated the reduction of total soluble solids and ascorbic acid content in fresh-cut *Typha latifolia* L.	Li, Lin, and Chen (2014)
Cysteine	Glucose	The cysteine-glucose MRPs and their ultrafiltrate exhibited inhibitory effects similar to those of sodium metabisulfite for up to 6 h of banana slice storage.	Yuan et al. (2015)
Cysteine	Glucose	When mushroom tyrosinase was pre-incubated with MRPs, a decrease in enzyme activity was observed as the reaction time increased, and the active site of mushroom tyrosinase is irreversibly blocked by the MRPs.	Xu et al. (2016)

glutathione-glucose conjugates, which had strong inhibitory effects and low color intensity. They assessed the potential inhibitory effects of these MRPs and compared their effectiveness with metabisulfite, a typical antibrowning agent, in apple purée, eggplant, mushrooms, and sliced apples. MRPs performed quite well in apple slices, particularly cysteine-xylose MRPs. Slices of eggplant did not brown when metabisulfite and MRPs were present. MRPs had the same impact on mushroom slices as did metabisulfite. Both cysteine-glucose MRPs and metabisulfite significantly reduced browning in apple purée. It was suggested that MRPs derived from thiols could be used as a replacement for sulfites in raw or lightly processed fruits and vegetables (Billaud et al., 2005). In a different study, Xu et al. (2016) assessed how MRPs made from glucose and L-cysteine inhibited the enzymatic browning process that was facilitated by mushroom tyrosinase. They demonstrated that MRPs have a direct and irreversible effect on the inactivation of tyrosinase, which was dependent on time. The chelation of copper ions in the enzyme's active center by these compounds contributes to their inhibitory effect.

Similar MRPs are produced during food processing like baking and roasting, so they are thought to be regular food constituents as PPO inhibitors, which may facilitate their approval as antibrowning agents. However, to ensure that the mixture of MRPs used for anti-browning purposes is safe, a thorough investigation should be conducted to confirm that there are no harmful substances present in the mixture at the concentration used. Moreover, to prevent any negative impact on the sensory properties of foods, it may be desirable to remove any aromatic compounds that may be present in the MRP mixture produced during the Maillard reaction.

5. Deleterious effect of MRPs as glycation end products in biological systems

A variety of advanced glycation end-products (AGEs) including crossline, pentosidine, pyrraline, N^ϵ (carboxymethyl)lysine (CML), and imidazolones are produced through additional rearrangement, oxidation, and reduction of initial MRPs (Van Nguyen, 2006). Certain studies have indicated that AGEs, particularly those derived from diet, are associated with oxidative stress and inflammation, leading to conditions like diabetes and cardiovascular diseases. Furthermore, AGEs play a role in food allergies and excessive consumption can cause significant cellular dysfunction. This includes alterations in protein structures, disruptions in lipid metabolism, vascular inflammation, and thrombogenesis (Henning & Glomb, 2016; Li et al., 2022). It has been observed that the daily consumption of AGEs CML and pyrraline in a typical Western diet ranges from 25 to 75 mg (Schwarzenbolz, Hofmann, Sparmann, & Henle, 2016). It's noteworthy that HMW AGEs are absorbed more slowly and less efficiently than their LMW counterparts. HMW AGEs can be partially broken down by gut proteases, and their bioavailability is influenced by factors such as diet type, gut environment, associated peptide size, and the duration of their presence in the gut (Poulsen et al., 2013). There exist several synthetic carbonyl scavengers like aminoguanidine and pyridoxamine, as well as natural AGE inhibitors such as phenolic compounds and plant extracts, which can prevent the formation of AGEs (Jia, Guo, Zhang, & Shi, 2023).

6. Conclusions

Many different compounds known as MRPs, which are produced by the Maillard reaction, may have antioxidant properties. The antioxidative assays have proposed various mechanisms for the antioxidant potency of MRPs, such as scavenging of free radicals, metal chelation, and breakdown of radical chains and hydrogen peroxide. Furthermore, MRPs have been successfully used to increase the oxidative stability of a variety of foods, including dairy, meat, pasta, bread, and edible oil. They may also be employed to replace sulfite compounds as antibrowning agents to prevent enzymatic browning in fruits and vegetables. However, additional research is required to characterize the substances in charge of MRPs' antioxidant activity. Additionally, MRPs should be tested for toxicity before being used as antioxidants in food systems.

Acknowledgments

The authors thank Ferdowsi University of Mashhad for financial support of this work (Grant No. 3/48147).

References

Anik, M. I., Mahmud, N., Masud, A. A., Khan, M. I., Islam, M. N., Uddin, S., & Hossain, M. K. (2022). Role of reactive oxygen species in aging and age-related diseases: A review. *ACS Applied Bio Materials, 5*(9), 4028–4054. https://doi.org/10.1021/acsabm.2c00411.

Augustyniak, A., Bartosz, G., Čipak, A., Duburs, G., Horáková, L. U., Łuczaj, W., ... Žarković, N. (2010). Natural and synthetic antioxidants: An updated overview. *Free Radical Research, 44*(10), 1216–1262. https://doi.org/10.3109/10715762.2010.508495.

Begum, N., Raza, A., Shen, D., Song, H., Zhang, Y., Zhang, L., & Liu, P. (2020). Sensory attribute and antioxidant capacity of Maillard reaction products from enzymatic hydrolysate of bovine bone marrow extract. *Journal of Food Science and Technology, 57*(5), 1786–1797. https://doi.org/10.1007/s13197-019-04212-8.

Billaud, C., Brun-Merimee, S., Louarme, L., & Nicolas, J. (2004). Effect of glutathione and Maillard reaction products prepared from glucose or fructose with glutathione on polyphenoloxidase from apple—I: Enzymatic browning and enzyme activity inhibition. *Food Chemistry, 84*(2), 223–233. https://doi.org/10.1016/S0308-8146(03)00206-1.

Billaud, C., Maraschin, C., Chow, Y.-N., Chériot, S., Peyrat-Maillard, M.-N., & Nicolas, J. (2005). Maillard reaction products as "natural antibrowning" agents in fruit and vegetable technology. *Molecular Nutrition & Food Research, 49*(7), 656–662. https://doi.org/10.1002/mnfr.200400101.

Billaud, C., Roux, E., Brun-Mérimee, S., Maraschin, C., & Nicolas, J. (2003). Inhibitory effect of unheated and heated d-glucose, d-fructose and l-cysteine solutions and Maillard reaction product model systems on polyphenoloxidase from apple. I. Enzymatic browning and enzyme activity inhibition using spectrophotometric and polarographic methods. *Food Chemistry, 81*(1), 35–50. https://doi.org/10.1016/S0308-8146(02)00376-X.

Borrelli, R. C., Visconti, A., Mennella, C., Anese, M., & Fogliano, V. (2002). Chemical characterization and antioxidant properties of coffee melanoidins. *Journal of Agricultural and Food Chemistry, 50*(22), 6527–6533. https://doi.org/10.1021/jf025686o.

Cao, J., Yan, H., & Liu, L. (2022). Optimized preparation and antioxidant activity of glucose-lysine Maillard reaction products. *LWT, 161*, 113343. https://doi.org/10.1016/j.lwt.2022.113343.

Chailangka, A., Seesuriyachan, P., Wangtueai, S., Ruksiriwanich, W., Jantanasakulwong, K., Rachtanapun, P., ... Phimolsiripol, Y. (2022). Cricket protein conjugated with different degrees of polymerization saccharides by Maillard reaction as a novel functional ingredient. *Food Chemistry, 395*, 133594. https://doi.org/10.1016/j.foodchem.2022.133594.

Chen, Z., Liu, Q., Zhao, Z., Bai, B., Sun, Z., Cai, L., ... Xi, G. (2021). Effect of hydroxyl on antioxidant properties of 2,3-dihydro-3,5-dihydroxy-6-methyl-4H-pyran-4-one to scavenge free radicals. *RSC Advances, 11*(55), 34456–34461. https://doi.org/10.1039/D1RA06317K.

Chiu, W. K., Tanaka, M., Nagashima, Y., & Taguchi, T. (1991). Prevention of sardine lipid oxidation by antioxidative Maillard reaction products prepared from fructose-tryptophan. *Nippon Suisan Gakaishi, 57*, 1773–1781.

Chung, S. Y., Lee, Y. K., Han, S. H., Lee, S. W., & Rhee, C. (2012). The inhibition effects of reactive oxygen species by Maillard reaction products in Caco-2 cells. *Starch-Stärke, 64*(11), 921–928. https://doi.org/10.1002/star.201200047.

Delgado-Andrade, C., Rufián-Henares, J. A., & Morales, F. J. (2005). Assessing the antioxidant activity of melanoidins from coffee brews by different antioxidant methods. *Journal of Agricultural and Food Chemistry, 53*(20), 7832–7836. https://doi.org/10.1021/jf0512353.

Dittrich, R., Dragonas, C., Kannenkeril, D., Hoffmann, I., Mueller, A., Beckmann, M. W., & Pischetsrieder, M. (2009). A diet rich in Maillard reaction products protects LDL against copper induced oxidation ex vivo, a human intervention trial. *Food Research International, 42*(9), 1315–1322. https://doi.org/10.1016/j.foodres.2009.04.007.

Dong, X., Du, S., Deng, Q., Tang, H., Yang, C., Wei, F., ... Liu, L. (2020). Study on the antioxidant activity and emulsifying properties of flaxseed gum-whey protein isolate conjugates prepared by Maillard reaction. *International Journal of Biological Macromolecules, 153*, 1157–1164. https://doi.org/10.1016/j.ijbiomac.2019.10.245.

Elizalde, B. E., Rosa, M. D., & Lerici, C. R. (1991). Effect of maillard reaction volatile products on lipid oxidation. *Journal of the American Oil Chemists' Society, 68*(10), 758–762. https://doi.org/10.1007/BF02662167.

Feng, J., Berton-Carabin, C. C., Fogliano, V., & Schroën, K. (2022). Maillard reaction products as functional components in oil-in-water emulsions: A review highlighting interfacial and antioxidant properties. *Trends in Food Science & Technology, 121*, 129–141. https://doi.org/10.1016/j.tifs.2022.02.008.

Feng, J., Berton-Carabin, C. C., Guyot, S., Gacel, A., Fogliano, V., & Schroën, K. (2023). Coffee melanoidins as emulsion stabilizers. *Food Hydrocolloids, 139*, 108522. https://doi.org/10.1016/j.foodhyd.2023.108522.

Fernández, C. L., Fogar, R. A., Doval, M. M., Romero, A. M., & Judis, M. A. (2016). Antioxidant effect of bovine plasma proteins modified via Maillard reaction on n3 fortified beef patties. *Food and Nutrition Sciences, 7*(8), 671–681. https://doi.org/10.4236/fns.2016.78068.

Giroux, H. J., Houde, J., & Britten, M. (2010). Use of heated milk protein–sugar blends as antioxidant in dairy beverages enriched with linseed oil. *LWT—Food Science and Technology, 43*(9), 1373–1378. https://doi.org/10.1016/j.lwt.2010.05.001.

Grzesik, M., Naparło, K., Bartosz, G., & Sadowska-Bartosz, I. (2018). Antioxidant properties of catechins: Comparison with other antioxidants. *Food Chemistry, 241*, 480–492. https://doi.org/10.1016/j.foodchem.2017.08.117.

Gu, F.-L., Kim, J. M., Abbas, S., Zhang, X.-M., Xia, S.-Q., & Chen, Z.-X. (2010). Structure and antioxidant activity of high molecular weight Maillard reaction products from casein–glucose. *Food Chemistry, 120*(2), 505–511. https://doi.org/10.1016/j.foodchem.2009.10.044.

Halliwell, B. (2006). Reactive species and antioxidants. Redox biology is a fundamental theme of aerobic life. *Plant Physiology, 141*(2), 312–322. https://doi.org/10.1104/pp.106.077073.

Han, J. R., Han, Y. T., Li, X. W., Gu, Q., Li, P., & Zhu, B. W. (2021). Antioxidant activity of Yesso scallop (Patinopecten yessoensis) female gonad hydrolysates-ribose Maillard reaction products extracted with organic reagents, before and after in vitro digestion. *Food Bioscience, 43*, 101262. https://doi.org/10.1016/j.fbio.2021.101262.

Henning, C., & Glomb, M. A. (2016). Pathways of the Maillard reaction under physiological conditions. *Glycoconjugate Journal, 33*, 499–512. https://doi.org/10.1007/s10719-016-9694-y.

Hodge, J. E. (1953). Dehydrated foods, chemistry of browning reactions in model systems. *Journal of Agricultural and Food Chemistry, 1*(15), 928–943. https://doi.org/10.1021/jf60015a004.

Jia, W., Guo, A., Zhang, R., & Shi, L. (2023). Mechanism of natural antioxidants regulating advanced glycosylation end products of Maillard reaction. *Food Chemistry, 404*, 134541. https://doi.org/10.1016/j.foodchem.2022.134541.

Jiang, Y., Sun, J., Yin, Z., Li, H., Sun, X., & Zheng, F. (2020). Evaluation of antioxidant peptides generated from Jiuzao (residue after Baijiu distillation) protein hydrolysates and their effect of enhancing healthy value of Chinese Baijiu. *Journal of the Science of Food and Agriculture, 100*(1), 59–73. https://doi.org/10.1002/jsfa.9994.

Jiang, Y., Zang, K., Yan, R., Sun, J., Zeng, X.-A., Li, H., ... Xu, L. (2022). Modification of Jiuzao glutelin with pullulan through Maillard reaction: Stability effect in nano-emulsion, in vitro antioxidant properties, and interaction with curcumin. *Food Research International, 161*, 111785. https://doi.org/10.1016/j.foodres.2022.111785.

Jiang, Z., Huangfu, Y., Jiang, L., Wang, T., Bao, Y., & Ma, W. (2023). Structure and functional properties of whey protein conjugated with carboxymethyl cellulose through maillard reaction. *LWT, 174*, 114406. https://doi.org/10.1016/j.lwt.2022.114406.

Jing, H., & Kitts, D. D. (2002). Chemical and biochemical properties of casein–sugar Maillard reaction products. *Food and Chemical Toxicology, 40*(7), 1007–1015. https://doi.org/10.1016/S0278-6915(02)00070-4.

Jing, H., & Kitts, D. D. (2004). Antioxidant activity of sugar–lysine Maillard reaction products in cell free and cell culture systems. *Archives of Biochemistry and Biophysics, 429*(2), 154–163. https://doi.org/10.1016/j.abb.2004.06.019.

Kim, J.-S., & Lee, Y.-S. (2009). Antioxidant activity of Maillard reaction products derived from aqueous glucose/glycine, diglycine, and triglycine model systems as a function of heating time. *Food Chemistry, 116*(1), 227–232. https://doi.org/10.1016/j.foodchem.2009.02.038.

Kitts, D. D. (2021). Antioxidant and functional activities of MRPs derived from different sugar–amino acid combinations and reaction conditions. *Antioxidants, 10*(11), 1840. https://doi.org/10.3390/antiox10111840.

Kitts, D. D., Chen, X. M., & Jing, H. (2012). Demonstration of antioxidant and anti-inflammatory bioactivities from sugar–amino acid Maillard reaction products. *Journal of Agricultural and Food Chemistry, 60*(27), 6718–6727. https://doi.org/10.1021/jf2044636.

Lavaei, Y., Varidi, M., & Nooshkam, M. (2022). Gellan gum conjugation with soy protein via Maillard-driven molecular interactions and subsequent clustering lead to conjugates with tuned technological functionality. *Food Chemistry: X, 15*, 100408. https://doi.org/10.1016/j.fochx.2022.100408.

Lee, M.-K., & Park, I. (2005). Inhibition of potato polyphenol oxidase by Maillard reaction product. *Food Chemistry, 91*(1), 57–61. https://doi.org/10.1016/j.foodchem.2004.05.046.

Li, M., Shen, M., Lu, J., Yang, J., Huang, Y., Liu, L., ... Xie, M. (2022). Maillard reaction harmful products in dairy products: Formation, occurrence, analysis, and mitigation strategies. *Food Research International, 151*, 110839. https://doi.org/10.1016/j.foodres.2021.110839.

Li, S. L., Lin, J., & Chen, X. M. (2014). Effect of chitosan molecular weight on the functional properties of chitosan-maltose Maillard reaction products and their application to fresh-cut Typha latifolia L. *Carbohydrate Polymers, 102*, 682–690. https://doi.org/10.1016/j.carbpol.2013.10.102.

Li, Y., Jiang, B., Zhang, T., Mu, W., & Liu, J. (2008). Antioxidant and free radical-scavenging activities of chickpea protein hydrolysate (CPH). *Food Chemistry.* https://doi.org/10.1016/j.foodchem.2007.04.067.

Lingnert, H. (1980). Antioxidative maillard reaction products iii. Application in cookies. *Journal of Food Processing and Preservation, 4*(4), 219–233. https://doi.org/10.1111/j.1745-4549.1980.tb00608.x.

Liu, H., Jiang, Y., Guan, H., Li, F., Sun-Waterhouse, D., Chen, Y., & Li, D. (2020). Enhancing the antioxidative effects of foods containing rutin and α-amino acids via the Maillard reaction: A model study focusing on rutin-lysine system. *Journal of Food Biochemistry, 44*(1), e13086. https://doi.org/10.1111/jfbc.13086.

Liu, H., Zhu, X., Jiang, Y., Sun-Waterhouse, D., Huang, Q., Li, F., & Li, D. (2021). Physicochemical and emulsifying properties of whey protein isolate (WPI)-polydextrose conjugates prepared via Maillard reaction. *International Journal of Food Science & Technology, 56*(8), 3784–3794. https://doi.org/10.1111/ijfs.14994.

Luo, Y., Tu, Y., Ren, F., & Zhang, H. (2022). Characterization and functional properties of Maillard reaction products of β-lactoglobulin and polydextrose. *Food Chemistry.* https://doi.org/10.1016/j.foodchem.2021.131749.

Maillard, L. C. (1912). Action des acides amines sur les sucres; formation des melanoidies par voie methodique. *Comptes Rendus de l'Académie Des Sciences, 154*, 66–68.

Martins, S. I. F. S., Jongen, W. M. F., & van Boekel, M. A. J. S. (2000). A review of Maillard reaction in food and implications to kinetic modelling. *Trends in Food Science & Technology, 11*(9), 364–373. https://doi.org/10.1016/S0924-2244(01)00022-X.

Michalska, A., Amigo-Benavent, M., Zielinski, H., & del Castillo, M. D. (2008). Effect of bread making on formation of Maillard reaction products contributing to the overall antioxidant activity of rye bread. *Journal of Cereal Science, 48*(1), 123–132. https://doi.org/10.1016/j.jcs.2007.08.012.

Miranda, L. T., Rakovski, C., & Were, L. M. (2012). Effect of Maillard reaction products on oxidation products in ground chicken breast. *Meat Science, 90*(2), 352–360. https://doi.org/10.1016/j.meatsci.2011.07.022.

Mogol, B. A., Yıldırım, A., & Gökmen, V. (2010). Inhibition of enzymatic browning in actual food systems by the Maillard reaction products. *Journal of the Science of Food and Agriculture, 90*(15), 2556–2562. https://doi.org/10.1002/jsfa.4118.

Morales, F. J., & Jiménez-Pérez, S. (2004). Peroxyl radical scavenging activity of melanoidins in aqueous systems. *European Food Research and Technology, 218*, 515–520. https://doi.org/10.1007/s00217-004-0896-3.

Morales, F. J., Fernandez-Fraguas, C., & Jiménez-Pérez, S. (2005). Iron-binding ability of melanoidins from food and model systems. *Food Chemistry, 90*(4), 821–827. https://doi.org/10.1016/j.foodchem.2004.05.030.

Moss, B. W. (2002). The chemistry of food colour. In D. B. MacDougall (Ed.). *Colour in food: Improving quality* (pp. 145–178). Cambridge: Woodhead Publishing Limited and CRC Press LLC. https://doi.org/10.1533/9781855736672.2.145.

Mshayisa, V. V., & Van Wyk, J. (2021). Hermetia illucens protein conjugated with glucose via maillard reaction: Antioxidant and techno-functional properties. *International Journal of Food Science, 2021*, 5572554. https://doi.org/10.1155/2021/5572554.

Nicoli, M. C., Elizalde, B. E., Pitotti, A., & Lerici, C. R. (1991). Effect of sugars and maillard reaction products on polyphenol oxidase and peroxidase activity in food. *Journal of Food Biochemistry, 15*(3), 169–184. https://doi.org/10.1111/j.1745-4514.1991.tb00153.x.

Nasrollahzadeh, F., Varidi, M., Koocheki, A., & Hadizadeh, F. (2017). Effect of microwave and conventional heating on structural, functional and antioxidant properties of bovine serum albumin-maltodextrin conjugates through Maillard reaction. *Food Research International, 100*, 289–297. https://doi.org/10.1016/j.foodres.2017.08.030.

Nishimura, K., Suzuki, M., & Saeki, H. (2019). Glucose-conjugated chicken myofibrillar proteins derived from random-centroid optimization present potent hydroxyl radical scavenging activity. *Bioscience, Biotechnology, and Biochemistry, 83*(12), 2307–2317. https://doi.org/10.1080/09168451.2019.1662276.

Nooshkam, M., & Madadlou, A. (2016a). Maillard conjugation of lactulose with potentially bioactive peptides. *Food Chemistry, 192*, 831–836. https://doi.org/10.1016/j.foodchem.2015.07.094.

Nooshkam, M., & Madadlou, A. (2016b). Microwave-assisted isomerisation of lactose to lactulose and Maillard conjugation of lactulose and lactose with whey proteins and peptides. *Food Chemistry, 200*, 1–9. https://doi.org/10.1016/j.foodchem.2015.12.094.

Nooshkam, M., & Varidi, M. (2020a). Maillard conjugate-based delivery systems for the encapsulation, protection, and controlled release of nutraceuticals and food bioactive ingredients: A review. *Food Hydrocolloids, 100*, 105389. https://doi.org/10.1016/j.foodhyd.2019.105389.

Nooshkam, M., & Varidi, M. (2020b). Whey protein isolate-low acyl gellan gum Maillard-based conjugates with tailored technological functionality and antioxidant activity. *International Dairy Journal, 109*, 104783. https://doi.org/10.1016/j.idairyj.2020.104783.

Nooshkam, M., & Varidi, M. (2021). Physicochemical stability and gastrointestinal fate of β-carotene-loaded oil-in-water emulsions stabilized by whey protein isolate-low acyl gellan gum conjugates. *Food Chemistry, 347*, 129079. https://doi.org/10.1016/j.foodchem.2021.129079.

Nooshkam, M., Babazadeh, A., & Jooyandeh, H. (2018). Lactulose: Properties, techno-functional food applications, and food grade delivery system. *Trends in Food Science & Technology, 80*, 23–34. https://doi.org/10.1016/j.tifs.2018.07.028.

Nooshkam, M., Falah, F., Zareie, Z., Tabatabaei Yazdi, F., Shahidi, F., & Mortazavi, S. A. (2019). Antioxidant potential and antimicrobial activity of chitosan–inulin conjugates obtained through the Maillard reaction. *Food Science and Biotechnology, 28*(6), 1861–1869. https://doi.org/10.1007/s10068-019-00635-3.

Nooshkam, M., Varidi, M., & Bashash, M. (2019). The Maillard reaction products as foodborn antioxidant and antibrowning agents in model and real food systems. *Food Chemistry, 275*, 644–660. https://doi.org/10.1016/j.foodchem.2018.09.083.

Nooshkam, M., Varidi, M., & Verma, D. K. (2020). Functional and biological properties of Maillard conjugates and their potential application in medical and food: A review. *Food Research International, 131*, 109003. https://doi.org/10.1016/j.foodres.2020.109003.

Nooshkam, M., Varidi, M., Zareie, Z., & Alkobeisi, F. (2023). Behavior of protein-polysaccharide conjugate-stabilized food emulsions under various destabilization conditions. *Food Chemistry: X, 18*, 100725. https://doi.org/10.1016/j.fochx.2023.100725.

Pan, Y., Wu, Z., Xie, Q.-T., Li, X.-M., Meng, R., Zhang, B., & Jin, Z.-Y. (2020). Insight into the stabilization mechanism of emulsions stabilized by Maillard conjugates: Protein hydrolysates-dextrin with different degree of polymerization. *Food Hydrocolloids, 99*, 105347. https://doi.org/10.1016/j.foodhyd.2019.105347.

Parveen, I., Threadgill, M. D., Moorby, J. M., & Winters, A. (2010). Oxidative phenols in forage crops containing polyphenol oxidase enzymes. *Journal of Agricultural and Food Chemistry, 58*(3), 1371–1382. https://doi.org/10.1021/jf9024294.

Pokorný, J. (2007). Are natural antioxidants better – and safer – than synthetic antioxidants? *European Journal of Lipid Science and Technology, 109*(6), 629–642. https://doi.org/10.1002/ejlt.200700064.

Poulsen, M. W., Hedegaard, R. V., Andersen, J. M., de Courten, B., Bügel, S., Nielsen, J., ... Dragsted, L. O. (2013). Advanced glycation endproducts in food and their effects on health. *Food and Chemical Toxicology, 60*, 10–37. https://doi.org/10.1016/j.fct.2013.06.052.

Puangmanee, S., Hayakawa, S., Sun, Y., & Ogawa, M. (2008). Application of whey protein isolate glycated with rare sugars to ice cream. *Food science and technology research, 14*(5), 457. https://doi.org/10.3136/fstr.14.457.

Ren, S.-T., Fu, J.-J., He, F.-Y., Chai, T.-T., Yu-Ting, L., Jin, D.-L., & Chen, Y.-W. (2023). Characteristics and antioxidant properties of Harpadon nehereus protein hydrolysate-xylose conjugates obtained from the Maillard reaction by ultrasound-assisted wet heating in a natural deep eutectic solvents system. *Journal of the Science of Food and Agriculture, 103*(5), 2273–2282. https://doi.org/10.1002/jsfa.12436.

Schwarzenbolz, U., Hofmann, T., Sparmann, N., & Henle, T. (2016). Free Maillard reaction products in milk reflect nutritional intake of glycated proteins and can be used to distinguish "organic" and "conventionally" produced milk. *Journal of Agricultural and Food Chemistry, 64*(24), 5071–5078. https://doi.org/10.1021/acs.jafc.6b01375.

Seiquer, I., Ruiz-Roca, B., Mesías, M., Muñoz-Hoyos, A., Galdó, G., Ochoa, J. J., & Navarro, M. P. (2008). The antioxidant effect of a diet rich in Maillard reaction products is attenuated after consumption by healthy male adolescents. In vitro and in vivo comparative study. *Journal of the Science of Food and Agriculture, 88*(7), 1245–1252. https://doi.org/10.1002/jsfa.3213.

Shahidi, F., & Ambigaipalan, P. (2018). *Antioxidants in oxidation control. Measurement of antioxidant activity & capacity*, 287–320. https://doi.org/10.1002/9781119135388.ch14.

Shang, J., Zhong, F., Zhu, S., Wang, J., Huang, D., & Li, Y. (2020). Structure and physiochemical characteristics of whey protein isolate conjugated with xylose through Maillard reaction at different degrees. *Arabian Journal of Chemistry, 13*(11), 8051–8059. https://doi.org/10.1016/j.arabjc.2020.09.034.

Singh, B., Suri, K., Shevkani, K., Kaur, A., Kaur, A., & Singh, N. (2018). Enzymatic browning of fruit and vegetables: A review. In M. Kuddus (Ed.). *Enzymes in food technology: Improvements and innovations* (pp. 63–78). Springer Singapore. https://doi.org/10.1007/978-981-13-1933-4_4.

Song, R., Shi, Q., Yang, P., & Wei, R. (2018). In vitro membrane damage induced by halffin anchovy hydrolysates/glucose Maillard reaction products and the effects on oxidative status in vivo. *Food & Function, 9*(2), 785–796. https://doi.org/10.1039/C7FO01459G.

Steiner, B. M., Shukla, V., McClements, D. J., Li, Y. O., Sancho-Madriz, M., & Davidov-Pardo, G. (2019). Encapsulation of lutein in nanoemulsions stabilized by resveratrol and maillard conjugates. *Journal of Food Science, 84*(9), 2421–2431. https://doi.org/10.1111/1750-3841.14751.

Su, Y., & Li, L. (2020). Structural characterization and antioxidant activity of polysaccharide from four auriculariales. *Carbohydrate Polymers, 229*, 115407. https://doi.org/10.1016/j.carbpol.2019.115407.

Suzuki, M., Matsumiya, K., Saeki, H., Matsumura, Y., & Nishimura, K. (2022). Development of glucose-conjugated chicken myofibrillar protein with the strongest superoxide anion radical scavenging activity using random-centroid optimization and maltotriose-conjugated ones. *Food science and technology research, 28*(6), 501–511. https://doi.org/10.3136/fstr.FSTR-D-22-00062.

Tanaka, M., Kuei, C. W., Nagashima, Y., & Taguchi, T. (1988). Application of antioxidative Maillard reaction products from histidine and glucose to sardine products. *Nippon Suisan Gakkaishi, 54*(8), 1409–1414.

Temiño, V., Gerardi, G., Cavia-Saiz, M., Diaz-Morales, N., Muñiz, P., & Salazar, G. (2023). Bioaccessibility and genoprotective effect of melanoidins obtained from common and soft bread crusts: Relationship between melanoidins and their bioactivity. *Foods, 12*(17), 3193. https://doi.org/10.3390/foods12173193.

Ullah, S. F., Khan, N. M., Ali, F., Ahmad, S., Khan, Z. U., Rehman, N., ... Muhammad, N. (2019). Effects of Maillard reaction on physicochemical and functional properties of walnut protein isolate. *Food Science and Biotechnology, 28*(5), 1391–1399. https://doi.org/10.1007/s10068-019-00590-z.

Van Nguyen, C. (2006). Toxicity of the AGEs generated from the Maillard reaction: On the relationship of food-AGEs and biological-AGEs. *Molecular Nutrition & Food Research, 50*(12), 1140–1149. https://doi.org/10.1002/mnfr.200600144.

Vhangani, L. N., & Van Wyk, J. (2016). Antioxidant activity of Maillard reaction products (MRPs) in a lipid-rich model system. *Food Chemistry, 208*, 301–308. https://doi.org/10.1016/j.foodchem.2016.03.100.

Walker, J. R. L. (1995). *Enzymatic browning in fruits. Enzymatic browning and its prevention, Vol. 600*, American Chemical Society, 8–22 https://doi.org/doi:10.1021/bk-1995-0600.ch002.

Wijewickreme, A. N., & Kitts, D. D. (1998). Metal chelating and antioxidant activity of model maillard reaction products. In F. Shahidi, C.-T. Ho, & N. Van Chuyen (Eds.). *Process-induced chemical changes in food* (pp. 245–254). Springer US. https://doi.org/10.1007/978-1-4899-1925-0_20.

Wijewickreme, A. N., Krejpcio, Z., & Kitts, D. D. (1999). Hydroxyl scavenging activity of glucose, fructose, and ribose-lysine model maillard products. *Journal of Food Science, 64*(3), 457–461. https://doi.org/10.1111/j.1365-2621.1999.tb15062.x.

Wu, J. J., Cheng, K. W., Li, E. T., Wang, M., & Ye, W. C. (2008). Antibrowning activity of MRPs in enzyme and fresh-cut apple slice models. *Food Chemistry, 109*(2), 379–385. https://doi.org/10.1016/j.foodchem.2007.12.051.

Xu, H., Zhang, X., & Karangwa, E. (2016). Inhibition effects of Maillard reaction products derived from l-cysteine and glucose on enzymatic browning catalyzed by mushroom tyrosinase and characterization of active compounds by partial least squares regression analysis. *RSC Advances, 6*(70), 65825–65836. https://doi.org/10.1039/C6RA15769F.

Yan, X., Gong, X., Zeng, Z., Wan, D., Xia, J., Ma, M., ... Gong, D. (2023). Changes in structure, functional properties and volatile compounds of *Cinnamomum camphora* seed kernel protein by Maillard reaction. *Food Bioscience, 53*, 102628. https://doi.org/10.1016/j.fbio.2023.102628.

Yang, S., Fan, W., & Xu, Y. (2022). Melanoidins present in traditional fermented foods and beverages. *Comprehensive Reviews in Food Science and Food Safety, 21*(5), 4164–4188. https://doi.org/10.1111/1541-4337.13022.

Yang, S., Zhang, G., Chu, H., Du, P., Li, A., Liu, L., & Li, C. (2023). Changes in the functional properties of casein conjugates prepared by Maillard reaction with pectin or arabinogalactan. *Food Research International, 165*, 112510. https://doi.org/10.1016/j.foodres.2023.112510.

Yilmaz, Y., & Toledo, R. (2005). Antioxidant activity of water-soluble Maillard reaction products. *Food Chemistry, 93*(2), 273–278. https://doi.org/10.1016/j.foodchem.2004.09.043.

Yuan, D., Xu, Y., Wang, C., Li, Y., Li, F., Zhou, Y., ... Jiang, Y. (2015). Comparison of anti-browning ability and characteristics of the fractionated Maillard reaction products with different polarities. *Journal of Food Science and Technology, 52*, 7163–7172. https://doi.org/10.1007/s13197-015-1869-1.

Zhang, H., Nakamura, S., & Kitts, D. D. (2020). Antioxidant properties of casein phosphopeptides (CPP) and maillard-type conjugated products. *Antioxidants, 9*(8), 648. https://doi.org/10.3390/antiox9080648.

Zhang, X., Li, X., Liu, L., Wang, L., Massounga Bora, A. F., & Du, L. (2020). Covalent conjugation of whey protein isolate hydrolysates and galactose through Maillard reaction to improve the functional properties and antioxidant activity. *International Dairy Journal, 102*, 104584. https://doi.org/10.1016/j.idairyj.2019.104584.

Zhang, Z., Holden, G., Wang, B., & Adhikari, B. (2023). Maillard reaction-based conjugation of Spirulina protein with maltodextrin using wet-heating route and characterisation of conjugates. *Food Chemistry, 406*, 134931. https://doi.org/10.1016/j.foodchem.2022.134931.

Zhao, Y., Ye, S., Wan, H., Zhang, X., & Sun, M. (2021). Characterization and functional properties of conjugates of rice protein with exopolysaccharides from Arthrobacter ps-5 by Maillard reaction. *Food Science & Nutrition, 9*(9), 4745–4757. https://doi.org/10.1002/fsn3.2336.

Zhong, S., Li, L., Shen, X., Li, Q., Xu, W., Wang, X., Tao, Y., & Yin, H. (2019). An update on lipid oxidation and inflammation in cardiovascular diseases. *Free Radical Biology and Medicine, 144*, 266–278. https://doi.org/10.1016/j.freeradbiomed.2019.03.036.

Zhu, K.-X., Li, J., Li, M., Guo, X.-N., Peng, W., & Zhou, H.-M. (2013). Functional properties of chitosan–xylose Maillard reaction products and their application to semi-dried noodle. *Carbohydrate Polymers, 92*(2), 1972–1977. https://doi.org/10.1016/j.carbpol.2012.11.078.

Zhu, Z., Chen, J., Chen, Y., Ma, Y., Yang, Q., Fan, Y., ... Liao, W. (2022). Extraction, structural characterization and antioxidant activity of turmeric polysaccharides. *LWT, 154*, 112805. https://doi.org/10.1016/j.lwt.2021.112805.

CHAPTER THIRTEEN

Vitamin B6 and diabetes and its role in counteracting advanced glycation end products

F. Vernì*
Department of Biology and Biotechnology "Charles Darwin" Sapienza University of Rome, Rome, Italy
*Corresponding author. e-mail address: iammetta.verni@uniroma1.it

Contents

1. Vitamin B6	402
1.1 Structure and biosynthesis of vitamin B6	402
1.2 Distribution, homeostasis, and catabolism of vitamin B6	404
1.3 Vitamin B6 functions	406
1.4 Causes and consequences of vitamin B6 deficiency	407
1.5 Diseases associated to *salvage pathway* enzymes	407
2. Vitamin B6 and diabetes	409
2.1 Diabetes mellitus	409
2.2 Relationship between vitamin B6 and diabetes	410
2.3 Diabetic complications	414
3. Advanced glycation end products	415
3.1 Discovery and metabolism of AGEs	415
3.2 Action mechanism of AGEs	418
3.3 AGEs and diabetes	419
4. Vitamin B6, diabetic complications and AGEs	421
4.1 Mechanisms through which vitamin B6 counteracts AGE accumulation	423
4.2 Vitamin B6, diabetes and cancer risk	425
5. Conclusions	427
References	428

Abstract

Naturally occurring forms of vitamin B6 include six interconvertible water-soluble compounds: pyridoxine (PN), pyridoxal (PL), pyridoxamine (PM), and their respective monophosphorylated derivatives (PNP, PLP, and PMP). PLP is the catalytically active form which works as a cofactor in approximately 200 reactions that regulate the metabolism of glucose, lipids, amino acids, DNA, and neurotransmitters. Most of vitamers can counteract the formation of reactive oxygen species and the advanced glycation end-products (AGEs) which are toxic compounds that accumulate in diabetic patients due to prolonged hyperglycemia. Vitamin B6 levels have been inversely associate with diabetes, while vitamin B6 supplementation reduces diabetes onset and its vascular complications.

The mechanisms at the basis of the relation between vitamin B6 and diabetes onset are still not completely clarified. In contrast more evidence indicates that vitamin B6 can protect from diabetes complications through its role as scavenger of AGEs. It has been demonstrated that in diabetes AGEs can destroy the functionality of macromolecules such as protein, lipids, and DNA, thus producing tissue damage that result in vascular diseases. AGEs can be in part also responsible for the increased cancer risk associated with diabetes. In this chapter the relationship between vitamin B6, diabetes and AGEs will be discussed by showing the acquired knowledge and questions that are still open.

1. Vitamin B6
1.1 Structure and biosynthesis of vitamin B6

Vitamin B6 term refers to a group of six interconvertible water-soluble chemical compounds (vitamers) all containing a pyridine ring in the center: pyridoxine (PN), pyridoxamine (PM), pyridoxal (PL) and the respective 5′-phosphorylated forms (PNP, PMP, PLP). These compounds differ for the functional group at the position 4′ (Fig. 1). PLP is the catalytically active form, used as a cofactor in numerous biochemical transformations. In few enzymes, PMP also plays a catalytic role (Di Salvo, Contestabile, & Safo, 2011).

The formula of vitamin B6 was first published in 1932 by Ohdake (1932). The crystalline vitamin B6 was isolated from yeast in 1938 from five separate groups of researchers and, a year later, after determination of its structure, György named the vitamin PN due to its structural homology to pyridine (György & Eckardt, 1939). In the same year, Stanton & Folkers (1939) accomplished the synthesis of vitamin B6.

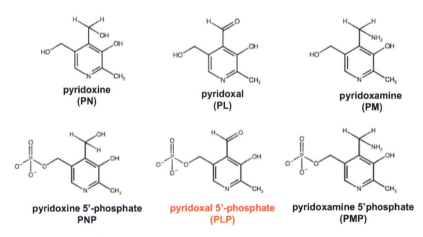

Fig. 1 Structures of the six B6 vitamers.

The food sources of vitamin B6 include beef liver and other organ meats, fish, cereals, potatoes, vegetables, and fruits such as bananas papaya and oranges. The vitamin B6 Recommended Dietary Allowance (RDA) is about 2 mg per day for adults (Hellmann & Mooney, 2010). Vitamin B6 is required for survival of all living organisms, but only microorganisms and plants can synthesize *de novo* this vitamin. The other organisms, including human, acquire vitamin B6 from exogenous sources and interconvert the vitamers through a biochemical pathway named *salvage pathway* (Mooney, Leuendorf, Hendrickson, & Hellmann, 2009). Vitamin B6 *de novo* synthesis is sustained by two pathways: the deoxyxylose 5′-phosphate (DXP)-dependent pathway and the DXP-independent pathway. The first is limited to a small number of bacteria and has been characterized in *Escherichia coli* (Hill et al., 1996). The second route is widespread among archaea, plants, fungi, and most bacteria. In the DXP-dependent pathway (Fig. 2) PNP is synthesized from the condensation of deoxyxylulose 5-phosphate and 4-phosphohydroxy-L-threonine catalysed by the concerted action of PdxA and PdxJ enzymes. Then, PdxH converts PNP into PLP. The DXP-independent pathway relies instead on the activity of the Pdx1 and Pdx2 proteins which produce PLP from a pentose

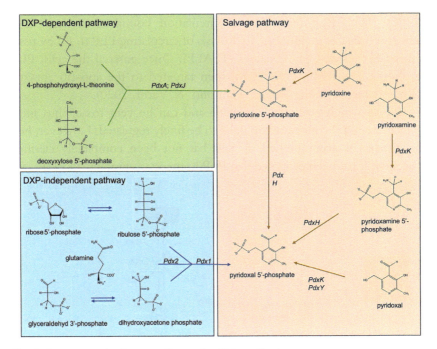

Fig. 2 *De-novo* and *salvage pathways* for PLP synthesis.

(ribose 5-phosphate or ribulose 5-phosphate) and a triose (glyceraldehyde 3-phosphate or dihydroxyacetone phosphate) in the presence of glutamine (Fitzpatrick et al., 2007) (Fig. 2). In addition to the *de novo* pathway, bacteria can synthesize PLP in the *salvage pathway* where PL, PN, and PM are phosphorylated by PdxK or PdxY kinases, and the PMP and PNP are oxidized to PLP by PdxH activity (Fig. 2) (Mukherjee, Hanes, Tews, Ealick, & Begley, 2011).

In humans, vitamin B6 is produced only through the *salvage pathway* where PL, PN and PM are first phosphorylated by a single ATP-dependent PL kinase (PDXK or PLK), and then oxidized to PLP by a FMN-dependent PN/PM [Pyridoxine/pyridoxamine 5′-phosphate oxidase (PNPO)] which are the orthologs of *PdxK* and *PdxH* bacteria genes respectively (Di Salvo, Safo, Musayev, Bossa, & Schirch, 2003)(Fig. 3).

1.2 Distribution, homeostasis, and catabolism of vitamin B6

In mammals, once ingested, phosphorylated B6 vitamers are hydrolyzed to PL, PM and PN by intestinal phosphatases (IP)(Fig. 4). Through blood circulation the absorbed vitamers reach the liver, where they are phosphorylated by PDXK; PNP and PMP are then further oxidized to PLP by PNPO. Then PLP re-enters the circulation bound to albumin. PLP delivery to tissues requires the hydrolysis of circulating PLP to PL by tissue nonspecific alkaline phosphatases (TSALP). Once entered the cells, PL is re-phosphorylated by PDXK and then targeted to the apo-B6 enzymes (Contestabile, di Salvo, Bunik, Tramonti, & Vernì, 2020).

PLP is a highly reactive molecule and can be very toxic, thus intracellular-free PLP concentration need to be finely regulated. However, how this process is accomplished in cells has not been completely clarified. Recently, it has been proposed that PLP homeostasis protein (PLPHP), also

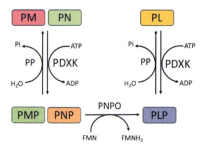

Fig. 3 Metabolism of B6 vitamers (Salvage pathway).

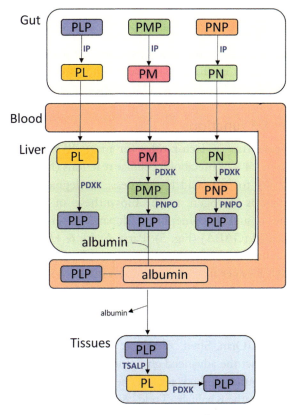

Fig. 4 Absorption, distribution, and interconversion of B6 vitamers in human tissues. *IP*, Intestinal phosphatases; *TSALP*, tissue nonspecific alkaline phosphatases; *PDXK*, pyridoxal kinase; *PNPO*, pyridoxine/pyridoxamine 5′-phosphate oxidase.

called PROSC in humans, can play a crucial role in regulating PLP levels. Mutations in the *PLPHP* human gene have been associated to a rare and severe form of epilepsy responsive to PN and/or PLP. Given that PLP is the cofactor of some enzymes involved in neurotransmitter metabolism which is altered in epilepsy, this finding suggested the involvement of PLPHP in PLP homeostasis (Wilson, Plecko, Mills, & Clayton, 2019). Studies carried out in *E. coli* to elucidate the mechanism through which the PLPHP counterpart (Yggs) regulates PLP levels revealed the involvement of a cluster of lysine residues located at the entrance of the active site, should be important for PLP binding regulation (Tramonti et al., 2022).

Little is known about PLP catabolism in humans or other mammals. In humans and mammals, the primary product of the degradation of PLP (and all other vitamin B6 vitamers) is 4-pyridoxic acid (4-PA). This compound is excreted in urine and is produced in two steps. In the first, PLP is hydrolyzed to PL by the intracellular enzyme PLP phosphatase (PLPase). In the second step, PL is oxidized to 4-PA by a non-specific aldehyde oxidase (AOX) or aldehyde dehydrogenase (Mukherjee et al., 2011).

1.3 Vitamin B6 functions

Vitamin B6 performs co-factor roles for many PLP-dependent enzymes involved in about 4% of metabolic cellular reactions (Mooney et al., 2009), thus contributing to fatty acid biosynthesis, to degradation of stored carbohydrates such as glycogen, to the biosynthesis of plant hormones and neurotransmitters such as epinephrine, dopamine, serotonin, and gamma-aminobutyric acid (GABA) (Fig. 5) (Parra, Stahl, & Hellmann, 2018). PLP is also important for the biosynthesis of tetrapyrroles such as heme, cobalamin and chlorophyl (Parra et al., 2018) and is also involved in the folding of PLP-dependent enzymes (Cellini, Montioli, Oppici, Astegno, & Voltattorni, 2014). In addition, PLP modulates the expression and action of steroid hormone receptors (Tully, Allgood, & Cidlowski, 1994) and influences the immune response (Qian, Shen, Zhang, & Jing, 2017). Moreover, PLP and PMP can counteract the advanced glycation end products (AGEs) which are genotoxic compounds linked to senescence and associated to diabetes complications (Booth, Khalifah, Todd, & Hudson, 1997) (Fig. 5). All B6

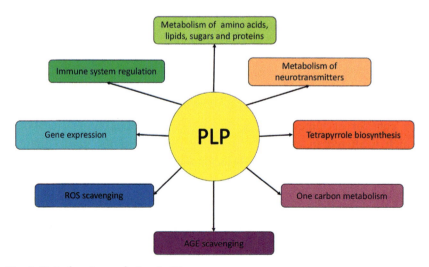

Fig. 5 Main functions of vitamin B6.

vitamers can quench reactive oxygen species (ROS) (Fig. 5). The antioxidant properties rely on the ability of compounds to directly react with the peroxyl radicals with both their hydroxyl (–OH) and amine (–NH2) substituents on the pyridine ring (Contestabile et al., 2020). In addition, PLP can protect from oxidative stress also as a cofactor in the pathway for the formation of cysteine, which is the rate-limiting precursor in glutathione synthesis (Contestabile et al., 2020).

1.4 Causes and consequences of vitamin B6 deficiency

Vitamin B6 deficiency due to insufficient dietary supply is rare because of the presence of this vitamin in almost all foods (Contestabile et al., 2020). However, a secondary vitamin B6 deficiency can be caused by absorption disorders, genetic factors, interactions with drugs or elevated requirements. Reduced PLP levels were associated to kidney diseases and malabsorption syndromes including celiac disease and inflammatory bowel diseases (Merrill & Henderson, 1987). In addition, low PLP levels are common in people with alcohol dependence (Cravo & Camilo, 2000), in pregnant women, obese individuals and diabetic patients (Ferro et al., 2017; Mascolo & Vernì, 2020; Merrill & Henderson, 1987). Reduced PLP availability may also result as a side effect of some commonly used drugs such as isoniazid, cycloserine and penicillamine (Lainé-Cessac, Cailleux, & Allain, 1997). Decreased PLP levels can also result from mutations in *PDXK* and *PNPO* genes involved in vitamin B6 synthesis and cause neurological diseases (Contestabile et al., 2020). Because of the pleiotropic role of vitamin B6, clinical vitamin B6 deficiency results in a broad spectrum of impaired features including anaemia (Linkswiler, 1967), impaired immune response (Qian et al., 2017), diabetes (Mascolo & Vernì, 2020) and cancer (Galluzzi et al., 2013), although molecular mechanisms underlying need to be still clarified. An excessive PLP intake is equally harmful as it produces signs of toxicity mostly affecting the peripheral nervous system, as shown in studies carried out in experimental animals and humans (Krinke & Fitzgerald, 1988).

1.5 Diseases associated to *salvage pathway* enzymes
1.5.1 Pyridoxal kinase (PDXK o PLK)
Human *PDXK* gene is located on the chromosome 21 (q22.3). PM kinase catalyses the phosphorylation of PN, PM and PL, in the presence of MgATP. The mechanism of this reaction has been elucidated in sheep and *E. coli*, showing that it occurs by a random sequential substrate addition. However, it has not clarified yet whether this mechanism also occurs in

humans (di Salvo et al., 2011). PDXK is a member of the ribokinase superfamily and the crystal structure of the human enzyme indicates that it functions as a homodimer (Cao, Gong, Tang, Leung, & Jiang, 2006).

Mutations in *PDXK* gene can affect different pathologies. Consistently with the role of PLP in the metabolism of neurotransmitters, it has been proposed that *PDXK* could be involved into Parkinson disease (Elstner et al., 2009) although this association is still controverse (Guella et al., 2010). In addition, *PDXK* mutations have been also recently associated to polyneuropathy, one of most common genetic neuromuscular disorders (Chelban et al., 2019).

In *Drosophila* mutations in *PDXK* gene increase the glucose content in the larval haemolymph (the blood of flies), thus establishing a condition of insulin resistance (IR) (Marzio, Merigliano, Gatti, & Vernì, 2014). Accordingly, metabolomic studies on human adipose tissue samples revealed a role of *PDXK* in adipogenesis and systemic insulin sensitivity (Moreno-Navarrete et al., 2016).

PDXK is also correlated to DNA damage as *PDXK* mutations cause chromosome aberrations in yeast, *Drosophila* and human cells (Kanellis et al., 2007; Marzio et al., 2014).

Altered *PDXK* expression has been associated with cancer. It has been shown that low *PDXK* expression within malignant cells indicates poor outcome in patients with non-small cell lung cancer (Galluzzi et al., 2012). In contrast, *PDXK* is abundantly expressed in myeloid leukaemia cells, where *PDXK* depletion has an antiproliferative effect (Chen et al., 2020). *PDXK* is also upregulated in serous ovarian cancer cells, and the knockdown of the gene reduces proliferation in vitro (Tan, Liu, & Ling, 2020). To reconcile these findings, it has been proposed that the effect of PDXK expression may vary depending on the type of cancer and/or the stage of the tumour (Galluzzi et al., 2013) (Contestabile et al., 2020).

1.5.2 Pyridoxine/pyridoxamine 5′-phosphate oxidase

Human *PNPO* gene is located on the chromosome 17 (q21.32). PNPO is a dimeric protein and each monomer uses an FMN molecule as a cofactor. In the salvage pathway PNPO catalyzes the oxidation of the 4′-hydroxyl group of PNP or the 4′-amino group of PMP into the aldehyde group of PLP (di Salvo et al., 2003).

PNPO mutations have been associated with a severe form of neonatal encephalopathy (Mills et al., 2005). Genome-wide association studies (GWAS) showed that different mutations in the gene can produce a large spectrum of phenotypes including cases able to respond to PN but not to PLP (di Salvo et al., 2017; Mills et al., 2014). The mechanisms through which the *PNPO* mutations cause epilepsy have not been fully elucidated, but it is conceivable that they are linked to the role of PLP as cofactor of enzymes implicated in the synthesis of some neurotransmitters. Likewise mutations in the *Drosophila sgll* gene (the *PNPO* counterpart), cause epilepsy (Chi et al., 2019). Moreover, like *PDXK* mutations, *sgll* mutations causes diabetes, accumulation of AGEs, DNA damage and impaired lipid metabolism (Mascolo, Amoroso, Saggio, Merigliano, & Vernì, 2020).

PNPO is overexpressed in several cancers and it has been included in a panel of seven genes that can predict overall survival of patients with colorectal cancer (Chen et al., 2017). *PNPO* was found overexpressed in epithelial ovarian cancer while in contrast the gene knockdown was found to decreases tumor cell proliferation and migration (Zhang et al., 2017). Computational analysis showed that at least 21 tumour types overexpress *PNPO* at mRNA and protein level, sometimes with a prognostic significance. Genomic studies showed that *PNPO* gene is altered in about 1.3% of all tumours. In most of them, *PNPO* is amplified, and missense, truncating mutations or deep deletions are also present (Zhang, Li, Zhang, & Xu, 2021).

2. Vitamin B6 and diabetes
2.1 Diabetes mellitus

Diabetes mellitus (DM) is a group of metabolic disorders characterized by hyperglycemia caused by defects in insulin secretion and/or insulin action. Type 1 (T1D), and type 2 (T2D) represent the most common forms of diabetes. T1D is a multifactorial disease which accounts for 5%–10% of diabetes; it is caused by autoantibody-mediated destruction of pancreatic beta cells which impairs insulin secretion (Katsarou et al., 2017). T2D accounts for 90%–95% of all diabetes and is typical of mature age, though nowadays is growing the number of young people affected. T2D is the results of an interaction between environmental factors and a strong hereditary component and is characterized by IR which is a reduced response of tissues to insulin (Galicia-Garcia et al., 2020). Linkage analysis, candidate gene approaches and GWAS discovered a wide number of T2D associated variants. However, these mutations can explain only a small proportion (~10%) of the heritability of T2D (Liguori, Mascolo, & Vernì, 2021). More common risk factors associated to T2D are obesity, hypertension, high concentration of HDL cholesterol and triglycerides and reduced physical activity. Gestational diabetes mellitus (GDM) is a common complication which affects about 7% of pregnancies and disappears after childbirth, although GDM women remain at risk of developing T2D after pregnancy (Plows, Stanley, Baker, Reynolds, & Vickers, 2018). The causes of GDM are not so far completely understood. It has been hypothesized that placental hormones combined to other factors may sometimes interfere with the action of insulin, causing IR. Fetal exposure to maternal hyperglycemia leads to fetal hyperglycemia providing excess nutrition that in turn accelerates fetal growth leading to macrosomia and neonatal disturbance in glucose metabolism (Plows et al., 2018).

2.2 Relationship between vitamin B6 and diabetes

Several studies carried out in both humans and animal models associated reduced levels of vitamin B6 with diabetes. Population studies revealed that plasmatic PLP levels are low in T1D and T2D patients as well as in GDM women (Ahn, Min, & Cho, 2011; Bennink & Schreurs, 1975; Nix et al., 2015; Satyanarayana et al., 2011). Moreover, vitamin B6 levels seem to be inversely related to the progression of diabetes (Ellis et al., 1991; Nix et al., 2015). In line with these findings Okada, Shibuya, Yamamoto, & Murakami (1999) and Rogers, Higgins, & Kline (1986) highlighted an inverse relationship between vitamin B6 levels and diabetes also in streptozotocin-induced diabetic rats.

Although it is known that PLP levels are reduced in diabetes, it is not clear whether this occurs because diabetes decreases PLP availability or whether, in contrast, reduced PLP levels trigger diabetes onset. In literature there are reports for both scenarios. The earliest evidence that diabetes decreases PLP levels was provided by Leklem & Hollenbeck (1990) in a study showing that glucose ingestion caused a reduction of PLP levels in healthy subjects. Okada and coworkers proposed that vitamin B6 deficiency in diabetes might be the consequence of an enhanced protein metabolism, due to the diet low in carbohydrates and rich in proteins (Okada et al., 1999). Since PLP is cofactor for many enzymes involved in protein metabolism, an increased PLP demand would decrease PLP levels. It is thought that inflammation may be another mechanism through which diabetes causes PLP depletion. An inverse relationship between plasma PLP levels and inflammation markers has been reported in diabetic patients (Nix et al., 2015). However, the exact mechanism by which vitamin B6 plays the anti-inflammatory role is still unclear. It has been recently proposed that

PLP may be involved in inflammation acting as a cofactor of sphingosine 1-phosphate lyase which regulates S1P levels in macrophages which promote signaling pathways correlated to inflammation (Du et al., 2020).

On the other hand, there is also evidence that reduced PLP levels can induce diabetes. Toyota et al. (1981) were the first to show that PN deficiency can impair insulin secretion in rats, establishing a diabetic condition. It has been also proposed that low vitamin B6 levels might contribute to the appearance of pancreatic islet autoimmunity in T1D. This because PLP is a cofactor for glutamic acid decarboxylase (GAD-65) which represents an important autoantigen implicated in the pathogenesis of T1D. Thus, reduced levels of the coenzyme may trigger autoimmunity by altering stability, tridimensional conformation, or antigenicity of GAD-65 (Rubí, 2012). More direct evidence on the causative effect of vitamin B6 deficiency in diabetes onset comes from studies in *Drosophila* showing that mutations in *dPdxk* or *sgll/PNPO* genes, involved in vitamin B6 biosynthesis, can establish a diabetic condition characterized by increased hemolymph glucose content, impaired lipid metabolism and reduced body size (Marzio et al., 2014; Mascolo et al., 2020). The same diabetes hallmarks were induced by treating wild type flies with the PLP inhibitor 4-deoxypyridoxine (4DP) (Merigliano, Mascolo, La Torre, Saggio, & Vernì, 2018). Evidence correlating the expression of *PDXK (PLK)* or *PNPO* human genes with diabetes are still scarce, however Moreno-Navarrete and coworkers (2016) demonstrated that reduced *PDXK* expression affects lipid metabolism (see Section 2.2.2), raising the possibility that vitamin B6 can protect from IR in obesity. In addition, we demonstrated that expressing the human wild-type PDXK protein in *dPdxk^1* mutant flies rescued hyperglycemia; in contrast the expression of four PDXK human variants, with impaired catalytic activity or affinity for substrates, was unable to restore normal glycemic values (Mascolo et al., 2019).

2.2.1 Vitamin B6 supplementation ameliorates diabetes and prevents the onset of the disease

Supplementation of vitamin B6 resulted effective in reversing diabetic hallmarks in humans and experimental animals. Bennink & Schreurs (1975) demonstrated that PN administration to GDM women with low blood vitamin B6 levels ameliorated glucose tolerance. Similar results were reported in a study from Spellacy, Buhi, & Birk (1977) showing that PN therapy increased the biologic activity of endogenous insulin in GDM women. Vitamin B6 supplementation can also prevent diabetes. Liu et al. (2016) demonstrated that Vitamin B6 prevents endothelial dysfunction, IR, and hepatic lipid accumulation in *Apoe* (−/−) mice fed with a high fat diet, which have an increased risk to develop diabetes due to impaired lipid metabolism. In line with this finding, a study carried out by Haidari, Mohammadshahi, Zarei, Haghighizadeh, & Mirzaee (2021) showed that vitamin B6 supplementation may be effective in reducing body mass index and improving biochemical parameters associated with obesity, which is one of the main risk factors for T2D. Conversely, Zhu et al. (2020) found that B6 supplementation was not inversely associated with diabetes incidence in a large US cohort examined for 30 years. However, the authors explained this result by hypothesizing that nutrient supplementation may only benefit individuals with insufficient dietary intakes.

2.2.2 Mechanisms linking vitamin B6 to diabetes

Based on the notion that PLP works as a coenzyme in a plethora of metabolic reactions and that it possesses antioxidant properties, it is conceivable that reduced vitamin B6 levels can affect diabetes through different mechanisms. However, two routes seem to mainly correlate diabetes to vitamin B6: the tryptophan (TRP) pathway and the lipid metabolism. TRP metabolism is often impaired in diabetes (Bennink & Schreurs, 1975; Connick & Stone, 1985). TRP is an essential amino acid, crucial for serotonin, N-acetylserotonin, and melatonin biosynthesis. About 95% of TRP is metabolized through the kynurenine pathway (Fig. 6) where work PLP-dependent enzymes such as the aminotransferases (KAT) and the kynureninase (KYNU). KYNU is more sensitive to PLP deficiency with respect to KAT thus, decreased PLP availability results in the accumulation of kynurenic acid (KYNA) and xanthurenic acid (XA) (Bender, Njagi, & Danielian, 1990; Takeuchi, Tsubouchi, Izuta, & Shibata, 1989). Evidence suggests that these compounds can interfere with insulin biological activity in different ways (Oxenkrug, 2007): (1) by forming chelate complexes with insulin (XA–In), which have reduced activity compared to pure insulin; (2) by forming Zn^{++} ion–insulin complexes that cause toxic effects on pancreatic beta cells; (3) by inhibiting insulin release from pancreas and (4) by inducing pathological apoptosis in pancreatic beta cells. The involvement of TRP metabolism also emerged in a recent study on a GDM mouse model revealing that vitamin B6 deficiency can induce GDM by perturbing the catabolism of TRP in the pancreatic islets, thus decreasing serotonin levels in maternal pancreatic islets, and reducing β-cell proliferation. However, these changes result in glucose intolerance and IR but do not alter insulin secretion (Fields, Welle, Ho, Mesaros, & Susiarjo, 2021).

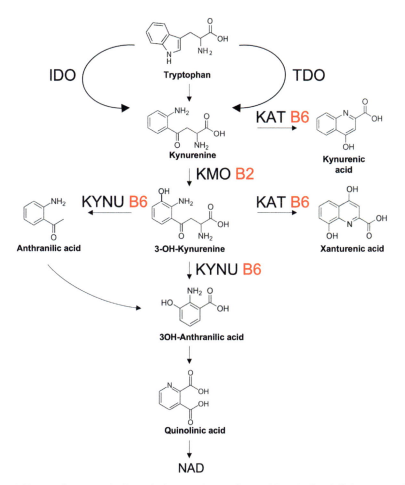

Fig. 6 Tryptophan metabolism via kynurenine pathway. Vitamin B6 deficiency mostly affects KYNU activity, thus resulting in the shift from 3OH-kynurenine metabolism and NAD formation to the production of kynurenic acid and xanthurenic acid. *IDO*, Indoleamine 2,3-dioxygenase; *TDO*, tryptophan 2,3-dioxygenase; *KAT*, kynurenine aminotransferase; *KMO*, kynurenine 3-monooxygenase; *KYNU*, kynureninase; *3OH-kynurenine*, 3-hydroxy kynurenine; *3OH-anthranilic acid*, 3-hydroxyanthranilic acid; *B6*, vitamin B6; *B2*, vitamin B2.

Decreased PLP levels can contribute to diabetes and to IR also by impairing lipid metabolism. Evidence indicates that PLP is a regulator of genes involved in adipogenesis and systemic insulin sensitivity (Moreno-Navarrete et al., 2016). It has been shown that vitamin B6-deficient diet results in a significant reduction in adipose tissue and lipogenesis in rat models (Radhakrishnamurty, Angel, & Sabry, 1968). Conversely, vitamin B6 administration increased intracellular lipid accumulation in 3T3-L1 adipocytes and decreased macrophage infiltration and adipose tissue inflammation in mice (Sanada et al., 2014). Accordingly, obese people have low circulating levels of vitamin B6 (Aasheim, Hofsø, Hjelmesaeth, Birkeland, & Bøhmer, 2008). Despite these findings, the exact mechanisms through which PLP influences lipid metabolism leading to diabetes are so far not completely clarified. It has been proposed that PLP might activate peroxisome proliferator-activated receptor- (PPAR), one of the master nuclear receptors involved in the expression of adipogenesis genes (Yanaka, Kanda, Toya, Suehiro, & Kato, 2011). Alternatively, PLP might conjugate with RIP140, a nuclear transcription factor, by enhancing its co-repressive activity and its physiological function in adipocyte differentiation (Huq, Tsai, Lin, Higgins, & Wei, 2007). Moreover, based on the finding that an altered DNA methylation is associated with adipose tissue dysfunction in T2D patients (Nilsson et al., 2014), given that PLP is a coenzyme for serine hydrossymethiltranferase, vitamin B6 might contribute to maintain the correct methylation pattern.

2.3 Diabetic complications

T1D and T2D can lead to long term complications affecting different organs classified as microvascular or macrovascular (Brownlee, 2001). Microvascular complications include nervous system damage (neuropathy), renal system damage (nephropathy) and eye damage (retinopathy). Macrovascular complications include cardiovascular diseases, stroke, and peripheral vascular disease. Peripheral vascular disease may lead to gangrene and ultimately, amputation. Diabetes complications are a significant cause of increased morbidity and mortality among people with diabetes, and result in a heavy economic burden on the health care system. With advances in treatment for diabetes and its associated complications, people with diabetes are living longer with their condition. However, this longer life span contributes to further increase the morbidity associated with diabetes, primarily in elderly people and in minority racial or ethnic groups.

The complex cascade of events which leads to the diabetes complications in response to high levels of glucose is not fully understood. However growing evidence pinpoints the formation of AGEs as one of the main events (Brownlee, 2001; Nin et al., 2011; Tan et al., 2002).

3. Advanced glycation end products

The AGEs are a heterogeneous group of compounds that form by the reaction of reducing sugars or reactive aldehydes with the free amino groups of proteins, lipids, and nucleic acids (Sergi, Boulestin, Campbell, & Williams, 2021). Thus, the rate of their endogenous formation is markedly enhanced in diabetes. They also form externally during the heat processing of food and it has become apparent that dietary AGEs contribute significantly to the body's AGE pool and may play a significant pathogenic role in several diseases (Uribarri et al., 2015). AGEs play crucial role in the pathogenesis of cardiovascular, kidney, Alzheimer's, neurological, diabetes and joints diseases and aging process.

3.1 Discovery and metabolism of AGEs

AGEs were discovered the early '900 as yellowish-brown color products generated during a chemical reaction in which amino acids were heated with reducing sugars (Maillard, 1912). Further studies elucidated the AGE formation through Maillard reaction, demonstrating that reducing sugars (i.e., glucose, fructose, pentoses, galactose, mannose, xylulose) react non-enzymatically with amino groups of proteins, lipids, and nucleic acids to form Schiff bases, early glycation products, which are converted into more stable Amadori products which progress to covalent adducts and accumulate on proteins giving rise to the formation of carboxymethyl-lysine (CML), carboxyethyl-lysine (CEL), pyrraline, 3-deoxyglucosone lysine dimer (DOLD), and glyoxal-lysine-dimer (GOLD) (Fig. 7). The Maillard reaction generates highly reactive dicarbonyl compounds which include methylglyoxal (MG), glyoxal (GO) and 3-deoxyglucosone (3-DG) (Fig. 7). These molecules can be also generated from other pathways such as glucose autoxidation, lipid peroxidation, and polyol pathway (Sergi et al., 2021). After further reacting with the amino compound, the dicarbonyl compounds undergo dehydration, condensation, cyclisation, and intermolecular crosslinking to form stable AGEs in the advanced stage.

Reactive carbonyl species produced by lipid peroxidation and lipid metabolism react nonenzymatically with the nucleophilic residues of macromolecule such as proteins, DNA, and aminophospholipids, leading to their irreversible modification and generating the advanced lipid peroxidation end products (ALEs). Some AGEs and ALEs have the same structure, since they arise from common precursors, as in the case of CML which is generated by glyoxal, which in turn is formed by both lipid and sugar oxidative degradation pathways (Fu et al., 1996).

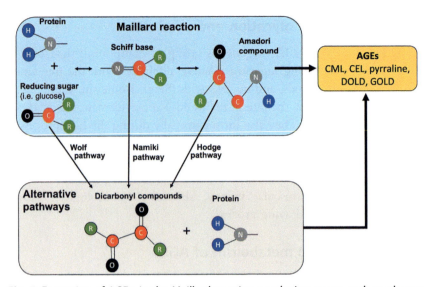

Fig. 7 Formation of AGEs. In the Maillard reaction a reducing sugar, such as glucose, reacts non-enzymatically with amino groups of proteins to form Schiff bases, which are converted into Amadori products, which progress to covalent adducts and give rise to AGEs. AGEs can also be produced through alternative pathways when highly reactive dicarbonyl compounds such as methylglyoxal (MG), glyoxal (GO) glycolaldehyde (GLA) and 3-deoxyglucosone (3-DG) react with proteins. The dicarbonyl compounds are in turn generated through the Wolff pathway by metal ion-catalyzed glucose autoxidation, or alternatively by Schiff bases and Amadori products through the Namiki pathway and Hodge pathway respectively.

3.1.1 Classification of AGEs

AGEs have been classified in several ways based on their properties or origin of formation (Table 1). One classification is based on their ability to create cross-links on proteins and to show fluorescence. This classification identifies three groups: (1) fluorescent cross-linking AGEs such as pentosidine and crossline; (2) non-fluorescent cross-linking AGEs such as glyoxal lysine dimer (GOLD), methylglyoxal-lysine dimer (MOLD), 3-deoxyglucosone lysine dimer (DOLD), methylglyoxal dimer imidazolone crosslink (MODIC); (3) fluorescent non-crosslinking including argpyrimidine; (4) non-fluorescent non-cross-linking such as pyrraline, CML and CEL (Perrone, Giovino, Benny, & Martinelli, 2020) (Fig. 8 and Table 1).

Another classification based on their origin identifies seven categories: (1) glucose-derived (Glc-AGE, or AGE-1); (2) glyceraldehyde-derived (Glycer-AGEs, or AGE-2); (3) glycol aldehyde-derived (Glycol-AGEs, or AGE-3); (4) methylglyoxal-derived (MGO-AGEs, or AGE-4); (5) glyoxal

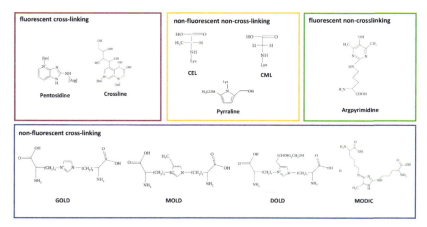

Fig. 8 AGE classification with respect to fluorescence and cross-linking. *CML*, Carboxymethyl-lysine; *CEL*, carboxyethyl-lysine; *GOLD*, glyoxal lysine dimer; *MOLD*, methylglyoxal-lysine dimer; *DOLD*, 3-deoxyglucosone lysine dimer; *MODIC*, methylglyoxal dimer imidazolone crosslink.

Table 1 AGE classification.

Sources	Fluorescence/ crosslinking	Origin	Toxicity
Exogenous (dietary AGEs)	Fluorescent cross-linking	Glucose-derived (Glc-AGEs or AGE-1)	Non-toxic AGEs
Endogenous	Non-fluorescent cross-linking	Glyceraldehyde-derived (Glycer-AGEs, or AGE-2)	Toxic AGEs
	Fluorescent non-crosslinking	Glycol aldehyde-derived (Glycol-AGEs, or AGE-3)	
	Non-fluorescent non-cross-linking	Methylglyoxal-derived (MGO-AGEs, or AGE-4)	
		Glyoxal (GO-AGEs, or AGE-5)	
		3-deoxyglucosone-derived (3DG-AGEs, or AGE-6)	
		Acetaldehyde-derived AGEs (AA-AGEs, or AGE-7)	

(GO-AGEs, or AGE-5); (6) 3-deoxyglucosone-derived (3DG-AGEs, or AGE-6); (7) acetaldehyde-derived AGEs (AA-AGEs, or AGE-7) (Kuzan, 2021) (Table 1).

Takeuchi, (2020) proposed a classification of AGEs into non-toxic AGEs, such as CML, CEL, pentosidine, pyrraline, methylglyoxal hydroimidazolone (MG-H1), MOLD, and GOLD and toxic AGEs (TAGEs) such as AGE-2, AGE-3 and AA-AGE (Table 1). TAGEs interact with the receptor of AGEs (RAGE), and exert their toxic effects in the blood vessels, and can also promote the development of several types of cancer as well as infertility. Regarding non-toxic AGEs, Takeuchi (2020) postulated that they may have a role in preventing the potentially damaging consequences of the advanced glycation process. However, it seems that the effect of each type of AGE may vary based on the specific conditions. There is considerable research showing that non-TAGE molecules, such as CML, pentosidine, pyrraline and crossline may be likewise dangerous for cells (Kuzan, 2021).

3.2 Action mechanism of AGEs

Two main mechanisms are at the basis of the pathogenic role of AGEs: the covalent crosslinking of proteins and DNA, and the interaction of AGEs with their receptors. Glycation of proteins interferes with their normal functions by disrupting molecular conformation, thus altering enzymatic activity or impairing receptor functioning (Singh, Bali, Singh, & Jaggi, 2014). Glycation of DNA gives rise mainly to N2-(1-carboxyethyl)-2'-deoxyguanosine (CEdG) and to cyclic dihydroimidazolone 1,N2-(1,2-dihydroxy-2-methyl)ethano-dG (cMG-dG). The CEdG adduct leads to G to T and G to C transversions and can cause single strand breaks, thus promoting genome instability (Ciminera, Shuck, & Termini, 2021). AGEs can produce genotoxicity also indirectly through oxygen free radicals (ROS) generated during the various phases of the glycation process (Wolff & Dean, 1987).

The most well-studied AGE receptor is the Receptor of Advanced Glycation End products (RAGE), which is the main up-regulator of cell activation in response to the AGE load. RAGE is a multi-ligand receptor, belonging to the immunoglobulin superfamily and has a highly charged cytoplasmic domain. AGE–RAGE interaction triggers a series of cascade reactions and signaling pathways, proliferation, autophagy, and apoptosis (Del Turco & Basta, 2012).

Several other AGE receptors were recently identified, including the AGE receptor complex (AGE-R1/OST-48, AGE-R2/80k-H, AGE-R3/galectin-3) and scavenger receptor family (SR-A, SR-B, SR-1, SR-E, LOX-1, FEEL-1, FEEL-2 and CD36). However, they have opposite functions to RAGE being involved in AGE homeostasis (Vlassara, Uribarri, Cai, & Striker, 2008).

3.3 AGEs and diabetes

Chronic complications of DM are caused by structural or functional modification of the vascular system. It has been shown that after long-term exposure to hyperglycemia, AGEs are responsible for these damages. Consistently, AGEs are found in serum, vasculature, retina, and various renal compartments of diabetic patients (Hammes et al., 1999). AGEs promote diabetes complications through both receptor-independent and receptor-dependent mechanisms. In the first case, AGEs form cross-links between key molecules in the basement membrane of the extracellular matrix (ECM) proteins, permanently altering cellular structure. In the second case, AGEs can interact with RAGE on cell surfaces, thus inducing signaling pathways that promote the expression of cytokine and growth factors and stimulate ROS production (Vlassara & Uribarri, 2014).

3.3.1 The role of AGEs in microvascular diseases

Diabetic nephropathy is characterized by glomerular hypertrophy, renal oxidative stress, and fibrosis. It has been shown that glomerular changes including thickening of tubular basement membranes, mesangial hypertrophy, and loss of podocytes, are provoked by AGEs. Tubular cells are exposed to a large amount of AGEs which increase the activation of intracellular signaling pathways via their RAGEs (Sun, Yuan, & Sun, 2016). AGE-RAGE interaction increases ROS and activates the NF-kB transcription factor (nuclear factor kappa-light-chain-enhancer of activated B cells). ROS in turn enhances the JAK-STAT signaling which plays a crucial role in glomerular hypertrophy by inducing growth factors. Moreover, NF-κB increases the expression of adhesion molecules and proinflammatory cytokines such as interleukin (IL)-6, tumor necrosis factor (TNF)-α, and monocyte chemoattractant protein (MCP)-1, all contributing to development of nephropathy (Matoba et al., 2019).

AGEs can also form cross-links with collagen, thus modifying its physical properties (e.g., elasticity) and its interactions with other molecules such as proteoglycans (PG), enzymes (e.g., collagenase) and cell integrins.

The consequent alteration of vascular permeability compromises the structural integrity of the endothelium (Rabbani & Thornalley, 2018).

Diabetic retinopathy is a serious disease characterized by abnormal vascular proliferation which accompanies hemorrhage and ischemia in the retina. AGEs were found in retinal vessels, and the levels were positively correlated with the disease (Stitt, 2001). It has been shown that the AGE–RAGE interaction induces the apoptosis of pericytes and increases oxidative stress via NF-κB production. Pericytes play an important role in the maintenance of microvascular homeostasis and thus, their loss could predispose the vessels to angiogenesis, thrombogenesis, and endothelial cell injury. Increased levels of NF-κB upregulate vascular endothelial growth factor (VEGF) and affect endothelial permeability (Wu, Yiang, Lai, & Li, 2018). ROS generation, as in nephropathy, can exacerbate angiogenesis and vascular permeability. All these pathologic changes can damage subretinal membrane and microvasculature.

Diabetic neuropathy is a life-threatening complication involving both peripheral and autonomic nerves. Peripheral neuropathy increases the risks of developing foot ulceration and subsequent necrosis that necessitates lower limb amputation. The levels of AGEs were increased in the serum and also in the peripheral nerves of diabetic patients (El-Mesallamy, Hamdy, Ezzat, & Reda, 2011). Accumulation of AGEs in the endothelium of the vasa nervorum causes damage to the vascular structure and ischemia or occlusion. AGEs reduce the conduction of sensory and motor nerves and nerve blood flow (Jack & Wright, 2012). Glycation of structural and functional proteins of nerves result in impaired nerve function and characteristic pathologic alterations. In addition, interaction between AGEs and RAGEs by activating the NF-kB cascade response induces pro-inflammatory cytokines such as IL-6 and TNF-α (Bierhaus et al., 2004). Moreover, also in nerves the AGE–RAGE pathway promotes the intracellular activation of NADPH oxidase and the production of ROS. Like glycating agents, excessive ROS alter proteins, lipids, and DNA causing damage to peripheral neurons (Vincent et al., 2007).

3.3.2 The role of AGEs in macrovascular complications

Macrovascular complications include coronary heart disease, chronic heart failure and stroke. The most important mechanisms through which AGEs contribute to these diseases are: (i) the interaction between AGEs and RAGE; (ii) the crosslinking of elastin and collagen; (iii) the inhibition of nitric oxide activity and the reduction of endothelial nitric oxide synthase activity; (iv) an increase in vascular permeability (Twarda-Clapa, Olczak, Białkowska, & Koziołkiewicz, 2022).

The AGEs/RAGE axis activates MAPK (Mitogen-activated protein kinase) and NF-kB cascades and promotes the accumulation of intracellular ROS. These pathways lead to the production of several inflammatory factors including vascular cell adhesion molecule-1, intercellular adhesion molecule-1, plasminogen activator inhibitor-1, MCP-1, and matrix metalloproteinase-2 protein whose increased expression causes arterial stiffness and vascular calcification (Fukami, Yamagishi, & Okuda, 2014). AGEs/RAGE axis also enhances the oxidation of low-density lipoprotein, which play a crucial role in the pathogenesis of CVD (cardio vascular disease) (Piarulli et al., 2005). The nonenzymatic modifications of collagen and lipoproteins by AGEs in large vessels result in increased collagen deposition, altering the structural integrity of arteries, disarray of elastic fibers, and the degeneration of smooth muscle tissue, which are key pathogenic factors in arteriosclerosis (Zhao, Randive, & Stewart, 2014).

4. Vitamin B6, diabetic complications and AGEs

In addition to protect from diabetes onset, vitamin B6 plays an important role in both micro and macrovascular diabetic complications. Low levels of PLP were found in serum from patients affected by diabetic neuropathy (McCann & Davis, 1978). In addition, alterations in vitamin B6 metabolism were associated with retinopathy and nephropathy (Ellis et al., 1991; Nix et al., 2015). Plasma levels of vitamin B6 were found lower in human subjects who had myocardial infarction than in controls (Verhoef et al., 1996). Moreover, a higher incidence of coronary artery disease was found in patients with lower vitamin B6 intake compared with those with higher intakes (Page et al., 2009; Rimm et al., 1998), thus suggesting that reduced PLP levels may potentially increase the predisposition to macrovascular diseases.

PLP supplementation has proven effective in preventing, as well as reducing, the symptoms of diabetes complications. For example, diabetic patients treated with vitamin B6 exhibited a reduced risk of retinopathy (Ellis et al., 1991; Horikawa et al., 2020). In a clinical trial PM ameliorated kidney function in T2D patients with overt nephropathy (Williams et al., 2007). Similar effects have been observed in animal models. Stitt et al. (2002)

showed that PM retarded the development of retinopathy in the STZ-diabetic rat, protecting against pericyte loss and formation of acellular capillaries. PM treatment also significantly protected Zucker obese rats, as well as alloxan-induced diabetic rats, against nephropathy (Alderson et al., 2003; Elseweidy, Elswefy, Younis, & Zaghloul, 2013).

In vitro and in vivo studies suggest that vitamin B6 may impact on the risk of diabetic complications mostly due to its capability of counteracting the AGEs. It is reported that PL inhibits glycation and AGE formation on bovine albumin at physiological concentrations (Vinson and Howard, 1996). Moreover, interventional studies demonstrated the B6 is able to modify the nonenzymatic glycosylation of hemoglobin in a cohort of male patients affected by T2D (Solomon & Cohen, 1989). Studies performed in diabetic rats revealed that PM administration is effective in reducing diabetes complications such as nephropathy and retinopathy, and concomitantly in decreasing plasma levels of GO, MGO, CML and CEL (Alderson et al., 2003; Stitt et al., 2002). Consistently, adducts of PM and carbonyl products were detected in urine of PM-treated animals (Metz, Alderson, Chachich, Thorpe, & Baynes, 2003). PM has also been tested in diabetic obese rats in combination with alpha lipoic acid, an antioxidant molecule that reduces the formation of AGE, and the double treatment proved to be more effective in reducing oxidative biomarkers of stress than the single treatments (Muellenbach et al., 2008). Vitamin B6 reduces also the formation of AGEs on DNA as PLP administration in STZ-induced diabetic rats significantly inhibited the accumulation of CEdG in glomeruli, thus preventing the progression of nephropathy (Nakamura, Li, Adijiang, Pischetsrieder, & Niwa, 2007).

There is also evidence that vitamin B6 treatment can reduce the ROS, which increase in diabetes because of AGE-RAGE interactions and also during the process of AGE formation. This topic has been addressed by Abdullah, Abul Qais, Hasan, & Naseem (2019) which demonstrated how PM supplementation is effective in reducing oxidative stress in alloxan-induced diabetic rats, enhancing the activity of antioxidative enzymes such as superoxide dismutase and catalase, and increasing the glutathione levels.

Also flies progressively accumulate AGEs during their lifespan (Oudes, Herr, Olsen, & Fleming, 1998). Like in mammals AGEs formation increases in diabetes and it is reduced by a PLP rich diet. This indicates that *Drosophila* can be an useful model organism also for diabetes complications (Marzio et al., 2014; Mascolo et al., 2020; Mascolo et al., 2021; Mascolo et al., 2022; Merigliano et al., 2018).

4.1 Mechanisms through which vitamin B6 counteracts AGE accumulation

It has been proposed that AGE formation is counteracted by mechanisms including competition for protein glycation sites, reduction of ROS levels, RAGE regulation, and trapping of dicarbonyl compounds and metal chelating ions.

Aminoguanidine (AG) is one of the most studied AGE inhibitors. AG is a nucleophilic hydrazine able to limit the formation of cross linking by trapping the reactive carbonyls, and by preventing free radical formation, lipid peroxidation and apoptosis induced by oxidative stress (Brownlee, Vlassara, Kooney, Ulrich, & Cerami, 1986). However, AG was retreated from phase III trials due to its various adverse effects (Borg & Forbes, 2016).

The identification of natural compounds such as polyphenols, polysaccharides, terpenoids, vitamins, and alkaloids, has therefore provided a good alternative to counteract AGE formation (Song, Liu, Dong, Wang, & Zhang, 2021). Among vitamins, B6 has been the most studied to clarify the mechanisms underlying its protective effect. These studies revealed that vitamin B6 inhibits AGE formation mainly by (i) blocking oxidative degradation of the Amadori intermediate of the Maillard reaction; (ii) scavenging toxic carbonyl products of glucose and lipid degradation; (iii) trapping ROS. The various B6 vitamers can act differently in consequence of their different chemical structure (Fig. 9). First studies, carried out on the nonenzymatic glycosylation of serum albumin, proposed that PLP can inhibit this process by forming a Schiff base with this protein as does glucose, thus preventing glycation through a competition mechanism (Khatami, Suldan, David, Li, & Rockey, 1988). Other studies on serum albumin, ribonuclease A, and human hemoglobin demonstrated that, among B6 vitamers, PM and PLP can strongly suppress AGE formation at lower concentrations, while PL and PN are effective at higher concentrations (Booth, Khalifah, & Hudson, 1996; Booth et al., 1997; Khalifah et al., 1996). These studies confirmed that PLP may prevent glycation by competitive mechanisms as previously proposed by Khatami et al., (1988). Furthermore, they provided evidence that PM (which lacks an aldehyde group) can inhibit late glycation reactions by forming an adduct with the Amadori intermediate and hence the production of AGEs (Booth et al., 1996; Booth et al., 1997; Khalifah et al., 1996) (Fig. 9). Subsequent ^{13}C NMR studies indicated, however, that PM does not directly interact with the carbonyl moiety of the Amadori intermediate by forming an adduct, but, more likely, it interferes

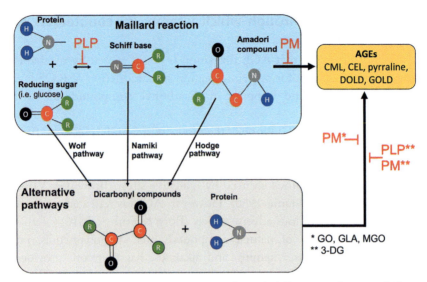

Fig. 9 Vitamin B6 counteracts AGE formation through different mechanisms. PLP may compete with glucose in protein binding, thus preventing the formation of Shiff base. Moreover, PLP may trap the 3-deoxyglucoson (3-DG), thus blocking AGE formation. PM prevents the transformation of Amadori compounds into AGEs through two possible mechanisms: by forming adducts with the Amadori products or by binding catalytic redox metal ions which transform Amadori compounds into AGEs. In addition, PM may scavenge the reactive carbonyl compounds such as glyoxal (GO), glycolaldehyde (GLA), methylglyoxal (MGO) and 3-deoxyglucoson (3-DG), thus preventing their transformation into AGEs. PM may also form a transient adduct with 3-DG, followed by irreversible PM-mediated 3-DG oxidative cleavage.

with post-Amadori oxidative reactions by binding catalytic redox metal ions which transform Amadori compounds into AGEs (Adrover, Vilanova, Frau, Muñoz, & Donoso, 2008) (Fig. 9).

PM can also act by scavenging reactive carbonyl compounds such as GO, GLA, and MGO formed by the degradation processes of sugars, lipids, and amino acids (Nagaraj et al., 2002; Voziyan, Metz, Baynes, & Hudson, 2002) (Fig. 9). Evidence from both vitro and in vivo studies give strong support to this hypothesis (Onorato, Jenkins, Thorpe, & Baynes, 2000; Voziyan & Hudson, 2005).

It has also been proposed that PLP may specifically trap the 3-deoxyglucoson (3-DG), a dicarbonyl compound synthesized via the Maillard reaction or the polyol pathway (Nakamura & Niwa, 2005) (Fig. 9). 3-DG serum concentration is elevated in diabetes. In addition 3-DG-derived protein adducts were found significantly increased in renal glomeruli and retinas of STZ-diabetic rats and in renal tissues and aortas of diabetic patients (Niwa, Katsuzaki, Ishizaki, et al., 1997; Niwa, Katsuzaki, Miyazaki, et al., 1997).

Chetyrkin, Zhang, Hudson, Serianni, & Voziyan (2008) proposed that also PM may protect against 3-DG-induced protein damage via a mechanism that consists in a transient adduction followed by irreversible PM-mediated oxidative cleavage of the 3-DG (Fig. 9). PM can also trap malondialdehyde (MDA) an important intermediate in the formation of ALEs (Kang, Li, Li, & Yin, 2006).

Vitamin B6 can counteract AGEs also indirectly by trapping ROS formed from (1) autoxidation of glucose, (2) oxidative degradation of Amadori compound and (3) interaction between AGEs with RAGE receptors (Mossine, Linetsky, Glinsky, Ortwerth, & Feather, 1999; Ott et al., 2014; Smith & Thornalley, 1992).

Major ROS generated during these reactions are superoxide anion radical ($O_2\bullet-$), hydrogen peroxide (H_2O_2), and hydroxyl radical (HO•). Other ROS, such as MGO radical, can also form during the reaction of carbonyl species with proteins (Yim, Yim, Lee, Kang, & Chock, 2001). In vitro PM can efficiently scavenge hydroxyl radical by its 3-hydroxyl group (Miyata et al., 2002). Moreover, in cell culture PM inhibits the formation of superoxide radical, nitric oxide and hydrogen peroxide (Jain & Lim, 2001; Kannan & Jain, 2004).

Ramis et al., (2019) applied the Density Functional Theory to determine the mechanisms underlying the ability of PM to scavenge ROS showing that PM traps the •OCH3 radical in both aqueous and lipidic media, whereas it can trap •OOH and •OOCH3 in aqueous media but not in lipidic media. The mechanism through which PM traps the •OCH3 radical may be a diffusion-controlled transfer of the hydrogen atoms from the protonated pyridine, from the protonated amino group, and from the phenolic oxygen atom. In contrast to scavenge •OOH and •OOCH3 radicals PM would form radical adducts on the aromatic carbon atoms adjacent to the pyridine nitrogen atom.

4.2 Vitamin B6, diabetes and cancer risk

People with diabetes are at significantly higher risk for many forms of cancer such as cancer of the pancreas, liver, bladder, colon, breast, ovary, and endometrium (Xu, Zhu, & Zhu, 2014). Diabetes and cancer are also related in *Drosophila*, as a diet rich in sucrose increases the malignancy of Ras/Src tumors (Hirabayashi, Baranski, & Cagan, 2013). However, despite the reported associations between diabetes and cancer risk the molecular bases remain still unclear. Plausible mechanisms are hyperinsulinemia, hyperglycemia, and inflammation (Zhu & Qu, 2022). Insulin is a potent growth factor that promotes cell proliferation and carcinogenesis directly and/or through insulin-like growth factor 1. High glucose may exert direct and indirect effects to promote proliferation of cancer cells. It is, indeed, known that glucose is specifically required to meet the metabolic demands of the fast proliferation cancer cell. In addition, it has been shown that high glucose concentrations can compromise DNA integrity causing mutations and DNA stand breaks in cultured human cells (Zhang, Zhou, Wang, & Cai, 2007). This raised the hypothesis that hyperglycemia in diabetic people can increase cancer risk also by triggering genome instability (Lee & Chan, 2015), which plays a significant role in the initiation and progression of cancer.

In support of this hypothesis, both T1D and T2D diabetic patients exhibit DNA strand breaks (Tatsch et al., 2012) and elevated levels of 8-Oxo-7,8-dihydro-20-deoxyguanosine (8-OHdG), a sensitive marker of DNA damage (Dandona et al., 1996; Goodarzi, Navidi, Rezaei, & Babahmadi-Rezaei, 2010; Hinokio et al., 1999). In addition they display a higher frequency of sister chromatid exchanges and micronuclei than healthy subjects (Binici, Karaman, Coşkun, Oğlu, & Uçar, 2013) and high levels of stable chromosomal aberrations in peripheral lymphocytes (Boehm et al., 2008).

One of the mechanisms that mediates the genotoxic effect of glucose in diabetes is the formation of AGEs on DNA. In support of this hypothesis, CEdG levels are increased in T1D and T2D animal models as well as in patients affected by T2D (Li et al., 2006). The AGE-DNA are difficult to repair in a hyperglycemic contexts because diabetes patients display decreased efficacy of DNA repair (Blasiak et al., 2004). Consistently, a recent study reported that chronic high glucose can inhibit the nucleotide excision repair (NER) system, thus causing CEdG accumulation and DNA strand breaks (Ciminera et al., 2021).

A bioinformatic analysis of a microarray dataset of normal and malignant breast cell lines revealed in contrast that high glucose triggers the overexpression of DNA damage response genes and causes significant DNA damage in cultured cells. This finding would indicate that an increased activity of the repair factors may be another possible mechanism that induces DNA damage in diabetes (Rahmoon et al., 2023).

The finding that vitamin B6 counteracts the formation of ROS and AGEs raises the hypothesis that reduced levels of PLP may contribute to increasing DNA damage and consequently the cancer risk in diabetics. Human trials proving this hypothesis are still missing however, it has been demonstrated in *Drosophila* that diabetic flies are more sensitive to DNA damage caused by PLP deficiency than wild-type individuals. The work of Merigliano et al. (2018) revealed, in fact, that larval brains from individuals treated with a PLP inhibitor displayed three times more chromosome damage with respect to wild type individuals which received the same treatment. Extrapolated to humans, these data may suggest that diabetic people need to keep under control their PLP levels to reduce the risk of DNA damage and consequently of cancer. In a hyperglycemic context per se characterized by oxidative stress, reduced antioxidant defenses, and weakened DNA repair systems, a reduced ability to counteract AGE and ROS such as that induced by a PLP depletion, may be further dangerous.

5. Conclusions

DM is a public health problem worldwide as many as 700 million people are expected to have DM by 2045. Diabetes-related complications—including cardiovascular disease, kidney disease, neuropathy, blindness, and lower-extremity amputation—are a significant cause of increased morbidity and mortality among people with diabetes, and result in a heavy economic burden on the national health care system.

Thus, earlier preventive therapies may help delay the development and progression of T2D complications. Compelling evidence associates vitamin B6 to diabetes. Depending on the contexts, vitamin B6 depletion can be a consequence of diabetes or can contribute to diabetes onset. Moreover, vitamin B6 has been also associated to diabetic complications. A crucial role in diabetes complications is played by AGEs that result from non-enzymatic glycation of proteins and DNA. These compounds, working through both receptor-dependent or independent mechanisms, destroy cellular functions, causes oxidative stress and DNA damage. The vitamin B6, besides to counteract ROS formation which increases because of AGE metabolism and action, can block AGE formation by several mechanisms including the formation of adducts with carbonyl compounds such as MO or 3-DG or by blocking the reactions that lead to AGEs from Amadori compounds. Vitamin B6 can also contribute to increase cancer risk in diabetic patients mostly by increasing genome instability.

Vitamin B6 entered clinical trials as adjuvant therapy in the prevention of complications in diabetic patients. Thus, improving in the knowledge of molecular mechanism linking B6 to AGEs and diabetes is important to better address the cares towards a precision medicine based on the use of this natural compound.

References

Aasheim, E. T., Hofsø, D., Hjelmesaeth, J., Birkeland, K. I., & Bøhmer, T. (2008). Vitamin status in morbidly obese patients: A cross-sectional study. *The American Journal of Clinical Nutrition,* 87(2), 362–369. https://doi.org/10.1093/ajcn/87.2.362.

Abdullah, K. M., Abul Qais, F., Hasan, H., & Naseem, I. (2019). Anti-diabetic study of vitamin B6 on hyperglycaemia induced protein carbonylation, DNA damage and ROS production in alloxan induced diabetic rats. *Toxicology Research,* 8(4), 568–579. https://doi.org/10.1039/c9tx00089e.

Adrover, M., Vilanova, B., Frau, J., Muñoz, F., & Donoso, J. (2008). The pyridoxamine action on Amadori compounds: A reexamination of its scavenging capacity and chelating effect. *Bioorganic & Medicinal Chemistry,* 16(10), 5557–5569. https://doi.org/10.1016/j.bmc.2008.04.002.

Ahn, H. J., Min, K. W., & Cho, Y.-O. (2011). Assessment of vitamin B(6) status in Korean patients with newly diagnosed type 2 diabetes. *Nutrition Research and Practice,* 5(1), 34–39. https://doi.org/10.4162/nrp.2011.5.1.34.

Alderson, N. L., Chachich, M. E., Youssef, N. N., Beattie, R. J., Nachtigal, M., Thorpe, S. R., & Baynes, J. W. (2003). The AGE inhibitor pyridoxamine inhibits lipemia and development of renal and vascular disease in Zucker obese rats. *Kidney International,* 63(6), 2123–2133. https://doi.org/10.1046/j.1523-1755.2003.00027.x.

Bender, D. A., Njagi, E. N., & Danielian, P. S. (1990). Tryptophan metabolism in vitamin B6-deficient mice. *The British Journal of Nutrition,* 63(1), 27–36. https://doi.org/10.1079/bjn19900089.

Bennink, H. J., & Schreurs, W. H. (1975). Improvement of oral glucose tolerance in gestational diabetes by pyridoxine. *British Medical Journal,* 3(5974), 13–15. https://doi.org/10.1136/bmj.3.5974.13.

Bierhaus, A., Haslbeck, K.-M., Humpert, P. M., Liliensiek, B., Dehmer, T., Morcos, M., ... Nawroth, P. P. (2004). Loss of pain perception in diabetes is dependent on a receptor of the immunoglobulin superfamily. *The Journal of Clinical Investigation,* 114(12), 1741–1751. https://doi.org/10.1172/JCI18058.

Binici, D. N., Karaman, A., Coşkun, M., Oğlu, A. U., & Uçar, F. (2013). Genomic damage in patients with type-2 diabetes mellitus. *Genetic Counseling (Geneva, Switzerland),* 24(2), 149–156.

Blasiak, J., Arabski, M., Krupa, R., Wozniak, K., Zadrozny, M., Kasznicki, J., ... Drzewoski, J. (2004). DNA damage and repair in type 2 diabetes mellitus. *Mutation Research,* 554(1–2), 297–304. https://doi.org/10.1016/j.mrfmmm.2004.05.011.

Boehm, B. O., Möller, P., Högel, J., Winkelmann, B. R., Renner, W., Rosinger, S., ... Brüderlein, S. (2008). Lymphocytes of type 2 diabetic women carry a high load of stable chromosomal aberrations: A novel risk factor for disease-related early death. *Diabetes,* 57(11), 2950–2957. https://doi.org/10.2337/db08-0274.

Booth, A. A., Khalifah, R. G., & Hudson, B. G. (1996). Thiamine pyrophosphate and pyridoxamine inhibit the formation of antigenic advanced glycation end-products: Comparison with aminoguanidine. *Biochemical and Biophysical Research Communications,* 220(1), 113–119. https://doi.org/10.1006/bbrc.1996.0366.

Booth, A. A., Khalifa, R. G., Todd, P., & Hudson, B. G. (1997). In vitro kinetic studies of formation of antigenic advanced glycation end products (AGEs). Novel inhibition of post-Amadori glycation pathways. *The Journal of Biological Chemistry, 272*(9), 5430–5437. https://doi.org/10.1074/jbc.272.9.5430.

Borg, D. J., & Forbes, J. M. (2016). Targeting advanced glycation with pharmaceutical agents: Where are we now? *Glycoconjugate Journal, 33*(4), 653–670. https://doi.org/10.1007/s10719-016-9691-1.

Brownlee, M. (2001). Biochemistry and molecular cell biology of diabetic complications. *Nature, 414*(6865), 813–820. https://doi.org/10.1038/414813a.

Brownlee, M., Vlassara, H., Kooney, A., Ulrich, P., & Cerami, A. (1986). Aminoguanidine prevents diabetes-induced arterial wall protein cross-linking. *Science (New York, N. Y.), 232*(4758), 1629–1632. https://doi.org/10.1126/science.3487117.

Cao, P., Gong, Y., Tang, L., Leung, Y.-C., & Jiang, T. (2006). Crystal structure of human pyridoxal kinase. *Journal of Structural Biology, 154*(3), 327–332. https://doi.org/10.1016/j.jsb.2006.02.008.

Cellini, B., Montioli, R., Oppici, E., Astegno, A., & Voltattorni, C. B. (2014). The chaperone role of the pyridoxal 5′-phosphate and its implications for rare diseases involving B6-dependent enzymes. *Clinical Biochemistry, 47*(3), 158–165. https://doi.org/10.1016/j.clinbiochem.2013.11.021.

Chelban, V., Wilson, M. P., Warman Chardon, J., Vandrovcova, J., Zanetti, M. N., Zamba-Papanicolaou, E., ... Houlden, H. (2019). PDXK mutations cause polyneuropathy responsive to pyridoxal 5′-phosphate supplementation. *Annals of Neurology, 86*(2), 225–240. https://doi.org/10.1002/ana.25524.

Chen, C.-C., Li, B., Millman, S. E., Chen, C., Li, X., Morris, J. P., 4th, ... Zhang, L. (2020). Vitamin B6 addiction in acute myeloid leukemia. *Cancer Cell, 37*(1), 71–84.e7. https://doi.org/10.1016/j.ccell.2019.12.002.

Chen, H., Sun, X., Ge, W., Qian, Y., Bai, R., & Zheng, S. (2017). A seven-gene signature predicts overall survival of patients with colorectal cancer. *Oncotarget, 8*(56), 95054–95065. https://doi.org/10.18632/oncotarget.10982.

Chetyrkin, S. V., Zhang, W., Hudson, B. G., Serianni, A. S., & Voziyan, P. A. (2008). Pyridoxamine protects proteins from functional damage by 3-deoxyglucosone: Mechanism of action of pyridoxamine. *Biochemistry, 47*(3), 997–1006. https://doi.org/10.1021/bi701190s.

Chi, W., Iyengar, A. S. R., Albersen, M., Bosma, M., Verhoeven-Duif, N. M., Wu, C.-F., & Zhuang, X. (2019). Pyridox (am) ine 5′-phosphate oxidase deficiency induces seizures in Drosophila melanogaster. *Human Molecular Genetics, 28*(18), 3126–3136. https://doi.org/10.1093/hmg/ddz143.

Ciminera, A. K., Shuck, S. C., & Termini, J. (2021). Elevated glucose increases genomic instability by inhibiting nucleotide excision repair. *Life Science Alliance, 4*(10), https://doi.org/10.26508/lsa.202101159.

Connick, J. H., & Stone, T. W. (1985). The role of kynurenines in diabetes mellitus. *Medical Hypotheses, 18*(4), 371–376. https://doi.org/10.1016/0306-9877(85)90104-5.

Contestabile, R., di Salvo, M. L., Bunik, V., Tramonti, A., & Vernì, F. (2020). The multifaceted role of vitamin B(6) in cancer: Drosophila as a model system to investigate DNA damage. *Open Biology, 10*(3), 200034. https://doi.org/10.1098/rsob.200034.

Cravo, M. L., & Camilo, M. E. (2000). Hyperhomocysteinemia in chronic alcoholism: Relations to folic acid and vitamins B(6) and B(12) status. *Nutrition (Burbank, Los Angeles County, Calif.), 16*(4), 296–302. https://doi.org/10.1016/s0899-9007(99)00297-x.

Dandona, P., Thusu, K., Cook, S., Snyder, B., Makowski, J., Armstrong, D., & Nicotera, T. (1996). Oxidative damage to DNA in diabetes mellitus. *Lancet (London, England), 347*(8999), 444–445. https://doi.org/10.1016/s0140-6736(96)90013-6.

Del Turco, S., & Basta, G. (2012). An update on advanced glycation endproducts and atherosclerosis. *BioFactors (Oxford, England), 38*(4), 266–274. https://doi.org/10.1002/biof.1018.

Di Salvo, M. L., Mastrangelo, M., Nogués, I., Tolve, M., Paiardini, A., Carducci, C., ... Leuzzi, V. (2017). Pyridoxine-5′-phosphate oxidase (Pnpo) deficiency: Clinical and biochemical alterations associated with the C.347g>A (P.·Arg116gln) mutation. *Molecular Genetics and Metabolism, 122*(1–2), 135–142. https://doi.org/10.1016/j.ymgme.2017.08.003.

Di Salvo, M. L., Safo, M. K., Musayev, F. N., Bossa, F., & Schirch, V. (2003). Structure and mechanism of *Escherichia coli* pyridoxine 5′-phosphate oxidase. *Biochimica et Biophysica Acta, 1647*(1–2), 76–82. https://doi.org/10.1016/s1570-9639(03)00060-8.

Di Salvo, M. L., Contestabile, R., & Safo, M. K. (2011). Vitamin B(6) salvage enzymes: Mechanism, structure and regulation. *Biochimica et Biophysica Acta, 1814*(11), 1597–1608. https://doi.org/10.1016/j.bbapap.2010.12.006.

Du, X., Yalong, Y., Xiaoxia, Z., Yulan, H., Yuling, F., ... Ma, L. (2020). Vitamin B6 prevents excessive inflammation by reducing accumulation of sphingosine-1-phosphate in a sphingosine-1-phosphate lyase-dependent manner. *Journal of Cellular and Molecular Medicine, 24*(22), 13129–13138. https://doi.org/10.1111/jcmm.15917 Epub 2020 Sep 23.

El-Mesallamy, H. O., Hamdy, N. M., Ezzat, O. A., & Reda, A. M. (2011). Levels of soluble advanced glycation end product-receptors and other soluble serum markers as indicators of diabetic neuropathy in the foot. *Journal of Investigative Medicine: The Official Publication of the American Federation for Clinical Research, 59*(8), 1233–1238. https://doi.org/10.2130/JIM.0b013e318231db64.

Ellis, J. M., Folkers, K., Minadeo, M., VanBuskirk, R., Xia, L. J., & Tamagawa, H. (1991). A deficiency of vitamin B6 is a plausible molecular basis of the retinopathy of patients with diabetes mellitus. *Biochemical and Biophysical Research Communications, 179*(1), 615–619. https://doi.org/10.1016/0006-291x(91)91416-a.

Elseweidy, M. M., Elswefy, S. E., Younis, N. N., & Zaghloul, M. S. (2013). Pyridoxamine, an inhibitor of protein glycation, in relation to microalbuminuria and proinflammatory cytokines in experimental diabetic nephropathy. *Experimental Biology and Medicine (Maywood, N. J.), 238*(8), 881–888. https://doi.org/10.1177/1535370213494644.

Elstner, M., Morris, C. M., Heim, K., Lichtner, P., Bender, A., Mehta, D., ... Turnbull, D. M. (2009). Single-cell expression profiling of dopaminergic neurons combined with association analysis identifies pyridoxal kinase as Parkinson's disease gene. *Annals of Neurology, 66*(6), 792–798. https://doi.org/10.1002/ana.21780.

Ferro, Y., Carè, I., Mazza, E., Provenzano, F., Colica, C., Torti, C., ... Montalcini, T. (2017). Protein and vitamin B6 intake are associated with liver steatosis assessed by transient elastography, especially in obese individuals. *Clinical and Molecular Hepatology, 23*(3), 249–259. https://doi.org/10.3350/cmh.2017.0019.

Fields, A. M., Welle, K., Ho, E. S., Mesaros, C., & Susiarjo, M. (2021). Vitamin B6 deficiency disrupts serotonin signaling in pancreatic islets and induces gestational diabetes in mice. *Communications Biology, 4*(1), 421. https://doi.org/10.1038/s42003-021-01900-0.

Fitzpatrick, T. B., Amrhein, N., Kappes, B., Macheroux, P., Tews, I., & Raschle, T. (2007). Two independent routes of de novo vitamin B6 biosynthesis: Not that different after all. *The Biochemical Journal, 407*(1), 1–13. https://doi.org/10.1042/BJ20070765.

Fu, M. X., Requena, J. R., Jenkins, A. J., Lyons, T. J., Baynes, J. W., & Thorpe, S. R. (1996). The advanced glycation end product, Nepsilon-(carboxymethyl)lysine, is a product of both lipid peroxidation and glycoxidation reactions. *The Journal of Biological Chemistry, 271*(17), 9982–9986. https://doi.org/10.1074/jbc.271.17.9982.

Fukami, K., Yamagishi, S.-I., & Okuda, S. (2014). Role of AGEs-RAGE system in cardiovascular disease. *Current Pharmaceutical Design, 20*(14), 2395–2402. https://doi.org/10.2174/13816128113199990475.

Galicia-Garcia, U., Benito-Vicente, A., Jebari, S., Larrea-Sebal, A., Siddiqi, H., Uribe, K. B., ... Martín, C. (2020). Pathophysiology of type 2 diabetes mellitus. *International Journal of Molecular Sciences, 21*(17), https://doi.org/10.3390/ijms21176275.

Galluzzi, L., Vacchelli, E., Michels, J., Garcia, P., Kepp, O., Senovilla, L., ... Kroemer, G. (2013). Effects of vitamin B6 metabolism on oncogenesis, tumor progression and therapeutic responses. *Oncogene, 32*(42), 4995–5004. https://doi.org/10.1038/onc.2012.623.

Galluzzi, L., Vitale, I., Senovilla, L., Olaussen, K. A., Pinna, G., Eisenberg, T., ... Kroemer, G. (2012). Prognostic impact of vitamin B6 metabolism in lung cancer. *Cell Reports, 2*(2), 257–269. https://doi.org/10.1016/j.celrep.2012.06.017.

György, P., & Eckhardt, R. E. (1939). Vitamin B-6 and skin lesions in rats. *Nature, 144,* 512.

Goodarzi, M. T., Navidi, A. A., Rezaei, M., & Babahmadi-Rezaei, H. (2010). Oxidative damage to DNA and lipids: Correlation with protein glycation in patients with type 1 diabetes. *Journal of Clinical Laboratory Analysis, 24*(2), 72–76. https://doi.org/10.1002/jcla.20328.

Guella, I., Asselta, R., Tesei, S., Zini, M., Pezzoli, G., & Duga, S. (2010). *The PDXK rs2010795 variant is not associated with Parkinson disease in Italy.* Annals of Neurology, Vol. 67, 411–412. author reply 412. https://doi.org/10.1002/ana.21964.

Haidari, F., Mohammadshahi, M., Zarei, M., Haghighizadeh, M. H., & Mirzaee, F. (2021). The effect of pyridoxine hydrochloride supplementation on leptin, adiponectin, glycemic indices, and anthropometric indices in obese and overweight women. *Clinical Nutrition Research, 10*(3), 230–242. https://doi.org/10.7762/cnr.2021.10.3.230.

Hammes, H. P., Alt, A., Niwa, T., Clausen, J. T., Bretzel, R. G., Brownlee, M., & Schleicher, E. D. (1999). Differential accumulation of advanced glycation end products in the course of diabetic retinopathy. *Diabetologia, 42*(6), 728–736. https://doi.org/10.1007/s001250051221.

Hellmann, H., & Mooney, S. (2010). Vitamin B6: A molecule for human health? *Molecules (Basel, Switzerland), 15*(1), 442–459. https://doi.org/10.3390/molecules15010442.

Hill, R. E., Himmeldirk, K., Kennedy, I. A., Pauloski, R. M., Sayer, B. G., Wolf, E., & Spenser, I. D. (1996). The biogenetic anatomy of vitamin B6. A ^{13}C NMR investigation of the biosynthesis of pyridoxol in *Escherichia coli*. *The Journal of Biological Chemistry, 271*(48), 30426–30435. https://doi.org/10.1074/jbc.271.48.30426.

Hinokio, Y., Suzuki, S., Hirai, M., Chiba, M., Hirai, A., & Toyota, T. (1999). Oxidative DNA damage in diabetes mellitus: Its association with diabetic complications. *Diabetologia, 42*(8), 995–998. https://doi.org/10.1007/s001250051258.

Hirabayashi, S., Baranski, T. J., & Cagan, R. L. (2013). Transformed Drosophila cells evade diet-mediated insulin resistance through wingless signaling. *Cell, 154*(3), 664–675. https://doi.org/10.1016/j.cell.2013.06.030.

Horikawa, C., Aida, R., Kamada, C., Fujihara, K., Tanaka, S., Tanaka, S., ... Sone, H. (2020). Vitamin B6 intake and incidence of diabetic retinopathy in Japanese patients with type 2 diabetes: Analysis of data from the Japan Diabetes Complications Study (JDCS). *European Journal of Nutrition, 59*(4), 1585–1594. https://doi.org/10.1007/s00394-019-02014-4.

Huq, M. D. M., Tsai, N.-P., Lin, Y.-P., Higgins, L., & Wei, L.-N. (2007). Vitamin B6 conjugation to nuclear corepressor RIP140 and its role in gene regulation. *Nature Chemical Biology, 3*(3), 161–165. https://doi.org/10.1038/nchembio861.

Jack, M., & Wright, D. (2012). Role of advanced glycation endproducts and glyoxalase I in diabetic peripheral sensory neuropathy. *Translational Research: The Journal of Laboratory and Clinical Medicine, 159*(5), 355–365. https://doi.org/10.1016/j.trsl.2011.12.004.

Jain, S. K., & Lim, G. (2001). Pyridoxine and pyridoxamine inhibits superoxide radicals and prevents lipid peroxidation, protein glycosylation, and (Na+ + K+)-ATPase activity reduction in high glucose-treated human erythrocytes. *Free Radical Biology & Medicine, 30*(3), 232–237. https://doi.org/10.1016/s0891-5849(00)00462-7.

Kanellis, P., Gagliardi, M., Banath, J. P., Szilard, R. K., Nakada, S., Galicia, S., ... Durocher, D. (2007). A screen for suppressors of gross chromosomal rearrangements identifies a conserved role for PLP in preventing DNA lesions. *PLoS Genetics, 3*(8), e134. https://doi.org/10.1371/journal.pgen.0030134.

Kang, Z., Li, H., Li, G., & Yin, D. (2006). Reaction of pyridoxamine with malondialdehyde: Mechanism of inhibition of formation of advanced lipoxidation endproducts. *Amino Acids, 30*(1), 55–61. https://doi.org/10.1007/s00726-005-0209-6.

Kannan, K., & Jain, S. K. (2004). Effect of vitamin B6 on oxygen radicals, mitochondrial membrane potential, and lipid peroxidation in H2O2-treated U937 monocytes. *Free Radical Biology & Medicine, 36*(4), 423–428. https://doi.org/10.1016/j.freeradbiomed.2003.09.012.

Katsarou, A., Gudbjörnsdottir, S., Rawshani, A., Dabelea, D., Bonifacio, E., Anderson, B. J., ... Lernmark, Å. (2017). Type 1 diabetes mellitus. *Nature Reviews: Disease Primers, 3*, 17016. https://doi.org/10.1038/nrdp.2017.16.

Khalifah, R. G., Todd, P., Booth, A. A., Yang, S. X., Mott, J. D., & Hudson, B. G. (1996). Kinetics of nonenzymatic glycation of ribonuclease A leading to advanced glycation end products. Paradoxical inhibition by ribose leads to facile isolation of protein intermediate for rapid post-Amadori studies. *Biochemistry, 35*(15), 4645–4654. https://doi.org/10.1021/bi9525942.

Khatami, M., Suldan, Z., David, I., Li, W., & Rockey, J. H. (1988). Inhibitory effects of pyridoxal phosphate, ascorbate and aminoguanidine on nonenzymatic glycosylation. *Life Sciences, 43*(21), 1725–1731. https://doi.org/10.1016/0024-3205(88)90484-5.

Krinke, G. J., & Fitzgerald, R. E. (1988). The pattern of pyridoxine-induced lesion: Difference between the high and the low toxic level. *Toxicology, 49*(1), 171–178. https://doi.org/10.1016/0300-483x(88)90190-4.

Kuzan, A. (2021). Toxicity of advanced glycation end products (Review). *Biomedical Reports, 14*(5), 46. https://doi.org/10.3892/br.2021.1422.

Lainé-Cessac, P., Cailleux, A., & Allain, P. (1997). Mechanisms of the inhibition of human erythrocyte pyridoxal kinase by drugs. *Biochemical Pharmacology, 54*(8), 863–870. https://doi.org/10.1016/s0006-2952(97)00252-9.

Lee, S. C., & Chan, J. C. N. (2015). Evidence for DNA damage as a biological link between diabetes and cancer. *Chinese Medical Journal, 128*(11), 1543–1548. https://doi.org/10.4103/0366-6999.157693.

Leklem, J. E., & Hollenbeck, C. B. (1990). Acute ingestion of glucose decreases plasma pyridoxal 5′-phosphate and total vitamin B-6 concentration. *The American Journal of Clinical Nutrition, 51*(5), 832–836. https://doi.org/10.1093/ajcn/51.5.832.

Li, H., Nakamura, S., Miyazaki, S., Morita, T., Suzuki, M., Pischetsrieder, M., & Niwa, T. (2006). N2-carboxyethyl-2'-deoxyguanosine, a DNA glycation marker, in kidneys and aortas of diabetic and uremic patients. *Kidney International, 69*(2), 388–392. https://doi.org/10.1038/sj.ki.5000064.

Liguori, F., Mascolo, E., & Vernì, F. (2021). The genetics of diabetes: What we can learn from Drosophila. *International Journal of Molecular Sciences, 22*(20), https://doi.org/10.3390/ijms222011295.

Linkswiler, H. (1967). Biochemical and physiological changes in vitamin B6 deficiency. *The American Journal of Clinical Nutrition, 20*(6), 547–561. https://doi.org/10.1093/ajcn/20.6.547.

Liu, Z., Li, P., Zhao, Z.-H., Zhang, Y., Ma, Z.-M., & Wang, S.-X. (2016). Vitamin B6 prevents endothelial dysfunction, insulin resistance, and hepatic lipid accumulation in apoe (−/−) mice fed with high-fat diet. *Journal of Diabetes Research, 2016*, 1748065. https://doi.org/10.1155/2016/1748065.

Maillard, L. C. (1912). Action des Acides Amines sur les Sucres: Formation des Melanoidines par voie Methodique. *Comptes rendus de l'Académie des Sciences (Paris), 154*, 66–68.

Marzio, A., Merigliano, C., Gatti, M., & Vernì, F. (2014). Sugar and chromosome stability: Clastogenic effects of sugars in vitamin B6-deficient cells. *PLoS Genetics, 10*(3), e1004199. https://doi.org/10.1371/journal.pgen.1004199.

Mascolo, E., Amoroso, N., Saggio, I., Merigliano, C., & Vernì, F. (2020). Pyridoxine/pyridoxamine 5′-phosphate oxidase (Sgll/PNPO) is important for DNA integrity and glucose homeostasis maintenance in Drosophila. *Journal of Cellular Physiology, 235*(1), 504–512. https://doi.org/10.1002/jcp.28990.

Mascolo, E., Barile, A., Stufera Mecarelli, L., Amoroso, N., Merigliano, C., Massimi, A., ... Vernì, F. (2019). The expression of four pyridoxal kinase (PDXK) human variants in Drosophila impacts on genome integrity. *Scientific Reports, 9*(1), 14188. https://doi.org/10.1038/s41598-019-50673-4.

Mascolo, E., Liguori, F., Stufera Mecarelli, L., Amoroso, N., Merigliano, C., Amadio, S., ... Vernì, F. (2021). Functional Inactivation of Drosophila GCK orthologs causes genomic instability and oxidative stress in a fly model of MODY-2. *International Journal of Molecular Sciences, 22*(2), https://doi.org/10.3390/ijms22020918.

Mascolo, E., & Vernì, F. (2020). Vitamin B6 and diabetes: Relationship and molecular mechanisms. *International Journal of Molecular Sciences, 21*(10), https://doi.org/10.3390/ijms21103669.

Mascolo, E., Liguori, F., Merigliano, C., Schiano, L., Gnocchini, E., Pilesi, E., ... Vernì, F. (2022). Vitamin B6 rescues insulin resistance and glucose-induced DNA damage caused by reduced activity of Drosophila PI3K. *Journal of Cellular Physiology*. https://doi.org/10.1002/jcp.30812.

Matoba, K., Takeda, Y., Nagai, Y., Kawanami, D., Utsunomiya, K., & Nishimura, R. (2019). Unraveling the role of inflammation in the pathogenesis of diabetic kidney disease. *International Journal of Molecular Sciences, 20*(14), https://doi.org/10.3390/ijms20143393.

McCann, V. J., & Davis, R. E. (1978). Serum pyridoxal concentrations in patients with diabetic neuropathy. *Australian and New Zealand Journal of Medicine, 8*(3), 259–261. https://doi.org/10.1111/j.1445-5994.1978.tb04520.x.

Merigliano, C., Mascolo, E., La Torre, M., Saggio, I., & Vernì, F. (2018). Protective role of vitamin B6 (PLP) against DNA damage in Drosophila models of type 2 diabetes. *Scientific Reports, 8*(1), 11432. https://doi.org/10.1038/s41598-018-29801-z.

Merrill, A. H. J., & Henderson, J. M. (1987). Diseases associated with defects in vitamin B6 metabolism or utilization. *Annual Review of Nutrition, 7*, 137–156. https://doi.org/10.1146/annurev.nu.07.070187.001033.

Metz, T. O., Alderson, N. L., Chachich, M. E., Thorpe, S. R., & Baynes, J. W. (2003). Pyridoxamine traps intermediates in lipid peroxidation reactions in vivo: Evidence on the role of lipids in chemical modification of protein and development of diabetic complications. *The Journal of Biological Chemistry, 278*(43), 42012–42019. https://doi.org/10.1074/jbc.M304292200.

Mills, P. B., Camuzeaux, S. S. M., Footitt, E. J., Mills, K. A., Gissen, P., Fisher, L., ... Clayton, P. T. (2014). Epilepsy due to PNPO mutations: Genotype, environment and treatment affect presentation and outcome. *Brain: A Journal of Neurology, 137*(Pt 5), 1350–1360. https://doi.org/10.1093/brain/awu051.

Mills, P. B., Surtees, R. A. H., Champion, M. P., Beesley, C. E., Dalton, N., Scambler, P. J., ... Clayton, P. T. (2005). Neonatal epileptic encephalopathy caused by mutations in the PNPO gene encoding pyridox(am)ine 5'-phosphate oxidase. *Human Molecular Genetics, 14*(8), 1077–1086. https://doi.org/10.1093/hmg/ddi120.

Miyata, T., Van Ypersele de Strihou, C., Ueda, Y., Ichimori, K., Inagi, R., Onogi, H., ... Kurokawa, K. (2002). Angiotensin II receptor antagonists and angiotensin-converting enzyme inhibitors lower in vitro the formation of advanced glycation end products: Biochemical mechanisms. *Journal of the American Society of Nephrology: JASN, 13*(10), 2478–2487. https://doi.org/10.1097/01.asn.0000032418.67267.f2.

Mooney, S., Leuendorf, J.-E., Hendrickson, C., & Hellmann, H. (2009). Vitamin B6: A long known compound of surprising complexity. *Molecules (Basel, Switzerland), 14*(1), 329–351. https://doi.org/10.3390/molecules14010329.

Moreno-Navarrete, J. M., Jove, M., Ortega, F., Xifra, G., Ricart, W., Obis, È., ... Fernández-Real, J. M. (2016). Metabolomics uncovers the role of adipose tissue PDXK in adipogenesis and systemic insulin sensitivity. *Diabetologia, 59*(4), 822–832. https://doi.org/10.1007/s00125-016-3863-1.

Mossine, V. V., Linetsky, M., Glinsky, G. V., Ortwerth, B. J., & Feather, M. S. (1999). Superoxide free radical generation by Amadori compounds: The role of acyclic forms and metal ions. *Chemical Research in Toxicology, 12*(3), 230–236. https://doi.org/10.1021/tx980209e.

Muellenbach, E. A., Diehl, C. J., Teachey, M. K., Lindborg, K. A., Archuleta, T. L., Harrell, N. B., ... Henriksen, E. J. (2008). Interactions of the advanced glycation end product inhibitor pyridoxamine and the antioxidant alpha-lipoic acid on insulin resistance in the obese Zucker rat. *Metabolism: Clinical and Experimental, 57*(10), 1465–1472. https://doi.org/10.1016/j.metabol.2008.05.018.

Mukherjee, T., Hanes, J., Tews, I., Ealick, S. E., & Begley, T. P. (2011). Pyridoxal phosphate: Biosynthesis and catabolism. *Biochimica et Biophysica Acta, 1814*(11), 1585–1596. https://doi.org/10.1016/j.bbapap.2011.06.018.

Nagaraj, R. H., Sarkar, P., Mally, A., Biemel, K. M., Lederer, M. O., & Padayatti, P. S. (2002). Effect of pyridoxamine on chemical modification of proteins by carbonyls in diabetic rats: Characterization of a major product from the reaction of pyridoxamine and methylglyoxal. *Archives of Biochemistry and Biophysics, 402*(1), 110–119. https://doi.org/10.1016/S0003-9861(02)00067-X.

Nakamura, S., Li, H., Adijiang, A., Pischetsrieder, M., & Niwa, T. (2007). Pyridoxal phosphate prevents progression of diabetic nephropathy. *Nephrology, Dialysis, Transplantation: Official Publication of the European Dialysis and Transplant Association—European Renal Association, 22*(8), 2165–2174. https://doi.org/10.1093/ndt/gfm166.

Nakamura, S., & Niwa, T. (2005). Pyridoxal phosphate and hepatocyte growth factor prevent dialysate-induced peritoneal damage. *Journal of the American Society of Nephrology: JASN, 16*(1), 144–150. https://doi.org/10.1681/ASN.2004020120.

Nilsson, E., Jansson, P. A., Perfilyev, A., Volkov, P., Pedersen, M., Svensson, M. K., ... Ling, C. (2014). Altered DNA methylation and differential expression of genes influencing metabolism and inflammation in adipose tissue from subjects with type 2 diabetes. *Diabetes, 63*(9), 2962–2976. https://doi.org/10.2337/db13-1459.

Nin, J. W., Jorsal, A., Ferreira, I., Schalkwijk, C. G., Prins, M. H., Parving, H.-H., ... Stehouwer, C. D. (2011). Higher plasma levels of advanced glycation end products are associated with incident cardiovascular disease and all-cause mortality in type 1 diabetes: A 12-year follow-up study. *Diabetes Care, 34*(2), 442–447. https://doi.org/10.2337/dc10-1087.

Niwa, T., Katsuzaki, T., Ishizaki, Y., Hayase, F., Miyazaki, T., Uematsu, T., ... Takei, Y. (1997). Imidazolone, a novel advanced glycation end product, is present at high levels in kidneys of rats with streptozotocin-induced diabetes. *FEBS Letters, 407*(3), 297–302. https://doi.org/10.1016/s0014-5793(97)00362-1.

Niwa, T., Katsuzaki, T., Miyazaki, S., Miyazaki, T., Ishizaki, Y., Hayase, F., ... Takei, Y. (1997). Immunohistochemical detection of imidazolone, a novel advanced glycation end product, in kidneys and aortas of diabetic patients. *The Journal of Clinical Investigation, 99*(6), 1272–1280. https://doi.org/10.1172/JCI119285.

Nix, W. A., Zirwes, R., Bangert, V., Kaiser, R. P., Schilling, M., Hostalek, U., & Obeid, R. (2015). Vitamin B status in patients with type 2 diabetes mellitus with and without incipient nephropathy. *Diabetes Research and Clinical Practice, 107*(1), 157–165. https://doi.org/10.1016/j.diabres.2014.09.058.

Ohdake, S. (1932). Isolation of "Oryzanin" (antineuritic vitamin) from rice-polishings. *Bulletin of the Agricultural Chemical Society of Japan, 8*, 11–46. https://doi.org: 10.1271/bbb1924.8.11.

Okada, M., Shibuya, M., Yamamoto, E., & Murakami, Y. (1999). Effect of diabetes on vitamin B6 requirement in experimental animals. *Diabetes, Obesity & Metabolism, 1*(4), 221–225. https://doi.org/10.1046/j.1463-1326.1999.00028.x.

Onorato, J. M., Jenkins, A. J., Thorpe, S. R., & Baynes, J. W. (2000). Pyridoxamine, an inhibitor of advanced glycation reactions, also inhibits advanced lipoxidation reactions. Mechanism of action of pyridoxamine. *The Journal of Biological Chemistry, 275*(28), 21177–21184. https://doi.org/10.1074/jbc.M003263200.

Ott, C., Jacobs, K., Haucke, E., Navarrete Santos, A., Grune, T., & Simm, A. (2014). Role of advanced glycation end products in cellular signaling. *Redox Biology, 2*, 411–429. https://doi.org/10.1016/j.redox.2013.12.016.

Oudes, A. J., Herr, C. M., Olsen, Y., & Fleming, J. E. (1998). Age-dependent accumulation of advanced glycation end-products in adult Drosophila melanogaster. *Mechanisms of Ageing and Development, 100*(3), 221–229. https://doi.org/10.1016/s0047-6374(97)00146-2.

Oxenkrug, G. F. (2007). Genetic and hormonal regulation of tryptophan kynurenine metabolism: Implications for vascular cognitive impairment, major depressive disorder, and aging. *Annals of the New York Academy of Sciences, 1122*, 35–49. https://doi.org/10.1196/annals.1403.003.

Page, J. H., Ma, J., Chiuve, S. E., Stampfer, M. J., Selhub, J., Manson, J. E., & Rimm, E. B. (2009). Plasma vitamin B(6) and risk of myocardial infarction in women. *Circulation, 120*(8), 649–655. https://doi.org/10.1161/CIRCULATIONAHA.108.809038.

Parra, M., Stahl, S., & Hellmann, H. (2018). Vitamin B_6 and its role in cell metabolism and physiology. *Cells, 7*(7), https://doi.org/10.3390/cells7070084.

Perrone, A., Giovino, A., Benny, J., & Martinelli, F. (2020). Advanced glycation end products (AGEs): Biochemistry, signaling, analytical methods, and epigenetic effects. *Oxidative Medicine and Cellular Longevity, 2020*, 3818196. https://doi.org/10.1155/2020/3818196.

Piarulli, F., Lapolla, A., Sartore, G., Rossetti, C., Bax, G., Noale, M., ... Fedele, D. (2005). Autoantibodies against oxidized LDLs and atherosclerosis in type 2 diabetes. *Diabetes Care, 28*(3), 653–657. https://doi.org/10.2337/diacare.28.3.653.

Plows, J. F., Stanley, J. L., Baker, P. N., Reynolds, C. M., & Vickers, M. H. (2018). The pathophysiology of gestational diabetes mellitus. *International Journal of Molecular Sciences, 19*(11), https://doi.org/10.3390/ijms19113342.

Qian, B., Shen, S., Zhang, J., & Jing, P. (2017). Effects of vitamin B6 deficiency on the composition and functional potential of T cell populations. *Journal of Immunology Research, 2017*, 2197975. https://doi.org/10.1155/2017/2197975.

Rabbani, N., & Thornalley, P. J. (2018). Advanced glycation end products in the pathogenesis of chronic kidney disease. *Kidney International, 93*(4), 803–813. https://doi.org/10.1016/j.kint.2017.11.034.

Radhakrishnamurty, R., Angel, J. F., & Sabry, Z. I. (1968). Response of lipogenesis to repletion in the pyridoxine-deficient rat. *The Journal of Nutrition, 95*(3), 341–348. https://doi.org/10.1093/jn/95.3.341.

Rahmoon, M. A., Elghaish, R. A., Ibrahim, A. A., Alaswad, Z., Gad, M. Z., El-Khamisy, S. F., & Elserafy, M. (2023). High glucose increases DNA damage and elevates the expression of multiple DDR genes. *Genes, 14*(1), https://doi.org/10.3390/genes14010144.

Ramis, R., Ortega-Castro, J., Caballero, C., Casasnovas, R., Cerrillo, A., Vilanova, B., ... Frau, J. (2019). How does pyridoxamine inhibit the formation of advanced glycation end products? The role of its primary antioxidant activity. *Antioxidants (Basel, Switzerland), 8*(9), https://doi.org/10.3390/antiox8090344.

Rimm, E. B., Willett, W. C., Hu, F. B., Sampson, L., Colditz, G. A., Manson, J. E., ... Stampfer, M. J. (1998). Folate and vitamin B6 from diet and supplements in relation to risk of coronary heart disease among women. *JAMA: The Journal of the American Medical Association, 279*(5), 359–364. https://doi.org/10.1001/jama.279.5.359.

Rogers, K. S., Higgins, E. S., & Kline, E. S. (1986). Experimental diabetes causes mitochondrial loss and cytoplasmic enrichment of pyridoxal phosphate and aspartate aminotransferase activity. *Biochemical Medicine and Metabolic Biology, 36*(1), 91–97. https://doi.org/10.1016/0885-4505(86)90111-8.

Rubí, B. (2012). Pyridoxal 5′-phosphate (PLP) deficiency might contribute to the onset of type I diabetes. *Medical Hypotheses, 78*(1), 179–182. https://doi.org/10.1016/j.mehy.2011.10.021.

Sanada, Y., Kumoto, T., Suehiro, H., Yamamoto, T., Nishimura, F., Kato, N., & Yanaka, N. (2014). IκB kinase epsilon expression in adipocytes is upregulated by interaction with macrophages. *Bioscience, Biotechnology, and Biochemistry, 78*(8), 1357–1362. https://doi.org/10.1080/09168451.2014.925776.

Satyanarayana, A., Balakrishna, N., Pitla, S., Reddy, P. Y., Mudili, S., Lopamudra, P., ... Reddy, G. B. (2011). Status of B-vitamins and homocysteine in diabetic retinopathy: Association with vitamin-B12 deficiency and hyperhomocysteinemia. *PLoS One, 6*(11), e26747. https://doi.org/10.1371/journal.pone.0026747.

Sergi, D., Boulestin, H., Campbell, F. M., & Williams, L. M. (2021). The role of dietary advanced glycation end products in metabolic dysfunction. *Molecular Nutrition & Food Research, 65*(1), e1900934. https://doi.org/10.1002/mnfr.201900934.

Singh, V. P., Bali, A., Singh, N., & Jaggi, A. S. (2014). Advanced glycation end products and diabetic complications. *The Korean Journal of Physiology & Pharmacology: Official Journal of the Korean Physiological Society and the Korean Society of Pharmacology, 18*(1), 1–14. https://doi.org/10.4196/kjpp.2014.18.1.1.

Smith, P. R., & Thornalley, P. J. (1992). Mechanism of the degradation of non-enzymatically glycated proteins under physiological conditions. Studies with the model fructosamine, N epsilon-(1-deoxy-D-fructos-1-yl)hippuryl-lysine. *European Journal of Biochemistry, 210*(3), 729–739. https://doi.org/10.1111/j.1432-1033.1992.tb17474.x.

Solomon, L. R., & Cohen, K. (1989). Erythrocyte O2 transport and metabolism and effects of vitamin B6 therapy in type II diabetes mellitus. *Diabetes, 38*(7), 881–886. https://doi.org/10.2337/diab.38.7.881.

Song, Q., Liu, J., Dong, L., Wang, X., & Zhang, X. (2021). Novel advances in inhibiting advanced glycation end product formation using natural compounds. *Biomedicine & Pharmacotherapy = Biomedecine & Pharmacotherapie, 140*, 111750. https://doi.org/10.1016/j.biopha.2021.111750.

Spellacy, W. N., Buhi, W. C., & Birk, S. A. (1977). Vitamin B6 treatment of gestational diabetes mellitus: Studies of blood glucose and plasma insulin. *American Journal of Obstetrics and Gynecology, 127*(6), 599–602. https://doi.org/10.1016/0002-9378(77)90356-8.

Stanton, A. H., & Folkers, K. (1939). Synthesis of vitamin B6. *Journal of the American Chemical Society, 61*(5), 1245–1247.

Stitt, A., Gardiner, T. A., Alderson, N. L., Canning, P., Frizzell, N., Duffy, N., ... Thorpe, S. R. (2002). The AGE inhibitor pyridoxamine inhibits development of retinopathy in experimental diabetes. *Diabetes, 51*(9), 2826–2832. https://doi.org/10.2337/diabetes.51.9.2826.

Stitt, A. W. (2001). Advanced glycation: An important pathological event in diabetic and age related ocular disease. *The British Journal of Ophthalmology, 85*(6), 746–753. https://doi.org/10.1136/bjo.85.6.746.

Sun, H., Yuan, Y., & Sun, Z. (2016). Update on mechanisms of renal tubule injury caused by advanced glycation end products. *BioMed Research International, 2016*, 5475120. https://doi.org/10.1155/2016/5475120.

Takeuchi, F., Tsubouchi, R., Izuta, S., & Shibata, Y. (1989). Kynurenine metabolism and xanthurenic acid formation in vitamin B6-deficient rat after tryptophan injection. *Journal of Nutritional Science and Vitaminology, 35*(2), 111–122. https://doi.org/10.3177/jnsv.35.111.

Takeuchi, M. (2020). Toxic AGEs (TAGE) theory: A new concept for preventing the development of diseases related to lifestyle. *Diabetology & Metabolic Syndrome, 12*(1), 105. https://doi.org/10.1186/s13098-020-00614-3.

Tan, K. C. B., Chow, W.-S., Ai, V. H. G., Metz, C., Bucala, R., & Lam, K. S. L. (2002). Advanced glycation end products and endothelial dysfunction in type 2 diabetes. *Diabetes Care, 25*(6), 1055–1059. https://doi.org/10.2337/diacare.25.6.1055.

Tan, W., Liu, B., & Ling, H. (2020). [Pyridoxal kinase (PDXK) promotes the proliferation of serous ovarian cancer cells and is associated with poor prognosis]. *Xi Bao Yu Fen Zi Mian Yi Xue Za Zhi = Chinese Journal of Cellular and Molecular Immunology, 36*(6), 542–548.

Tatsch, E., Bochi, G. V., Piva, S. J., De Carvalho, J. A. M., Kober, H., Torbitz, V. D., ... Moresco, R. N. (2012). Association between DNA strand breakage and oxidative, inflammatory and endothelial biomarkers in type 2 diabetes. *Mutation Research, 732*(1–2), 16–20. https://doi.org/10.1016/j.mrfmmm.2012.01.004.

Toyota, T., Kai, Y., Kakizaki, M., Ohtsuka, H., Shibata, Y., & Goto, Y. (1981). The endocrine pancreas in pyridoxine deficient rats. *The Tohoku Journal of Experimental Medicine, 134*(3), 331–336. https://doi.org/10.1620/tjem.134.331.

Tramonti, A., Ghatge, M. S., Babor, J. T., Musayev, F. N., di Salvo, M. L., Barile, A., ... Contestabile, R. (2022). Characterization of the Escherichia coli pyridoxal 5'-phosphate homeostasis protein (YggS): Role of lysine residues in PLP binding and protein stability. *Protein Science: A Publication of the Protein Society, 31*(11), e4471. https://doi.org/10.1002/pro.4471.

Tully, D. B., Allgood, V. E., & Cidlowski, J. A. (1994). Modulation of steroid receptor-mediated gene expression by vitamin B6. *FASEB Journal: Official Publication of the Federation of American Societies for Experimental Biology, 8*(3), 343–349.

Twarda-Clapa, A., Olczak, A., Białkowska, A. M., & Koziołkiewicz, M. (2022). Advanced Glycation End-Products (AGEs): Formation, chemistry, classification, receptors, and diseases related to AGEs. *Cells, 11*(8), https://doi.org/10.3390/cells11081312.

Uribarri, J., del Castillo, M. D., de la Maza, M. P., Filip, R., Gugliucci, A., Luevano-Contreras, C., ... Garay-Sevilla, M. E. (2015). Dietary advanced glycation end products and their role in health and disease. *Advances in Nutrition (Bethesda, Md.), 6*(4), 461–473. https://doi.org/10.3945/an.115.008433.

Verhoef, P., Stampfer, M. J., Buring, J. E., Gaziano, J. M., Allen, R. H., Stabler, S. P., ... Willett, W. C. (1996). Homocysteine metabolism and risk of myocardial infarction: Relation with vitamins B6, B12, and folate. *American Journal of Epidemiology, 143*(9), 845–859. https://doi.org/10.1093/oxfordjournals.aje.a008828.

Vincent, A. M., Perrone, L., Sullivan, K. A., Backus, C., Sastry, A. M., Lastoskie, C., & Feldman, E. L. (2007). Receptor for advanced glycation end products activation injures primary sensory neurons via oxidative stress. *Endocrinology, 148*(2), 548–558. https://doi.org/10.1210/en.2006-0073.

Vinson, J. A., & Howard, T. B. (1996). Inibition of protein glycation and avanced glycation end products by ascorbic acid and other vitamins and nutrients. *The Journal of Nutritional Biochemistry, 7*(12), 659–663. https://doi.org/10.1016/S0955-2863(96)00128-3.

Vlassara, H., & Uribarri, J. (2014). Advanced glycation end products (AGE) and diabetes: Cause, effect, or both? *Current Diabetes Reports, 14*(1), 453. https://doi.org/10.1007/s11892-013-0453-1.

Vlassara, H., Uribarri, J., Cai, W., & Striker, G. (2008). Advanced glycation end product homeostasis: Exogenous oxidants and innate defenses. *Annals of the New York Academy of Sciences, 1126*, 46–52. https://doi.org/10.1196/annals.1433.055.

Voziyan, P. A., & Hudson, B. G. (2005). Pyridoxamine as a multifunctional pharmaceutical: Targeting pathogenic glycation and oxidative damage. *Cellular and Molecular Life Sciences: CMLS, 62*(15), 1671–1681. https://doi.org/10.1007/s00018-005-5082-7.

Voziyan, P. A., Metz, T. O., Baynes, J. W., & Hudson, B. G. (2002). A post-Amadori inhibitor pyridoxamine also inhibits chemical modification of proteins by scavenging carbonyl intermediates of carbohydrate and lipid degradation. *The Journal of Biological Chemistry, 277*(5), 3397–3403. https://doi.org/10.1074/jbc.M109935200.

Williams, M. E., Bolton, W. K., Khalifah, R. G., Degenhardt, T. P., Schotzinger, R. J., & McGill, J. B. (2007). Effects of pyridoxamine in combined phase 2 studies of patients with type 1 and type 2 diabetes and overt nephropathy. *American Journal of Nephrology, 27*(6), 605–614. https://doi.org/10.1159/000108104.

Wilson, M. P., Plecko, B., Mills, P. B., & Clayton, P. T. (2019). Disorders affecting vitamin B(6) metabolism. *Journal of Inherited Metabolic Disease, 42*(4), 629–646. https://doi.org/10.1002/jimd.12060.

Wolff, S. P., & Dean, R. T. (1987). Glucose autoxidation and protein modification. The potential role of "autoxidative glycosylation" in diabetes. *The Biochemical Journal, 245*(1), 243–250. https://doi.org/10.1042/bj2450243.

Wu, M. Y., Yiang, G. T., Lai, T. T., & Li, C. J. (2018). The oxidative stress and mitochondrial dysfunction during the pathogenesis of diabetic retinopathy. *Oxidative Medicine and Cellular Longevity, 2018*, 3420187. https://doi.org/10.1155/2018/3420187.

Xu, C. X., Zhu, H. H., & Zhu, Y. M. (2014). Diabetes and cancer: Associations, mechanisms, and implications for medical practice. *World Journal of Diabetes, 5*(3), 372–380. https://doi.org/10.4239/wjd.v5.i3.372.

Yanaka, N., Kanda, M., Toya, K., Suehiro, H., & Kato, N. (2011). Vitamin B6 regulates mRNA expression of peroxisome proliferator-activated receptor-γ target genes. *Experimental and Therapeutic Medicine, 2*(3), 419–424. https://doi.org/10.3892/etm.2011.238.

Yim, M. B., Yim, H. S., Lee, C., Kang, S. O., & Chock, P. B. (2001). Protein glycation: Creation of catalytic sites for free radical generation. *Annals of the New York Academy of Sciences, 928*, 48–53.

Zhang, L., Li, X., Zhang, J., & Xu, G. (2021). Prognostic implication and oncogenic role of PNPO in pan-cancer. *Frontiers in Cell and Developmental Biology, 9*, 763674. https://doi.org/10.3389/fcell.2021.763674.

Zhang, L., Zhou, D., Guan, W., Ren, W., Sun, W., Shi, J., ... Xu, G. (2017). Pyridoxine 5′-phosphate oxidase is a novel therapeutic target and regulated by the TGF-β signalling pathway in epithelial ovarian cancer. *Cell Death & Disease, 8*(12), 3214. https://doi.org/10.1038/s41419-017-0050-3.

Zhang, Y., Zhou, J., Wang, T., & Cai, L. (2007). High level glucose increases mutagenesis in human lymphoblastoid cells. *International Journal of Biological Sciences, 3*(6), 375–379. https://doi.org/10.7150/ijbs.3.375.

Zhao, J., Randive, R., & Stewart, J. A. (2014). Molecular mechanisms of AGE/RAGE-mediated fibrosis in the diabetic heart. *World Journal of Diabetes, 5*(6), 860–867. https://doi.org/10.4239/wjd.v5.i6.860.

Zhu, B., & Qu, S. (2022). The relationship between diabetes mellitus and cancers and its underlying mechanisms. *Frontiers in Endocrinology, 13*, 800995. https://doi.org/10.3389/fendo.2022.800995.

Zhu, J., Chen, C., Lu, L., Yang, K., Reis, J., & He, K. (2020). Intakes of folate, Vitamin B (6), and Vitamin B(12) in relation to diabetes incidence among american young adults: A 30-year follow-up study. *Diabetes Care, 43*(10), 2426–2434. https://doi.org/10.2337/dc20-0828.